Intensification of Liquid–Liquid Processes

Explore and review novel techniques for intensifying transport and reaction in liquid–liquid and related systems with this essential toolkit. Topics include discussion of the principles of process intensification, the nexus between process intensification and sustainable engineering, and the fundamentals of liquid–liquid contacting, from an expert with over 45 years' experience in the field. Providing promising directions for investment and for new research in process intensification, in addition to a unique review of the fundamentals of the topic, this book is the perfect guide for senior undergraduate students, graduate students, developers, and research staff in chemical engineering and biochemical engineering.

Laurence R. Weatherley is the Albert P Learned Distinguished Professor of Chemical Engineering and Department Chair at the University of Kansas. Prior to joining the University of Kansas, he served as the Chaired Professor of Chemical Engineering, and Head of the Department of Chemical and Process Engineering, at the University of Canterbury in New Zealand. He has served as the Executive Co-Editor of the *Chemical Engineering Journal*. Dr. Weatherley is a Fellow of the Institution of Chemical Engineers, United Kingdom, and a Fellow of the Institution of Professional Engineers of New Zealand.

Cambridge Series in Chemical Engineering

BOOKS IN THE SERIES

Baldea and Daoutidis, *Dynamics and Nonlinear Control of Integrated Process Systems*
Chamberlin, *Radioactive Aerosols*
Chau, *Process Control: A First Course with MATLAB*
Cussler, *Diffusion: Mass Transfer in Fluid Systems, Third Edition*
Cussler and Moggridge, *Chemical Product Design, Second Edition*
De Pablo and Schieber, *Molecular Engineering Thermodynamics*
Deen, *Introduction to Chemical Engineering Fluid Mechanics*
Denn, *Chemical Engineering: An Introduction*
Denn, *Polymer Melt Processing: Foundations in Fluid Mechanics and Heat Transfer*
Dorfman and Daoutidis *Numerical Methods with Chemical Engineering Applications*
Duncan and Reimer, *Chemical Engineering Design and Analysis: An Introduction, Second Edition*
Fan *Chemical Looping Partial Oxidation Gasification, Reforming, and Chemical Syntheses*
Fan and Zhu, *Principles of Gas–Solid Flows*
Fox, *Computational Models for Turbulent Reacting Flows*
Franses, *Thermodynamics with Chemical Engineering Applications*
Leal, *Advanced Transport Phenomena: Fluid Mechanics and Convective Transport Processes*
Lim and Shin, *Fed-Batch Cultures: Principles and Applications of Semi-Batch Bioreactors*
Litster, *Design and Processing of Particulate Products*
Marchisio and Fox, *Computational Models for Polydisperse Particulate and Multiphase Systems*
Mewis and Wagner, *Colloidal Suspension Rheology*
Morbidelli, Gavriilidis, and Varma, *Catalyst Design: Optimal Distribution of Catalyst in Pellets, Reactors, and Membranes*
Nicoud, *Chromatographic Processes*
Noble and Terry, *Principles of Chemical Separations with Environmental Applications*
Orbey and Sandler, *Modeling Vapor–Liquid Equilibria: Cubic Equations of State and Their Mixing Rules*
Pfister, Nicoud, andMorbidelli, *Continuous Biopharmaceutical Processes: Chromatography, Bioconjugation, and Protein Stability*
Petyluk, *Distillation Theory and Its Applications to Optimal Design of Separation Units*
Ramkrishna and Song, *Cybernetic Modeling for Bioreaction Engineering*
Rao and Nott, *An Introduction to Granular Flow*
Russell, Robinson, and Wagner, *Mass and Heat Transfer: Analysis of Mass Contactors and Heat Exchangers*
Schobert, *Chemistry of Fossil Fuels and Biofuels*
Shell, *Thermodynamics and Statistical Mechanics*
Sirkar, *Separation of Molecules, Macromolecules and Particles: Principles, Phenomena and Processes*
Slattery, *Advanced Transport Phenomena*
Varma, Morbidelli, and Wu, *Parametric Sensitivity in Chemical Systems*
Weatherley, *Intensification of Liquid-Liquid Processes*

Intensification of Liquid–Liquid Processes

LAURENCE R. WEATHERLEY
University of Kansas

CAMBRIDGE
UNIVERSITY PRESS

CAMBRIDGE
UNIVERSITY PRESS

University Printing House, Cambridge CB2 8BS, United Kingdom

One Liberty Plaza, 20th Floor, New York, NY 10006, USA

477 Williamstown Road, Port Melbourne, VIC 3207, Australia

314–321, 3rd Floor, Plot 3, Splendor Forum, Jasola District Centre, New Delhi – 110025, India

79 Anson Road, #06–04/06, Singapore 079906

Cambridge University Press is part of the University of Cambridge.

It furthers the University's mission by disseminating knowledge in the pursuit of education, learning, and research at the highest international levels of excellence.

www.cambridge.org
Information on this title: www.cambridge.org/9781108421010
DOI: 10.1017/9781108355865

First published 2020

Printed in the United Kingdom by TJ International Ltd, Padstow Cornwall

A catalogue record for this publication is available from the British Library.

Library of Congress Cataloging-in-Publication Data
Names: Weatherley, Laurence R., author.
Title: Intensification of liquid-liquid processes / Laurence R. Weatherley, University of Kansas.
Description: First edition. | New York, NY : Cambridge University Press, 2020. | Series: Cambridge series in chemical engineering | Includes bibliographical references and index.
Identifiers: LCCN 2019043644 (print) | LCCN 2019043645 (ebook) | ISBN 9781108421010 (hardback) | ISBN 9781108355865 (epub)
Subjects: LCSH: Chemical processes. | Liquids. | Diffusion. | Liquid-liquid equilibrium. | Catalysts. | Drops.
Classification: LCC TP155.7 .W35 2020 (print) | LCC TP155.7 (ebook) | DDC 660/.28–dc23
LC record available at https://lccn.loc.gov/2019043644
LC ebook record available at https://lccn.loc.gov/2019043645

ISBN 978-1-108-42101-0 Hardback

Contents

1 Introduction

1.1 Process Intensification

Process intensification (PI) is the term used to describe the means by which a process manufacturing stage can be rendered more compact than the conventional standard. This implies a lower inventory and a corresponding increase in manufacturing rate compared with the standard for that stage. Process intensification is important because its application can lead to: novel or enhanced products, better use of chemistry, improved processing (higher efficiency), distributed manufacturing; energy and environmental benefits; process flexibility, improved product quality, reduced footprint, improved inherent safety and energy efficiency, capital cost reduction, and reduced material inventories. Process intensification also encapsulates a novel design philosophy that aims to revolutionize process engineering by revisiting the fundamentals of fluid dynamics and transport phenomena.

The concept of process intensification has been around for a number of years and the adoption of ground-breaking applications of recent, innovative process intensification techniques in industrial processing has been relatively slow. One of the principal reasons is the risk-averse approach taken by manufacturing companies (Tsouris and Porcelli, 2003) toward the application of entirely new technologies. However, as greater understanding of the fundamentals is forthcoming and with the refinement of modeling techniques, some of the major risks can be reduced at relatively low cost. In addition, by its very nature process intensification often involves scale-up using multiple equipment modules, which singly or in small numbers can be rigorously tested and evaluated prior to implementation at full scale.

Intensification technologies and the associated research and development focus mostly on multiphase processes: gas/liquid, gas/solid, liquid/solid, and liquid/liquid. In the case of both gas/liquid and liquid/liquid, the fluid mechanics are the most complex because of the fact that gas bubbles and liquid droplets are dynamic entities whose shapes and sizes are time dependent. For example, gas bubbles oscillate, which influences internal circulation, exhibiting variable shear conditions at the gas–liquid boundary and time-dependent interfacial area values. Gas bubbles in a swarming system may collide and coalesce, thus experiencing large hydrodynamic disturbance, and experiencing a change in shape and size. This not only complicates the understanding of the fluid mechanics, but also complicates our understanding of the interrelated processes of heat and mass transfer across the phase boundary. Recently developed concepts in intensification inevitably include novel equipment designs, some with complex geometries that further complicate rigorous description and fundamental understanding of the fluid mechanics of the system. The key to the success of the majority of intensification techniques, which will be discussed in later chapters of this book, is the control of the fluid mechanics.

In the case of liquid–liquid processes, complexities similar to those already briefly mentioned for gas–liquid systems also apply, including complex drop size behavior, shape dynamics, and coalescence phenomena. In addition, knowledge of the physics of the drop surface and the interface is also a vital consideration for understanding intensification of liquid–liquid contacting.

A number of definitions of intensification have been articulated. An overall definition may be described as follows: "The reduction in size of a process or element of a process by improved design, choice of better reagents or materials, and improved control of fluid mechanics." The above overarching result may be achieved in many different ways. These include achievement of higher heat and mass transfer fluxes, improved reaction selectivity, development of new solvents and materials for improving separation efficiency, and improved physical geometries of reactors and contactors. There are a number of important underpinning reasons why intensification techniques are of on-going interest among designers, researchers, and innovators. Energy saving is a major reason why intensification techniques are considered (Baird and Rao, 1995; Baird and Stonestreet, 1995; Qiu, Zhao, and Weatherley, 2010; Ramshaw, 1999). There is substantial evidence of demonstrated savings in energy utilization through the adoption of process intensification principles through improvements in mechanical efficiency, through the application of better catalysts, through the application of intensification principles to heat exchanger design, and through the use of reaction and separation media that enable more efficient and easier product separation. A second basic reason for the adoption of process intensification is the reduction in size and process plant footprint that the application to equipment design affords. The intensification of reaction rates and heat and mass transfer fluxes results in reduction of equipment size for a given throughput and thus the space required to accommodate the plant hardware is correspondingly reduced. Another potential benefit of reduction in plant size is the possibility of a distributed manufacturing approach being adopted. Distributed manufacturing involves organizing production in a larger number of smaller plants that are located close to sources of raw materials and the market, rather than in a series of larger plants in a central location. There are economic and environmental benefits to be gained from distributed manufacturing due to reduction of transportation costs of both raw materials and products. A further positive feature of intensification is the reduction in inventory due to smaller equipment size and reduced residence times. Inventory reduction is not only economically beneficial but also is a huge safety feature of PI, especially when processing highly toxic, explosive, or flammable materials.

Modularization of process plant is also made possible by reducing equipment size resulting from more intensive reaction rates and higher material and energy fluxes. Building plants in modular form can be achieved off-site with cost benefits and improved flexibility for the configuration of multipurpose plants. Modularization is also a major consideration in the design of offshore oil and gas processing facilities where space costs are extremely high, with a necessity for plant of the smallest possible footprint.

Traditionally, process intensification has been strongly identified with the development of new or improved devices for processing. These include devices whose design exploits phenomena such as cavitation, centrifugal forces, electrical fields, magnetic fields, creation of thin films, high shear, cyclic dispersion and coalescence, pulsatile flows, and others. Another emerging view of process intensification is strongly based on the development of new materials and fluids. These include new catalysts, new reaction media and solvents for separation, and the application of biological processes.

Of special relevance to liquid–liquid systems is the development of two new families of solvents, ionic liquids and solvents based on near-critical carbon dioxide. Much of the interest in these groups of solvents has focused on the possibilities they offer as reaction and extraction media in terms of higher rates of reaction, high reactant solubilities, and easier separations. The availability of ionic liquid media (ILM) at near-ambient temperatures has stimulated many possibilities for their use as media for organic reactions involving gas, liquid, and solid reaction substrates. The principal process advantages of ILM center on their very low vapor pressure and relatively low toxicity. These address the major concerns surrounding volatile organic compound (VOC) emissions and toxic exposure within the chemical and process industries, which currently use very large amounts of conventional organic solvents both as reaction media and as extractants. Another key advantage of ILM is their excellent performance as solvents for a wide range of compounds. The extensive work of Seddon and co-workers and others (Earle and Seddon, 2000; Holbrey, Rooney, and Seddon, 1999; Welton, 1999; Wilkes, 2004) has confirmed the potential of ILM for catalysis at laboratory scale for a range of systems. However, there has been less attention paid thus far to the issues of contacting, mixing, and phase separation in liquid–liquid systems involving ionic liquids. A detailed understanding of these key process phenomena is required if a fuller appreciation of the performance of ionic liquids as solvents in liquid–liquid processes is to be obtained.

Another "clean" solvent of importance to intensification is carbon dioxide (CO_2), which may be used either as a subcritical or supercritical fluid functioning as a benign solvent or as a reaction medium. The ability to control solvent properties, both the solvation properties and the physical properties, offers significant flexibility for process applications. Advantages include: relatively low cost; low toxicity; properties that are well understood; for separations, product isolation is straightforward; wide control of solvation and selectivity behavior (via temperature and pressure control, and addition of entrainers); high diffusion coefficients that offer potential for increased reaction rates; potential for homogeneous catalytic reactions; high gas miscibility; nonoxidizable; nonflammable; excellent medium for oxidation and reduction reactions; and further tunability can be obtained by combination of near-critical CO_2 with conventional solvents (CXLs) and entrainers. The potential for using CO_2-expanded liquids (CXLs) as reaction media has received increased attention in recent times. These are compressible media that are synthesized by mixing dense-phase CO_2 with an organic solvent to produce a liquid having properties that are substantially different from either of the constituents. The physical and chemical properties of the mixture are highly tunable through variation of the chemical composition and the temperature and pressure of the system. One of the significant features of CXLs is the intensification of gas solubility (Bezanehtak, Dehghani, and Foster, 2004; Hert et al., 2005) as this accrues potential enhancements in reaction conditions. Similarly, some catalytically significant transition metal complexes, many of which are only sparingly soluble in conventional organic solvents, also demonstrate enhanced solubility, resulting in improved reaction conditions.

Knez et al. (2014) highlight the interesting possibility of conducting environmentally benign enzymatic catalytic reactions in a liquid–liquid system comprising an ionic liquid and a supercritical fluid. It is shown that reactions may be successfully conducted in the resulting two-phase liquid–liquid environment that effectively facilitates simultaneous reaction and separation (Fan and Qi, 2010; Paljevac, Knez, and Habulin, 2009). In a significant review article, Keskin et al. (2007) showed how ionic liquid solvents have become the "partner" solvents with supercritical CO_2 in many applications, based on a significant literature focus on the interaction of these two green solvents.

The drawbacks of CO_2 as a processing fluid include the constraints associated with high-pressure processing, high capital cost for large-scale applications, and lack of good design data.

Developments in these areas are part of the strong nexus between intensification technologies, "green engineering" and "green chemistry," sometimes also referred to as subsets of "sustainable engineering." Sustainable engineering could be defined as the multidisciplinary integration of sciences and engineering for the restoration and preservation of the environment and for the sustainable use of resources. The associated process engineering is crucial for the success of clean technology. The principles of waste minimization which have been articulated for several years have a direct bearing on the goals of sustainable engineering. The continuum of goals may be put simply as: treatment > reduction > minimization > zero emission.

It is useful to quote the 1995 definition of sustainable development from the UK Engineering Council:

> Sustainable development [is] ... development, which has as its goal the maximization of human well-being within the constraints necessary for conserving the integrity of the environment.
>
> Making the transition to a sustainable model for human activity will be an extremely complex and long term task ... It will require considerable technological innovation and many other skills which engineers are uniquely placed to offer ...

It may be argued that sustainable development can be described as the "goal" – and sustainable engineering as the "means." We can consider some modified definitions of sustainable engineering and we can use analogies with some of the definitions of process intensification, for example:

> The term used to describe the means by which a process manufacturing element, or entire process can be rendered more sustainable than the conventional standard. This implies a lower inventory and usage of hazardous and non-renewable material in manufacturing compared with the standard. (Jones, 1996)

Sustainable engineering may be considered to be the adoption of a novel approach to process development, plant design, operation, and control that aims to minimize environmental impact at every stage. A related goal is a sustainable environment for chemical processing that results in safer and less polluting products in compact, highly efficient equipment at low cost.

The sustainable engineering approach, in addition to leading to the replacement of inefficient processes, can also provide opportunities for new chemical products, distributed manufacturing, and higher profitability. A design philosophy underpinned by the principles of sustainability can result in improved energy efficiency, lower carbon footprint, reduced hazard, lower capital and inventory costs, and reduction in plant size and footprint.

In considering the actualities of moving toward sustainable engineering, there are a number of key questions to be examined. Can the goals behind sustainable engineering be achieved and delivered through: (i) novel or enhanced products; (ii) better use of chemistry; (iii) reduced inventories; (iv) enhanced safety and control systems; (v) improved processing (higher efficiency); (vi) distributed manufacturing; (vii) energy and environmental benefits; or (viii) capital cost reduction? It can be argued that most of the above include the design and operational goals of good process engineering and have done so for many years. It can also be argued that there are a host of specific examples already in place, such as recycle reactors, renewable energy (wave, wind, solar, tidal), waste water treatment and recycling, and biofuels (biodiesel, bioethanol, gasified wood and other cellulose). A difference observed in recent developments is the importance of fundamental knowledge in addressing the challenge of sustainable development. New science and deeper understanding of engineering principles are making a positive impact and contributing to the implementation of sustainable engineering. Process intensification is a crucial part of that implementation, bringing not only economic advantages but also environmental benefits to manufacturing processes. This applies to processes involving liquid–liquid mixtures as much as to any other multiphase system.

The evolution in the design of liquid–liquid processes has occurred over many years, as shown in the improvements in equipment performance and reliability, and in the development of solvents with improved chemical and physical properties. Deeper understanding of hydrodynamics, advances in computational modeling of equipment behavior, novel approaches to equipment geometry, and innovations in solvent chemistry have been the main driving forces behind development of improved liquid–liquid processes.

The design of liquid–liquid contacting equipment has evolved over many years, covering a wide range of applications. The major application that has underpinned the majority of developments in design has been for extractive separations, product purification, and recovery. These range from purification and separation processes in high-volume industries, such as petrochemicals, to separations in the metals and mining industry, to the stringent separation requirements of the nuclear industry and the pharmaceutical industry. The underpinning principles in all cases are concerned with ensuring high mass transfer efficiency, tight control of hydraulic conditions during contact, and effective phase separation. In the case of certain industries other factors also have influenced design. These include product stability, which is a major factor in certain pharmaceutical separations where rapid extraction and product stabilization are paramount requirements for process efficiency and economic operation. Another industry-specific factor is the need for hygienic process conditions and the avoidance of contamination, the ease of cleaning, and sterilization, which is again the case in the pharmaceutical industry but also in the food, perfumery, and ingredients industries. The nuclear industry also has additional special conditions required for the design of liquid–liquid contact equipment. Here the need for maintenance-free equipment throws up challenges for equipment designers, added to which is the need to ensure equipment design takes into account criticality safety criteria in cases where fissile nuclear fuel compounds are processed.

In general the design of separation process equipment is governed largely by requirements to achieve the desired separation reliably, safely, and economically, requirements that continue today to dominate decision making regarding equipment design and choice for a particular application. Other considerations have also become important with the emergence of green chemistry, the drive for greater process efficiencies and product yield, and increased emphasis on process safety and inventory reduction. Exploitation of new catalysts and innovative solvent media will depend in some cases on the development of new equipment technology in order to make the application of new materials feasible at industrial scale. The high cost of homogeneous catalysts and novel solvent media such as ionic liquids, aqueous-based solvents, and complexing agents such liquid ion-exchangers, have led to new developments in equipment design and operations. The reasons include the need to maximize solvent recovery and recycle, to ensure phase separations are highly efficient, and that extractions are conducted to minimize solvent degradation and loss. As discussed earlier, these are major driving forces behind process intensification.

Liquid–liquid processes involving interfacial transport must ensure that the hydrodynamic conditions of the contact create an environment for high mass transfer rates. For all types of equipment this involves breaking up or dispersing one phase into the second phase, generating small but unstable drops, achieving high rates of interfacial shear, and minimizing back-mixing, dead space, and short-circuit flows. These considerations are paramount in stagewise equipment, column contactors, batch or continuous contact, and in high-intensity contactors such as centrifugal devices. The fundamentals of the liquid–liquid behaviors that govern the meeting of these requirements are considered in Chapters 2 and 3.

The overarching requirement can be simply described in two steps; contacting followed by phase separation. Figure 1.1(a)–(f) shows a sequence of shots in which the static liquid–liquid mixture proceeds to a complete dispersion after the onset of agitation. Frames (b) and (c) show the onset of interfacial breakup, followed by the dispersion of the droplet phase as discrete drops in frame (d), and then to complete dispersion in frame (e). Frame (f) show the progress toward phase disengagement after approximately 20 s from the discontinuation of agitation.

Frame (f) illustrates one of the fundamental conflicts in liquid–liquid mixing and the design and operation of mixer settlers. On the one hand, high levels of drop break-up,

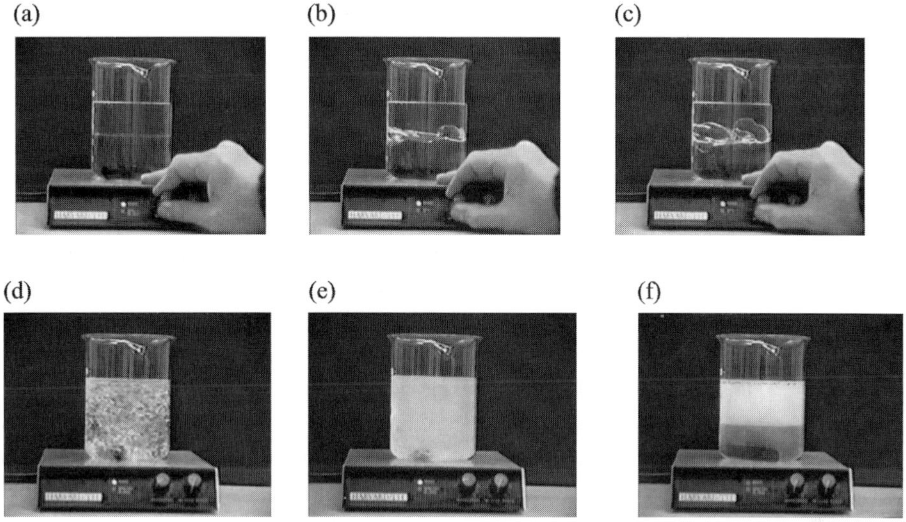

Figure 1.1 Stages of mechanical dispersion of a liquid–liquid mixture.

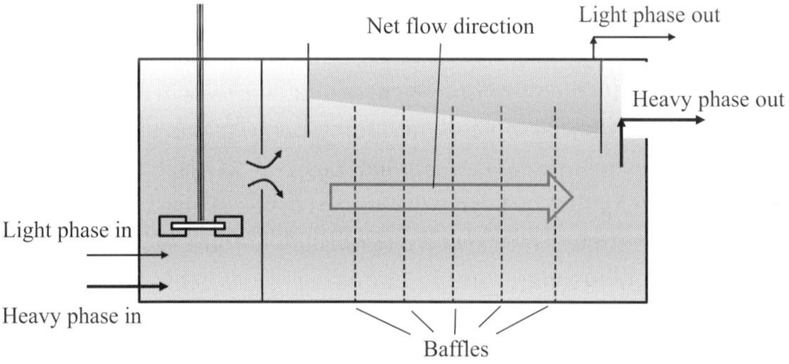

Figure 1.2 Mixer settler arrangement with baffled settling chamber.

mixing, and dispersion are required for high rates of mass transfer and efficient mixer compartment utilization. On the other hand, high rates of phase disengagement and coalescence are also an essential requirement. As can be seen in frame (f), in the light phase there is strong evidence of fine drops of the heavy phase remaining that may be slow to settle on account of their very small size. Therefore the degree of agitation must be carefully controlled to ensure that settling can be realized efficiently while at the same time maintaining high mass transfer rates.

Innovations in liquid–liquid contacting equipment design are mainly focused on enhancing the fluid dynamics and interfacial phenomena that govern mass and energy transport, and that govern physical phase separation. To recap, Figure 1.2 shows the principle of mixing and settling for liquid–liquid extraction or solvent extraction. This may be defined as the selective transfer of a desired solute for recovery and purification from a feed liquid phase into a second partially or totally immiscible liquid phase whose physical and chemical thermodynamic properties are designed for the separation required. There are many texts that describe liquid–liquid extraction technologies in detail, for example Logsdail and Slater (1993). The principle features of the process are physical contacting of the two liquid feed phases to achieve high interfacial area and efficient interfacial mass transfer between the phases. The promotion of interfacial shear between the surface of droplets of one phase and the surrounding continuum of the second phase together with droplet instability are important for good mass transfer. The second major part of liquid–liquid processes is the efficient separation of the liquid phases after mass transfer has approached completion. This is important in order to ensure high efficiency and to avoid contamination of each exiting liquid stream with traces of the other phase as much as possible.

These fundamental requirements strongly impact the design and operation of industrial liquid–liquid contacting equipment. The intensification of liquid–liquid processes and future developments of this approach require not only clear understanding of the relevant fluid mechanics and transport phenomena in existing equipment designs but also innovation in equipment design and geometries. The intensification of liquid–liquid processes through improved equipment design therefore must be based on better understanding of the fluid mechanics and how these may be better controlled. Improved equipment performance through intensification opens up avenues for the viable commercialization of new chemistry and biotechnology for new and improved chemical products, and for the manufacture of existing products with lower environmental impacts. The nexus between equipment innovation and new chemistry, for example in the development of new pathways, solvent media, and catalysts, is an essential component driving development of new equipment technologies.

The technologies for industrial-scale liquid–liquid contacting processes are traditionally categorized as follows:

(i) mixer settler (stagewise) equipment;
(ii) continuous column (differential) contact equipment;
(iii) centrifugal contact equipment.

The majority of modern, high-performance contact devices use externally applied mechanical energy to achieve good dispersion and use turbulence of the liquid–liquid mixture in order to ensure high rates of mass transfer. The proportion of total energy supplied to the contactor that is actually utilized to enhance mass transfer rates is small in the majority of cases and a major proportion is dissipated through frictional loss. Thus there is significant scope for improvements in efficiency.

As we review both traditional liquid–liquid contacting technologies and new, emerging technologies we see that a number of well-established equipment designs already embrace many of the tenets of process intensification. As already stated, liquid–liquid contacting equipment design falls into three broad categories: (i) stagewise equipment, for example mixer settler trains; (ii) continuous countercurrent column contactors; and (iii) centrifugal contactors. Equipment that falls into any of these three categories to a greater or lesser degree contain design features to increase kinetics, improve phase separation, and enhance hydraulic capacity. In new and emerging equipment technology we see a move toward miniaturization with equipment having smaller footprints, modular design, and incorporation of novel methods of intensification such as microwaves, ultrasonic fields, and electrical fields. Another factor that has enabled improved optimization of equipment design is the availability of highly accurate mathematical modeling tools that allow us to study the performance of new equipment geometries without the expense of building and experimental testing until design choices are significantly narrowed down. Such tools also provide valuable insights into the fluid dynamics inside liquid–liquid contacting equipment hitherto unavailable to designers.

1.2 Review of Current Equipment Technologies

1.2.1 Mixer Settlers

Mixer settlers have provided the mainstay of many large-scale industrial liquid–liquid extraction processes. A typical arrangement is shown in Figure 1.2.

The two immiscible liquid phases are fed to the mixing compartment at the left-hand end, where mixing of the phases is conducted. The main objectives required in the mixing compartment are as follows:

- To increase the interfacial area per unit volume in order to enhance overall mass transfer rates. This is achieved by efficient breakup and dispersion of one phase into the other phase.
- To reduce the diffusion boundary layer around individual droplets to minimize mass transfer resistances in the continuous phase surrounding the dispersed droplets.
- To promote good mass transfer within the droplet phase by reducing internal concentration gradients. Promotion of cyclic dispersion and coalescence processes to enhance convection inside individual droplets within the mixing chamber can result in significant internal hydrodynamic disturbances thus intensifying mass transfer.
- To ensure that the entire volume of the mixing chamber is used for effective mixing of the two phases, avoiding dead space and short-circuit unmixed flow of either of the phases.
- To produce a mixed phase that can be readily separated in the coalescence section of the mixer settler.

The above requirements have a fundamental influence on the design and operating conditions of the mixing chamber. Understanding the interactions between the fluid dynamics, geometry, and transport rates is critical for intensification of the liquid–liquid interaction.

With reference to Figure 1.2, the mixing chamber is shown on the left-hand end of the unit and is shown as being equipped with an impeller on a vertical drive shaft. The speed, depth, and shape of impeller, together with the volume of the mixing compartment, are all considerations in determining the optimum mixing and dispersion conditions as listed. Early work on liquid–liquid mixing focused on establishing the minimum agitator speed required to obtain a well-mixed dispersion. For example, Skelland and Seksaria (1978) proposed the following correlation for the minimum agitator speed N_{cd} for complete dispersion in terms of physical properties (viscosities of continuous and dispersed phases μ_c and μ_d respectively, interfacial tension σ, and density difference $\Delta\rho$)

$$N_{cd} = C_0 D^{\alpha_0} \mu_C^{1/9} \mu_d^{-1/9} \sigma^{0.3} \Delta\rho^{0.25} \tag{1.1}$$

where C_0 and α_0 are constants depending on the type of impeller, the location of the impeller, and dimensions (height : width ratio).

In later work, Skelland developed another correlation for minimum agitator speed $(N_{Fr})_{min}$ in baffled vessels (Skelland and Ramsey, 1987):

$$(N_{Fr})_{min} = C^2 \left(\frac{T}{D}\right)^{2\alpha} \phi^{0.106} (N_{Ga} N_{Bo})^{-0.084} \tag{1.2}$$

$$N_{Bo} = \frac{D^2 g \Delta\rho}{\sigma} \quad \text{(Bond number)}$$

$$N_{Ga} = \frac{D^3 \rho_M g \Delta\rho}{\mu_M^2} \quad \text{(Gallileo number)}$$

C = a shape factor
T = tank diameter (m)
D = impeller diameter (m)
ϕ = volume fraction of the dispersed phase
μ_m = mean viscosity (N s m^{-2})
ρ_m = mean density (kg m^{-3})

Another major requirement of mixer design is the need to ensure well-mixed conditions in the mixing chamber while achieving uniform flow. Here the design of mixing impeller, the rotation speed, the positioning of the impeller relative to the base of the mixing chamber, the design and positioning of baffles in the mixing chamber, and conditions of operation are critical to achieving uniform mixing. One of the conflicts in liquid–liquid mixing is, on the one hand a need to generate interfacial area for high mass transfer through dispersion, on the other hand avoidance of stable emulsion formation. There is a significant literature on the subject of mixing, so the coverage here is necessarily concise and we will review the main concepts and governing relationships relevant to liquid–liquid systems, and in particular the nexus between understanding mixing and intensification.

The type of impeller exerts a significant influence on the nature of the flows within the mixing chamber, although all three regimes are present to a greater or lesser extent (Figure 1.3). Flat-bladed turbine impellers are associated with strong radial flows, where angled turbine impellers tend to be associated with dominant axial flows. Flat-bladed paddle impellers tend to be associated with more dominant tangential flows. The latter

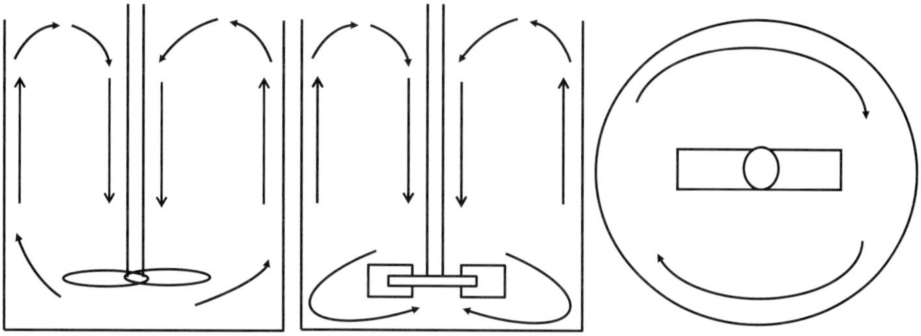

Figure 1.3 Simple flow regimes in a liquid–liquid mixer.

Figure 1.4 Examples of impellers for mixer settler applications.
With permission from www.postmixing.com/mixing%20forum/impellers/impellers.htm#r300

are generally undesirable since tangential flows tend to generate stratification that is inconsistent with good mixing and good mass transfer conditions. Figure 1.4 shows these three generic types.

1.2.2 Mixer Settler Design

1.2.2.1 Mixing

Impeller design is one consideration, another is the positioning of inlet and outlet ports from the mixing chamber. The positioning of inlet port and outlet port may have an important influence on the overall residence time distribution. Classical studies (Cloutier and Chollette, 1968) show the importance of the positioning of inlet and outlet ports in continuous stirred tank systems, in order to avoid short-circuit and by-pass flows. Such flows result in inadequate mixing time for elements of fluid thus leading to inefficient contact and low stage efficiency.

The design of stagewise mixer settler equipment for liquid–liquid contacting processes has relied upon prior experience and incremental practical development based on operating experience and pilot plant studies. In the context of intensification, the development of formal design and performance equations provides valuable tools for optimization and efficient design. The summary that follows is based on a paper by Post

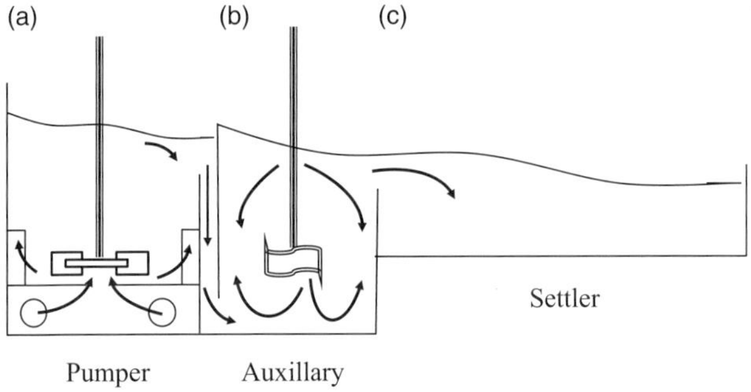

Figure 1.5 Arrangement of a mixer settler stage showing phase inlets, (a) pumper/mixer compartment, (b) auxiliary mixer compartment, and (c) settler (Post, 2003).

(2003) presented at the 5th International Hydrometallurgy Conference in 2003, which significantly advanced design methodology for mixer settlers.

The design analysis of Post (2003) is based on a mixer settler unit stage shown in Figure 1.5.

The first compartment labeled "pumper" (a) itself is divided into two sections. The lower section shows the inlet ports for the light phase and heavy phase. The lower sction communicates with the main mixing section above through an orifice located below the stirrer. The two liquid phases flow into the upper section of (a) and experience mixing. The mixed phase then overflows into the auxiliary mixer (b) and thence into the settling section (c). The mixing conditions need to be carefully controlled to ensure good conditions for mass transfer while at the same time avoiding the creation of either a stable liquid–liquid emulsion or a slow settling liquid–liquid mixture. The design of the pumper compartment must achieve three goals. Firstly, good mixing and conditions for interfacial mass transfer are essential. Secondly, the overall residence time must be sufficient for mass transfer to be close to completion prior to the liquids entering the auxiliary compartment. Thirdly, the mixing impeller design and speed must be matched to the size of the compartment to ensure the generation of sufficient hydraulic head for the transfer of the liquid mixture to the next compartment. A further requirement of the design is that air entrainment through the free surface of the liquid must be avoided.

With reference to Figure 1.6, Post defined three terms to characterize the conditions in the pump mixer compartment:

- the flow number N_q

$$N_q = \frac{Q_{total}}{ND_p^3}$$

- the power number N_p

$$N_p = \frac{P_{pumper}}{\rho_d N^3 D_p^5}$$

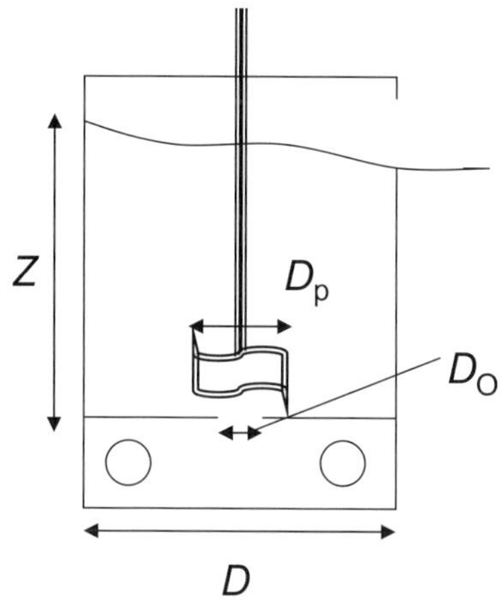

Figure 1.6 Dimensioned sketch of pump mixer compartment.

- the head number N_h

$$N_h = \frac{2gH}{(TS)^2} = \frac{2gH}{\pi^2 N^2 D_p^2}$$

where:

D_p = impeller diameter, m
D = tank diameter, m
g = acceleration due to gravity, m s^{-2}
H = head developed, m
N = rotational speed of impeller, s^{-1}
P_{pumper} = power to the impeller, W
Q_{total} = total liquid flow rate, $m^3 s^{-1}$
TS = tip speed of the impeller, m s^{-1}
Z = tank height, m
ρ_d = mean density of the dispersion, kg m^{-3}

Empirically based relationships such as in Figure 1.7 allow the determination of hydraulic performance of the mixer settler.

Comparison of different mixer impellers and operating conditions, requires the power-flow and head-flow curves for each type. Such data are usually available from the manufacturers, for example as cited in the reference by Giralico, Post, and Preston (1995).

Various suitable impellers are reviewed at www.postmixing.com/mixing%20forum/impellers/impellers.htm#r300 and include the following, as shown in Figures 1.8, 1.9, and 1.10.

The impellers shown in Table 1.1 are tried and tested designs and there is significant experience and data available for design and evaluations. The parameters required for the hydraulic performance evaluations are available at the www.postmixing.com website and from the manufacturers.

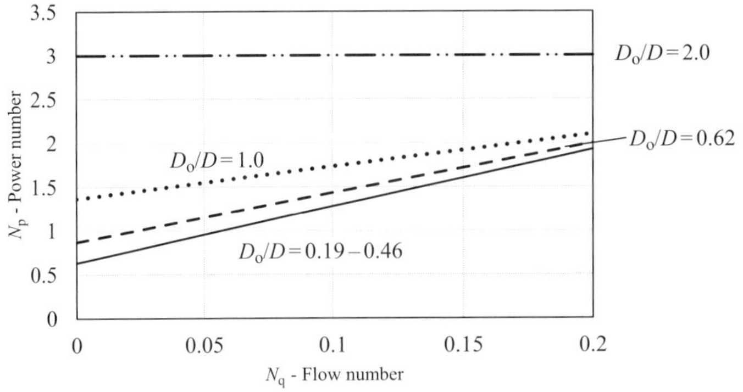

Figure 1.7 Flow number vs power number.
Data from Post (2003)

Figure 1.8 Mixer with impellor in place – arrangement view.
Courtesy of MGT Mixing

There are a number of other innovations in impeller design that are relevant to liquid–liquid processes.

Examples highlighted here include the MGT range of rotor/stator impellers which are designed to intensify mass transfer rates but avoid generation of stable emulsions. These high shear mixers use a high-speed rotor/stator generator to apply intense mechanical

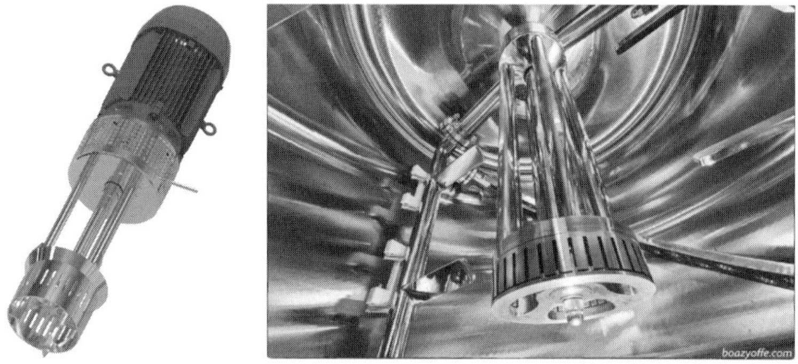

Figure 1.9 Photographs of MGT impellors and drive shafts – in situ in mixing chamber.
Courtesy of MGT Mixing

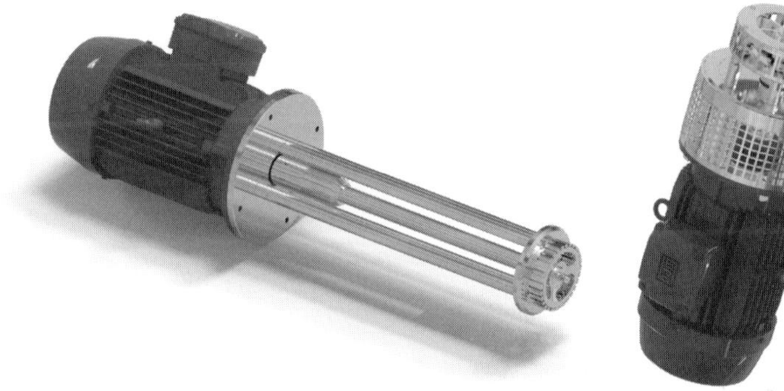

Figure 1.10 Rotor stator impellers.
Courtesy of MGT Mixing

and hydraulic shear with the blades of the rotor running at peripheral speeds of in the range 15–30 m s^{-1} within a fixed stator. As the blades rotate past each opening in the stator (see Figures 1.8–1.10), they shear particles and droplets, expelling material at high velocity into the surrounding mass. Figure 1.8 shows the general arrangement of one of the novel impellors located at the lower end of the mixing tank. Figure 1.9 shows a view of the drive motor attached to the upper end of the drive shaft with the impellor located at the lower end. There are two views, one showing the impeller and shaft in place, the other with the impeller and drive shaft removed. These are high shear mixers involving a high-speed rotor/stator arrangement that allows the generation of intensive hydraulic shear. Typical tip speeds are in the range 15–30 m s^{-1}. The mode of action of these types of impellers is such that the fluid phases are drawn from below into the high shear zone above the impellors. As fast as material is expelled, more is drawn from beneath into the high shear zone of the rotor/stator, promoting continuous flow and fast droplet/particle size reduction. The manufacturers confirm that the specification of rotor/stator depends on balancing shear and flow. The design of the stator head can be varied, as shown in Figure 1.10, and aims to provide effective liquid–liquid mixing for a range of shear to flow ratios.

Table 1.1 Impellers types for mixer settler applications (with permission Tom Post 2018) http://www.postmixing.com/mixing%20forum/impellers/impellers.htm#a310

Radial Flow Impellers	Axial Flow Impellers
 Rushton 4 bladed impeller	 Propellor type
 Paddle with 4 blades	 4 bladed pitched turbine impeller
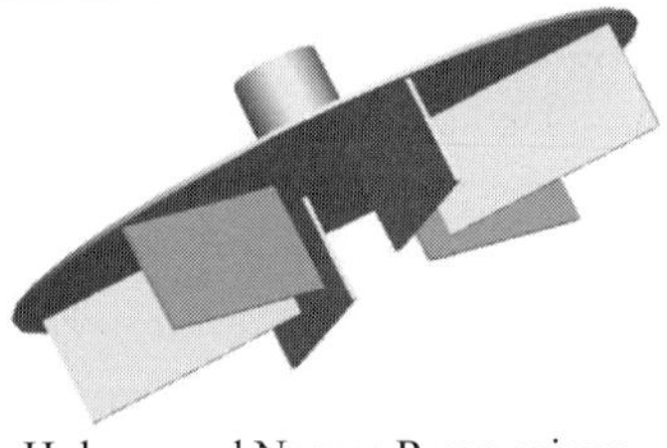 Holmes and Narver Pump mixer	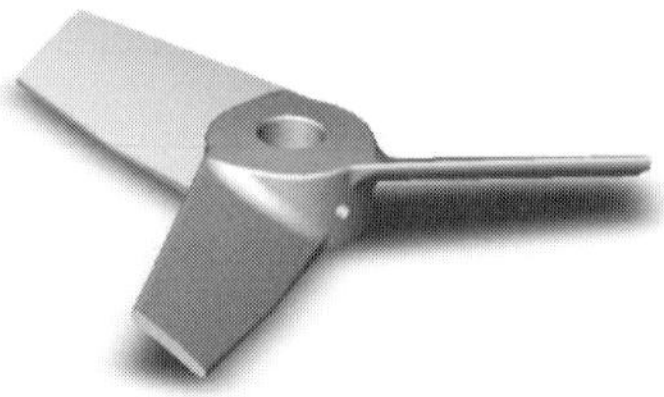 Lightnin downpumper 3 bladed impeller
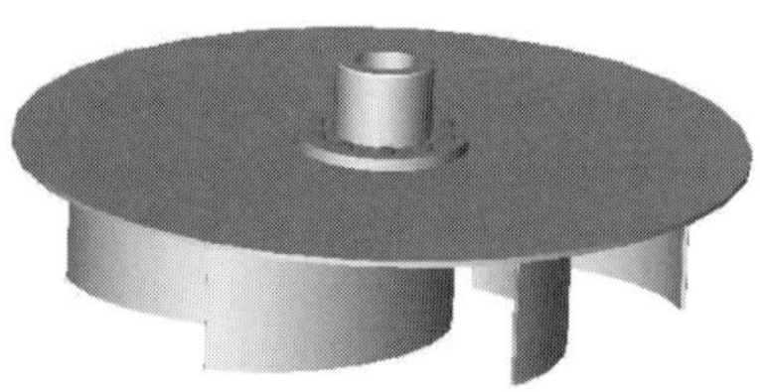 Curved bladed pumper	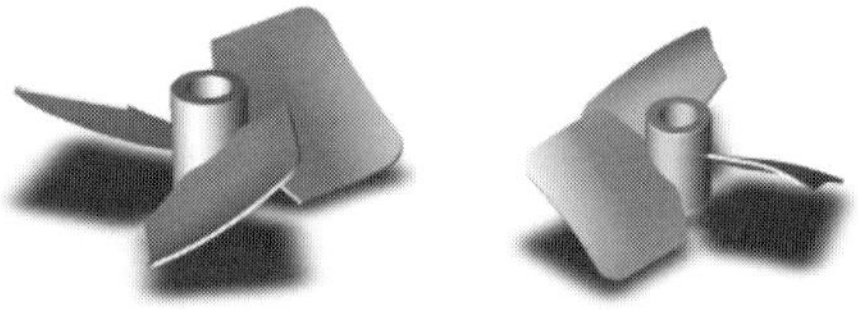 Lightnin impellers A340 up-pumper A320 down-pumper
 RS6 Impeller	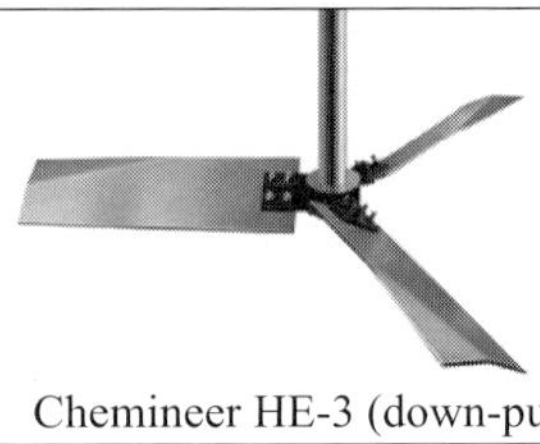 Chemineer HE-3 (down-pumper)

According to the manufacturer's information, the wetting head can come with different shapes of perforations. Those with larger round holes appear to work well for general purposes, generating vigorous flow, rapidly reducing the size of large drops. The slotted heads, for example shown in Figures 1.9 and 1.10, provide for an effective combination of high shear and efficient flow rate. It is reported as ideal for emulsions and medium viscosity materials.

There are many proprietary designs of mixers and impellers and the final choice is highly dependent upon the physical properties of the two liquids involved, especially viscosity, density difference, and interfacial tension.

1.2.2.2 Settling

Settling, or liquid–liquid phase separation, is of equal importance for efficient deployment of liquid–liquid systems. Ideally, residence time and the hydrodynamic environment in the settling chamber, Figure 1.11, should be optimized to provide complete separation of the light and heavy phases of the mixture entering the settling chamber from the mixer. Presence of entrained liquid of light phase in heavy phase and vice-versa easily reduces overall efficiency of an extraction process. Contamination of either of the exiting liquid phases will result in off-specification product, loss of product, and loss of solvent. If the raffinate stream is contaminated this may also lead to problems with downstream environmental clean up.

Settling may be achieved in a number of ways, which include: simple gravity settling, enhanced coalescence, centrifugal separation, deployment of ultrasonic fields, and electrostatically enhanced coalescence. There are also circumstances where the addition of chemical de-emulsifiers may also be used to promote phase separation but these are comparatively rare because of the reluctance to add chemical compounds that must be removed at a later stage, requiring additional separation technology and increased costs. The fundamentals of settling, phase separation and coalescence are dealt with in

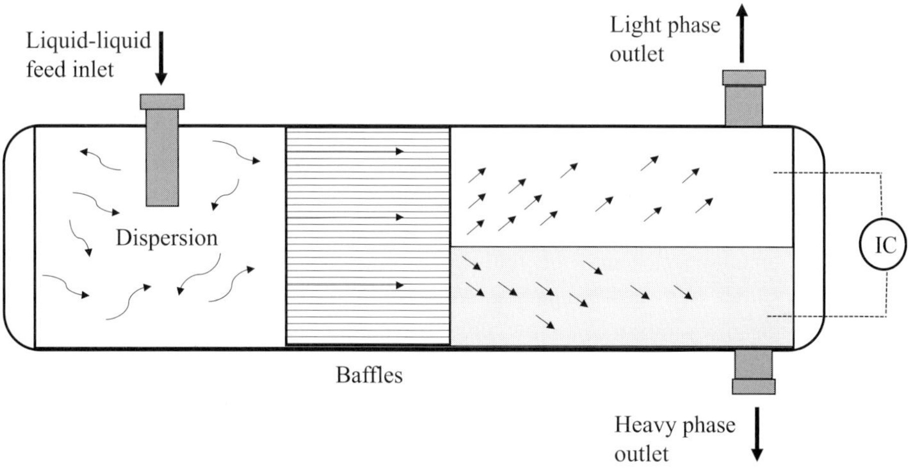

Figure 1.11 Two-phase horizontal settler.
Adapted from Koch-Glitsch

Chapter 7. In practical terms, gravity settling is the mainstay technique adopted in many industrial liquid–liquid processes based on mixer settler technology. This is an "extensive" technique requiring laminar flow conditions and minimization of hydrodynamic mixing. The basic principle is based on dispersed droplets rising or falling according to Stokes law and density difference and coalescing into a planar interface between the two phases. Density difference, viscosity, and interfacial tension represent the important fluid properties that will dictate the rate of phase separation. Thus flow lengths tend to be large, flow velocities tend to be low, and therefore equipment of correspondingly large size with high cross-sectional flow area. Enhancements in gravity settling may be achieved by the incorporation of baffles into the settling chamber, as shown in the right-hand settling chamber shown in Figure 1.2. In this case, two vertical coalescence plates are shown. These provide surfaces that are preferentially wetted by the dispersed phase thus facilitating drop breakdown and coalescence between adjoining elements of spreading liquid derived from the drops. Also the positioning of baffles may increase path length thus providing additional time for phase separation to occur.

The presence of fine solids, finely dispersed gas bubbles, or macromolecular contaminants such as proteins and other surfactants generally have a negative effect upon phase separation.

Some of the fundamental aspects of coalescence phenomena are discussed in Chapter 7. In this introductory chapter, the current technology and equipment types are briefly discussed.

Cusack (2009) lists the main types of phase separators for liquid–liquid systems:

- two-phase horizontal settler
- two-phase horizontal coalescer/settler
- three-phase horizontal coalescer/settler
- three-phase vertical separator
- two-phase vertical coalescer.

Cusak (2009) describe these different types, which are summarized briefly below.

Figure 1.11 shows a diagram of a two-phase horizontal settler.

The two-phase horizontal separator shown in Figure 1.11 comprises four sections. At the left-hand end is the inlet section, shown here with a mixed liquid–liquid feed entering from the top. The section may be equipped with internals or an impeller to promote mixing and turbulence. To the right of this section is a coalescing section that communicates into the settling chamber that is equipped with an interface detector, shown here as a gauge glass. The residence time in this section is critical to achieve clean separation of the two phases, with the light phase leaving through a top port, and the heavy phase leaving through a port in the base of the vessel. An important feature common to most types of liquid–liquid contacting equipment is the section at the right-hand end of the unit, which is for control of the interface. The interface controller (IC) in principle comprises a detector that measures the position of the interface relative to the total overall height of the liquids. The signal from the measuring element determines deviation from a set point and translates that into a feedback signal that will promote adjustment of flowrate. In the event that interface control cannot be established then an alarm would be activated, leading to manual control or shutdown to avoid loss of solvent and product due to entrainment.

Figures 1.13 and 1.14 show a development of the two-stage unit of Figure 1.11, whereby the internals of the separator incorporate two sections equipped with coalescing media of the type shown in Figure 1.12. The first coalescing section is shown to the right of the turbulence isolation plate. The majority of the coalescence occurs here, mainly between the larger dispersed phase drops. This section is followed by a primary settling section, also filled with structured media to intensify the rate of settling. To the right of this section are shown further two sections for "polishing," which involves coalescence of very fine, more stable droplets of dispersed phase, and final settling. The structure and materials of construction of the coalescence and settling media are critical to successful performance, especially the effective voidage, hydraulic mean diameter of the flow spaces, and the wetting characteristics of the media. The wetting characteristics of the media are especially important. If the dispersed phase preferentially wets the packing material it will tend to coalesce; thus, as a general rule, media material that preferentially wet the droplet phase should be employed.

Figure 1.14 shows a further development designed for three-phase (i.e. gas/liquid/liquid) phase separation, such as would be encountered in three-phase homogeneous catalytic reactors. This is a three-phase horizontal coalescer. In this design, the obvious difference compared with the two-phase device is the presence of a significant gas space and mist eliminator in the upper section of the vessel. The lower section for the

Figure 1.12 Structured packing media for high efficiency liquid–liquid separations. With permission from Sulzer

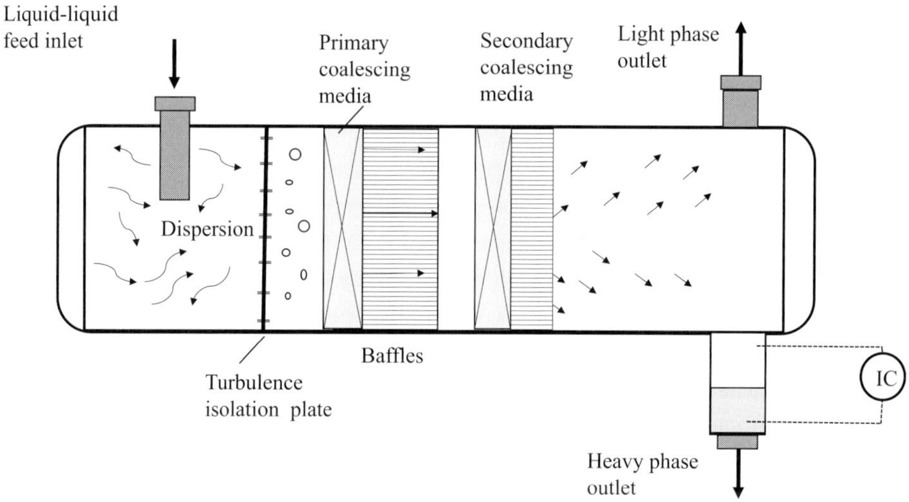

Figure 1.13 Two-phase horizontal settler coalescer.
Adapted from Koch-Glitsch

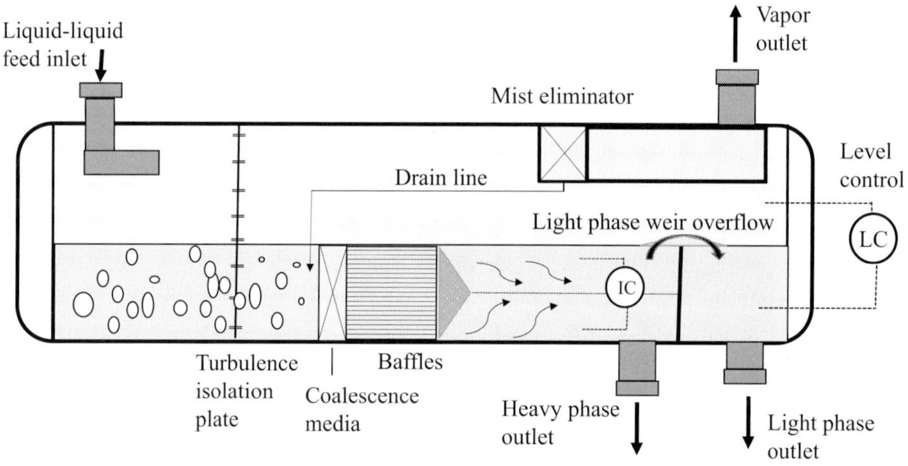

Figure 1.14 Arrangement of a three-phase horizontal coalescer settler.
Adapted from Koch-Glitsch

liquid–liquid coalescence and separation mirrors the arrangement of the two-phase horizontal coalescer settler of Figures 1.11 and 1.13.

Figure 1.15 shows an example of a vertical three-phase separator that operates on the same principles as the horizontal device.

Each of the devices described above are excellent examples of equipment developed and designed to intensify liquid–liquid operations. The results of these developments include smaller equipment size for phase separation, an ability to operate the phase contacting at smaller drop sizes and higher turbulence, thus intensifying mass transfer

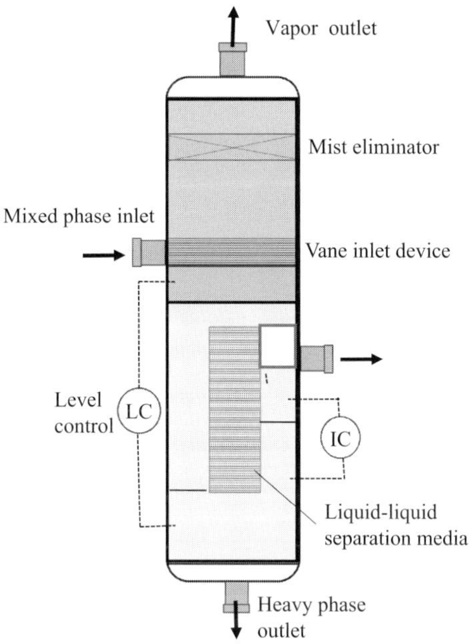

Figure 1.15 Three-phase vertical separator.
Adapted from Koch-Glitsch

rates, and lower incidence of entrainment with corresponding environmental and economic benefits.

Sulzer also describe a range of coalescence equipment based on similar principles. These also include vertical and horizontal types, using a variety of materials that may be tailored to the specific properties of the liquid–liquid system both in terms of corrosion and in terms of interfacial properties. These proprietary designs are claimed to reduce size through the improved phase separation efficiencies. An important performance criterion is the lower limit of the cut-off drop diameter that can in the range of 30–40 μm. Other advantages include application to primary dispersions. Specific applications listed by Sulzer include separation of dispersions remaining after water washing stages; entrainment reduction of both phases in liquid–liquid extraction columns such as LPG amine treaters and hydrogen peroxide and caustic washers; separation of dispersions formed by condensation following azeotropic distillation, as in butanol/water distillation; and separation of liquids following steam stripping. For example, Figure 1.16 shows the arrangement of the horizontal Sulzer Mellaplate coalescer.

Another development is based on the principle described in Chapter 7, where the intensification of coalescence is achieved by contacting the liquid–liquid dispersion with fiber bundles (see Figures 7.6 and 7.7 in Chapter 7). The Dusec™ Coalescers by Sulzer are particularly valuable for the break up of secondary dispersions, where droplets are so small (typically in the range 1–30 μm) that they do not readily wet a surface or settle under gravity. The Dusec Coalescer series comprise an array of fiber-filled

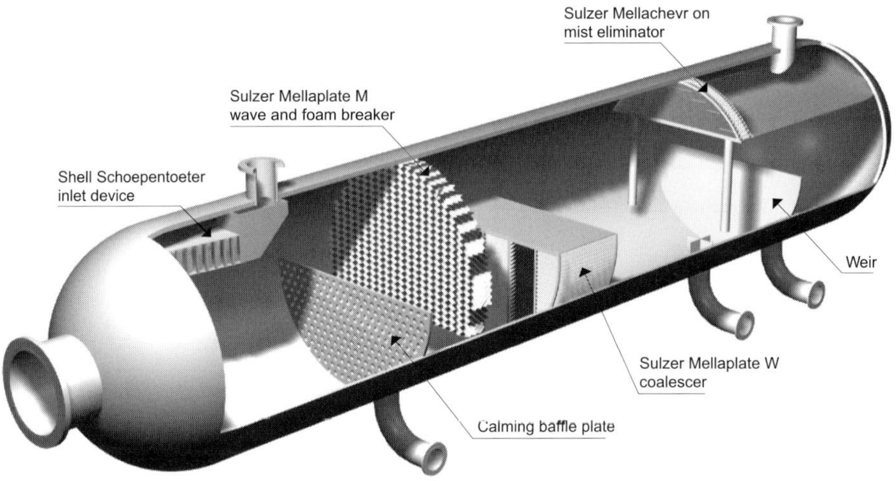

Figure 1.16 Sulzer Mellaplate coalescer.
With permission from Sulzer

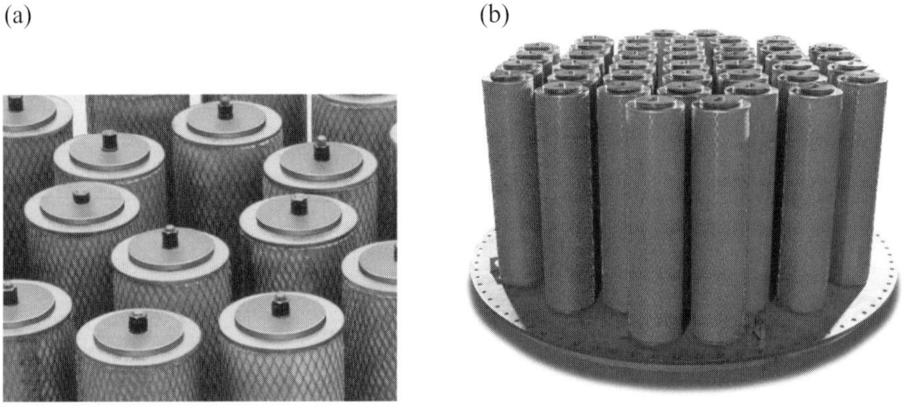

Figure 1.17 The Dusec™ Coalescer: (a), (b) modules.
With permission from Sulzer

cylinders, as shown in Figure 1.17. The cylindrical modules are constructed in cartridge form with the liquid flowing from the center radially outwards. The process of coalescence is enhanced as the liquid mixture encounters resistance in the fiber bundle with the droplets attaching to the fiber surfaces, clustering together, and coalescing. The larger drops resulting from the coalescence experience viscous drag forces resulting in separation from the fiber surface and transport to another part of the fiber bed where the process is repeated until drop sizes greater than 60 μm are achieved. This type of coalescer can be designed either in vertical or horizontal arrangement. Another claim is that effective performance may be achieved for both light phase dispersed and for heavy phase dispersed. The overall arrangement of the coalescer is shown in Figure 1.18.

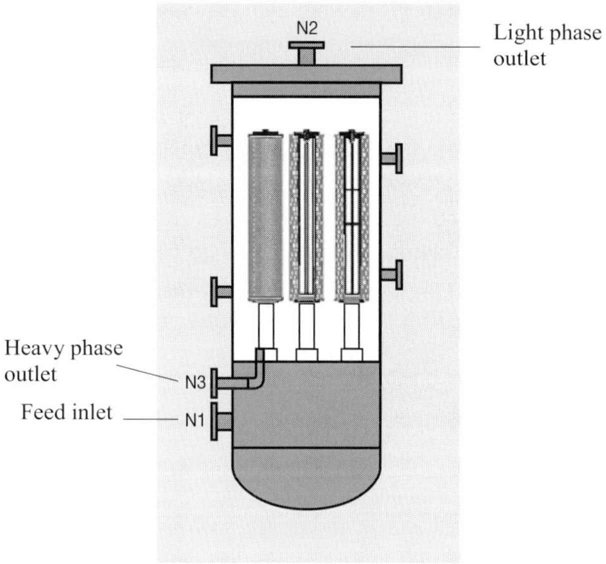

Figure 1.18 The Dusec™ Coalescer with modules in place.
With permission from Sulzer

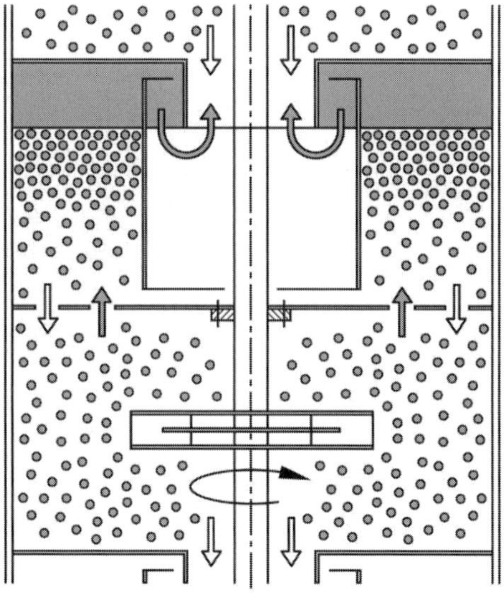

Figure 1.19 The mixer settler column.
With permission from Sulzer

1.3 Mixer Settler Columns

A recent innovative design of mixer settler units, the mixer settler column, is described by Sulzer (Figure 1.19). The design is based on a series of mixer settler units stacked on top of each other to form a column unit. Unlike other liquid–liquid column contactors, the mixer settler column features stages that are hydraulically separated from each other, each with its mixing compartment and settling compartment. A significant feature is that the turbine mixers in each compartment only function as mixers and do not transport the liquid from stage to stage. Therefore mixing conditions can be controlled independently to achieve optimum droplet size. In the Sulzer literature the device is described as similar to the ECR Kühni column that allows for the long residence times that are needed for slow mass transfer systems and for reacting systems with slow reaction kinetics. A further claim is the ability to operate over a very wide range of phase ratios, though specific flow rates through the entire device may be lower than in a truly continuous column contactor due to the need for settling in each stage prior to redispersion.

1.4 Continuous Column Contactors

Continuous column liquid–liquid contactors form a major class of equipment which are ubiquitous in chemical processes and in industries such as the nuclear industry. The development of column contactor design has reached an advanced stage of maturity and recent innovations are confined to incremental improvement of existing designs. This section will present a brief review of column contactors and summarize their evolution to the current state of the art. Some recent innovations and new concepts that have come forward will be reviewed.

One of the earliest designs of continuous column liquid–liquid contactors was the spray column, Figure 1.20; these are described extensively in the literature and standard texts on liquid–liquid extraction (for example, Logsdail and Slater, 1993). The spray column is the simplest and one of the earliest designs of countercurrent liquid extraction device and comprises a column with few internals, and the drops of the dispersed phase rise or fall through the continuous phase according to a balance of gravitational and drag forces. While the spray column provided the advantages of simplicity of construction, and of continuous processing, the main weakness is the presence of back-mixing of both dispersed phase and continuous phases, which leads to major loss of mass transfer efficiency and hydraulic load capacity. The large upward facing arrows in Figure 1.20 indicate the direction of induced axial flows in the continuous phase due to the downward motion of dispersed droplet of the heavy phase. To reduce these effects and to reduce entrainment, large diameter columns are required. Because of the relatively poor mass transfer even in the absence of axial back-mixing, taller columns are required relative to more intensified designs.

The development of packed columns was one of the earliest examples of intensification of liquid–liquid contacting processes in continuous columns. A simple mechanistic description of the enhancement in performance that may be observed in a packed extraction column is depicted in Figure 1.21 (Batey and Thornton, 1989). Batey and Thornton state that the conventional explanation for the improved performance of packed towers over spray columns is in terms of a stabilized, equilibrium droplet size distribution and a reduction in back-mixing as a result of the baffling effect of the packing. They also point out that while these factors are important, effects such as

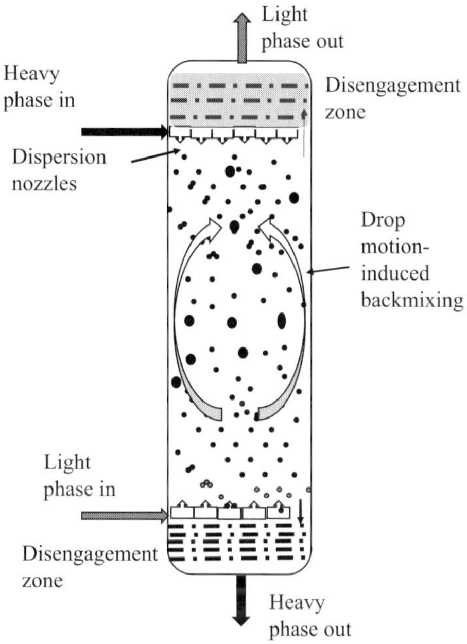

Figure 1.20 Countercurrent spray extraction column, heavy phase dispersed.

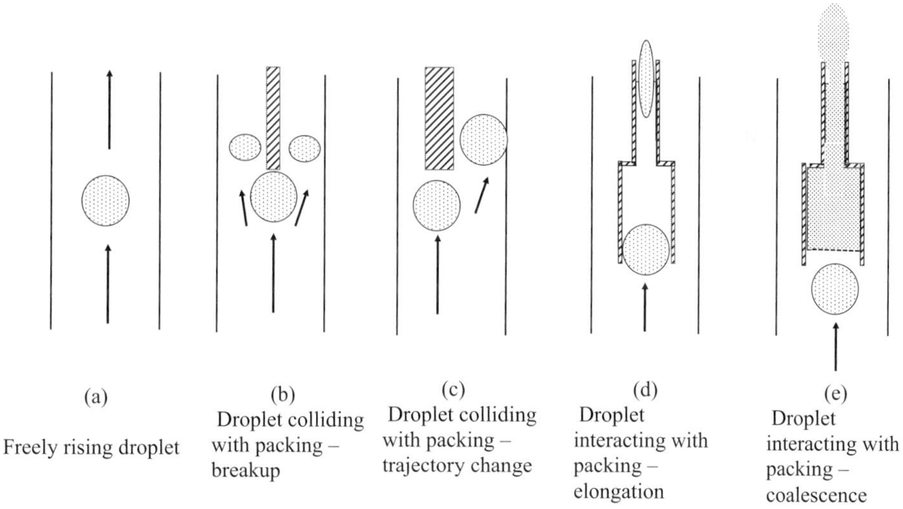

Figure 1.21 Visual simulation of drop behavior in a packed tower.
Reprinted (adapted) with permission from Thornton, J.D. and Batey, W. (1989). Partial mass-transfer coefficients and packing performance in liquid-liquid extraction. *Industrial and Engineering Chemistry Research*, **28**, 1096–1101. Copyright (1989) American Chemical Society

droplet collision, drop-splitting, and coalescence coupled with redispersion play a key role in the enhancement of mass transfer rates.

These positive effects on mass transfer are explained with reference to the diagrams in Figure 1.21 which are based on a droplet phase rising through a more dense continuous phase. Diagram (a) depicts a single drop rising freely, for example in a spray column or in the void space of a packed column. Diagrams (b) and (c) represent the situation where a drop collides "head-on" with a solid packing particle and either splits into smaller drops, or experiences a substantial change in trajectory and possibly some deformation. Diagrams (d) and (e) show droplet elongation and coalescence, either of which may occur as the dispersed phase flows through a restriction in the packing array. In cases (b)–(e) the hydrodynamic disturbances will have a positive effect on mass transfer.

Table 1.2 lists some of the significant full commercial-scale liquid–liquid columns that are available from process equipment manufacturers Koch Modular Process Systems and Sulzer.

The table lists the type of column, gives information taken from the manufacturer's website, and gives the URL address. Each type of column is illustrated in Figures 1.20 and 1.22–1.29.

Packed columns (Figure 1.22) for liquid–liquid operations are in two categories: (i) columns that contain particulate packings such as rings, saddles or spheres; and (ii) columns that contain structured packings, such as is shown in Figure 1.12.

There are many types of packing particles that may be used in liquid–liquid contacting columns. There are a number of general criteria that should be considered, which include mass transfer performance, pressure drop, hydraulic performance, and corrosion resistance. Packed extraction columns are of longstanding and there are many standard works on the design and operation of this equipment. Koch Modular Process Systems, for example, make some general statements for randomly packed columns for liquid–liquid extraction at www.modularprocess.com. These include capacity range of 20–30 $m^3\ m^{-2}$, noting poor efficiency due to back-mixing and wetting, limited turndown flexibility, performance affected by wetting characteristics, and limitations on the choice of dispersed phase.

The more recent developments in packed column design for liquid–liquid systems have focused on so-called structured packings; see Figure 1.12 which shows an example by Sulzer. Structured packings are truly modular in design and comprise a rigid grid assembly or series of grid assemblies that can be installed or removed from the column as one piece or in sections, making installation and replacement much easier compared with random packings. In addition to the advantages of modular design, the geometry of the packing assemblies have been optimized to combine high mass transfer efficiency with much increased hydraulic capacity.

Koch Modular Process Systems also highlight features of structured packings for liquid–liquid extraction that are greatly enhanced, including hydraulic capacities that can be in the range 40–80 $m^3\ m^{-2}$, which is more than double that for random packings. Nevertheless, structured packings can suffer from similar drawbacks to randomly packed systems, including limited turndown, sensitivity to wetting properties of the liquids to be processed, and limitations on choice of dispersed phase.

The ECP packed column by Sulzer consists of a packed bed equipped with distributors. Pulsation may also be incorporated to achieve a degree of mechanical intensification. The design of the packing modules is such that back-mixing in the continuous phase is much reduced. Thus conditions close to plug flow are claimed

Table 1.2 Examples of current commercial intensified continuous liquid–liquid extraction columns.

Column type	Features	Figure number and source
Spray column	Simple construction, few internals, low hydraulic capacity, high risk of back mixing and entrainment	Figure 1.20
Packed column	High capacity: 20–30 $m^3\ m^{-2}\ h^{-1}$ Poor efficiency due to back-mixing and wetting Limited turndown flexibility Affected by changes in wetting characteristics Limited as to which phase can be dispersed	Figure 1.22
Packed column, structured packing	High capacity: 40–80 $m^3\ m^{-2}\ h^{-1}$ Poor efficiency due to back-mixing and wetting Limited turndown flexibility Affected by changes in wetting characteristics Limited as to which phase can be dispersed	Figure 1.22
Schiebel column	Reasonable capacity: 15–25 $m^3\ m^{-2}\ h^{-1}$ High efficiency due to internal baffling Good turndown capability (25%) Best suited when many stages are required Not recommended for highly fouling systems or systems that tend to emulsify	Figure 1.23 https://kochmodular.com/liquid-liquid-extraction/extraction-column-types/scheibel-columns/
Karr column	Highest capacity: 30–60 $m^3\ m^{-2}\ h^{-1}$ Good efficiency Good turndown capability (25%) Uniform shear mixing Best suited for systems that emulsify	Figure 1.24 https://kochmodular.com/liquid-liquid-extraction/extraction-column-types/karr-columns/
Rotating disk contactor	Reasonable capacity: 20–30 $m^3\ m^{-2}\ h^{-1}$ Limited efficiency due to axial back-mixing Suitable for viscous materials Suitable for fouling materials Sensitive to emulsions due to high shear mixing Reasonable turndown (40%)	Figure 1.25 https://kochmodular.com/liquid-liquid-extraction/extraction-column-types/rdc/
Pulsed column	Reasonable capacity: 20–30 $m^3\ m^{-2}\ h^{-1}$ Best suited for nuclear applications due to lack of seal Suited for corrosive applications when constructed out of nonmetals	Figure 1.26 https://kochmodular.com/liquid-liquid-extraction/extraction-column-types/other-columns/

Table 1.2 *(cont.)*

Column type	Features	Figure number and source
	Limited stages due to back-mixing Limited diameter/height due to pulse energy required	
Sieve tray column	High capacity: 30–50 $m^3\ m^{-2}\ h^{-1}$ Good efficiency due to minimum back-mixing Multiple interfaces can be a problem Limited turndown flexibility Affected by changes in wetting characteristics Limited as to which phase can be dispersed	Figure 1.27 https://kochmodular.com/liquid-liquid-extraction/extraction-column-types/other-columns/
Kühni column	Universal use High flexibility for varying process parameters and physical properties Reliable scale-up More than 30 theoretical stages in one single column Standard turndown 1:3 Diameter: 30 mm to 3.5 m Physical properties Density difference >50 kg m^{-3} Viscosity: up to 500 mPa s Interfacial tension: >1mN m^{-1} Phase ratio up to 70:1 Materials Stainless steel and higher alloys, PP/PVDF/PFA	Figure 1.28 www.sulzer.com/en/Products-and-Services/Separation-Technology/Liquid-Liquid-Extraction/Kuehni-Agitated-Columns-ECR

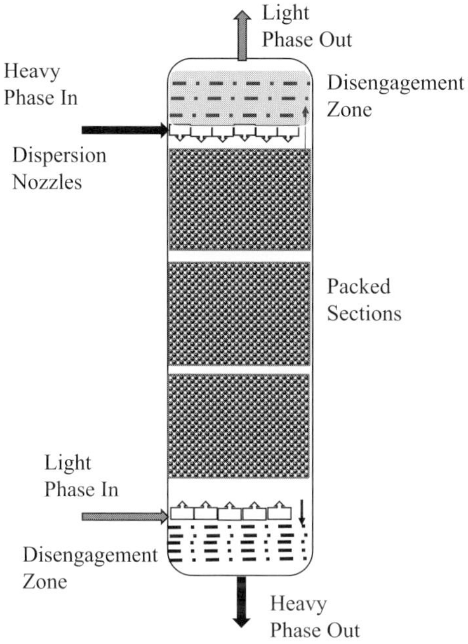

Figure 1.22 Diagrammatic arrangement of a packed column for liquid–liquid operations.

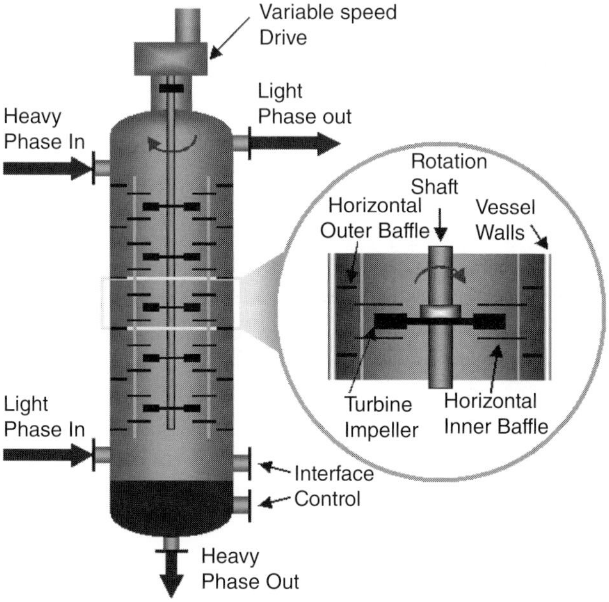

Figure 1.23 Scheibel column.
Adapted from Koch Modular

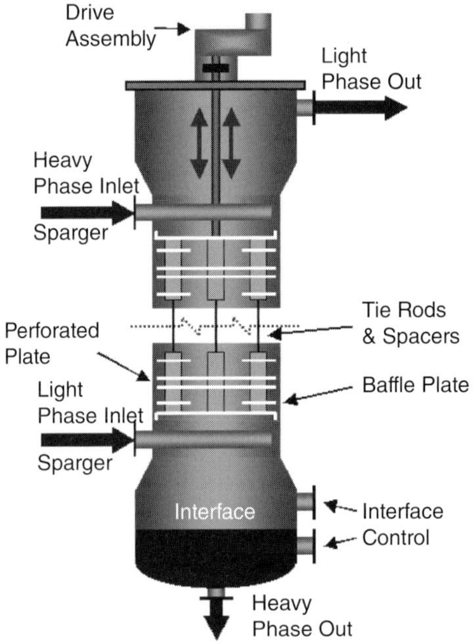

Figure 1.24 Karr column.
Adapted from Koch Modular

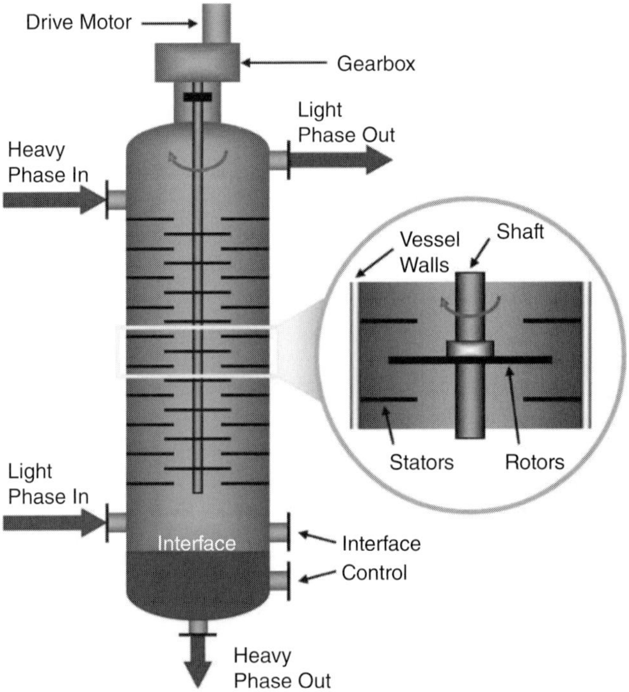

Figure 1.25 Rotating disk column.
Adapted from Koch Modular

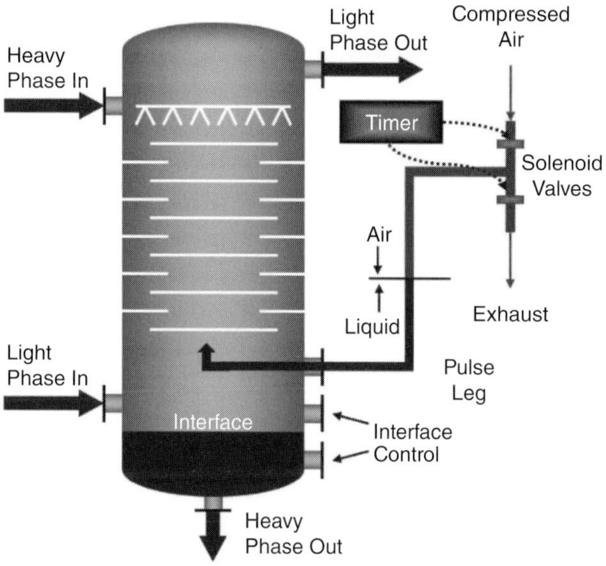

Figure 1.26 Pulsed column.
Adapted from Koch Modular

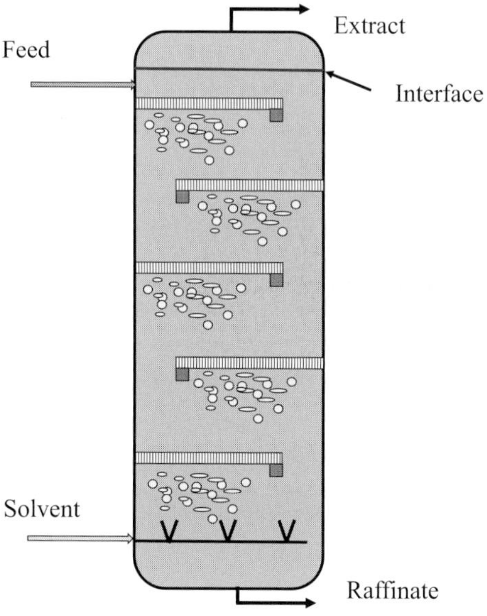

Figure 1.27 Sieve tray liquid–liquid column contactor.
Adapted from Koch Modular

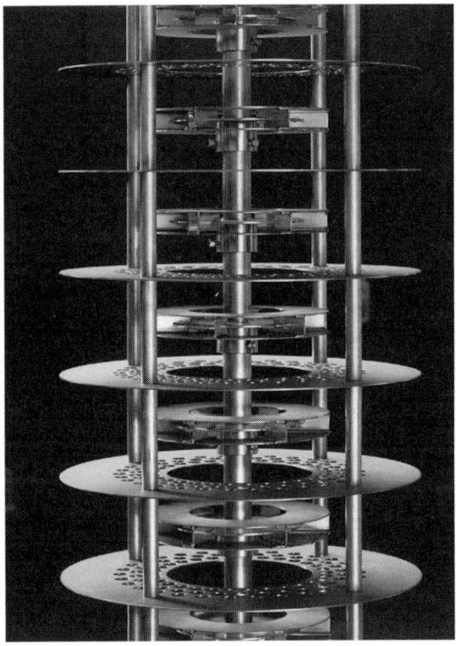

Figure 1.28 Kühni column – internals.
With permission from Sulzer

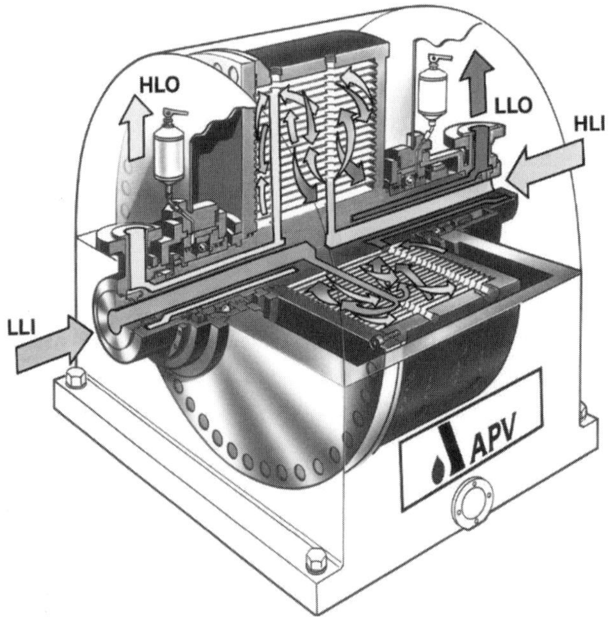

Figure 1.29 Cutaway of the Podbielniak contactor. LLO, light liquid out; LLI, light liquid in; HLO, heavy liquid out; HLI, heavy liquid in.
Courtesy of APV Chemical Machinery

together with achievement of narrow drop size distributions. Further increase in the separation performance is reported through the addition of dual flow perforated plates inserted between the packing elements. The distributors even out the velocity profiles within the column. The manufactures report that the ECP packed column may be operated to handle systems having interfacial tension of less than 2 mN m^{-1} and density differences of less than 50 kg m^{-3}. The design is also such that the mechanical drives can be added to further improve performance.

The agitated Kühni column is another example of an effective mechanically intensified continuous contactor based on a relatively simple design that features flexibility in terms of drive units and features low maintenance.

1.5 Rotary Contactors

Rotary liquid–liquid contactors represent an important class of intensified devices for extractive separations. The main feature of rotary contactors is that the contact and separation of the two liquid phases are conducted in a high gravity centrifugal field. All types involve a cylindrical geometry, exploiting density difference to manage the net flow of each phase radially across the contactor, usually in a countercurrent configuration. The features that distinguish rotary contactors from either mixer settler or column contactors are as follows:

- significant introduction of mechanical energy to create a high gravity environment;
- contact and phase separation integrated into a single device;
- achievement of very short residence times suitable for the recovery of unstable products;
- small plant footprint;
- the ability to process liquids containing solid materials without significant pretreatment.

There are significant differences in the design and operational conditions of rotary contactors, but all have the above features in common. The development of rotary contactors has significantly advanced the application of principles of process intensification to liquid–liquid operations with low solvent and product inventory, increased yields, and smaller equipment size. In some cases there have been significant improvements in process design with fewer unit operations required and corresponding enhancements in product yield. In this section we will focus on reviewing some of the proprietary designs and describe their main features and applications. Fundamental analysis, modeling, and new developments in rotary devices (including high gravity systems) are discussed in Chapter 5.

The technology used in rotary contactor designs is well established and we will review here a selection of the main types used in current industrial liquid–liquid contacting operations. The arrangement, the mode of operation, and some applications of the Podbielniak centrifugal contactors and the Westfalia decanter extractors, will be briefly reviewed. Each group of devices uses centrifugal force to achieve intensified performance and much of the technology specific to extraction emerged from the development of

Table 1.3 Applications of rotating machinery for intensified liquid–liquid processes.

Type of process	Application
Chemical	Herbicides; insecticides; acrylics; lube oil acid treatment; acid or alkali washing; aromatics
Food industry	Vegetable oil degumming; lecithin separation
Liquid ion exchange	Dye separations; citric acid manufacture; hydrometallurgical processes
Simultaneous extract and strip	naphthenic acid recovery; continuous soap-stock acidulation; paraffin neutralization and washing; polyol washing, wastewater treatment, extractive fermentations
Supercritical fluid extractions	Flavor extraction; fragrances

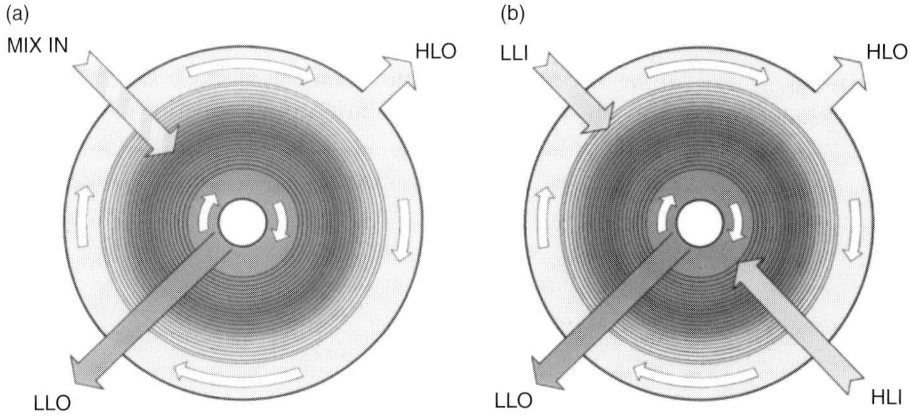

Figure 1.30 Flow directions in the Podbielniak contactor.
Courtesy of APV Chemical Machinery

simple centrifuge designs initially developed for solid–liquid and liquid separation as opposed to continuous extractive separations.

Some well-known applications of rotary contactors for intensified liquid–liquid contacting are listed in Table 1.3.

The Podbielniak contactor was one of the first commercially applied high speed rotating liquid–liquid centrifugal extractors (see Figures 1.29 and 1.30) and has been in use at industrial scale since the 1950s. The contactor comprises a rotor assembly that is made up of a series of concentric perforated cylinders, the entirety of which is spun at high speed on a rotating shaft. The rotor assembly is located within a cylindrical containment. The image shown in Figure 1.30(a) shows the Podbielniak operating purely as a liquid–liquid separator, with the feed stream comprising a liquid–liquid mixture or emulsion. The principle of the Podbielniak for continuous extraction is shown in Figure 1.30(b), with the promotion of countercurrent flow of the two liquid phases in a hydrodynamically intensive environment created by the rotation of the rotor drum. The g forces that are present at high rotational speeds intensify dispersion formation, resulting in large specific areas for mass transfer. This results in highly efficient mass transfer across a short distance. With reference to Figure 1.30, the heavy (more dense) liquid is fed to the central region of the rotor in the stream labeled HLI, while the light phase, labeled LLI, is fed to the periphery of the drum. Due to the density difference and the high g forces to which the two liquids are subjected, countercurrent flow of the two phases is achieved. The heavy liquid progresses to the outer region of the rotor and exits the device in the stream labeled HLO. The light liquid flows toward the center of the rotor, is collected and transported out of the device through conduits located radially across the drum. This flow regime provides for differential countercurrent extraction with up to five theoretical stages in a single unit. The ability to handle liquids of very small density difference is one of the significant features of the Podbielniak and liquids having a density difference of less than 10 kg m^{-3} have been successfully processed.

Podbielniak technology shows numerous advantages, not least of which is that the devices are positioned horizontally, providing greater stability from vibration compared with vertical geometry. Another advantage is that the vessel volume is fully utilized for the liquid–liquid contacting, since the drive machinery is external to the operating volume on account of the availability of reliable seals. The ability to handle liquid rates of up to 150 m^3 h^{-1} is reported.

In common with the other types of rotary contactors, the Podbielniak incorporates advantages consistent with the definition of process intensification – including low material inventory with lower costs and reduced hazard on account of the sealed nature of the unit. Short residence time contact to equilibration is an advantage exploited to significant benefit for the separation and purification of unstable molecules. The recovery of penicillin G is a classic example of one of the early applications of Podbielniak-based technology. Other documented pharmaceutical applications include vaccines, hormones, vitamins, and other pharmaceuticals.

1.5.1 GEA Westfalia Rotary Extractors

GEA in their product literature (GEA – Liquids to Value) list the major advantages of centrifugal extraction with reference to their range of decanter extractors and mixers as shown in Table 1.4.

The range of applications for this technology is quite wide and GEA cite examples of industrial processes for production of antibiotics, statins, hormones, plant extracts, polycarbonates, and pectins. The production of each of these important well-known products either would not be possible, or at least would be very expensive and impractical, in the absence of appropriate intensive centrifugal liquid–liquid contactors.

The GEA decanter extractor (Figure 1.31) is a type of rotary contactor that cleverly combines the advantages of centrifugal intensification of liquid–liquid contact with efficient phase separation, both integrated into a single device. With this design Westfalia Separator has succeeded in developing a centrifuge which combines the two essential process stages in solvent extraction. The design also meets the very stringent hygiene and safety requirement for applications in the food, pharmaceutical, and biotechnology industries.

Table 1.4 Advantages of centrifugal extraction (GEA)

Advantages of centrifugal extraction, Westfalia separator	Results
Low phase hold-up	Low solvent requirement, thus lower costs
Short contact time	Reduce product degradation, hence enhanced product yield
High stage efficiency	Reduce number of stages and hence reduced capital costs
High load range and optimum throughput capacity with minimum space requirement	Reduced capital costs
Narrow residence time distribution	Improved mass transfer, reduced back-mixing and thus higher efficiency
Can handle low density difference liquids and high viscosity liquids	Extends the range of solvents that may be used and improves mass transfer
Efficient phase separation	Achieve higher yields and reduce solvent losses
Optimum mass transfer with small drop size	Higher process yield
Operation not affected by presence of solids	Improves process security, avoids product losses in solids removal in pre-treatment

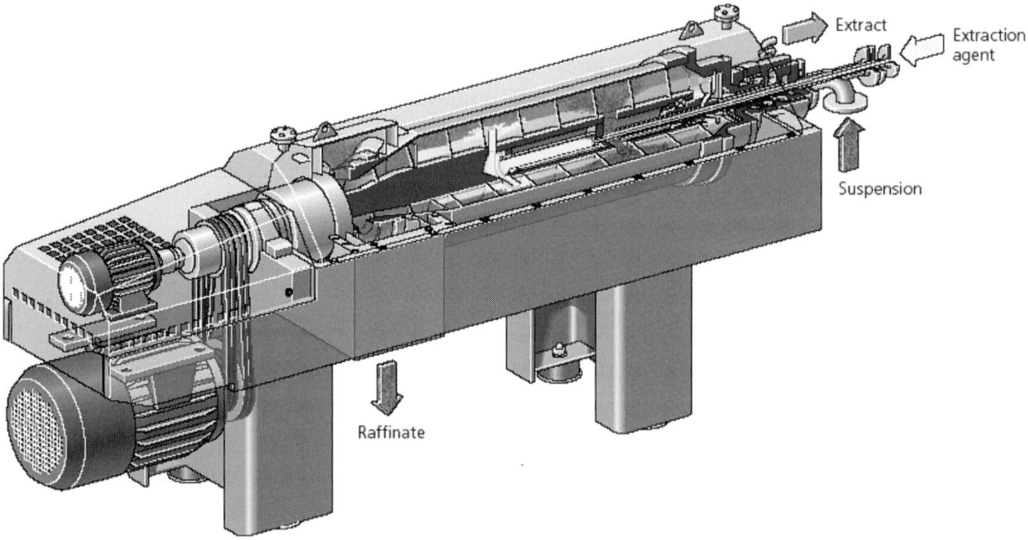

Figure 1.31 The decanter extractor.
With permission from GEA

This decanter extractor allows suspensions with a high solids content to be processed, making it suitable for application to direct extraction of compounds such as antibiotics directly from unfiltered fermentation broth with high biomass content, including mycelium, whole cells, biomass filaments, cell debris, and biopolymer by-products. From a process design angle the need for prefiltration of the feed liquor can be avoided, saving on capital and at the same time improving overall product yield. It is claimed that for some pharmaceutical fermentation products yield values of up to 98% can be achieved. Another advantage of direct extraction compared with conventional countercurrent extraction is that apart from avoiding the need for filtration equipment, filter aids such as precoat are no longer required, thus effecting further savings. Finally, since prefiltration is eliminated, the fermentation broth is not diluted by filter wash water, reducing the solvent requirement and the risk of infection and product contamination. The amount of waste water is also reduced.

These decanters are designed to perform direct extractions from feed liquid containing significant suspended solids. Direct extraction of antibiotics from fermentation broth is a well-established example. The principle of operation of the decanter-extractor is that feed suspension (or broth) to be processed is fed through a series of distributor slots into the scroll assembly located inside the bowl. The feed liquid then enters the countercurrent extraction zone of the bowl, where it contacts the extracting solvent. This is introduced close at the opposite end of the bowl to (the right-hand end in Figure 1.31) and then flows right to left followed by reversal of the flow to be discharged as the extract. The rotating scroll conveys the solid phase along the axis of the contactor and is discharged at the same point as the liquid raffinate phase. The extracting solvent flows countercurrent to the liquid/solid suspension following entry into the contacting zone, through a series of apertures located at intervals along the axis of the contactor. Efficient phase separation is achieved in clarification zones located close to the outer vessel containment. The contactors are designed to be gas-tight thus reducing the risk of leakage of flammable or toxic solvent vapors. In addition, the gas-tight containment also allows inert gas blanketing to reduce fire risk.

Another type of intensified liquid–liquid contactor, also by GEA, is the disc-type separator; these are equipped either with a solid-wall bowl or a self-cleaning disc-type bowl. Both types are used for liquid–liquid extraction processes.

Figure 1.32 shows a separator with a solid-wall disc-type bowl. The mode of operation, which applies either to solid–liquid mixtures or to liquid–liquid mixtures, involves a feed arrangement that minimizes the shear forces, which is important for shear-sensitive products. The mixture is separated into light phase and heavy phase in the disc stack and are is discharged under pressure by means of the centripetal pumps located at the outlets. The use of a disc stack separator provides area for coalescence and separation that is much greater than the equivalent area available in a conventional single-chamber centrifugal device. The bowl can be equipped with discharge bore holes so that residual material can be drained from the bowl when operation is halted. The advantage of this arrangement is that the exposure to process media can be minimized during maintenance. Cleaning in place (CIP) is also possible without the bowl being opened. The choice of centripetal pumps also enables the separating zone to be optimized even when there is a substantial density difference between the products. The separating zone can be adjusted by means depending on the volumetric flowrate of feed to be handled. This permits efficient separation even with significant fluctuations in the composition of the phases. There are also gas-tight versions for processes involving flammable or potentially explosive liquids, the separator is equipped with

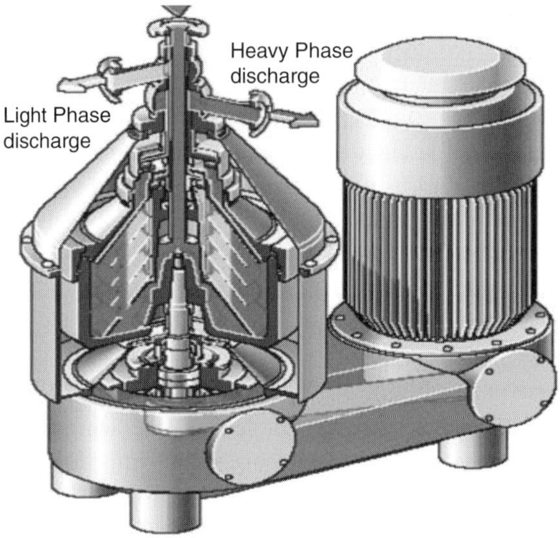

Figure 1.32 Disc-type separator for liquid–liquid extraction with solid-wall bowl. With permission from GEA

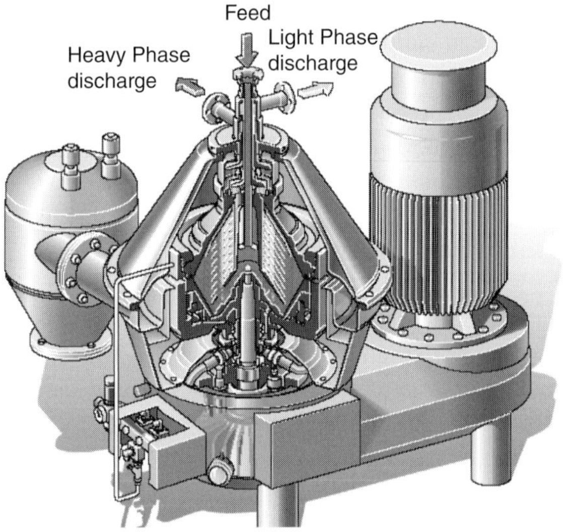

Figure 1.33 GEA separator with self-cleaning disc-type bowl for liquid–liquid extraction. With permission from GEA

explosion-protected components, and provision of flame-proofing where necessary is possible. The solid-wall disc-type separator is used primarily for separating liquid mixtures with zero or minimal solid contents (less than $<0.1\%$ by volume).

Self-cleaning separators (Figure 1.33) are able to discharge the separated solids at full bowl speed. They are able to process feeds with a solids content of up to 7% (by volume) although the primary function of this type of separator is for liquid–liquid separation and not primarily for removal of solids.

1.6 Oscillatory Flow Contactors

The oscillating flow baffled reactor (OFBR) is another concept that has been successfully demonstrated for the intensification of liquid–liquid contacting operations. The basic principle of oscillatory flow is similar to that deployed in the pulsed column; however, the control of the hydrodynamics is achieved in a different manner. Whereas the pulsed plate column comprises a series of perforated horizontal plates fitted along the axis of the column, the OFBR achieves optimal mixing with a series of circular baffles fitted to the interior wall of the column, as shown in Figure 1.34 (Mignard et al., 2006). The oscillatory flow is achieved by pulsation of the two liquid phases, which are held up in the column at any point in time. The diagram in Figure 1.34(a) shows the arrangement of the OFBR. It shows that the main contactor does not rely on a vertical configuration but rather that the reactor can be designed as a series of reversed column sections, which results in significant space saving, and avoiding the need for high bay space. The example shown is for continuous mode operation, with the dispersed phase and continuous phase feeds shown. The flow arrangement illustrated in Figure 1.34(b,c) shows in outline the flow patterns. Figure 1.34(b) shows the pattern in the upward cycle of oscillation, noting the vortex formation on the trailing edge of the baffle and the secondary vortex formation in the compartment itself. Both these led to high degrees of mixing that is confined to the section of column between each pair of baffles. Figure 1.34 (c) shows the corresponding flow patterns in the downward cycle of oscillation.

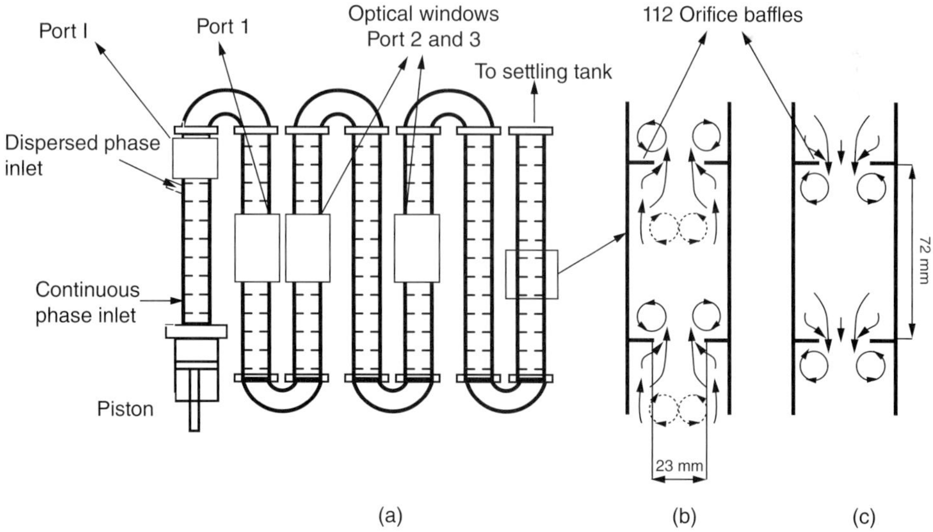

Figure 1.34 (a) Arrangement of the OFBR. (b) Enlarged view, showing the outline flow profiles during the upward cycle of oscillation. (c) Enlarged view, showing the outline flow profiles during the downward cycle of oscillation.

Reprinted (adapted) with permission from Mignard, D., Lekhraj P., Amin, L. P., and Ni, X. (2006). Determination of breakage rates of oil droplets in a continuous oscillatory baffled tube. *Chemical Engineering Science*, **61**, 6902–6917.

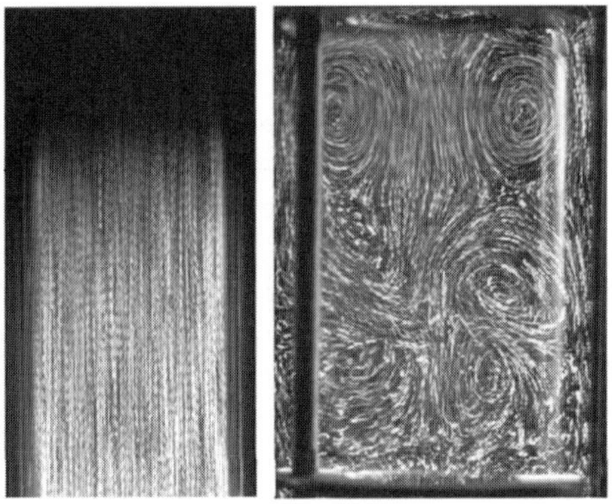

(a) (b)

Figure 1.35 (a) Visualization of flow in a smooth walled column. (b) Visualization of flow in an oscillatory baffled column.
Reprinted with permission from Ni, X., Mackley, M. R., Harvey, A. P., et al. (2003). Mixing through oscillations and pulsations, a guide to achieving process enhancements in the Chemical and Process industries. *Transactions of the Institution of Chemical Engineers*, Part A 81, 373–383. Copyright (2003) Elsevier B.V. All Rights Reserved

The significance of the flow patterns is that each section of the column between each pair of baffles may be considered as a single stirred tank contactor with controlled mass transfer conditions. In the column shown, there are a number of such sections comprising the whole length of the contactor, thus the complete assembly can be considered as a sequence of stirred tanks in series. Classical chemical reactor theory shows that increasing number of stirred tanks in series tends to approach plug flow behavior for the entire unit. Unlike a conventional plug flow system, high rates of heat and mass transfer may be promoted by the localized mixing that is promoted in each "tank."

Images shown in Figure 1.35 show typical flow patterns in an OFBR during the imposition of oscillatory motion (Ni et al., 2003). The patterns observed in an unbaffled tube, Figure 1.35(a) are compared with the patterns in the OFBR, with baffles, as shown in Figure 1.35(b). The image in Figure 1.35(a) clearly shows no evidence of enhanced mixing and thus provides no advantage. On the other hand, the image in Figure 1.35(b) shows strong evidence of recirculatory flows and vortex formation under moderate oscillatory conditions. The authors claim that in this type of device control of mixing to a very high degree may be achieved. Mixing regimes ranging from plug flow through to intense mixed flow conditions are possible. The application of the OFBR concept stretches beyond liquid–liquid extraction into other multiphase processes requiring controlled shear conditions such as in polymerization (Gaidhani, McNeil, and Ni, 2002) and crystallization (Chew et al., 2004). Application to gas–liquid systems has also been demonstrated.

The design of scaled-up OFBR devices is claimed to be straightforward using linear geometric scaling, taking account of dynamic similarity based on oscillatory Reynolds number and linear flow Reynolds number (Smith, 2000). Work by Smith showed that axial dispersion in an OFBR was unaffected at different scales if geometric and dynamic similarity is maintained. This approach was validated in the design of industry-scale contactors described by Harvey and Stonestreet (2002).

References

Baird, M. H. I. and Rao, N. V. R. (1995). Power dissipation and flow patterns in reciprocating baffle-plate columns. *Canadian Journal of Chemical Engineering*, **73**(4), 417–425.

Baird, M. H. I. and Stonestreet, P. (1995). Energy dissipation in oscillatory flow within a baffled tube. *Chemical Engineering Research and Design*, **73**, 503–511.

Batey, W. and Thornton, J. D. (1989). Partial mass-transfer coefficients and packing performance in liquid-liquid extraction. *Industrial and Engineering Chemistry Research*, **28**, 1096–1101.

Bezanehtak, K., Dehghani, F., and Foster, N. R. (2004). Vapor-liquid equilibrium for the carbon dioxide + hydrogen + methanol ternary system. *Journal of Chemical Engineering Data*, **49**, 430–434.

Chew, C. M., Ristic, R. I., Dennehy, R. D., and De Yoreo, J. J. (2004). Crystallization of paracetamol under oscillatory flow mixing conditions. *Crystal Growth & Design*, **4**(5), 1045–1052.

Cloutier, L. and Cholette, A. (1968). Effect of various parameters on level of mixing in continuous flow systems. *Canadian Journal of Chemical Engineering*, **46**(2), 82–88.

Cusack R. (2009). Rethink your liquid-liquid separations. *Hydrocarbon Processing*. **June edition**, 53–60.

Earle, M. J. and Seddon, K. R. (2000). Ionic liquids; Green solvents for the future. *Pure and Applied Chemistry*, **72**(7), 1391–1398.

Fan, Y. and Qi, J. (2010). Lipase catalysis in ionic liquids/supercritical carbon dioxide and its applications. *Journal of Molecular Catalysis B: Enzymatic*, **66**, 1–7.

Gaidhani, H. K., McNeil, B., and Ni, X. (2002). Production of pullulan using an oscillatory baffled bioreactor. *Journal of Chemical Technology and Biotechnology*, **78**, 260–264.

Giralico, M. A., Post, T. A., and Preston, M. J. (1995). Improve the performance of your copper solvent extraction process by optimizing the design and operation of your pumper and auxiliary impellors. *SME Annual Meeting (March 6–9)*, preprint number 95-189, Denver, Colorado.

Harvey, A. P. and Stonestreet, P. (2002). A mixing-based design methodology for continuous oscillatory flow reactors. *Transactions of the Institution of Chemical Engineers, Part A*, **80**, 31–44.

Hert, D. G., Anderson, J. L., Aki, S. N. V. K., and Brennecke, J. F. (2005). Enhancement of oxygen and methane solubility in 1-hexyl-3-methylimidazolium bis(trifluoromethylsulfonyl) imide using carbon dioxide. *Chemical Communications*, 20, 2603–2605.

Holbrey, J. D., Rooney, D., and Seddon, K. R. (1999). Use of ambient-temperature ionic liquids as reaction media for clean synthesis. *Abstracts of Papers of the American Chemical Society*, **218**, U800–U800.

Jones, D. O. (1996). Process Intensification of Batch Exothermic Reactors. HSE Report Research Contract Report 105/1996. Health and Safety Executive, United Kingdom.

Keskin, S., Kayrak-Talay, D., Akman, U., and Hortacsu, O. (2007). A review of ionic liquids towards supercritical fluid applications. *Journal of Supercritical Fluids*, **43**, 150–180.

Knez, Z., Markocic, E., Leitgeb, M., et al. (2014). Industrial applications of supercritical fluids: A review. *Energy*, **77**, 235–243.

Logsdail, D. H. and Slater, M. J. (1993). *Solvent Extraction in the Process Industries: ISEC 93*. London and New York: SCI and Elsevier Applied Science.

Mignard, D., Lekhraj P., Amin,L. P., and Ni, X. (2006). Determination of breakage rates of oil droplets in a continuous oscillatory baffled tube. *Chemical Engineering Science*, **61**, 6902–6917.

Ni, X., Mackley, M. R., Harvey, A. P., et al. (2003). Mixing through oscillations and pulsations, a guide to achieving process enhancements in the Chemical and Process industries. *Transactions of the Institution of Chemical Engineers*, Part A **81**, 373–383.

Paljevac, M., Knez, Z., and Habulin, M. (2009). Lipase-catalyzed transesterification of (*R,S*)-1-phenylethanol in SC CO_2 and in SC CO_2/ionic liquid systems. *Acta Chimica Slovenica*, 56, 399–409.

Post T. A. (2003). Scale-up of pumpers & mixers for solvent extraction. Proceedings of Hydrodmetallurgy 2003, Proceedings of the 5th International Symposium. The Minerals, Materials and Metals Society, Warrendale, PA, 1037–1052.

Qiu, Z., Zhao, L., and Weatherley, L. R. (2010). Process intensification technologies in continuous biodiesel production. *Chemical Engineering & Processing: Process Intensification*, **49**(4), 323–330.

Ramshaw C. (1999). Process intensification and green chemistry. *Green Chemistry*, February, G15–G17.

Skelland, A. H. P. and Ramsay, G. G. (1987). Minimum agitator speeds for complete liquid-liquid dispersion. *Industrial and Engineering Chemistry Research*, **26**(1), 77–81.

Skelland, A. H. P. and Seksaria, R. (1978). Minimum impeller speeds for liquid–liquid dispersion in baffled vessels. *Industrial and Engineering Chemistry Process Design and Development*, **17**(**1**), 56–61.

Smith, K. B. (2000). The scale-up of oscillatory flow mixing. Ph.D. Thesis, Cambridge University, UK.

Tsouris, C. and Porcelli, J. V. (2003). Process intensification – has its time finally come? *Chemical Engineering Progress*, October, 50–55.

Welton, T. (1999). Room-temperature ionic liquids. Solvents for synthesis and catalysis. *Chemical Reviews*, **99**, 2071–2083.

Wilkes, J. S. (2004). Properties of ionic liquid solvents for catalysis. *Journal of Molecular Catalysis A: Chemical*, **214**, 11–17.

2 Droplets and Dispersions

2.1 Introduction

The importance of droplet behavior in determining the performance of liquid–liquid extraction equipment is very well known. The size and holdup of the droplet phase have a profound influence on the mass transfer rate in both batch and continuous contactors, and they also dictate the hydraulic requirements of the contactor for the efficient handling of both dispersed and continuous phase throughputs. The size distribution of drops and the dynamic (as opposed to the static) dispersed phase holdup directly affect the interfacial area per unit volume available for mass transfer. Another significant factor is the frequency at which drops coalesce and break up during their passage through the contactor. A high frequency of breakup/coalescence/breakup generally enhances mass transfer due to the macro disturbances that are imparted to the liquid phase comprising the interior of each drop. The frequency of drop breakup and coalescence may also affect the effective size distribution of drops within the contactor and thus the interfacial area. A further factor that affects the mass transfer rate is the linear velocity of the drop phase relative to the continuous phase, since the higher the velocity, the higher the interfacial shear rate. This will enhance the mass transfer rate in the continuous phase surrounding the drop. Another important factor affecting the overall performance of the contactor is the degree of mixing (and back-mixing) present in the continuous phase, and the extent to which macro-scale back-mixing and forward-mixing of individual drops occur during their passage through the contactor. In this respect the residence time distributions of the drops and of the continuous phase are of some significance as is the case in continuous chemical reactor systems. Secondary flows in the nondispersed phase created by the motion of the dispersed phase are also an important factor in determining the overall mixing and velocity distribution in the liquid–liquid continuum.

The study of droplet behavior (size and motion) in liquid–liquid systems falls into three area: (i) discrete droplet behavior; (ii) swarming droplets (referred to here as dispersions); (iii) emulsions. Understanding of all three regimes is essential in the context of application, process design, and process intensification of liquid–liquid operations. Early research on liquid–liquid systems focused on establishing understanding of the factors that determine drop size of newly formed drops at a nozzle, detaching into the continuous phase (Figure 2.1) (Harkins and Brown, 1919; Hayworth and Treybal, 1950; Meister and Scheele, 1969; Scheele and Meister, 1968; Skelland

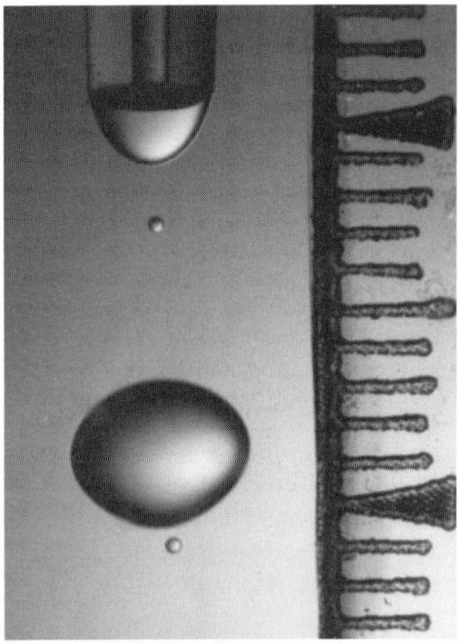

Figure 2.1 Photograph of a newly formed drop of water in n-decanol.

and Wellek, 1964; Wellek, Agrawal, and Skelland 1966). The work was largely empirical but established many useful correlations based on dimensional analysis, with valid relationships between drop size, physical properties of both phases, and dimensions of the nozzle at which the drops are forming. Knowledge of the size and motion of single discrete droplets is essential for establishing interfacial mass transfer rates. Knowledge of discrete drop motion is also essential for providing insights into coalescence phenomenon in systems comprising swarms of droplets where data on collision frequency may be an important requirement. Understanding of drop behavior is also a requirement for predicting hydraulic performance (e.g. in continuous column equipment).

In the context of intensification of liquid–liquid processes, understanding the motion of droplets in hydrodynamic environments involving high gravity fields, electrical fields, and oscillatory fields is a prerequisite, and requires an initial understanding of the mechanics of discrete drop motion.

In industrial liquid–liquid extractions systems, for example those involving mixer settler equipment and continuous columns, the main interest is in swarms of droplets in which the motion of each individual drop is almost certainly influenced by the presence of the other droplets (Figure 2.2). Collisions between droplets may either result in coalescence or at least impact the motion of both drops in a significant manner. Secondary flows promoted in the continuous phase due to shear forces and due to volume displacement are also significant factors in swarming droplet systems, impacting hydraulic design and phenomena such as entrainment and axial back-mixing.

Efficient performance of industrial liquid–liquid extraction equipment is highly dependent on controlling droplet size, minimizing back-mixing, and ensuring efficient

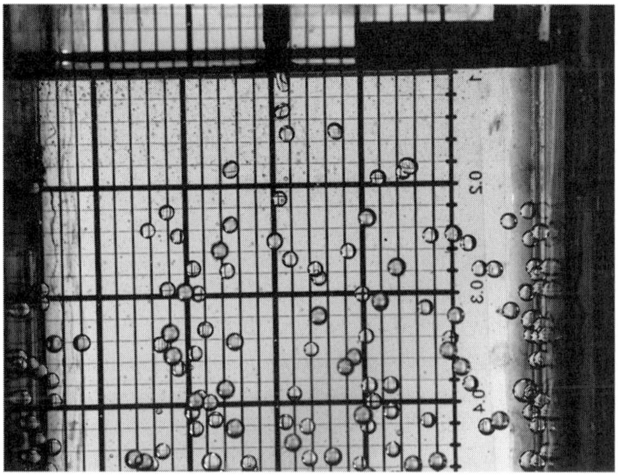

Figure 2.2 Photograph of a swarming droplet system.

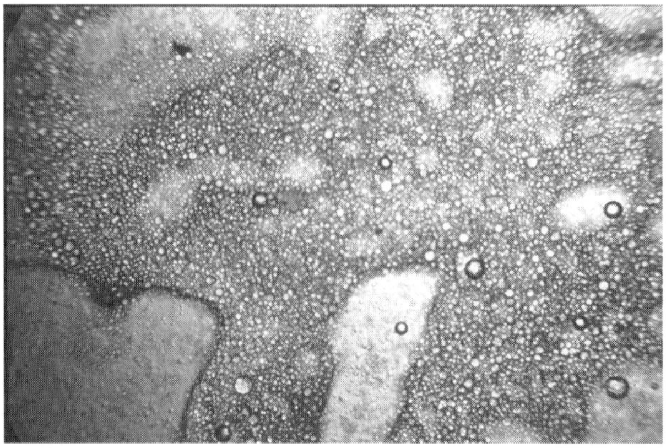

Figure 2.3 Micrograph of a stable emulsion.

phase separation (coalescence) post extraction/reaction for product recovery. All of the fore-mentioned require an understanding of drop motion and size behavior. The issue of coalescence is very important and highlights one of the main optimization issues in liquid–liquid system design for extraction. On the one hand, maximization of specific interfacial area by the production of multiple very fine droplets and the promotion of highly turbulent conditions are desirable for high rates of interfacial mass transfer. On the other hand, the liquid–liquid mixture must be readily separable into two very discrete phases if the process is to be efficient. As will be seen, many factors influence the balance between two apparently conflicting requirements and the choice of operating conditions.

The third major category of liquid–liquid system is that of emulsions (Figure 2.3).

Emulsions are important in many parts of the food and pharmaceutical industry where a primary interest is in the generation of stable emulsions that meet food and drug safety criteria, and exhibit controlled release performance. The criteria in this context are somewhat different from those relevant to chemical engineering separations. Emulsion stability, viscosity, density, flavor release rate, and compatibility with other food ingredients are significant factors in food industry applications of emulsions. In the pharmaceutical industry, emulsions are used either directly for drug delivery or are used in the production of other drug delivery vehicles such as microcapsules. Control of drop size distribution is of importance, strongly relating to rates of release and production efficiency. The fundamental phenomena remain the same, thus considerations of mass transfer and understanding those factors that influence droplet size, coalescence, and interfacial stability are of interest.

Another application of emulsions relevant to process intensification is for the production of emulsion liquid membranes (ELM) and micellar systems for separations and bioseparations. Bioseparations in particular require separation systems that are relatively benign to biomass and enzymes, are highly selective, and that provide efficient separation power. The factors that control drop size during emulsion formation, including hydrodynamic conditions, surfactant dosing, and density difference, must be understood. As is discussed in Chapter 7, understanding the conditions under which emulsion breakage occurs is also essential. The development of microreactors and microcontactors involving liquid–liquid systems as a component of process intensification, also requires understanding of droplet and dispersion behavior. For example, Cherlo, Kariveti, and Pushpavanam (2010) discuss the development of microfluidic devices for liquid–liquid contact in which the contact is achieved in a slug flow environment in microchannels (Figure 2.4).

The development of microfluidic devices for liquid–liquid extraction and as reactors for liquid–liquid reactions is a relatively new emerging area. The fundamental factors that control droplet (slug) size, shape, and motion are not well understood and is an area requiring further research. The impact of successful microfluidic and microchannel technology as a vehicle for process intensification should not be understated.

All of the above categories of liquid–liquid system involve the formation and motion of droplets in which the droplet surface, be it in an emulsion, in a stirred tank, in a centrifuge, or in an extraction column, is surrounded almost entirely by a continuous phase.

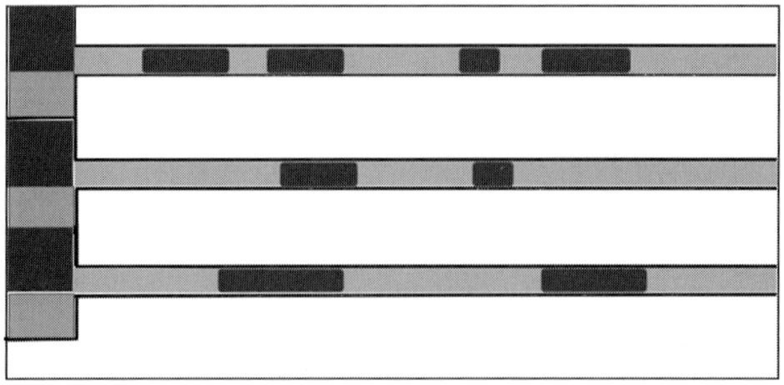

Figure 2.4 Representation of slug flow in a microchannel contactor. Adapted after Cherlo et al. (2010)

In this chapter the mechanisms of drop formation are reviewed with detailed presentation from the literature of the controlling relationships expressed in terms of physical properties, system geometry, and phase flow conditions. The relationships governing the size of discrete drops are reviewed in some detail. The governing equations for the motion of discrete drops in unhindered flow are then presented, followed by consideration of motion of swarming drops and of dispersions. The problem of prediction of drop size and of drop size distribution is considered with a review of the relevant literature relating to dispersion from nozzles and drop size behavior in stirred tanks and in mechanical column contactors.

2.2 Drop Size: Discrete Drops

The design and optimization of equipment used for liquid–liquid processes requires a knowledge of drop sizes and holdup in order to determine values of interfacial area for mass transfer calculations. A further prime consideration is the need for drop size data and holdup for the determination of hydraulic conditions, including pressure drop, flooding, and entrainment, for example in column contactors. Drop size is central to successful understanding of the internal hydrodynamics in contactors, that includes not only mass transfer but also mixing regime and coalescence phenomena.

A great deal of experimental work has been done in the study of motion of bubbles, drops, and particles (Clift, Grace, and Weber, 1978; Grace, Wairegi, and Nguyen, 1976). The most frequent interpretations are based upon the correlation of experimental data using the force balance acting on single droplets, and accounting for the effects of viscous, gravitational, interfacial, and drag forces. Approximate theoretical solutions have been obtained for the case of drops moving with high Reynolds numbers (Harper and Moore, 1968; Moore, 1959, 1963; Parlange, 1970) and for the case of drops moving with low Reynolds numbers (Taylor and Acrivos, 1964). However, the motion of drops and bubbles with a significant degree of deformation in the presence of mass transfer has not been solved thoroughly. The main reason is the problem of dealing with an unknown shape of the free surface and allowing for the dynamic nature of the interface. In general, the early theoretical analysis of the motion of deforming drops and bubbles has only been done in "ideal" situations, such as with inviscid fluids or with those comprising a highly viscous liquid and with droplets close to ideal spherical shape.

Early work on the prediction of drop size was based on the mechanics of single drop formation at a nozzle and subsequent motion in an unhindered environment with the assumption of zero interaction with other neighboring droplets. In this highly controlled scenario and assuming spherical drops, drop size on detachment is predictable based on physical properties, nozzle geometry, and flow rate. Even so, empirical correction factors are required for accuracy of prediction (Harkins and Brown, 1919; Hayworth and Treybal, 1950; Letan and Kehat, 1967; Scheele and Meister, 1968).

Harkins and Brown's (1919) approach assumed that the drop volume at detachment was determined by the force balance of gravitational forces and interfacial tension forces, thus, the interfacial tension force, F_σ, is given by:

$$F_\sigma = \pi D_N \sigma \quad (2.1)$$

where D_N is the nozzle diameter and σ the interfacial tension.

The volume of the drop to overcome the interfacial tension force is given by:

$$V_\sigma = \frac{\pi D_N \sigma}{\Delta \rho g} \quad (2.2)$$

where $\Delta\rho$ = the density difference between the two phases and g is acceleration due to gravity.

Harkins and Brown (1919) first identified the nonideal breakoff of a newly formed single drop at a nozzle and established a relationship for determining the fraction of the drop volume that did not detach and thus the actual volume of the detached drop.

This gave rise to the Harkins correction factor $\psi\left(\frac{D_N}{D_S}\right)$ where $\psi\left(\frac{D_N}{D_S}\right)$ is correlated as function of the volume of the drop and the nozzle diameter.

Application of the Harkins correction factor thus results in a modified drop volume prior to detachment:

$$V_S = \frac{\pi D_N \sigma}{\Delta \rho g} \psi \frac{D_N}{D_S} \quad (2.3)$$

Figure 2.5 shows the four principle stages of formation and detachment of a single drop from the nozzle. The semiempirical models of Scheele and Meister (1968) and others utilize the concept here for expressing the various forces upon which the size at detachment is contingent. Figure 2.6 shows in reality the phenomenon of drop

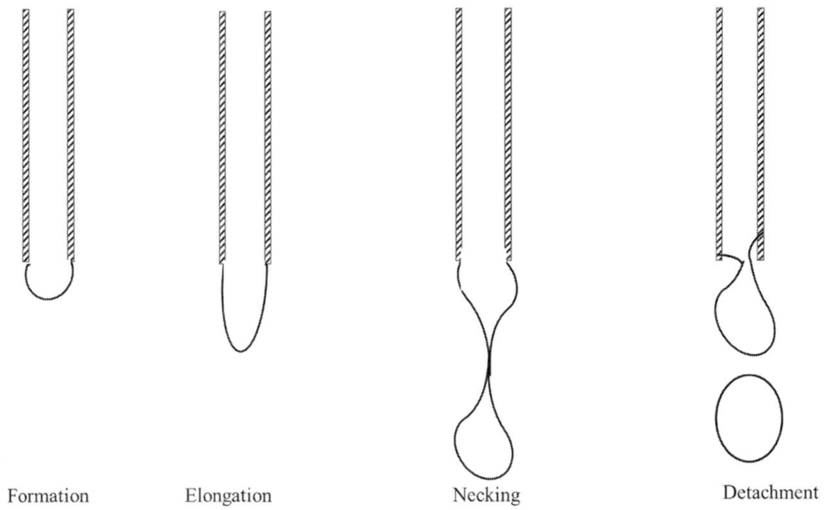

Figure 2.5 The four principle stages of formation and detachment of a single drop from a nozzle.

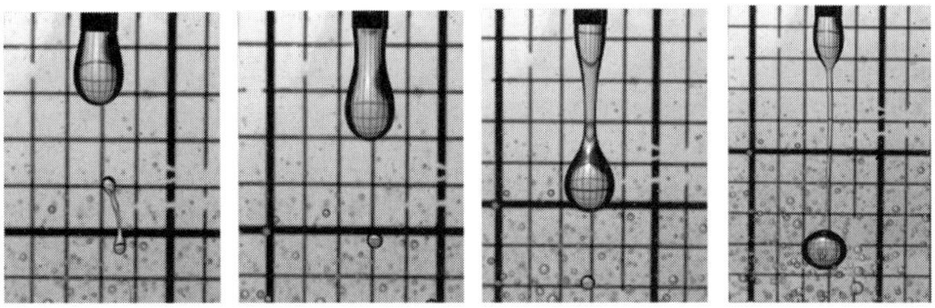

(a) (b) (c) (d)

Figure 2.6 Time sequence of formation and detachment of a discrete drop from a single nozzle: drop phase is water/carboxy methyl cellulose (cmc); continuous phase is mineral oil/ polyisobutylene.

formation, necking, and detachment. Figure 2.6(a) shows the formation of the pendent drop, retained on the nozzle by interfacial forces between the drop phase and the nozzle circumference. Figure 2.6(b) shows the onset of "necking," which is followed by "necking flow" shown in Figure 2.6(c). Figure 2.6(d) shows the instant immediately prior to full detachment of the spherical drop. An important feature of these images is in Figure 2.6(d), which shows the substantial retention of dispersed phase on the nozzle. This shows an extreme example of the liquid retention after the force balance for detachment has been met, and thus the importance of the Harkins correction factor when predicting drop size.

The approach of Harkins and Brown was developed further by Hayworth and Treybal (1950) who presented a more accurate equation, also semiempirical, as shown in equation 2.4:

$$V_F + 4.11\left(10^{-4}\right)V_F^{2/3}\left(\frac{\rho_D v^2}{\Delta\rho}\right) = 21\left(10^{-4}\right)\left(\frac{\sigma D_N}{\Delta\rho}\right) + 1.069\left(10^{-2}\right)\left(\frac{D_N^{0.747} v^{0.365} \mu_C^{0.188}}{\Delta\rho}\right)^{3/2} \tag{2.4}$$

where V_F is the drop volume upon formation at a single nozzle.

The expression is based upon the assumption that droplet detachment does not take place until the rise velocity of the droplet exceeds the velocity of the dispersed phase through the nozzle and until the buoyancy force equals the force due to interfacial tension.

In both cases an iterative procedure is required to calculate the droplet volume. Hayworth and Treybal also suggested that under certain circumstances the Harkins factor can be replaced by the constant 0.655 without appreciable error. Their treatment also incorporated a kinetic force term into the force balance equation, thus taking into account the streaming force due to flow of the dispersed phase through the nozzle. The final semiempirical correlation for droplet volume was expressed as shown above.

Scheele and Meister (1968) proposed additional refinements to the approach of Hayworth and Treybal by considering two additional aspects of the droplet formation process. Firstly, the drag force acting on the droplet during formation was taken into account by considering the forming droplet as a rigid sphere with a velocity close to that when the net force acting on the droplet has reached zero. The drag force term was thus incorporated into the force balance equation at the nozzle to obtain the droplet volume at breakoff. A further refinement took into account the additional flow through the nozzle due to the process of necking immediately prior to droplet breakaway. Finally, it was assumed that both components of flow contributing to the final droplet volume (the main contribution due to nozzle flow and the contribution due to necking flow), are not 100% efficient and that some proportion of the total material is retained at the nozzle. Correction for this was included in the final correlation using the Harkins and Brown correction factor. The final form of correlation proposed by Scheele and Meister is:

$$V_{\mathrm{F}} = F\left[\frac{\pi\sigma D_{\mathrm{N}}}{g\Delta\rho} + \frac{20\mu Q D_{\mathrm{N}}}{{D_{\mathrm{F}}}^{2} g\Delta\rho} - \frac{4\rho Q U_{\mathrm{N}}}{3g\Delta\rho} + 4.5\left(\frac{Q^{2} D_{\mathrm{N}}^{2}\rho\sigma}{(g\Delta\rho)^{2}}\right)^{1/3}\right] \tag{2.5}$$

where $g\ \Delta\rho$ is the buoyancy force term,

F is the Harkins correction factor,

$\pi\sigma D_{\mathrm{N}}$ represents the contribution of interfacial tension forces,

$$\frac{20\mu Q D_{\mathrm{N}}}{{D_{\mathrm{F}}}^{2} g\Delta\rho} \tag{2.6}$$

represents the contribution of drag forces to drop breakaway,

$$\frac{4\rho' Q U_{\mathrm{N}}}{3g\Delta\rho} \tag{2.7}$$

represents the contribution of kinetic forces to drop breakaway, and

$$4.5\left(\frac{Q^{2} D_{\mathrm{N}}^{2}\rho'\sigma}{(g\Delta\rho)^{2}}\right)^{1/3} \tag{2.8}$$

represents the contribution of necking flow.

The drop volume correlation of Scheele and Meister was claimed to be more accurate than previous correlations over a wider range of physical properties.

The correlation was validated experimentally for a wide range of water/organic systems in which the organic phase was dispersed through a range of nozzle diameters (range 0.813–6.688 mm) and at a range of nozzle velocities (0–30 cm s^{-1}).

The Harkins and Brown correction factor F was used in the form of a correlation of F versus $D_{\mathrm{N}}/(F/V_{\mathrm{f}})^{1/3}$, see Figure 2.7.

The Scheele and Meister model was significantly better at prediction of droplet size across the whole range of velocities compared to those using the Hayworth and Treybal correlation. Figure 2.8 shows data from Scheele and Meister for two very different ranges. Figure 2.8(a) shows drop volume data versus nozzle flow velocity for dispersion of n-heptane into glycerol through a small nozzle of diameter 0.613 mm.

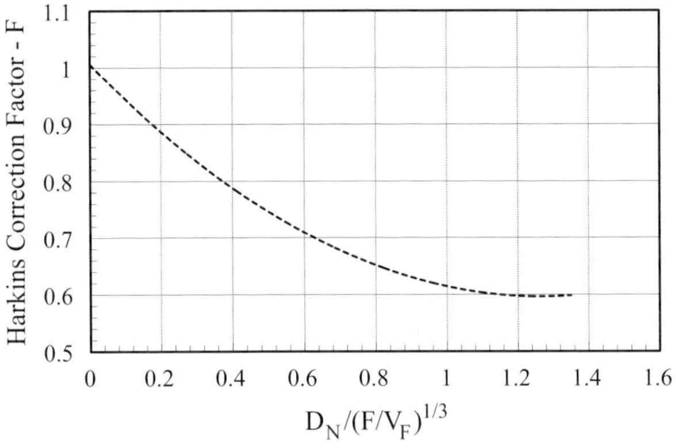

Figure 2.7 Harkins and Brown correction factor chart.

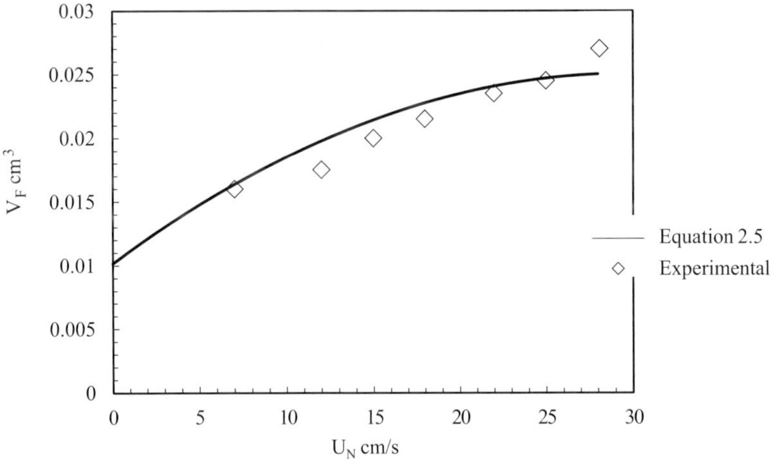

(a) Heptane drops dispersing into glycerol - nozzle diam = 0.0613 cm

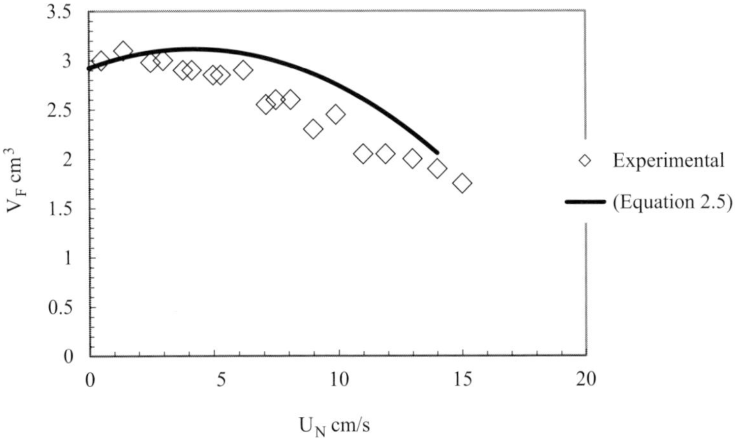

(b) 55% carbon tetrachloride / 45% heptane drops dispersing into water - nozzle diam = 0.254 cm

Figure 2.8 (a) and (b) Drop volume (V_F) vs nozzle velocity (U_N) for discrete drops showing comparison with of experimental drop volume and predictions based on equation 2.5. Data from Scheele and Meister (1968)

The drop volume range is 0.015–0.03 cm^3, which corresponds to equivalent spherical drop diameters in the range 3–4 mm. Figure 2.8(b) shows drop volume data versus nozzle flow velocity for dispersion of a mixture of carbon tetrachloride/heptane into glycerol through a nozzle of diameter 2.54 mm. The drop volume range is 1.5–3 cm^3, which corresponds to equivalent spherical drop diameters in the range 14–18 mm. Both these data sets show good agreement. Comparison of the Scheele and Meister model with the Hayworth and Treybal's correlation and with the empirical model of Null and Johnson (1958) shows a much more consistent accuracy for the theoretically based Scheele and Meister model. This is particularly marked for systems involving a high viscosity continuous phase in which the effect of drag force upon droplet size will be substantial.

Takamatsu et al. (1981, 1982), however, point out that the Scheele and Meister model in its original form was specifically validated using experiments in which a light organic phase was dispersed in water. For the case of water dispersed in a light organic phase they proposed that more accurate prediction of droplet volume is achieved by reducing the coefficient for the "necking flow" term in the overall correlation from 4.5 to 1.15.

All models are based on a force balance acting on the droplets at the point of formation and attempt to predict the point of breakoff. However, the more refined approach adopted by Scheele and Meister included not only consideration of buoyancy, kinetic, and drag forces but also included a term to account for the contribution to droplet volume due to necking flow immediately prior to droplet breakaway. Scheele and Meister point out that the range of application is generally limited to the range of experimental parametric values upon which the correlation is based.

For pure systems in which the dispersed phase was the less dense and the continuous phase of low viscosity, this correlation showed a more consistent accuracy when compared to the models of either Harkins and Brown or Hayworth and Treybal. The correlation in its general form was also shown by Takamatsu et al. to be accurate for systems in which the dispersed phase was the more dense, although better predictions were obtained by reducing the value of the coefficient associated with the necking flow term from 4.5 to 1.15. In later work, Weatherley and Wilkinson (1988), examined the accuracy of the correlation for drop formation in systems in which the continuous organic phase is substantially more viscous than the dispersed phase. Weatherley and Wilkinson showed that a correlation of the general form of Scheele and Meister accurately predicted droplet size in the system n-decanol/water and n-decanol/filtered yeast broth solutions. In this case the additional drag force due to viscous drag of the continuous phase was allowed for by enhancing the relative magnitude of the drag term in the correlation and the best predictions were obtained by increasing the value of the drag term coefficient from 1.33 to 3.3. Other work by Kalyanasundaram, Kumar, and Kuloor (1968) showed that viscosity of the dispersed phase could also have an appreciable effect upon drop size behavior due the additional tensile forces exerted on the dispersed phase during drop formation. Chen et al. (2001) validated Scheele and Meister's equation for prediction of size of newly formed drops from a flat-tipped nozzle.

Klee and Treybal (1956) proposed a correlation for maximum drop size, d_{cr}, in which spherical shape may be assumed. Baumler et al. (2011) validated this for rising drops of toluene in water to accurately predict diameter of 4.45 mm.

$$d_{cr} = 1.77\rho_C^{-0.14}\Delta\rho^{-0.43}\mu_C^{0.3}\sigma^{0.24} \tag{2.9}$$

2.3 Drop Motion

In liquid systems comprising two immiscible liquids, the velocity profiles of both dispersed and continuous phases are of fundamental importance. Interfacial mass transfer, especially in the case of small drops, is strongly influenced by the slip velocity vector between the phases and by the degree of interfacial shear. The occurrence of back-mixing in either or both phases is very likely to be detrimental to the overall mass transfer efficiency, for example in continuous column contactors. Mass transfer rate may generally be considered a first-order process and therefore is likely to be negatively impacted by back-mixing on account of reduction in effective concentration driving force. Motion of droplets may convey significant secondary flows in the continuous phase due to momentum transfer. The majority of fundamental studies of drop motion in liquid systems have been conducted on freely rising or falling drops. This is in contrast to the situation in a process environment, such as extraction columns, where normally the dispersed phase moves through the continuous phase as a population of swarming drops that interact with each other and thus experience changes in velocity (magnitude and direction) due to droplet collision, coalescence, and encounters with the walls of the equipment. However, it is argued by Baumler et al. (2011) that, despite this: "The free rise or fall of single droplets in a fluid of unrestricted extent is the basis of all further modeling approaches and thus an important design parameter."

Generally, the prediction of the rise or fall velocity of droplets in unhindered motion is complicated by the dynamic nature of the shape of the droplet. The physical relationships that are valid for solid spheres are a classical starting point for the prediction of droplet velocity behavior, but that assumption is only valid for a small number of cases. These are cases where the drop size is very small and may be assumed to be spherical, or if the droplet viscosity is very large relative to that of the continuous phase. The terminal velocity of a falling droplet, for example, is determined by the balance of gravitational forces, viscous forces, and drag forces. In turn, these are a function of the droplet shape and diameter, density difference, and the physical properties of the two liquid phases. In the case of rigid spherical drops, if we consider the relationship between drop diameter and terminal velocity, initially as the drop diameter increases, the terminal velocity increases due to the dominant effect of gravitational forces on account of the increase in the effective mass of the drop. As the drop diameter increases further the drag forces on the drop start to dominate and therefore the terminal velocity declines. In this region, onset of drop oscillation is likely as the drop may assume an unstable prolate spheroidal shape that further increases the dominant drag forces and thus a further reduction in linear velocity will be observed.

The droplet shape is a critical factor that delineates the motion of liquid droplets from rigid spheres. In some cases the assumption of perfect sphericity may be assumed and the criterion for that assumption is generally expressed in terms of the Weber number:

$$\mathrm{We} = \frac{u_T^2 d_p \rho_c}{\sigma}$$

Systems with Weber numbers less than 3.58 are usually assumed to exhibit the motion behavior of a rigid spherical particle, which makes the prediction of drop motion relatively easy.

Bond and Newton (1928) showed the important effect of viscosities of each phase upon the condition for rigid sphere behavior. The terminal velocity u_T may be expressed by equation 2.10, in terms of density difference, continuous phase viscosity μ_c, drop diameter D, and gravitational constant g. The term k is defined in equation 2.11, and is a function of the viscosity ratio.

$$u_T = \frac{1}{k}\left[\frac{(\rho_d - \rho_C)gD^2}{18\mu_C}\right] \tag{2.10}$$

$$k = \frac{0.6667 + \mu_d/\mu_C}{1 + \mu_d/\mu_C} \tag{2.11}$$

Examination of equation 2.11 shows that if $\mu_d \gg \mu_c$ the drop will behave as a rigid sphere.

The stability of droplet shape and maintenance of sphericity is strongly influenced by the presence of impurities in the drop phase. Work of Clift et al. (1978), Loth (2008), Thorsen, Stordalen, and Terjesen (1968), and Weatherley and Turmel (1992) showed that the presence of surfactants and impurities influence drop motion. The main mechanism that is put forward is that impurities in the drop phase in many cases reduce internal circulation and thus promote stability of the drop shape. The converse is that highly pure systems exhibit drop motion behaviors that vary significantly from those that are contaminated. Clift maintained that small drops are especially influenced by the presence of contaminants, increasing their rigidity, and thus making prediction of terminal velocity straightforward in unhindered motion. Baumler et al. (2011) add that in carefully purified systems even small fluid particles have an inner circulation due to the momentum transfer exerted by the outer fluid and this can be further influenced by the presence of impurities.

The critical Weber number of around approximately 3.6 for the oscillation threshold, see above, is confirmed in a number of early papers (Hu and Kintner, 1955; Krishna et al.,1959; Winnikow and Chao,1966). Beyond the threshold, the droplet starts to oscillate periodically until the wobbling regime begins, where droplets exhibit irregular shape deformations, as confirmed by Clift et al. (1978). Rigorous solutions of the governing Navier–Stokes equations until recently were only possible for drops in the creeping flow regime (see Hadamard, 1911, Rybczynski, 1911).

In terms of correlations, there are several options:

Hamielec, Storey, and Whitehead (1963) proposed the following expression for drag coefficient:

$$C_D = \frac{3.05(783\mu^2 + 2142\mu + 1080)}{(60 + 29\mu)(4 + 3\mu)\,\mathrm{Re}^{0.74}} \tag{2.12}$$

Baumler et al. (2011) showed excellent agreement with this correlation for droplets up to 3 mm in diameter in the EFCE standard liquid–liquid systems.

Grace et al. (1976) correlated drop velocity for systems contaminated with surface-active agents and proposed the following:

$$u_T = \frac{\mu_C}{\rho_C d_P} \mathrm{Mo}^{-0.149}(J - 0.857) \tag{2.13}$$

$$J = 0.94H^{0.757} \; (2 < H \leq 59.3) \tag{2.14a}$$

and

$$J = 3.42H^{0.441} \; (H > 59.3) \tag{2.14b}$$

where H is defined as follows:

$$H = \frac{4}{3} E_0 M_0^{-0.149} \left(\frac{\mu_C}{\mu_W}\right)^{-0.14} \tag{2.15}$$

with $\mu_w = 0.9$ m Pa s.

Another model expression is by Waheed, Henschke, and Pfennig (2004) for terminal velocity, and is as follows:

$$u_T = \frac{u_{def,OS} u_{sph}}{\left(u_{def,OS}^{p3} + u_{sph}^{p3}\right)^{1/p3}} \tag{2.16}$$

where p_3 is a parameter and

$$u_{def,OS} = \left(u_{def}^{ps} + u_{OS}^{ps}\right)^{1/ps} \tag{2.17}$$

with

$$u_{def} = (d_p g \Delta\rho / 2\rho_C)^{1/2} \text{ and } u_{OS} = (2p_2\sigma / d_P \rho_C)^{1/2} \tag{2.18}$$

The terminal velocity of spherical drops is expressed as follows:

$$u_{sph} = \frac{\mu_C \mathrm{Re}_{sph}}{d_P \rho_C} \tag{2.19}$$

$$\text{with } \mathrm{Re}_{sph} = (1 - f)\,\mathrm{Re}_{rigid} + f\,\mathrm{Re}_{circ} \text{ where } f = 2(K_{HR} - 1) \tag{2.20}$$

where K_{HR} is the corrected Hadamard–Rybczynski factor:

$$K_{HR} = \frac{3(\mu_C + \mu_d/\beta)}{2\mu_C + 3\mu_d/\beta} \tag{2.21}$$

where $\beta = 1-1/(1+(d_P/p_1)^{p4})$

$$\mathrm{Re}_{\mathrm{circ}} = \frac{\mathrm{Ar}}{12(0.065\mathrm{Ar}+1)^{1/6}} \tag{2.22}$$

$$\mathrm{Re}_{\mathrm{rigid}} = \left(\frac{4\mathrm{Ar}}{3C_\mathrm{D}}\right)^{1/2} \tag{2.23}$$

where

$$\mathrm{Ar} = \frac{d_\mathrm{P}^3 g \rho_\mathrm{C} \Delta\rho}{\mu_\mathrm{C}^2}$$

Calculation of the two Reynolds numbers Re_{rigid} *and* Re_{circ} requires the Archimedes number Ar.

Weatherley and Turmel (1992) provide a summary of some of the first serious analysis of drop motion in liquid–liquid systems as follows.

Early analysis of the unhindered motion of drops formed in the non-jetting regime into an immiscible second phase is presented by Garner and Skelland (1955). The key factors influencing the motion comprise frictional drag, which may produce internal circulation in the drop that affects the drop velocity. A second factor is the effect of gravitational forces on drop shape. The third factor is oscillatory behavior, which results in time-dependent variation in the drag forces acting on the drop. Motion behavior can to some extent be predicted assuming rigid sphere behavior, especially for small drops. Hu and Kinter (1955) demonstrated a linear relationship between drag coefficient and Reynolds number analogous to rigid sphere behavior, providing that the Reynolds number (equation 2.24) for the drop did not exceed 300.

$$\mathrm{Re} = \frac{\rho U_0 d}{\mu} \tag{2.24}$$

where ρ is density of the continuous phase and μ is viscosity of the continuous phase.

Further studies showed that the peak terminal velocity for descending drops above which oscillation has significant influence on motion may be expressed by equations 2.25 and 2.26.

$$(U_0)_\mathrm{p} = 1.23\left(\frac{\sigma_\mathrm{i}}{\mu}\right)P^{-0.238} \tag{2.25}$$

$$P = \frac{\rho \sigma_\mathrm{i}{}^3}{g\mu^4}\frac{\rho}{\Delta\rho} \tag{2.26}$$

Above this maximum, oscillation results in large increase in drag coefficient. The Weber number, as defined in equation 2.27, provides the criterion for the onset of significant oscillation, where $\mathrm{We} < 3.58$.

$$\mathrm{We} = \frac{U_0{}^2 d\rho}{\sigma_\mathrm{i}} \tag{2.27}$$

Correlations for drag coefficient are presented by Hu and Kinter as shown in equations 2.28 and 2.29.

$$Y = 1.333\ X^{1.275} \quad \text{for } 2 < Y < 70 \tag{2.28}$$

$$Y = 0.045X^{2.37} \quad \text{for } Y > 70 \tag{2.29}$$

where

$$Y = \text{Cd. We. } P^{0.15}; \quad X = (\text{Re}/P^{0.15}) + 0.75 \tag{2.30}$$

$$C_\text{d} = \frac{4gD\Delta\rho}{3\rho_\text{C} u_\text{T}^2} \tag{2.31}$$

Another significant effect on drop motion is the presence of surfactants and other impurities (Skelland, Woo, and Ramsey, 1987; Thorsen et al., 1968). These studies surprisingly showed that the correlations presented by Hu and Kinter proved less accurate for ultra-pure liquid–liquid systems.

Internal drop circulation also has an important influence on drop behavior (Garner and Skelland, 1955) especially in regards to mass transfer and motion. The onset of circulatory flow depends on the drop Reynolds number, with viscosity and interfacial tension being the key variables.

The relationship between drop size and velocity is also important to analyze. There are two principle regimes: (i) in which velocity increases with diameter and (ii) in which velocity remained relatively invariant with diameter. In regime (ii) the increase in gravitational force due to increase in drop mass is largely offset by increase in drag force.

Controlling equations for each regime were proposed by Klee and Treybal (1956) as follows:

Region (i):

$$\text{Re} = 22.2 C_\text{d}^{-5.18} \text{We}^{-0.169} \tag{2.32}$$

Region (ii):

$$\text{Re} = 0.00418 C_\text{d}^{2.91} \text{We}^{-1.81} \tag{2.33}$$

Terminal velocity $u_{\text{T(I)}}$ and $u_{\text{T(II)}}$ in each regime may be written as shown in equation 2.34:

$$\begin{aligned} u_{\text{T(I)}} &= 38.3\rho_\text{C}^{-0.45}\Delta\rho^{0.58}\mu_\text{C}^{0.11}D^{0.7} \\ u_{\text{T(II)}} &= 17.6\rho_\text{C}^{-0.55}\Delta\rho^{0.28}\mu_\text{C}^{0.1}\sigma^{0.18} \end{aligned} \tag{2.34}$$

Later research by Skelland et al. (1987) and Thorsen et al. (1968) underlined the importance of impurities on terminal velocity behavior of single drops in unhindered motion. Thorsen presented a correlation for the drag coefficient as written in equation 2.35, which built on the correlation developed by Winnikow and Chao (1966) assuming complete internal circulation in the drop:

$$C_d = \frac{16}{\mathrm{Re}}\left[1 + \frac{0.814}{\mathrm{Re}^{1/2}}\right]\left[\frac{2 + 3\mu_d/\mu_c}{1 + \left[(\rho_d/\rho_c)(\mu_d/\mu_c)^{1/2}\right]}\right] \quad (2.35)$$

In the case of drops that experience significant oscillation, Thorsen presented the correlation in equation 2.36 for terminal velocity:

$$u_T = \frac{6.8}{1.65 - \Delta\rho/\rho_d}\left[\frac{\sigma}{3\rho_d + 2\rho_C}\right]^{1/2}\frac{1}{D^{1/2}} \quad (2.36)$$

In light of the huge growth in computational power, combined with the application of finite element techniques and computational fluid dynamics, the capability now exists to undertake modeling of drop behavior in liquid–liquid systems based on rigorous solution of the relevant controlling equations. Despite these developments, Baumler et al. (2011) highlight the fact that rigorous numerical simulations based on solution of the fundamental equations are comparatively rare. Several examples are reviewed by Baumler et al., including that of Deshpande and Zimmerman (2006) who adopted a level set approach to the study of drop motion covering wide range of Eötvös numbers, where the Eötvös number E_o is given by:

$$E_o = \frac{\Delta\rho g L^2}{\sigma} \quad 2.37$$

where E_o is the Eötvös number,
$\Delta\rho$ is the difference in density of the two phases, kg m^{-3},
g is the gravitational acceleration, m s^{-2},
L is the characteristic length, m,
σ is the surface tension, N m^{-1}.

Baumler confirmed the validity of the early equations of Grace et al. (1976) for rising single droplets as did Watanabe and Ebihara (2003). Baumler's simulations also took account of droplet shape, demonstrating the importance of drop shape on velocity. Petera and Weatherley (2001), as part of a study into mass transfer from falling droplets, also conducted quantitative numerical simulations of drop shape and terminal velocity using finite element methods in a Lagrangian framework assuming axisymmetric flow. Their results showed excellent agreement with experimental measurements.

2.4 Dispersions and Swarming Drops

Many industrial liquid–liquid processes are conducted in equipment involving stirred vessels, where the hydrodynamic conditions may be significantly more complex than is the case for drops emerging from a nozzle. Understanding the behavior of dispersions in stirred vessels is critical in the context of process intensification. Simply increasing energy input to a mixing vessel may not necessarily lead to higher efficiency or better overall performance. On the one hand, reduction in drop size and high rates of interfacial shear may lead to high mass transfer rates, but on the other hand, increasing mixing and turbulence may lead to slow phase separation. The impact on overall fluid mixing is also an important consideration so that operational conditions and equipment design ensure minimization of dead space and short-circuit flows.

In this section we will briefly review some of the correlations developed over many years for mechanically agitated liquid–liquid contacting equipment. These include stirred vessels, such as are used in multi-stage mixer settler units, and a range of mechanically agitated columns. There are many papers on drop size in mechanically agitated systems going back to the 1950s and beyond and none could be classed as very accurate, with minimum errors reported at around ±10% but in many cases errors are in the range 20–30%. While this level of uncertainty may be taken into account in equipment design and design margins, it does demonstrate the great degree of complexity and difficulty in accurately predicting drop size and, as will be discussed later, holdup and mass transfer rate. The next part will examine drop size behavior in intensive liquid–liquid contacting equipment such as centrifugal contactors, microchannel contactors, oscillatory baffled contactors, and spinning disks.

In all situations, the prediction of drop size, holdup, and interfacial area is critical for effective design. Ultimately, the drop size and drop size distribution is a function of the balance between rate of drop creation via breakup due to shear forces, and coalescence of drops occurring due to drop–drop encounters. Much of the early work on predicting drop size and distribution in mechanically agitated vessels focused on an empirical approach in which the measured drop size data are correlated to independent variables, which include impellor size and speed, phase ratio, and property data such as density, viscosity, and interfacial tension.

2.5 Drop Size in Stirred Tanks

There are many correlations proposed for prediction of drop size in stirred tanks. Singh et al. (2008) have presented an excellent review, in which they compare and comment on some 20 papers over the period 1955–2006. Some of the literature highlighted included the following proposed relationships for drop size in stirred tanks.

Vermeulen, Williams, and Langlois (1955) proposed the following:

$$\frac{N^2 d^{5/3} D^{4/3} \rho_{\mathrm{m}}}{\sigma f_{\sigma}^{5/3}} = 0.016 \tag{2.38}$$

$$f_{\sigma} = \frac{d}{d_{\sigma=0.1}} \tag{2.39}$$

$$\rho_{\mathrm{m}} = 0.6\rho_{\mathrm{d}} + 0.4\rho_{\mathrm{c}} \tag{2.40}$$

The work of Hinze (1955) was among the earliest to develop a relationship between the maximum stable drop size and energy dissipation rate ε_{T} in a stirred vessel containing a liquid–liquid mixture. The basis of the relationship is the balance between interfacial tension stresses and turbulence induced stresses at the liquid–liquid interface.

$$d_{\max} = K_1 (\varepsilon_T)_{\max}^{-0.4} \left(\frac{\sigma}{\rho_c} \right)^{0.6} \tag{2.41}$$

Rodger, Trice, and Rushton (1956) proposed the following correlation for interfacial area of agitated dispersions:

$$\bar{a} = \frac{K}{D} \left[\frac{D^3 N^2 \rho_C}{\sigma} \right]^{0.36} \left(\frac{D}{T} \right)^k \left(\frac{v_d}{v_C} \right)^{1/5} \left(\frac{t}{t_o} \right)^{1/6} \exp \left[3.6 \frac{\Delta\rho}{\rho_C} \right] \Psi \tag{2.42}$$

where ψ is a scale up factor, K is a constant, D is the impeller diameter, T is the tank diameter, and t and t_o are settling times as defined in the paper.

Weinstein and Treybal (1973) presented a number of correlations for Sauter mean drop size in agitated vessels based on the study of eight different liquid–liquid systems covering a range of density differences, interfacial tension values, vessel sizes, and rotational speeds.

Equation 2.43 is based on data collected for dispersion in batch vessels. Equation 2.44 is for continuous stirred tanks.

$$d_{32} = 10^{(-2.316+0.672\phi)} v_C^{0.0722} \varepsilon^{-0.194} \left(\frac{\sigma}{\rho_C} \right)^{0.196} \tag{2.43}$$

$$d_{32} = 10^{(-2.066+0.732\phi)} v_C^{0.047} \varepsilon^{-0.204} \left(\frac{\sigma}{\rho_C} \right)^{0.274} \tag{2.44}$$

Another well known early correlation for drop size in agitated liquid–liquid mixtures is that proposed by Shinnar (1961):

$$\frac{D_{32}}{D\ \alpha \mathrm{We}^{-0.6}} = \frac{C_1 \rho_c N^2 D^3}{\sigma} \tag{2.45}$$

where C_1 is determined experimentally.

This is for a dilute system and for liquids of low viscosity.

A development of the original correlation of Shinnar that does account for viscous forces and their influence on drop size was proposed in three papers by Calabrese, Wang and others, as follows.

Calabrese, Chang, and Dang (1986a) present a correlation based on work done at low dispersed phase holdup (0.0015) in the silicone oil–water system to establish the effect of dispersed phase viscosity on drop size.

Equations 2.46 and 2.47 were found to provide accurate predictions for dispersed phase viscosities in the range 0.1–10 Pa s.

$$\frac{d_{32}}{d_o} = \left[1 + 11.5 \left(\frac{\rho_C}{\rho_d} \right)^{1/2} \frac{\mu_d \varepsilon^{1/3} d_{32}^{1/3}}{\sigma} \right]^{5/3} \tag{2.46}$$

$$\frac{d_{32}}{D} = 2.1 \left(\frac{\mu_d}{\mu_C} \right)^{3/8} \left(\frac{\rho_C N D^2}{\mu_C} \right)^{-3/4} \tag{2.47}$$

Equation 2.46 correlates mean drop size d_{32} implicitly in terms of power dissipation ε, whereas equation 2.47 more usefully correlates mean drop size explicitly in terms of impeller speed and diameter.

In a second paper, Wang and Calabrese (1986) present further correlations developed to establish the effect of both viscosity and interfacial tension on drop size, incorporating the Weber number as follows:

$$\frac{d_{32}}{D} = C_4 \mathrm{We}^{-0.6}\left[1 + C_5 V_\mathrm{i}\left(\frac{d_{32}}{D}\right)^{1/3}\right]^{0.6} \tag{2.48}$$

$$\frac{d_{32}}{D} = 0.066 \mathrm{We}^{-0.66}\left[1 + 13.8 V_\mathrm{i}^{0.82}\left(\frac{d_{32}}{D}\right)^{1/3}\right]^{0.59} \tag{2.49}$$

$$V_\mathrm{i} = \frac{\mu_\mathrm{d} N D}{\sigma}\left(\frac{\rho_\mathrm{C}}{\rho_\mathrm{d}}\right)^{1/2} \tag{2.50}$$

$$\mathrm{We} = \frac{u_\mathrm{t}^2 d_\mathrm{p} \rho_\mathrm{c}}{\sigma}$$

Then, in a third paper by the same authors (Calbrese, Wang, and Bryner 1986b), an overarching correlation is proposed, as shown in equation 2.51:

$$\frac{d_{32}}{D} = 0.054(1 + 3\phi)\mathrm{We}_\mathrm{I}^{-0.6}\left[1 + 4.42(1 - 2.5\phi)V_\mathrm{i}\left(\frac{d_{32}}{D}\right)^{1/3}\right]^{3/5} \tag{2.51}$$

Other early correlation techniques for Sauter mean drop size in turbulent liquid–liquid dispersions include that of Hughmark (1971) (equation 2.52):

$$1.69 d_\mathrm{m} = \frac{\sigma}{\rho_\mathrm{c} u^2} + \frac{1.3 \mu_\mathrm{d}}{(\rho_\mathrm{c}\rho_\mathrm{d})^{1/2} u} \tag{2.52}$$

A further development and more accurate correlation was proposed by Pacek, Man, and Nienow (1998), but it is also reported that this correlation is valid only for situations in which viscous forces are small, as shown in equations 2.53 and 2.54:

$$\frac{d_{32}}{D} = 0.052(1 + 22.8\Phi_\mathrm{d})\mathrm{We}^{-0.6} \tag{2.53}$$

$$\frac{d_{32}}{D} = 0.0221(1 + 23.3\Phi_\mathrm{d})\mathrm{We}^{-0.43} \tag{2.54}$$

The two equations are based on extensive drop size data determined in the system chlorobenzene/water and chlorobenzene/sodium chloride solution.

Laso, Steiner, and Hartland (1987) proposed the following for CCl_4/n-heptane in water and 1-octanol in water, and MIBK in water:

$$\frac{d_{32}}{D} = 0.118 \mathrm{We}_\mathrm{I}^{-0.4}\phi^{0.27}\left(\frac{\mu_\mathrm{d}}{\mu_\mathrm{c}}\right)^{-0.056} \tag{2.55}$$

Singh et al. (2008) proposed one of the most accurate correlations when compared with the twenty previous correlations that they reviewed. Their measurements were conducted on a system comprising an organic phase of n-paraffin, di-2-ethylhexyl phosphoric acid (D2EHPA)/TBP and an aqueous phase of 30% phosphoric acid using a turbine pump mix impeller. The correlation is as follows:

$$\frac{d_{32}}{D} = 1.849 \times 10^{-3} N^{-1.7025} \times \left(1 + 0.392\phi_{\mathrm{f}} + 3.2435\phi_{\mathrm{f}}^2\right) \exp\left(0.4302\tau\right) \qquad (2.56)$$

where the holdup term ϕ is expressed as follows:

$$\phi = 2.1214\phi_{\mathrm{f}}^{0.4012} N^{-1.097} \tau^{-0.0631} \qquad (2.57)$$

Singh et al. show the parity plots for their correlation in Figure 2.9. These predictions are also compared with a number of other correlations and show significant improvement.

A different approach to the prediction of droplet size distributions in turbulent systems was adopted by Skartlien, Sollum, and Schumann (2013), based on Kolmogorov microscale comparisons. This is appropriate for systems where the viscosity of the fluids dominate the breakup and in which turbulent energy is dissipated as heat. These workers conducted Lattice Boltzmann simulations of water-in-oil dispersions in steady-state, homogeneous turbulent flow. They showed that where the internal viscous stresses in the dispersion is comparable to the interfacial tension. the drop sizes equilibrate close to the Kolmogorov scale. The maximum drop size being proportional

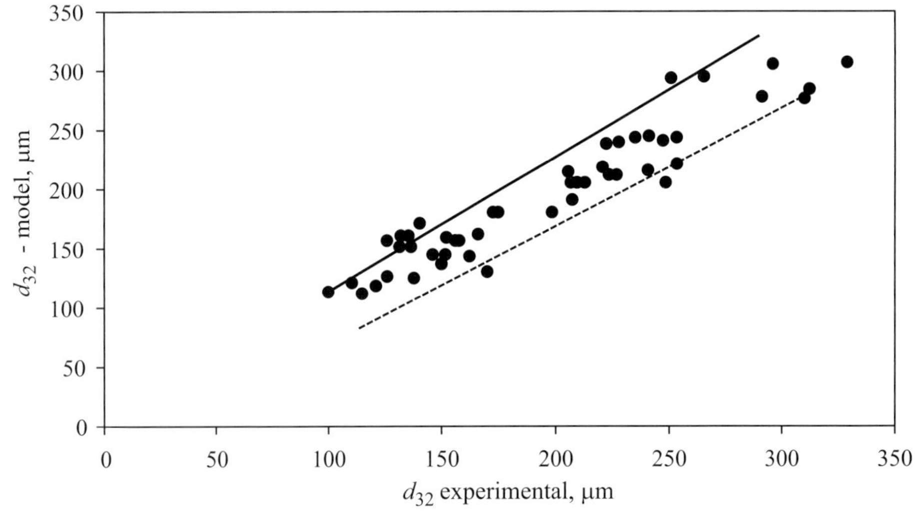

Figure 2.9 Validation plot for Sauter Mean Drop size for a stirred tank system based on equation 2.56

Points – based on model predictions
+15% experimental
–15% experimental

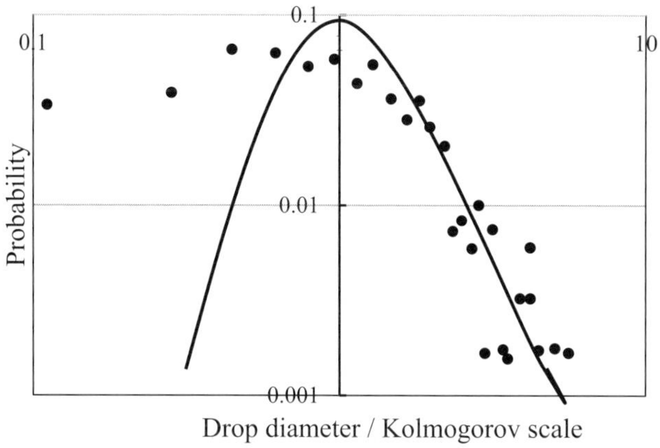

Figure 2.10 Drop size distribution for oil/water dispersions with high surface activity and intermediate energy input.
Reprinted (adapted) with permission from Skartlien, R., Sollum, E., and Schumann, H. (2013). Droplet size distributions in turbulent emulsions: breakup criteria and surfactant effects from direct numerical simulations. *Journal of Chemical Physics*, **139**, 174901. Copyright (2013) AIP Publishing

to that scale when viscous stress dominates. It was shown that the droplet size distribution in the inertial subrange scaled with the known power law $\sim d^{-10/3}$, as a consequence of breakup by external turbulent stress fluctuations.

The Kolmogorov length scale is defined in equation 2.58:

$$\eta = \left(\frac{\upsilon^3}{\varepsilon}\right)^{1/4} \tag{2.58}$$

where ν = viscosity; ε = average turbulent kinetic energy dissipation per unit mass.

Based on this analysis, dispersions in a turbulent flow environment were numerically simulated to determine the drop size distribution for different sets of conditions, and to calculate the maximum stable drop diameter. Simulations were undertaken successfully using a modified version of the lattice Boltzmann equation by incorporating a stochastic force term. The simulations were limited to systems of isotropic turbulence and assuming energy dissipation independent of time. A sample of the simulations based on this approach is shown in Figure 2.10.

2.6 Dispersions in Continuous Liquid–Liquid Columns

Many continuous liquid–liquid processes are conducted in continuous column contactors with countercurrent flow of the dispersed and continuous phases. The large variety of column internals, which may range from sieve plates, rotating discs, or packing pieces of many types, makes the general prediction of drop size challenging. In comparison to stirred tank systems, the hydraulic behavior of column contactors is also heavily influenced by the phase flow rates. Additionally, some column designs involve oscillatory flows imposed externally by pulsation or internally by reason of oscillating internals such as sieve tray stringers or baffle stringers that are mechanically pulsed. One of the mechanisms for improving mass transfer efficiency in column contactors is through the promotion of cyclic drop breakup and coalescence, which assists in the breaking down of internal mass transfer resistances in the drop phase. Another factor in column contactors is that most operate as differential contactors, which means that the concentrations in both phases vary with axial position. Therefore, the density difference, and other physical properties such as viscosity and interfacial tension, are position dependent in the presence of mass transfer. Such differences become important especially in sieve plate or pulsed plate contactors where there is a constant cycle of coalescence followed by creation of a fresh dispersion at every plate. The local drop size distribution at each point in the column is therefore prone to variation, given the importance of viscosity, density, and interfacial tension on drop formation as discussed in detail in Sections 2.2 and 2.3.

Studies of drop size and holdup in column contactors are legion, but fall into four broad categories: (i) in the absence of mass transfer; (ii) in the presence of mass transfer; (iii) nonmechanically agitated columns; and (iv) mechanically agitated columns.

A noteworthy early example of an accurate correlation for a column contactor was by Baird and Lane (1973) who proposed the following expression for Sauter mean drop size in the absence of mass transfer in a reciprocating plate column:

$$d_{32} = 0.357 \frac{\gamma^{0.6}}{\rho_{\mathrm{C}}^{0.2} \Psi^{0.4}} \tag{2.59}$$

where the power dissipation term was estimated from Kolmogoroff's theory of isotropic turbulence in terms of the turbulent energy dissipation rate per unit volume, and is expressed by:

$$\Psi = \frac{2\pi^2 \left(1 - S^2\right)}{3 h_{\mathrm{C}} C_{\mathrm{o}}^2 S^2} \rho_{\mathrm{m}} (Af)^3 \tag{2.60}$$

where A is the amplitude of oscillation, f is the frequency, S is the fractional free area of the plate, h_{c} is the center to center plate spacing, C_{o} is the discharge coefficient for plate perforations, and ρ_{m} is the mean dispersion density.

This was further developed by Kumar and Hartland (1988) for systems with low agitation rates and where gravitational forces become more dominant to give:

$$\Psi_{\mathrm{g}} = V_{\mathrm{s}} g \phi \Delta\rho = g \Delta\rho \left(V_{\mathrm{d}} + \frac{\phi}{\phi + 1} V_{\mathrm{c}} \right) \tag{2.61}$$

where V_{s} is the slip velocity, V_{d} is the linear velocity of the dispersed phase, and ϕ is the fractional holdup of the dispersed phase.

Thus a modified expression for the Sauter mean drop size was presented:

$$d_{32} = 0.357 \frac{\gamma^{0.6}}{\rho_{\mathrm{C}}^{0.2} \left(\Psi + \Psi_{\mathrm{g}}\right)^{0.4}} \tag{2.62}$$

Hartland and co-workers studied drop size and holdup extensively in a range of industrial-type mechanically enhanced liquid–liquid contactors and successfully generated accurate correlations based on many experimental data.

Kumar and Hartland (1996) present a detailed treatise on drop size behavior in liquid–liquid column contactors. The starting point is for nonagitated systems. As with single drops dealt with earlier, the drop size is controlled by the ratio of interfacial forces to gravitational forces see in equation 2.63. The constant C_1 is a function of the type of column, with differing values, for example, for rotating disk columns and for pulsed plate columns:

$$d_{32} = C_1(\gamma/\Delta\rho g)^{1/2} \tag{2.63}$$

In cases where agitation is present, which is most frequently the case for high efficiency contact, as will be discussed below, the application of fundamental theoretical models for the prediction of drop breakup and coalescence becomes complex. Therefore an empirical approach has traditionally been adopted and with some success.

For stirred tanks, Calabrese et al. (1986b) developed and validated the following correlation for Sauter mean drop diameter during turbulent liquid–liquid mixing:

$$d_{32} = C_2\varepsilon^{-0.4}\left(\frac{\gamma}{\rho_C}\right)^{0.6}\left[1 + C_4\left[\left(\frac{\rho_C}{\rho_d}\right)^{1/2}\frac{\mu_d\varepsilon^{1/3}D_R{}^{1/3}}{\gamma}\right]^{m_1}\right]^{3/5} \tag{2.64}$$

Coalescence is also a complex process to model from fundamentals. Coalescence is not only a function of holdup and drop size but as Kumar and Hartland clearly state, the direction of mass transfer and the interfacial tension gradients that exist close to the liquid–liquid interface also exert an important influence on coalescence. For systems where mass transfer is from continuous phase to dispersed phase, the solute concentration in the film between two approaching drops will be lower than in the bulk continuous phase and thus the interfacial tension will tend to be higher, thus tending to impede coalescence. In the case of mass transfer from dispersed to continuous phase the reverse tends to be true.

In an earlier paper, Kumar and Hartland (1996), the power dissipation per unit mass of the phases ε was proposed as a useful predictor for drop size, expressed as follows:

$$\varepsilon = \frac{4P}{\pi D_C^2 H\rho_C} \tag{2.65}$$

where P is the power dissipation per unit volume determined from N_p and Re_R.

$$P = \frac{N_p N^3 D^5 \rho}{g_c} \tag{2.66}$$

$$N_p = \frac{C_5}{Re_R} + C_6\left[\frac{1000 + 1.2Re_R{}^{m_2}}{1000 + 3.2Re_R{}^{m_2}}\right]^{m_3} \tag{2.67}$$

For rotating disk columns:

$$C_5 = 109.36,\ C_6 = 0.74,\ m_2 = 0.72,\ m_3 = 3.30 \tag{2.68}$$

For asymmetric rotating disk columns:

$$C_5 = 90.00,\ C_6 = 0.62,\ m_2 = 0.73,\ m_3 = 3.17 \tag{2.69}$$

For Kühni columns they propose:

$$N_{\mathrm{P}} = 1.08 + \frac{10.94}{\mathrm{Re_R}^{0.5}} + \frac{257.37}{\mathrm{Re_R}^{1.5}} \tag{2.70}$$

For pulsed plate columns and for Karr columns the following is proposed:

$$\varepsilon = \frac{2\pi^2(1-e^2)}{3HC_{\mathrm{o}}^2 e^2}(Af)^3 \tag{2.71}$$

where the value of C_{o} is set at 0.6.

Bensalem, Steiner, and Hartland(1986) also studied the effect of mass transfer on drop size in a Karr column and proposed two correlations based on experiments in the toluene-acetone-water system:

$$d_{32(d\,\rightarrow\,c)} = 2.774\times10^{-3}(Af)^{-0.400}(V_{\mathrm{C}})^{-0.232}(V_{\mathrm{d}})^{0.414} \tag{2.72}$$

$$d_{32(c\,\rightarrow\,d)} = 2.709\times10^{-4}(Af)^{-0.59} \tag{2.73}$$

V_{c} and V_{d} are the superficial velocities of continuous dispersed phases, respectively.

Kumar and Hartland presented a unified correlation as follows that is useful for a range of different mechanically agitated columns:

$$\frac{d_{32}}{H} = \frac{C_\Psi e^n}{\dfrac{1}{C_\Psi\left(\frac{\gamma}{\Delta\rho g H^2}\right)^{1/2}} + \dfrac{1}{C_\pi\left[\left(\frac{\varepsilon}{g}\right)\left(\frac{\rho_C}{g\gamma}\right)^{1/4}\right]^{n1}\left[H\left(\frac{\rho_C g}{\gamma}\right)^{1/2}\right]^{n2}}} \tag{2.74}$$

The parameters used in equation 2.74 are shown in Table 2.1.

Kumar and Hartland also present a suite of correlations for multistage columns and for packed extraction columns:

Table 2.1 Parameters for use in equation 2.74 to calculate drop size d_{32} for mechanically agitated columns (Kumar and Hartland, 1996).

Column type	Data points	Cψ cont- > disp	Cψ disp- > cont.	C_Ω	Cπ	n	n_1	n_2	% error
RDC	749	1	1.29	2.54	0.97	0.64	−0.45	−1.12	22.4
Kühni	702	1	3.04	1.6	0.034	0.45	−0.63	−0.38	22.4
Wirz	221	1	1	2.24	0.87	0	−0.47	−1.24	16.1
Pulsed Karr	1067	0.9	1.66	1.19	1.12	0.41	−0.62	−1.2	18.1
All data	2739	1	1.65	1.43	0.45	0.45	−0.57	−0.96	25.5

For Kühni columns:

$$d_{32} = \frac{C_{\psi} e^{0.64} S^{-0.10}}{\frac{1}{\left[3.42\left(\frac{\gamma}{\Delta\rho g}\right)^{0.5}\right]^{2}} + \frac{1}{\left[0.65\varepsilon^{-0.4}\left(\frac{\gamma}{\rho_{\mathrm{c}}}\right)^{0.6}\right]^{1/2}}} \tag{2.75}$$

For multistage Kühni columns:

$$\frac{d_{32}}{H} = \frac{C_{\psi} e^{0.56} S^{-0.10}}{\frac{1}{\left[2.13\left(\frac{\gamma}{\Delta\rho g H}\right)^{1/2}\right]} + \frac{1}{\left[7.08 \times 10^{-2}\left(\frac{\varepsilon}{g}\right)\left(\frac{\rho_{\mathrm{c}}}{g\gamma}\right)^{1/4}\right]^{-0.63}\left[H\left(\frac{\rho_{\mathrm{c}} g}{\gamma}\right)^{1/2}\right]^{-0.5}}} \tag{2.76}$$

where S = number of stages.

For the Wirz column:

$$d_{32} = \frac{(1 + 1.78\phi)}{\left[\frac{1}{\left[0.98\left(\frac{\gamma}{\Delta\rho g}\right)^{0.5}\right]^{2}} + \frac{1}{\left[0.27\varepsilon^{-0.4}\left(\frac{\gamma}{\rho_{\mathrm{c}}}\right)^{0.6}\right]^{2}}\right]^{1/2}} \tag{2.77}$$

This was further improved by incorporating a term for the holdup, ϕ, as follows:

$$\frac{d_{32}}{H} = \frac{(1 + 2\phi)}{\frac{1}{\left[2.22\left(\frac{\gamma}{\Delta\rho g H^{2}}\right)^{1/2}\right]} + \frac{1}{0.88\left[\left(\frac{\varepsilon}{g}\right)\left(\frac{\rho_{\mathrm{c}}}{g\gamma}\right)^{1/4}\right]^{-0.47}\left[H\left(\frac{\rho_{\mathrm{c}} g}{\gamma}\right)^{1/2}\right]^{-01.32}}} \tag{2.78}$$

For Karr columns:

$$\frac{d_{32}}{H} = \frac{C_{\psi} e^{0.32}}{\frac{1}{\left[1.55\left(\frac{\gamma}{\Delta\rho g H^{2}}\right)^{1/2}\right]} + \frac{1}{0.42\left[\left(\frac{\varepsilon}{g}\right)\left(\frac{\Delta\rho_{\mathrm{c}}}{g\gamma}\right)^{1/4}\right]^{-0.35}\left[H\left(\frac{\Delta\rho_{\mathrm{c}} g}{\gamma}\right)^{1/2}\right]^{-1.15}}} \tag{2.79}$$

For nonmechanically agitated columns, for example, a packed liquid–liquid extraction column, Kumar and Hartland present the following correlation for drop size:

$$d_{32} = 0.74 C_{\Psi}\left(\frac{\Delta\rho\rho_{\mathrm{d}}\gamma}{{\rho_{\mathrm{w}}}^{2}\gamma_{\mathrm{w}}}\right)^{-0.12}\left(\frac{\lambda\gamma}{\Delta\rho g}\right)^{1/2} \tag{2.80}$$

Finally, for columns in which droplets of the dispersed phase are formed at multi nozzles, Kumar and Hartland present a correlation that is applicable to both low speed (discrete drops) and high speed (jetting flow) drop formation as follows:

$$d_{32} = \frac{C_{\psi}}{\frac{1}{C_{\Omega}\left(\frac{6 D_{\mathrm{N}}\gamma}{\Delta\rho g}\right)^{1/3}} + \frac{1}{C_{\kappa}\left(\frac{12\gamma}{\rho_{\mathrm{d}} V_{\mathrm{N}}^{2}}\right)}} \tag{2.81}$$

This was further improved as follows:

$$\frac{d_{32}}{D_{\mathrm{N}}}=\frac{C_{\psi}}{\frac{1}{\left(\frac{6\gamma}{\Delta\rho g D_{\mathrm{N}}^{2}}\right)^{1/3}}+\frac{1}{4.15\left(\frac{12\gamma}{\rho_{\mathrm{d}} D_{\mathrm{N}} V_{\mathrm{N}}^{2}}\right)^{0.73}\left[D_{\mathrm{N}}\left(\frac{\rho_{\mathrm{d}} g}{\gamma}\right)^{1/2}\right]^{0.62}}} \tag{2.82}$$

Kumar and Hartland's paper (1996) on drop size in continuous extraction columns provides a comprehensive comparison and set of correlations for a number of mainstream types of continuous liquid–liquid contactors: Kühni, Wirz, Wirz pulsed, pulsed Karr, packed, spray column, rotating disk, and Karr.

For the packed columns and spray columns, the classical approach for prediction of drop size based on the balance of interfacial and buoyancy forces, and taking account of the balance of interfacial and kinetic forces, is satisfactory. In the case of externally agitated columns, the correlations take into account power dissipation per unit mass of fluid. Overall, a comprehensive two-term correlation for the agitated columns is presented and validated for each type and an average error deviation of $\pm 22\%$, which in this context is very good. The effect of holdup is considered to be minor.

Other more recent studies on a Schiebel column by Yuan et al. (2012) propose the following correlation for low interfacial tension systems and with a maximum relative error of $\pm 16\%$ and a mean relative error of $\pm 4.6\%$. This study was conducted using a n-heptane/glycerol/water system.

$$\frac{d_{32}}{D}=0.509\left(\frac{Q_{\mathrm{d}}}{ND^{3}}\right)^{0.093}\left(\frac{\mu_{\mathrm{c}}}{\mu_{\mathrm{d}}}\right)^{0.246}\left(\frac{\Delta\rho}{\rho_{\mathrm{d}}}\right)^{-0.240}\left(\frac{ND^{2}\rho_{\mathrm{d}}}{\mu_{\mathrm{d}}}\right)^{-0.62}\left(\frac{\sigma}{ND\mu_{\mathrm{d}}}\right)^{0.456} \tag{2.83}$$

For a pulsed packed column, Samani et al. (2012) propose two correlations, the first not taking account of column height thus:

$$d_{32}=8.26\times 10^{-5}\left(\frac{(af)^{4}\rho_{\mathrm{c}}}{\sigma g}\right)^{-0.2304}\left(\frac{\mu_{\mathrm{c}}^{4} g}{\Delta\rho\sigma^{3}}\right)^{-0.0514}\left(1+\frac{Q_{\mathrm{c}}}{Q_{\mathrm{c}}}\right)^{-0.0321} \tag{2.84}$$

The second, which incorporates a term for column height:

$$d_{32}=7.1\times 10^{-5}\left(\frac{(af)^{4}\rho_{\mathrm{c}}}{\sigma g}\right)^{-0.239}\left(\frac{\mu_{\mathrm{c}}^{4} g}{\Delta\rho\sigma^{3}}\right)^{-0.0473}\left(\frac{h}{H_{\mathrm{o}}}\right)^{-0.1645} \tag{2.85}$$

where af = pulse frequency × amplitude, h = height coordinate relevant to d_{32}, and H_{o} = overall height of column.

Torab-Mostaedia, Ghaemi, and Asadollahzadeh (2011) evaluated Kumar and Hartland's drop-size correlation for a pulsed plate column (Kumar and Hartland, 1986) for disc and donut pulsed column as follows:

$$\frac{d_{32}}{\sqrt{\frac{\sigma}{\Delta\rho g}}} = C_1 e^{n_1}\left(h_c \frac{\sqrt{\rho_* g}}{\sigma_*}\right)^{n_2}\left(\frac{\mu_d g^4}{\rho_*^{1/4}\sigma_*^{3/4}}\right)^{n_3}\left(\frac{\sigma}{\sigma_*}\right)^{n_4}\left[C_2 + \exp\left(C_3 \frac{Af}{e\left(\frac{\sigma_* g}{\rho_*}\right)^{1/4}}\right)\right] \tag{2.86}$$

The parameters C_1, C_2, C_3, n_1, n_2, n_3, n_4 in the above correlation are 2.84, 0.16, −2.59, 0.3, 0.18, 0.14, and 0.06, respectively.

The authors comment that: "The experimental data are fitted reasonably with a relative deviation of 22.23%; however, a marked deviation was observed for the system of low interfacial tension. Moreover, the effect of dispersed phase velocity was not considered in this equation."

So a new correlation was developed, thus:

$$\frac{d_{32}}{\sqrt{\frac{\sigma}{\Delta\rho g}}} = 33.53 \times 10^{-3}\left(\frac{Af^4\rho_c}{g\sigma}\right)^{-0.283}\left(\frac{d_a\rho_c\sigma}{\mu_c^2}\right)^{0.29}\left(\frac{\sigma^4\rho_c}{\mu_c\psi}\right)^{-0.13}\left(\frac{\Delta\rho}{\rho_c}\right)^{2.86}$$
$$\left(\frac{\mu_d}{\mu_c}\right)^{0.085}\left(\frac{h_c}{d_a}\right)^{-0.734} \times (1+R)^{0.34}$$

Where

$$\psi = \frac{(2\pi^2(1-e^2))}{(3e^2C_0^2h_c)(Af)^3} \tag{2.87}$$

e = fractional free area in the column.

This correlation provided a fit to observed drop size with an average mean error of ±7.32%.

Syll et al. (2011) tested Kumar and Hartland's drop size correlation for pulsed columns during mass transfer of acetic acid from a dispersed phase of ethyl acetate to a continuous phase of water under standard conditions as follows, and determined good prediction of mean drop size ±10%.

$$d_{32} = \frac{C\psi e^{0.5}}{\left\{\frac{3}{5}\left(\frac{\Delta\rho g}{\gamma_{LV}}\right) + \frac{9}{4}\epsilon^{0.8}\left(\frac{\rho_c}{\gamma_{LV}}\right)^{1.2}\right\}^{0.5}} \tag{2.88}$$

Usman et al. (2009) also showed that Kumar and Hartland's (1986) drop size correlation for pulsed plate columns proved to be very appropriate for the benzoic acid/water/kerosene system in the region of practical interest close to the end of mixer settler regime and within the dispersion regime. However they observed that it failed in the lower end of the mixer settler regime where superficial velocities have a profound effect on diameters.

Al-Rahawi (2007) proposed an improved drop size correlation for RDC contactors, as follows:

The first correlation is mainly based upon the operating variables, i.e. dispersed phase, distributor hole, diameter, disc speed, column geometry, and the system physical properties:

$$d_{32} = 0.67\left(\frac{\sigma}{g\Delta\rho}\right)^{0.5}\frac{D_{\mathrm{h}}^{0.73}}{N^{0.15}}\left(\frac{Q_c/Q_d}{Q_c+Q_d}\right)^{0.11}(D_c - D)^{0.04}h_c^{\ 0.008} \tag{2.89}$$

The second is a simpler correlation and excludes the column diameter term:

$$d_{32} = 0.705\left(\frac{\sigma}{g\Delta\rho}\right)^{0.5}\frac{D_{\mathrm{h}}^{0.8}}{N^{0.185}}\frac{(Q_c/Q_d)^{0.15}}{(Q_c+Q_d)^{0.1}} \tag{2.90}$$

The correlations were evaluated on two standard systems: toluene/acetone/water, and n-butanol/succinic acid/water, showing a maximum absolute error of ±6.8%.

2.7 Dispersion and Coalescence Modeling: Quantitative Approach

The gaining of deeper understanding of those factors that determine drop size distribution in turbulent liquid–liquid systems required consideration of rates of drop breakup and rates of drop coalescence.

A more elegant approach to prediction of the behavior of liquid–liquid dispersions in turbulent flows is by the application of quantitative population balance methodology and by the application of computational fluid dynamics modeling. Luo and Svendsen (1996) reviewed drop breakup modeling and presented an overview of drop breakup rate models published at that time. The limitation of most models was the requirement of experimental data fitting involving at least two fitting parameters. The assumptions of the approach taken by Luo and Svendsen are as follows:

(i) turbulence is isotropic;

(ii) breakage of drops is binary, meaning that a drop breakup event results in two daughter drops;

(iii) the breakage volume is assumed to be a stochastic variable, where breakage volume f_{BV} is defined thus:

$$f_{\mathrm{BV}} = \frac{v_{\mathrm{I}}}{v} = \frac{d_{\mathrm{I}}^{\,3}}{d^3} \tag{2.91}$$

(iv) occurrence of breakup is determined by the energy level of the arriving fluid eddy;

(v) only eddies of length scale smaller or equal to drop diameter can induce drop oscillations; this is especially important since drop oscillation is one of the key mechanisms involved in drop breakup.

Prediction of droplet breakage rate (expressed as a frequency with units of time^{-1}) and prediction of coalescence rate are the keys to determining drop size distributions in turbulent liquid–liquid systems.

Breakup frequency may be defined as follows:

$$\Omega(V_i) = \frac{1}{\text{breakuptime}} \times \text{fractionofdropsbreaking} \tag{2.92}$$

where V_i = volume of the parent droplet.

There are many equations put forward to predict breakup frequency.

Chatzi and Lee (1987) proposed:

$$\frac{\Omega(d)}{n} = c_1 \left(\frac{\varepsilon}{d^2}\right)^{1/3} \Gamma\left(\frac{3}{2}\frac{c_2 \sigma}{\rho d \varepsilon^{2/3} d^{5/3}}\right) \tag{2.93}$$

Hesketh, Etchells, and Russell (1991) proposed:

$$\frac{\Omega(d)}{n} = 2.7\left(\frac{{\rho_c}^{0.1}{\rho_d}^{0.3}\varepsilon^{0.6}}{\sigma^{0.4}}\right) \tag{2.94}$$

Luo and Svendsen (1996) reviewed other older models for breakage and then presented their model that predicts breakage rate and the resulting drop size distribution as follows:

$$\Omega_B(v) = \frac{1}{2}\int_0^1 \Omega_B(v : vf_{BV})df_{BV} \tag{2.95}$$

where:

$$\frac{\Omega_B(v : vf_{BV})}{(1-\varepsilon_d)n} = c_4\left(\frac{\varepsilon}{d^2}\right)^{1/3} \int_{\xi_{min}}^{1} \frac{(1+\xi)^2}{\xi^{11/3}} \exp\left(-\frac{12 c_f \sigma}{\beta \rho_c \varepsilon^{2/3} d^{5/3} \xi^{11/3}}\right) d\xi$$

and where:

$$\xi_{min} = \frac{\lambda_{min}}{d}$$

Drop size distribution is given by:

$$\eta(v : vf_{BV}) = \frac{2\int_{\xi_{min}}^1 \frac{(1+\xi)^2}{\xi^{11/3}} \exp(-\chi_c) d\xi}{v\int_{\xi_{min}}^1 \frac{(1+\xi)^2}{\xi^{11/3}} \exp(-\chi_c) d\xi df_{BV}} \tag{2.96}$$

Tsouris and Tavlarides (1994) addressed the challenge of quantitative population balance methodology and their approach is summarized as follows:

$$\frac{\partial n(v,t)}{\partial t} + D - B = 0 \tag{2.97}$$

Where $n(v,t)$ is the number of drops of volume v at time t. D and B represent death and birth rates of drops, respectively. The death term D consists of breakage to smaller drops and coalescence to larger ones, while the birth term B consists of breakage of larger drops and coalescence of smaller ones.

$$D = n(v,t)\int_0^{\infty}\lambda(v,v')h(v,v')n(v,v')dv' + g(v)n(v,t) \tag{2.98}$$

$$\begin{aligned} B = & \int_0^{v/2}\lambda(v-v',v')h(v-v',v')n(v-v',t)n(v',t)dv' \\ & + \int_0^{\infty}\beta(v,v')\nu(v')g(v')n(v',t)dv' \end{aligned} \tag{2.99}$$

$$\begin{aligned} g(d) = 0.0118DF(\phi)\varepsilon_1^{1/3} & \int_{\frac{2}{d}}^{\frac{2}{d_{r,\,\min}}}\left(\frac{2}{k}+d\right)^2 \\ & \times\left(8.2k^{-2/3}+1.07d^{2/3}\right)^{1/2}k^2\exp\left[-\frac{E_c}{c_1 e}\right]dkn_d \end{aligned} \tag{2.100}$$

$$\beta(d,d') = \frac{\varepsilon_{\min}+[\varepsilon_{\max}-\varepsilon(d)]}{\int_0^{d'}[\varepsilon_{\min}+[\varepsilon_{\max}-\varepsilon(d)]]\delta d} \tag{2.101}$$

$$h(d_i,d_j) = \frac{\pi}{4}\left(d_i+d_j\right)^2\left(\overline{u_i^2}+\overline{u_j^2}\right)^{1/2}n_i n_j \tag{2.102}$$

$$\overline{u^2} = 1.07\varepsilon^{2/3}d^{2/3} \tag{2.103}$$

$$\lambda(d_1,d_2) = \exp\left[-\frac{6\pi\mu_c c_2\zeta}{\rho_c\varepsilon^{*2/3}(d_1+d_2)^{2/3}}\frac{31.25ND_i}{\left(T^2H\right)^{1/3}}\right] \tag{2.104}$$

Using Equations 2.100–2.104, one can obtain drop breakage and coalescence rates that can be utilized by the population balance equation model, equation 2.97, for calculations of the transient drop size distribution in a batch stirred tank contactor. Discretization of equation 2.97 leads to a system of nonlinear ordinary differential equations that form an initial value problem.

Figure 2.11 shows the results of the simulations of Sauter mean diameter presented by Tsouris and Tavlarides (1994) for the system toluene (dispersed)/water (continuous) in a stirred tank. These are based on the above analysis and very good agreement between experimental data and simulations is shown.

More recent work described by Srilatha et al. (2010) confirmed this approach in which a quantitative population balance methodology is demonstrated to accurately model drop size distributions in a pump-mix mixer. CFD modeling was used to establish local hydrodynamic conditions and local energy dissipation rates in the mixer relating, these to rates of drop breakup and coalescence to determine a quasi-steady state drop size distribution for each of a range of conditions. The range of variables for which the predictions were accurately demonstrated included flow rate, impeller design, speed and location, and system properties. The experimental work described by Srilatha et al. (2010), and the modeling, confirmed the importance of convection and an

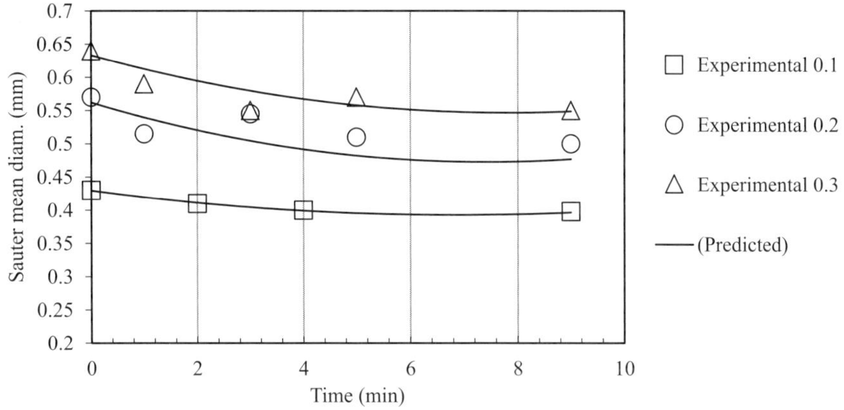

Figure 2.11 Calculated and experimental Sauter mean diameter in a stirred tank. System: toluene (dispersed)/water (continuous): 270 rpm.

Reprinted (adapted) with permission from Tsouris, C. and Tavlarides, L. L. (1994). Breakage and coalescence models for drops in turbulent dispersions. *American Institute of Chemical Engineers Journal*, **40**, 395–406.

understanding of energy dissipation in predicting drop size in turbulent dispersions. The nexus between CFD and population balance modeling was highlighted in this work.

In a comprehensive review of drop and bubble breakup phenomena, Liao and Lucas (2009) define four mechanisms that are involved in breakup:

- breakup due to turbulent fluctuation and collision;
- breakup due to the viscous shear forces;
- breakup due to interfacial instability;
- breakup due to shearing-off processes.

In the case of drop breakup due to turbulent fluctuations, Liao and Lucas (2009) proposed that the mechanisms at play include turbulent pressure fluctuations along the surface of the drop–continuous phase interface, and droplet–turbulent eddy collisions. It is suggested that the onset of drop breakup occurs when the amplitude of the fluctuations exceeds a certain value, causing the interface to disrupt. Thus the dominant forces at play are caused by the dynamic pressure differences at the interface. The extent of deformation of the drop is related to the Weber number for the drop, We.

The nature and cause of the turbulent pressure fluctuations has been the subject of much analysis by many researchers over an extended period of time. A number of papers refer to critical values of kinetic energy for the droplet or of the particle velocity value (Alopaeus et al., 1999, 2002; Chatzi, 1983; Chatzi and Lee, 1987; Chatzi, Gavrielides, and Kiparissides, 1989; Coulaloglou and Tavlarides, 1977; Lee, Erickson, and Glasgow, 1987a, 1987b; Luo and Svendsen, 1996; Martinez-Bazan et al., 1999a, 1999b; Narsimhan and Gupta, 1979; Prince and Blanch, 1990; Tsouris and Tavlarides, 1994). Others have suggested that breakup is possible when the inertial force of the hitting eddy is greater than the interfacial force of the smallest daughter particle (Lehr, Millies, and Mewes 2001, 2002) and others propose a combination of both possible mechanisms.

Much of the work reviewed above is focused on liquid systems of relatively normal viscosity, which reflect the solvents used in industrial separations where high mass transfer and ease of disengagement are critical requirements. In low viscosity systems, the surface force dominates drop stabilization and the influence of the internal viscous force is usually neglected. Quantitative analysis of systems involving dispersed liquids of higher viscosity is less well developed. Viscous forces can play an important role in droplet breakup. Such breakup occurs due to viscous shear forces being developed through velocity gradients around the drop caused by interfacial shear forces. The gradient causes deformation of the drop and ultimately may result in breakup. The occurrence of a wake behind a moving droplet also can create local viscous shear forces that contribute to drop elongation and possible breakup.

Viscous shear forces are also responsible for the mechanism of "shearing off." This is particularly the case as drop size increases, making the drop less stable and more prone to the shear-induced "tearing off" of smaller drops. The treatment of dispersion of droplets of higher viscosities in a system of complex hydrodynamics has seen some significant progress in recent years. Work by Abidin, Raman, and Nor (2014) is of particular note, where they develop an accurate theoretical analysis of drop size prediction for a stirred tank system involving viscous drops. The starting point is based on the long-established relationship between Sauter mean drop diameter and Weber number, as detailed in equation 2.105:

$$\frac{d_{32}}{D} \propto \mathrm{We}^{-0.6} \qquad (2.105)$$

This form of relationship has been used extensively for dilute systems involving dispersed liquids of low viscosity. The constant of proportionality is frequently based on experimental data. Abidin et al. incorporated a viscosity term into a form of equation 2.105 using a dimensionless viscosity number V_i, configured to represent the ratio of viscous to surface forces:

$$V_i = \left(\frac{\rho_c}{\rho_d}\right)^{0.5} \frac{\mu_d N D}{\sigma} \qquad (2.106)$$

The viscosity number is incorporated into the expression for d_{32} as follows:

$$\frac{d_{32}}{D} = 0.000014\left[1 + 1.428\left(\varepsilon_T N D^2\right)^{-0.7}\right] V_i^{0.27} \qquad (2.107)$$

The correlations were tested using data measured for mixtures of silicone oil in water in a flat-bottomed cylindrical baffled tank equipped with a six-bladed Rushton turbine. The main independent variable was the viscosity of the silicone oil (the droplet phase), three values being: employed 20 mPa s, 350 mPa s, and 500 mPa s.

The addition of this term accounts for the drop stabilization due to both surface forces and viscous forces. Two further correlations are presented by Abidin et al. as follows, shown in equations 2.108 and 2.109:

$$\frac{d_{32}}{D} = 0.285\left[1 + 0.0035\left(\frac{\mu_d}{\mu_c}\right)^{0.9}\right] \mathrm{We}^{-0.83} \qquad (2.108)$$

In equation 2.108, the effect of viscosity is represented by the viscosity ratio of the dispersed phase liquid to the continuous phase.

In equation 2.109, the effect of viscosity is represented by the viscosity number, V_i, as defined in equation 2.106.

$$\frac{d_{32}}{D} = 1.558\left[1 + 0.0227 V_i^{1.4}\right] \mathrm{We}^{-1.05} \qquad (2.109)$$

The difference in these equations is reflected in the degrees of accuracy with which they were shown to predict the mean drop size value in a stirred vessel as shown in Figures 2.12–2.14.

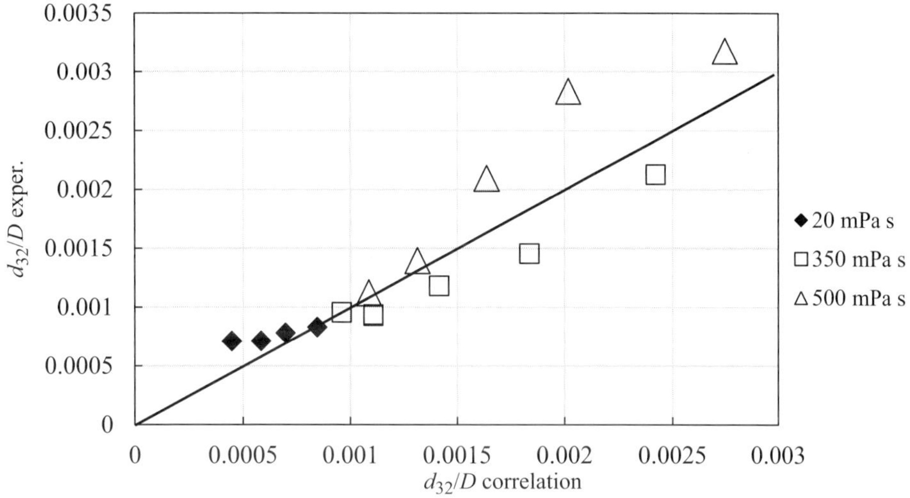

Figure 2.12 Comparison of experimental mean drop size vs correlation (equation 2.107). Reprinted (adapted) with permission from Abidin, M., Raman, A., and Nor, M. (2014). Experimental investigations in liquid–liquid dispersion system: effects of dispersed phase viscosity and impeller speed. *Industrial Engineering Chemistry Research*, **53**, 6554–6561. Copyright (2014) American Chemical Society

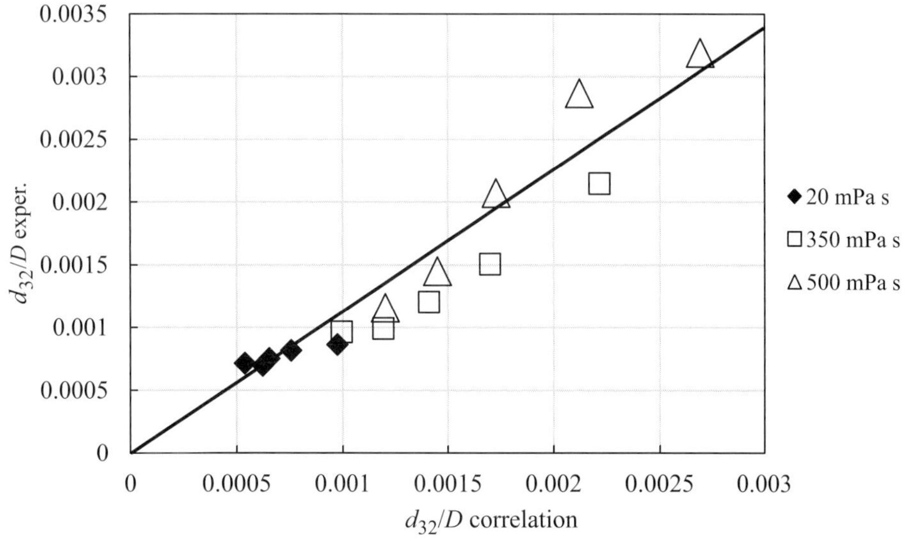

Figure 2.13 Comparison of experimental mean drop size vs correlation (equation 2.108). Reprinted (adapted) with permission from Abidin M., Raman, A., and Nor, M. (2014). Experimental investigations in liquid–liquid dispersion system: effects of dispersed phase viscosity and impeller speed. *Industrial Engineering Chemistry Research*, **53**, 6554–6561. Copyright (2014) American Chemical Society

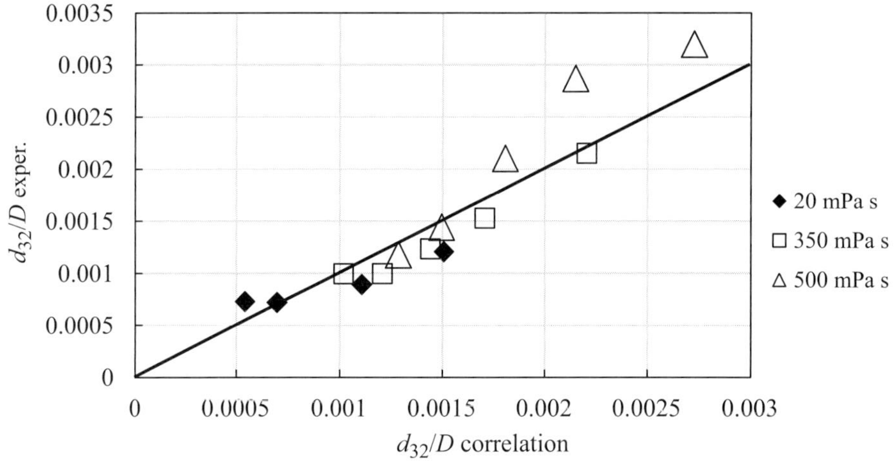

Figure 2.14 Comparison of experimental mean drop size versus correlation (equation 2.109). Reprinted (adapted) with permission from Abidin M., Raman, A., and Nor, M. (2014). Experimental investigations in liquid–liquid dispersion system: effects of dispersed phase viscosity and impeller speed. *Industrial Engineering Chemistry Research*, **53**, 6554–6561. Copyright (2014) American Chemical Society

Abidin et al. concluded that the predicted d_{32} values using either equation 2.108 or 2.109 correlated well with the experimental d_{32} values with a small degree of scatter at the highest value of viscosity. Equation 2.109 provides the best fit, although the differences are small.

References

Abidin, M., Raman, A., and Nor, M (2014). Experimental investigations in liquid–liquid dispersion system: effects of dispersed phase viscosity and impeller speed. *Industrial Engineering Chemistry Research*, **53**, 6554–6561.

Alopaeus, V., Koskinen, J., and Keskinen, K. I. (1999). Simulation of the population balances for liquid–liquid systems in a nonideal stirred tank, Part 1: Description and qualitative validation of the model. *Chemical Engineering Science*, 54, 5887–5899.

Alopaeus, V., Koskinena, J., Keskinena, K. I., and Majander, J. (2002). Simulation of the population balances for liquid–liquid systems in a nonideal stirred tank, Part 2: Parameter fitting and the use of the multiblock model for dense dispersions. *Chemical Engineering Science*, 57, 1815–1825.

Al-Rahawi, A. M. I. (2007). New predictive correlations for the drop size in a rotating disc contactor liquid–liquid extraction column. *Chemical Engineering & Technology*, **30**(2), 184–192.

Baird, M. H. I. and Lane, S. J. (1973) Drop size and holdup in a reciprocating plate extraction column. *Chemical Engineering Science*, **28**(3), 947–957.

Baumler, K., Wegener, M., Paschedag, A. R., and Bansch, E. (2011). Drop rise velocities and fluid dynamic behavior in standard test systems for liquid/liquid extraction – experimental and numerical investigations. *Chemical Engineering Science*, **66**, 426–439.

Bensalem, A., Steiner, L., and Hartland, S. (1986), Effect of mass transfer on drop size in a Karr column. *Chemical Engineering and Processing*, **20**(3), 129–135.

Bond, W. N. and Newton, D. A. (1928). Bubbles, drops and Stokes law. *Philosophical Magazine*, **5**(30), 794–800.

Calabrese, R. V., Chang, T. P. K., and Dang, P. T. (1986a). Drop breakup in turbulent stirred-tank contactors. Part I: Effect of dispersed-phase viscosity. *American Institute of Chemical Engineers Journal*, **32**, 657–666.

Calabrese, R. V., Wang, C. Y., and Bryner, N. P. (1986b) Drop breakup in turbulent stirred tank contactors. Part III: Correlation for mean size and drop size distributions. *American Institute of Chemical Engineers Journal,* **32**, 677–681.

Chatzi, E. (1983). Analysis of Interactions in Fluid–Fluid Dispersion Systems in Agitated Vessels. Cleveland, OH: Cleveland State University Press.

Chatzi, E. G. and Lee, J. M. (1987). Analysis of interactions for liquid–liquid dispersions in agitated vessels. *Industrial Engineering Chemistry Research*, 26, 2263–2267.

Chatzi, E. G., Gavrielides, A. D., and Kiparissides, C. (1989). Generalized model for prediction of the steady-state drop size distributions in batch stirred vessels. *Industrial Engineering Chemistry Research*, **28**, 1704–1711.

Chen, C. T., Maa, J., Yang, Y., and Chang, C. (2001). Drop formation from flat tip nozzles in a liquid–liquid system. *International Communications in Heat Mass Transfer*, **28**(5), 681–692.

Cherlo, S. K. R., Kariveti, S., and Pushpavanam, S. (2010). Experimental and numerical investigations of two-phase (liquid–liquid) flow behavior in rectangular microchannels. *Industrial Engineering Chemistry Research*, **49**, 893–899.

Clift, R,. Grace, J. R., and Weber, M. E. (1978). *Bubbles, Drops and Particles*. New York: Academic Press.

Coulaloglou, C. A. and Tavlarides, L. L. (1977). Description of interaction processes in agitated liquid–liquid dispersions. *Chemical Engineering Science*, **32**, 1289–1297.

Deshpande, K. B. and Zimmerman, W. B. (2006). Simulation of interfacial mass transfer by droplet dynamics using the level set method. *Chemical Engineering Science*, 61, 6486–6498.

Garner, F. H. and Skelland, A. H. P. (1955). Some factors affecting droplet behaviour in liquid–liquid systems. *Chemical Engineering Science*, **4**(4), 149–158.

Grace, J. R., Wairegi, T., and Nguyen, T. H. (1976). Shapes and velocities of single drops and bubbles moving freely through immiscible liquids. *Transactions of the Institution of Chemical Engineers*, **54**, 167.

Hadamard, J. S. (1911). Mouvement permanent lent d'une sphere liquide et visqueuse dans un liquide visqueux. *C. R. Acad. Sci.* (in French), 152, 1735–1738.

Hamielec, A. E., Storey, S. H., and Whitehead, J. M. (1963). Viscous flow around fluid spheres at intermediate Reynolds-numbers. *Canadian Journal of Chemical Engineering*, **12**, 246–251.

Harkins, W. D. and Brown, F. E. (1919). The determination of surface tension (free surface energy) and the weight of falling drops: the surface tension of water and benzene by the capillary height method. *Journal of the American Chemical Society*, **41**(4), 499–524.

Harper, J. F. and Moore, D. W. (1968). Motion of a spherical liquid drop at high Reynolds number. *Journal of Fluid Mechanics*, **32**, 367–391.

Hayworth, C. B. and Treybal, R. E. (1950). Drop formation in two-liquid phase systems. *Industrial and Engineering Chemistry Research*, **June**, 1174–1181.

Hesketh, R. P., Etchells, A. W., and Russell, T. W. F. (1991). Experimental observations of bubble breakage in turbulent flow. *Industrial and Engineering Chemistry Research*, 30, 835–841.

Hinze, J. O. (1955). Fundamentals of the hydrodynamic mechanism of splitting dispersion process. *American Institute of Chemical Engineers Journal*, **1**, 289–295.
Hu, S. and Kintner, R. C. (1955). The fall of single drops through water. *American Institute of Chemical Engineers Journal*, **1**(1), 42–48.
Hughmark, G. A. (1971). Drop breakup in turbulent pipe flow. *American Institute of Chemical Engineers Journal*, **17**(4), 1000.
Kalyanasundaram, C. V., Kumar, R., and Kuloor, N. R. (1968). Direct contact heat transfer between two immiscible phases during drop formation. *International Journal of Heat and Mass Transfer,* **11**(12), 1826–1830.
Klee, A. J. and Treybal, R. E. (1956). Rate of rise or fall of liquid drops. *American Institute of Chemical Engineers Journal*, **2**(4), 444–447.
Krishna, P. M., Venkateswarlu, D., and Narasimhamurty, G. S. R. (1959). Fall of liquid drops in water. Drag coefficients, peak velocities, and maximum drop sizes. *Journal of Chemical Engineering Data*, 4, 340–343.
Kumar, A. and Hartland, S. (1986). Prediction of drop size in rotating-disk extractors. *Canadian Journal of Chemical Engineers*, **64**(6), 915–924.
Kumar, A., and Hartland, S. (1988). Prediction of dispersed-phase holdup and flooding velocities in Karr reciprocating-plate extraction columns. *Industrial and Engineering Chemistry Research*, **27**(1), 131–138.
Kumar, A. and Hartland, S. (1996). Unified correlations for the prediction of drop size in liquid–liquid extraction columns. *Industrial and Engineering Chemistry Research*, **35**, 2682–2695.
Laso, M., Steiner, L., and Hartland, S. (1987). Dynamic simulation of agitated liquid–liquid dispersions: II. Experimental determination of breakage and coalescence rates in a stirred tank. *Chemical Engineering Science*, **42**(10), 2437–2445.
Lee, C. H., Erickson, L. E., and Glasgow, L. A. (1987a). Bubble breakup and coalescence in turbulent gas–liquid dispersions. *Chemical Engineering Communications*, **59**, 65–84.
Lee, C. H., Erickson, L. E., and Glasgow, L. A. (1987b). Dynamics of bubble size distribution in turbulent gas–liquid dispersions. *Chemical Engineering Communications*, **61**, 181–195.
Lehr, F., Millies, M., and Mewes, D. (2001). Coupled calculation of bubble size distribution and flow fields in bubble columns. *Chemie Ingenieur Technik*, 73, 1245–1259.
Lehr, F., Millies, M., and Mewes, D. (2002). Bubble-size distributions and flow fields in bubble columns. *AIChE Journal*, 48, 2426–2443.
Letan, R. and Kehat, E. (1967). The mechanics of a spray column. *American Institute of Chemical Engineers Journal*, **13**(3), 443–449.
Liao, Y. and Lucas, D. A. (2009). Literature review of theoretical models for drop and bubble breakup in turbulent dispersions. *Chemical Engineering Science*, **64**, 3389–3406.
Loth, E. (2008). Quasi-steady shape and drag of deformable bubbles and drops. *International Journal of Multiphase Flow,* **34**(6), 523–546.
Luo, H. and Svendsen, H. F.(1996). Theoretical model for drop and bubble breakup in turbulent dispersions. *American Institute of Chemical Engineers Journal*, **42**, 1225–1233.
Martinez-Bazan, C., Montanes, J. L., and Lasheras, J. C. (1999a). On the breakup of an air bubble injected into fully developed turbulent flow. Part 1. Breakup frequency. *Journal of Fluid Mechanics*, **401**, 157–182.
Martinez-Bazan, C., Montanes, J. L., and Lasheras, J. C. (1999b). On the breakup of an air bubble injected into fully developed turbulent flow. Part 2. Size PDF of the resulting daughter bubbles. *Journal of Fluid Mechanics*, **401**, 183–207.

Moore, D. W. (1959).The rise of a gas bubble in a viscous liquid. *Journal of Fluid Mechanics*, **6**(1), 113–130.

Moore, D. W. (1963). The boundary layer on a spherical gas bubble. *Journal of Fluid Mechanics*, **16**(2), 161–176.

Narsimhan, G. and Gupta, J. P. (1979). A model for transitional breakage probability of droplets in agitated lean liquid–liquid dispersions. *Chemical Engineering Science*, **34**, 257–265.

Null, H. R. and Johnson, H. F. (1958). Drop formation in liquid–liquid systems from single nozzles. *American Institute of Chemical Engineers Journal*, **4**(3), 273–281.

Pacek, A. W. Man, C. C., and Nienow, A. W. (1998). On the Sauter mean diameter and size distributions in turbulent liquid–liquid dispersions in a stirred vessel. *Chemical Engineering Science,* **53**(11), 2005–2011.

Parlange, J. Y. (1970). Motion of spherical drops at large Reynolds numbers. *Acta Mechanica*, **9**, (3–4), 323–328.

Petera, J. and Weatherley, L.R. (2001). Modelling of mass transfer from falling droplets. *Chemical Engineering Science*, 56, 4929–4947.

Prince, M. J. and Blanch, H. W. (1990). Bubble coalescence and break-up in air-sparged bubble columns. *American Institute of Chemical Engineers Journal*, **36**,1485–1499.

Rodger, W. A., Trice, V. G., and Rushton, J. H. (1956). Effect of fluid motion on interfacial area of dispersions. *Chemical Engineering Progress*, **52**(12), 515–520.

Rybczynski, W. (1911). Über die fortschreitende Bewegung einer flüssigen Kugel in einem zähen Medium. *Bull. Acad. Sci. Cracovie, A.* (in German), 40–46.

Samani, M. G., Asl, A. H., Safdari, J. M., and Torag-Moestaedi, M. (2012). Drop size distribution and mean drop size in a pulsed packed extraction column. *Chemical Engineering Research and Design*, **90**, 2148–2154.

Scheele, G. F. and Meister, B. J. (1968). Drop formation at low velocities in liquid–liquid systems: prediction of drop volume. *American Institute of Chemical Engineers Journal*, 1968, **14**(1), 9–15.

Shinnar, R. (1961). On the behaviour of liquid dispersions in mixing vessels. *Journal of Fluid Mechanics*, **10**, 259–275.

Singh, K. K., Mahajani, S. M., Shenoy, K. T., and Ghosh, S. K. (2008). Representative drop sizes and drop size distributions in A/O dispersions in continuous flow stirred tank. *Hydrometallurgy*, **90**, 121–136.

Skartlien, R., Sollum, E. and Schumann, H. (2013) Droplet size distributions in turbulent emulsions: breakup criteria and surfactant effects from direct numerical simulations. *Journal of Chemical Physics*, **139**, 174901.

Skelland, A. H. P. and Wellek, R. M. (1964). Resistance to mass transfer inside droplets. *American Institute of Chemical Engineers Journal*, **10**, 491–496.

Skelland, A. H. P., Woo, S., and Ramsay, G. G. (1987). Effects of surface-active agents on drop size, terminal velocity, and droplet oscillation in liquid–liquid systems. *Industrial & Engineering Chemistry Research*, 26, 907–911.

Srilatha, C., Morab, V. V., Mundada, T. P., and Patwardhan, A. W. (2010). Relation between hydrodynamics and drop size distributions in pump–mix mixer. *Chemical Engineering Science*, 65, 3409–3426.

Syll, O. S., Mabille, I., Moscosa-Santillan, M., Traore, M., and Amouroux, J. (2011). Study of mass transfer and determination of drop size distribution in a pulsed extraction column. *Chemical Engineering Research and Design,* **89**, 60–68.

Takamatsu, T., Hashimoto, Y., Yamaguchi, M., and Kataym a, T. (1981). Theoretical and experimental studies of charged drop formation in a uniform electric field. *Journal of Chemical Engineering of Japan*, **14**(3), 178–182.

Takamatsu, T., Yamaguchi, M., and Katayma, T. (1982). Formation of single charged drops in liquid media under a uniform electric field. *Journal of Chemical Engineering of Japan*, **15**(5), 349–355.

Taylor, T. D. and Acrivos, A. (1964). On the deformation and drag of a falling viscous drop at low Reynolds number. *Journal of Fluid Mechanics*, **18,** 466–476.

Thorsen, G., Stordalen, R. M., and Terjesen, S. G. (1968). On the terminal velocity of circulating and oscillating liquid drops. *Chemical Engineering Science*, 1968, **23**, 413–426.

Torab-Mostaedi, M., Ghaemi, A., and Asadollahzadeh, M. (2011). Flooding and drop size in a pulsed disc and doughnut extraction column. *Chemical Engineering Research and Design*, **89**, 2742–2751.

Tsouris, C. and Tavlarides, L. L. (1994). Breakage and coalescence models for drops in turbulent dispersions. *American Institute of Chemical Engineers Journal*, **40**, 395–406.

Usman, M. R., Sattar, H., Hussain, S. N., et al. (2009). Drop size in a liquid pulsed sieve-plate extraction column. *Brazilian Journal of Chemical Engineering*, **26(4)**, 677–683.

Vermeulen, T., Williams, G. M., and Langlois, G. E. (1955). Interfacial area in liquid–liquid and gas–liquid agitation. *Chemical Engineering Progress*, **51**(2), 85–94.

Waheed, M. A., Henschke, M., and Pfennig, A. (2004). Simulating sedimentation of liquid drops. *International Journal for Numerical Methods in Engineering*, **59**,1821–1837.

Wang, C. Y. and Calabrese, R. V. (1986). Drop breakup in stirred-tank contactors, Part II: Relative influence of viscosity and interfacial tension. *American Institute of Chemical Engineers Journal*, **32**(4), 667–676.

Watanabe, T. and Ebihara, K. (2003). Numerical simulation of coalescence and breakup of rising droplets. *Computers & Fluids*, 32, 823–834.

Weatherley, L. R. and Wilkinson, D. T. (1988). Droplet size studies in a whole broth liquid/liquid extraction system. *Process Biochemistry*, **23**(5), 149–156.

Weinstein, B. and Treybal, R. E. (1973). Liquid–liquid contacting in unbaffled agitated vessels. *American Institute of Chemical Engineers Journal*, **19**, 304–312.

Wellek, R. M., Agrawal, A. K., and Skelland, A. H. P. (1966). Shape of liquid drops moving in liquid media. *American Institute of Chemical Engineers Journal*, **12**(5), 854–862.

Winnikow, S. and Chao, B. T. (1966). Droplet motion in purified systems. *Physics of Fluids*, 9, 50–61.

Yuan, S., Shi, Y., Yin, H., Chen, Z., and Zhou, J. (2012). Correlation of the drop size in a modified Scheibel extraction column. *Chemical Engineering and Technology*, 35, 1810–1816.

3 Mass Transfer

3.1 Introduction

Knowledge of the rate of interfacial mass transfer between two immiscible liquid phases is of critical importance for the design of industrial liquid–liquid extraction equipment. The size of the equipment is highly dependent on the mass transfer kinetics and one of the key goals of intensification of liquid–liquid processes is the acceleration of mass transfer rates. The achievement of optimal mass transfer rates is fundamentally dependent upon the fluid dynamics during the contact. The important fluid dynamic effects include drop size, rate of drop breakup and coalescence, mixing phenomena in both phases, interfacial shear, and velocity profiles of both phases. Ultimately, a knowledge of the interfacial mass transfer rate is used in combination with hydraulic data to predict the performance of an entire unit. On account of the complexity of the fluid dynamics, and phenomena such as interfacial instabilities, for design purposes the interfacial mass transfer rate has been calculated using mass transfer coefficients. Mass transfer coefficients, as will be seen in this chapter, are derived through empirically based correlations that describe resistances to mass transfer independently in both phases, expressed as a function of physical properties (Pratt and Stevens, 1992). Use of mass transfer coefficients in most cases applies to steady-state conditions. Other related information required for design include volumetric holdup of the dispersed phase, drop size distribution, and axial and radial mixing. These are generally determined independently of mass transfer although there are many situations where the two sets of phenomena are closely interdependent. This interdependence occurs for example due to changes in physical properties on account of mass transfer and/or reaction resulting in changes in density, viscosity, and interfacial tension. Changes in any of these can impact hydraulic behavior, including drop size, coalescence rate, and relative velocity. The latter can be strongly influenced by changes in density and viscosity due to solute gain or loss by each of the liquid phases.

Hydraulic behavior, mass transport, and reaction rates are inherently linked, as we shall demonstrate. Developments in finite element mathematics have greatly improved our ability to quantitatively simulate the fluid dynamic behavior of liquid–liquid systems. Much of the early research into mass transfer in liquid–liquid systems focused on mass transfer to and from single droplets in unhindered flow, then proceeded to examine effects of droplet oscillation, interfacial disturbance, and the combined effects of external hydrodynamics and droplet oscillation on not only intra-droplet mass transfer but also external mass transfer in the continuous phase. Later work then focused on swarming droplet systems such as are encountered in liquid–liquid mixers, column contactors, and rotating contactors. Further developments examined mass transfer in systems involving viscous liquid phases, and systems containing impurities such as solids and surfactants that influence mass transfer and droplet hydrodynamics. In Chapters 5 and 6 the influence of externally applied fields, high gravity, and electrical fields on mass transfer in liquid–liquid systems is considered.

Understanding the interrelationships between drop size, drop hydrodynamics, drop stability, and mass transfer is fundamental to the intensification of kinetics of liquid–liquid processes. The chapter is divided into two principal sections. Firstly, we will review single droplet mass transfer and discuss the early fundamental work based on experiment and correlation based on dimensional analysis. Recent developments in the modeling of mass transfer in discrete droplet systems will be reviewed. Secondly, mass transfer in swarming droplet systems will be reviewed, examining the effects of hindered and unhindered droplet motion, and mass transfer in different types of industrial equipment.

3.2 Single Droplet Systems

It is quite widely accepted that understanding of mass transfer in single droplet systems can be valuable for equipment design and simulation with the important proviso that mass transfer data for single drop systems are used in combination with appropriate hydrodynamic models (Brodkorb et al., 2003). Examples include population balance models, statistical models such as those using Monte Carlo simulations, and, more recently, those using cloud models (Petera et al., 2009).

At the outset of this section it is useful to summarize the key mechanisms influencing mass transfer for a single drop, forming and detaching from a nozzle, subsequently moving through the continuous phase in a column, until coalescing into a layer of coalesced dispersed phase, as shown in Figure 3.1 (after Johnson and Hamielec, 1960). They defined three distinct mechanisms that are at play in such a system. Mass transfer occurs during drop formation and there is a significant body of research into description of mass transfer in forming droplets. Once the droplet has detached, mass transfer continues to occur but at a different rate, since the hydrodynamic conditions during unhindered motion are different from those experienced by the droplet during formation. Thirdly, as the droplet coalesces into the lower interface (as shown in Figure 3.1) a different set of hydrodynamic conditions are created due to the action of the drop hitting the interface and coalescing, thus changing the mass transfer mechanism. The mass transfer rates during droplet formation and during coalescence are generally accepted as being very significant, especially during formation when the concentration difference between the phases is at a maximum thus giving rise to interfacial disturbances due to local gradients of interfacial tension.

Fundamental description of the mass transfer phenomena associated with drop formation and coalescence is complex and is an area of continuing interest in the literature. In the case of mass transfer in the "free fall" zone, drop size and velocity in addition to physical properties of all components are the key parameters of interest. There is a substantial historical literature dealing with droplet size, velocity, and holdup in liquid–liquid systems, as discussed in detail in Chapter 2 (Garner and Skelland, 1955;

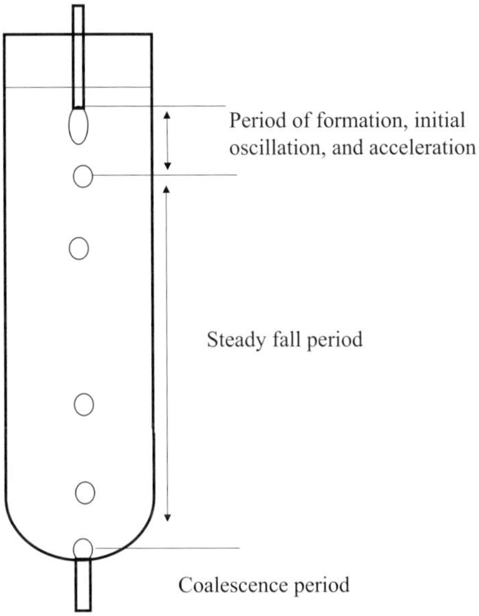

Figure 3.1 Drop mechanisms that influence mass transfer in a moving drop (Johnson and Hamielec, 1960).

Grace, Wairegi, and Nguyen, 1976; Hu and Kintner, 1955; Klee and Treybal, 1956; Scheele and Meister, 1968). Dimensional analysis is the main tool used in those papers to establish description of the observations. Kumar and Hartland (1999) present an excellent review of the literature relevant to single drop mass transfer and list published correlations for both circulating single droplets and for oscillating single droplets.

Before reviewing the correlations for single drop mass transfer in more detail, it is worth noting the extensive work of Wegener et al. (2009a) on the interrelationships between droplet instability and mass transfer. They present a useful diagrammatic summary of internal flow profiles in a single drop under different conditions (see Figure 3.2, taken from the reference). Diagram (c) shows the circulation pattern for a normal circulating droplet. The other diagrams, (a), (b1), and (b2), show the progressive impact of Marangoni instabilities on the internal flow profile relating to mass transfer enhancement. Further consideration of Marangoni effects appears in Section 3.5.

Figure 3.3 shows the development of oscillation for a detaching organic drop rising into a continuous phase of water. The onset of oblate spheroidal cyclic oscillation is illustrated.

Drop size alone has been shown to be very important. In a recent study by Zheng et al. (2014) for the extraction of acetic acid from water into n-propyl acetate, they present data for single drop mass transfer and demonstrate how this is influenced by drop size (see Figure 3.4).

These data show the large increase in mass transfer coefficient as the drop diameter increases from approximately 2.9 mm to 3.1 mm due to the onset of oscillation and thus internal instability within the drop.

The approach presented by Kumar and Hartland (1999) characterizes the relationships for mass transfer coefficients as follows:

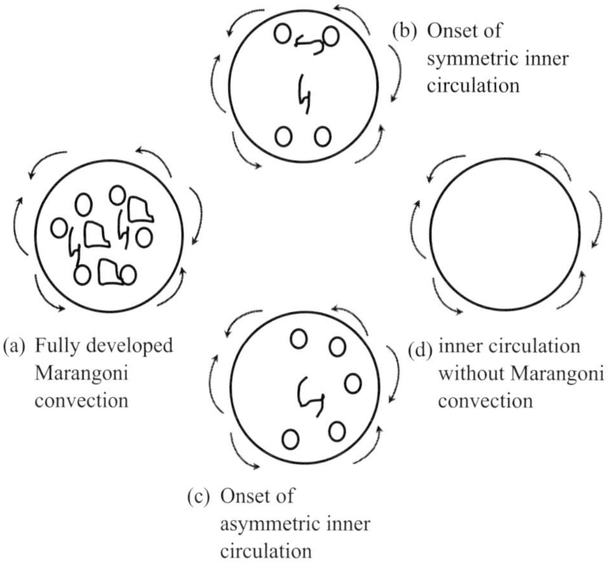

Figure 3.2 Scheme of inner circulation pattern: (a) fully developed Marangoni convection; (b) onset of symmetric inner circulation; (c) onset of asymmetric inner circulation; (d) inner circulation without Marangoni convection.

Reprinted (adapted) with permission from Wegener, M., Fevre, M., Paschedag, A. R., and Kraume, M. (2009). Impact of Marangoni instabilities on the fluid dynamic behaviour of organic droplets. *International Journal of Heat and Mass Transfer*, **52**, 2543–2551. Copyright (2009) Elsevier B.V. All Rights Reserved

(i) Continuous phase mass transfer coefficients, k_c, describing mass transfer resistance exterior to the dispersed drop for: (a) rigid drops, (b) circulating drops, (c) oscillating drops, and (d) contaminated systems.

(ii) Dispersed phase mass transfer coefficients, k_d, describing mass transfer resistance internal to the drop for: (a) rigid drops, (b) circulating drops, (c) oscillating drops, and (d) contaminated systems.

(iii) Overall mass transfer coefficients, k_{od}, k_{oc}, for: (a) rigid drops, (b) circulating drops, (c) oscillating drops, and (d) contaminated systems.

Kumar and Hartland (1999) provide a detailed summary and critical appraisal of correlations available from the literature for continuous phase side mass transfer coefficients for single droplets that indicate the range of applicability in terms of droplet Reynolds number.

Correlations for the case of single circulating droplets and for oscillating droplets are included. The dimensionless Sherwood number, Sh, is most often used as the dependent variable in each correlation, defined thus:

$$\mathrm{Sh} = \frac{kd}{D}$$

where k = mass transfer coefficient,

d = drop diameter, and
D = diffusion coefficient.

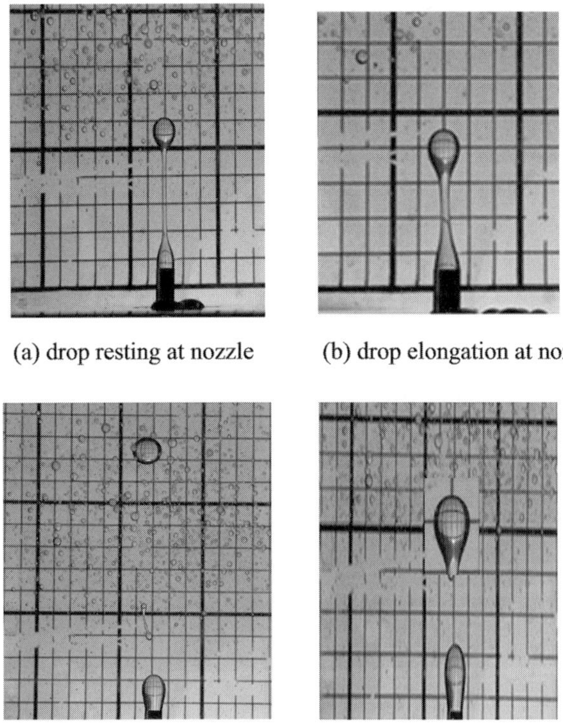

(a) drop resting at nozzle (b) drop elongation at nozzle

(c) oblate spheroid oscillating drop after detachment (d) prolate spheroid oscillating drop after detachment

Figure 3.3 The development of oscillation for a detaching organic drop rising into a continuous phase of water, showing the onset of oblate spheroidal cyclic oscillation (Gangu, 2013).

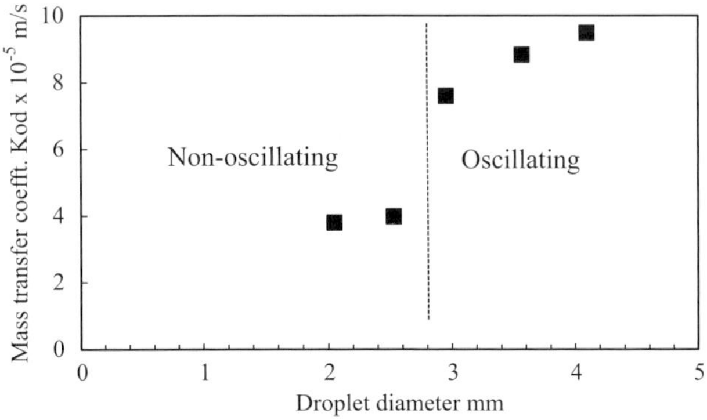

Figure 3.4 Influence of drop diameter on overall mass transfer coefficient for the extraction of acetic acid from water into n-propyl acetate.

For single circulating droplets, Kumar and Hartland list the following correlations, see equations 3.1–3.14 below.

$8 < \mathrm{Re} < 800$

(Garner, Foord, and Tayeban, 1959)

$$\mathrm{Sh_c} = -126 + 1.8\mathrm{Re}^{0.5}\mathrm{Sc}^{0.42} \quad (3.1)$$

(Garner and Tayeban, 1960)

$$\mathrm{Sh_c} = 0.6\mathrm{Pe_c}^{0.5} \quad (3.2)$$

$4 < \mathrm{Re} < 100$

Slater (1994)

$$\mathrm{Sh_c} = 2 + 1.13C_1^{0.5}\mathrm{Pe_c}^{0.5} \quad (3.3)$$

$$C_1 = 0.16 \ln(\mathrm{Re}^{0.5}/\kappa) + 0.35$$

$20 < \mathrm{Re} < 100$

Lochiel and Calderbank (1964)

$$\mathrm{Sh_c} = 2\left(\frac{\mathrm{Pe_c}}{\pi}\right)^{1/2}\left(1 - \frac{2 + 3\kappa}{1 + (\kappa\rho_d/\rho_c)^{0.5}}\frac{1.45}{\mathrm{Re}^{0.5}}\right)^{1/2} \quad (3.4)$$

$50 < \mathrm{Re} < 800$

Thorsen and Terjesen (1962)

$$\mathrm{Sh_c} = -178 + 3.62\mathrm{Re}^{0.5}\mathrm{Sc_c}^{1/3} \quad (3.5)$$

$4 < \mathrm{Re} < 1000;\ 130 < \mathrm{Sc_c} < 23\,600$

Brauer (1971), Ihme, Schmidtt, and Brauer (1972)

$$\mathrm{Sh_c} = 2 + 0.0511\mathrm{Re}^{0.724}\mathrm{Sc_c}^{0.70} \quad (3.6)$$

$\mathrm{Re} > 40,\ \kappa < 3$

Yamaguchi and Katayama (1977)

$$\mathrm{Sh_c} = 2 + C_1\,\mathrm{Pe}^{0.5} \quad (3.7)$$

Where C_1 is expressed as follows:

$$C_1 = \frac{1}{3(1+\kappa)} + \frac{(2+3\kappa)\mathrm{Re}}{3(1+\kappa)(16+16\kappa+\mathrm{Re})} - \frac{4}{3}\left(\frac{2+3\kappa}{1+\sqrt{\kappa\rho_\mathrm{d}/\rho_\mathrm{C}}}\right)\sqrt{\frac{2}{\pi\mathrm{Re}}} \tag{3.8}$$

Kumar and Hartland also then present and analyze correlations for dispersed phase mass transfer coefficients. Mass transfer modeling for single drop systems dates to early work by Newman (1931) who presented an expression for the dispersed phase mass transfer coefficient in the absence of mass transfer resistance in the continuous phase as follows:

$$k_\mathrm{d} = -\left(\frac{d}{6t}\right)\ln\left[\left(\frac{6}{\pi^2}\right)\sum_{n=1}^{\infty}\left(\frac{1}{n^2}\right)\exp\left(-\frac{4n^2\pi^2 D_\mathrm{OE}t}{d^2}\right)\right] \tag{3.9}$$

The Newman model was developed for solid spheres and has a fundamental basis using the solution of the partial differential equations governing transient diffusion through a rigid sphere. In the context of liquid droplets, the model has an acceptable degree of accuracy only for stagnant drops that do not experience significant oscillation or interfacial instability. Generally this is exclusive to small drop sizes.

This equation is a special case of an earlier expression by Grober (1925), which was developed for mass transfer resistance in both phases, hence k_od.

$$k_\mathrm{od} = -\frac{d}{6t}\ln\left[6\sum_{n=1}^{\infty}B_n\exp\left(-\frac{4\lambda_n^2 D_\mathrm{a}t}{d^2}\right)\right] \tag{3.10}$$

where B_n and λ_n are functions as listed by Elzinga and Banchero (1959).

Other correlations for dispersed phase mass transfer for stagnant droplets were developed as follows. By retaining only the first term of the series of the Newman equation, Kumar and Hartland showed that the dispersed phase mass transfer coefficient k_d for long contact times for stagnant droplets can be represented by equation 3.11:

$$k_\mathrm{d} \simeq \frac{2\pi^2 D_\mathrm{d}}{3d} \text{ or } \mathrm{Sh} \simeq 6.58 \tag{3.11}$$

In the case of circulating droplets, Kronig and Brink (1950) developed a model for mass transfer for the case of dispersed phase side mass transfer control and for droplets having Reynolds numbers less than unity and assuming that the internal circulation pattern conformed to that suggested by Hadamard (1911) and Rybczynski (1911). The result for the mass transfer coefficient is as follows:

$$k_\mathrm{d} = -\frac{d}{6t}\ln\left[\frac{3}{8}\sum_{n=1}^{\infty}B_n^2\exp\left(-\frac{64\lambda_n D_\mathrm{d}t}{d^2}\right)\right] \tag{3.12}$$

Values of the coefficients B_n and λ_n are published by Elzinga and Banchero (1959) as before.

Again, Kumar and Hartland claim that this can be simplified without significant loss of accuracy by only using the first term of the expansion, to yield:

$$k_{\mathrm{d}} \simeq 17.7\frac{D_{\mathrm{d}}}{d} \quad \text{or} \quad \mathrm{Sh}_{\mathrm{d}} \simeq 17.7 \tag{3.13}$$

This was further improved on in work by Skelland and Wellek (1964), who proposed a correlation as follows for mass transfer in single circulating droplets based on measurements in four different binary liquid–liquid systems:

$$\frac{k_{\mathrm{d}}d}{D_{\mathrm{d}}} = 31.4\left(\frac{4D_{\mathrm{d}}t}{d^2}\right)^{-0.338}\left(\frac{dV_{\mathrm{T}}^2\rho_{\mathrm{C}}}{\gamma}\right)^{0.371}\mathrm{Sc}^{-0.125} \tag{3.14}$$

3.3 Single Oscillating Droplets

For single oscillating droplets a different set of correlations has been developed. The onset of oscillation is defined by the Weber number, We, which is as follows:

$$\mathrm{We} = \frac{U_{\mathrm{o}}^2 d\rho}{\sigma_i} \tag{3.15}$$

As a general rule the onset of drop oscillation occurs when $\mathrm{We} > 3.58$ (Hu and Kintner, 1955).

The onset of droplet oscillation is critically important in relation to mass transfer since the presence of oscillation means that the surface of the droplet is constantly changing shape as it moves through the continuous phase. Change in shape, for example from spherical to oblate spheroid, fundamentally affects interfacial area, drop fall or rise velocity, and internal mixing within the droplet.

Work by Rose and Kintner (1966) analyzed the effect of drop oscillation by considering the time dependent but regular distortion of the stagnant film at the interface, also using mass transfer coefficients to achieve a reasonable fit with experimental data. Their work involved a geometrical analysis of an oscillating drop to establish the time dependence of drop shape and thus change in interfacial area and boundary layer thickness, as shown in Figure 3.5 (reproduced from the paper). The boundary layer thickness is assumed to change according to the surface position and the time in oscillation cycle. Their underpinning assumption is that the interior of the drop is well mixed and therefore of uniform concentration and thus the resistance to mass transfer is confined to a thin film at the interface. The resistance may comprise mass transfer resistances in both phases.

Their correlation is expressed in terms of the degree of extraction achieved as droplets traverse through a column of continuous phase. E = fractional extraction occurred during the passage of the drop through a column of continuous phase:

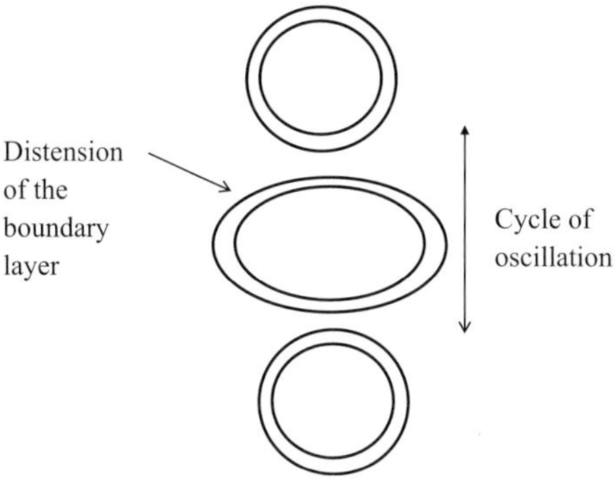

Figure 3.5 One period of mass transfer model showing the variation in boundary layer thickness with drop oscillation.

Reprinted (adapted) with permission from Rose, P. M. and Kintner, R. C. (1966). Mass transfer from large oscillating drops. *American Institute of Chemical Engineers Journal*, **12**(3), 530–534.

$$E = 1 - \exp\left[-\frac{2\pi D_E}{V}\int_{t_0}^{t_f}\frac{1}{f_1(t)}\left\{\left(\frac{3V}{4\pi\left(a_0 + a_p|\sin\omega' t|\right)^2}\right)^2\right.\right.$$

$$\left.\left.\times\frac{1}{2\alpha}\ln\frac{1+\alpha}{1-\alpha} + \left(a_0 + a_p|\sin\omega' t|\right)^2\right\}dt\right] \tag{3.16}$$

where α is given by:

$$\alpha^2 = \frac{\left(a_0 + a_p|\sin\omega' t|\right)^2 - \left(\frac{3V}{4\pi\left(a_0 + a_p|\sin\omega' t|\right)^2}\right)^2}{\left(a_0 + a_p|\sin\omega' t|\right)^2} \tag{3.17}$$

a_0 = initial radius,
a_p = amplitude of oscillation,
ω' = 0.5ω,
D_E = effective diffusivity,
V = volume.

The frequency of drop oscillation, ω, is expressed as a modified version of Lamb's equation:

$$\omega^2 = \frac{\sigma b}{a^3}\frac{n(n+1)(n-1)(n+2)}{((n+1)\rho_D + n\rho_C)} \tag{3.18}$$

b is an empirical amplitude coefficient for the droplet and was estimated using:

$$b = \frac{d_e^{0.225}}{1.242}$$

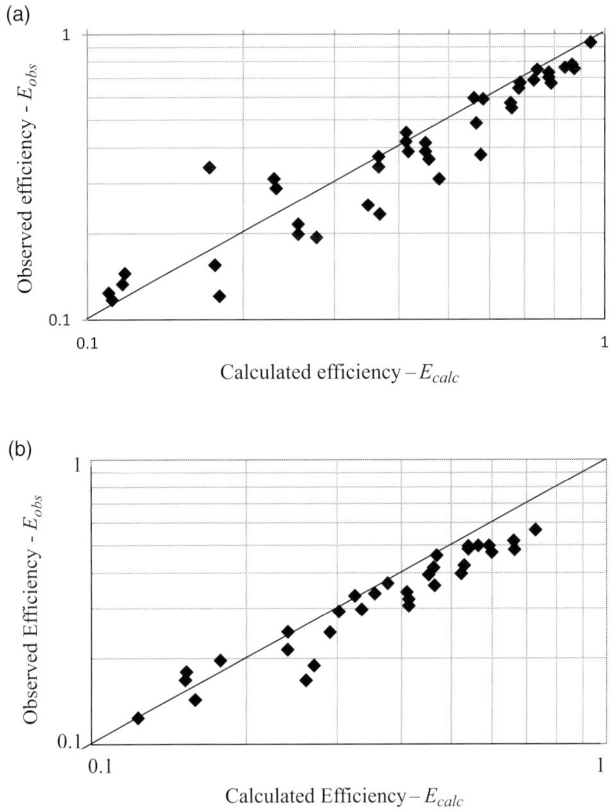

Figure 3.6 (a) A comparison of calculated and experimentally measured single drop mass transfer, resistance in drop phase. (b) Comparison of calculated and experimentally measured single drop mass transfer, resistance in drop phase combined data from Skelland and Wellek (1964), Johnson and Hamielec (1960), and Garner and Skelland (1955). Various systems: ethyl acetate/water, water/nitrobenzene/acetic acid.

Plotted data from Rose and Kintner (1966). Reprinted (adapted) with permission from Rose, P. M. and Kintner, R. C. (1966). Mass transfer from large oscillating drops. *American Institute of Chemical Engineers Journal*, **12**(3), 530–534.

Rose and Kintner present good comparison of calculated single drop mass transfer based on the above analysis versus experimental single drop mass transfer data, as shown in Figure 3.6.

Angelo, Lightfoot, and Howard (1966) analyzed the effect of the regular oscillatory deformation of drops upon mass transfer rate. Their analysis was mostly confined to the consequential effect of time-dependent deformation upon interfacial area and local convection though a useful model was developed using mass transfer coefficients. This reaffirmed the approach of Rose and Kinter by building into the expression for mass transfer coefficients, functions describing the time variation of the interface referred to as "surface stretch."

The comparison with the Rose and Kintner analysis is instructive.

Angelo et al. (1966) firstly define the mass transfer coefficient in dimensionless form. For example, the average dispersed phase mass transfer coefficient k_D is expressed in dimensionless form as:

$$\bar{\kappa} = \frac{k_D}{\sqrt{\frac{D_{AD}}{\pi t_0}}} \tag{3.19}$$

The time variance of the interface is incorporated as a function $S(\tau)$, where S is the interfacial area value. The function $S(\tau)$ may be expressed in different forms. Based on the Rose and Kinter approach, Angelo et al. firstly express $S(\tau)$ as follows:

$$S(\tau) = S_0 e^{\varepsilon \sin \tau} \tag{3.20}$$

where ε is an amplitude of oscillation expressed in dimensionless form, τ is dimensionless time.

The time-averaged dispersed phase mass transfer coefficient is presented as follows:

$$\bar{\kappa} = \frac{1}{\tau} \int_0^{\tau} \frac{(S/S_0)^2}{\sqrt{t}} dt \tag{3.21}$$

$$\bar{\kappa} = \frac{2}{\tau} \int_0^{\sqrt{\tau}} e^{2\varepsilon \sin(t^2)} dt \tag{3.22}$$

Angelo et al. developed this further with an alternative function for S(τ):

$$S = S_0(1 + \varepsilon \sin^2 \tau) \tag{3.23}$$

This function results in a simple expression for the dimensionless dispersed phase mass transfer coefficient, setting $\tau = \pi$, thus:

$$\bar{\kappa} = \sqrt{4(1 + \varepsilon_O)/\pi} \tag{3.24}$$

$$\varepsilon_O = \varepsilon + \frac{3}{8}\varepsilon^2 \tag{3.25}$$

$$t_O = 1/(\pi\omega) \tag{3.26}$$

$$k_D = \sqrt{\frac{4 D_{AD} \omega (1 + \varepsilon_O)}{\pi}} \tag{3.27}$$

where ω = oscillation frequency, s^{-1},
ε = dimensionless amplitude factor that is system specific,
k_D is the time-averaged dispersed phase mass transfer coefficient.

Using a similar approach, the overall mass transfer coefficient based on dispersed phased units is given as:

$$K_D = \sqrt{\frac{4D_{AD}\omega(1+\varepsilon_O)}{\pi}}\left[\frac{1}{1+m\sqrt{\frac{D_{AD}}{D_{AC}}}}\right] \tag{3.28}$$

Where $m = \frac{c_{A0} - c_A^*}{\left(c_{A0} - c_{A\infty}^*\right)c}$

c_{A0} = interfacial concentration of A,
$c_{A\sim}$ = bulk concentration of A,
c_A^* = equilibrium concentration of A,
D_{AD}, D_{AC} = binary diffusivity for A, dispersed phase, continuous phase respectively.

Their predictions are compared with those of the Rose and Kintner model based on a surface stretch approach. This entailed taking account of the unsteady oscillatory stretching of the droplet surface into the prediction of the mass transfer coefficient, and placing the comparison of the Rose and Kintner model, the penetration model, and the revised model of Angelo on a common basis. The comparison shows close agreement, with some minor deviation at lower mass transfer coefficient values (Figure 3.7).

Clift, Grace, and Weber (1978) present the following correlation for the dispersed phase mass transfer coefficient for oscillating drops.

$$\frac{k_d A}{A_O} = 2\sqrt{\frac{D_d f_O}{\pi}}\sqrt{1+\varepsilon+\frac{3}{8}\varepsilon^2} \tag{3.29}$$

$(1+\varepsilon)$ is the ratio of the maximum to minimum drop surface area and thus reflects the degree of surface deformation.

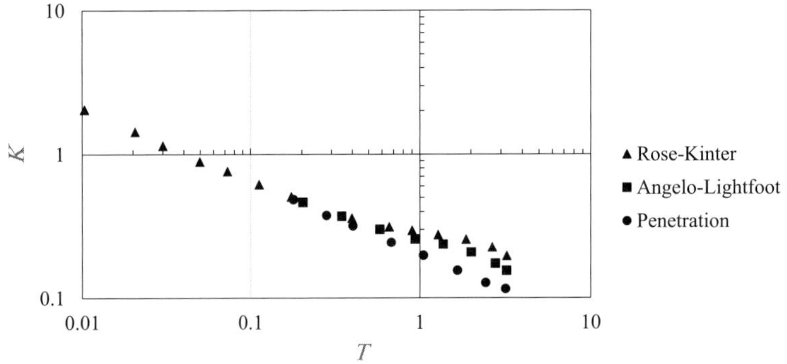

Figure 3.7 Comparison of the Rose-Kintner model with the development of Angelo et al. (1966) for prediction of time-average mass transfer coefficients with an assumed surface-time relation $S = S_O e^{\varepsilon \sin T}$ for $\varepsilon = 0.5$.

Reprinted (adapted) with permission from Angelo, J. B., Lightfoot, E. N., and Howard, D. W. (1966). Generalisation of the penetration theory for surface stretch: application to forming and oscillating drops. *American Institute of Chemical Engineers Journal*, **12**(4), 751–760. Copyright (1966) John Wiley and Sons – All Rights Reserved

f_o is the oscillation frequency according to the correlation of Schroeder and Kintner (1965) and this is reexpressed clearly by Kumar and Hartland in the following form:

$$f_O = \left[\frac{48\gamma C_1}{\pi^2 d^3 (2\rho_C + 3\rho_d)}\right]^{1/2} \tag{3.30}$$

where:

$$C_1 = 2.27 d^{0.225}$$

Kumar and Hartland proposed that a value of ε of 0.3 is valid for many systems and thus Clift's expression may be simplified using this value, reformulating the value of k_d into a Sherwood number to give:

$$\mathrm{Sh_d} = 1.30\left(\frac{d^2 f_O \rho_d}{\mu_d}\right)^{0.5} \mathrm{Sc_d^{0.5}} \tag{3.31}$$

This is similar to a correlation derived independently by Yamaguchi, Fujimoto, and Katayama (1975).

$$\mathrm{Sh_d} = 1.14\left(\frac{d^2 f_O \rho_d}{\mu_d}\right)^{0.56} \mathrm{Sc_d^{0.5}} \tag{3.32}$$

Kumar and Hartland also developed and validated mass transfer relationships for continuous phase mass transfer in the case of oscillating drops using Reynolds number, Schmidt number, capillary number, and viscosity ratio. Their analysis is proposed in equation 3.33:

$$\frac{\mathrm{Sh_c} - \mathrm{Sh_{c,rigid}}}{\mathrm{Sh_{c,\infty}} - \mathrm{Sh_c}} = C_2 \mathrm{Re}^{n_1} \mathrm{Sc}^{n_2} \left(\frac{V_T \mu_c}{\gamma}\right)^{n_3} \frac{1}{1+\kappa^{n_4}} \tag{3.33}$$

Comparison of experimental data with the predictions of equation 3.33 reproduced from Kumar and Hartland (1999) are shown in Figure 3.8. Figure 3.8(a) shows the variation of $\overline{\mathrm{Sh}}_c$ with Re for transfer of acetone in the system carbon tetrachloride/*o*-nitrophenol/water with water as the continuous phase (experimental data is from Thorsen and Terjesen, 1962). Figure 3.8(b) shows the variation of $\overline{\mathrm{Sh}}_c$ with Re for transfer of isobutyl alcohol in the system isobutyl alcohol/water (experimental data from Heertjes, Holve, and Talsma, 1954).

3.4 Single Drop Systems: Quantitative Approach

Developments in finite element mathematics together with huge increases in computer power have enabled the quantitative solution of transport problems for nonspherical drops (Petera and Weatherley, 2001). The high dependence of mass transfer rate on interfacial area dictates the need for quantitative description of interfacial area at any point in time if accurate rate predictions are to be made. It is noteworthy in particular that behavior at intermediate Reynolds numbers has been successfully predicted, for example by Petera and Weatherley (2001) and Dandy and Leal (1989). Others have developed and validated mathematical tools to provide robust tools for investigating

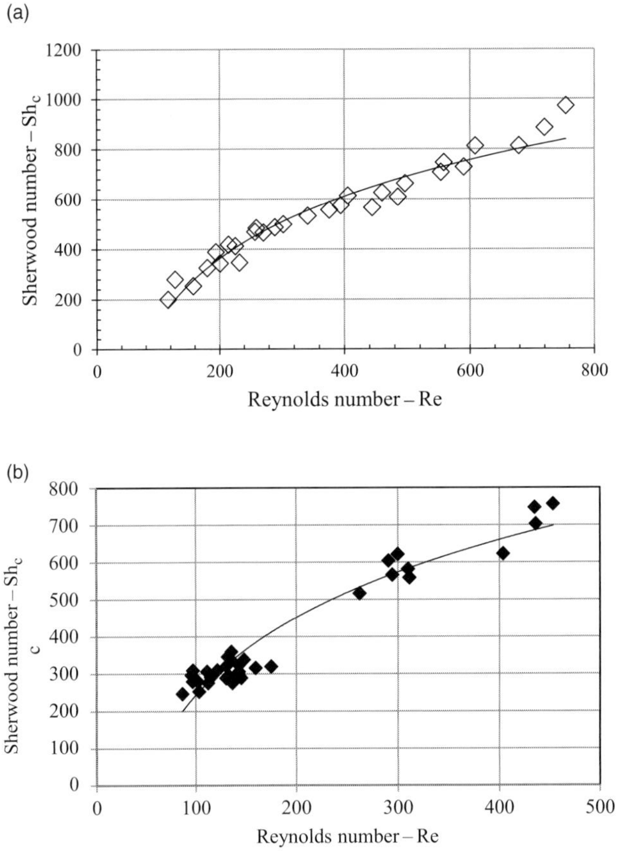

Figure 3.8 Comparison of experimental data with the predictions of equation 3.33. (a) Variation of Sh_c with Re for transfer of acetone in the system carbon tetrachloride/*o*-nitrophenol/water (continuous phase). (b) Variation of Sh_c with Re for transfer of isobutyl alcohol in the system isobutyl alcohol/water.

Reprinted (adapted) with permission from Kumar, A. and Hartland, S. (1999). Correlations for prediction of mass transfer coefficients in single drop systems and liquid–liquid extraction columns. *Transactions of the Institution of Chemical Engineers*, **77(A)**, 372–384.

secondary flows in the continuous phase external to moving drops and solutions using finite element techniques (Bozzi et al., 1997; Tsukada et al., 1984, 1990). This is also noteworthy since the fluid dynamics in the fluid phase external to the drop may also have a profound effect on the rate of mass transfer or reaction rate. Despite the dynamic nature of drop shape, steady state and pseudo-steady state approaches can be successful, with the final shape of the moving droplet determined using an iterative process at the nonpenetrable interface. Droplet shapes, as discussed in Chapter 2, are generally found to be in the form of flattened, oblate ellipsoids whose shape is determined by the balance of stresses that are normal to the interface, which include interfacial tension, hydrodynamic stresses, and hydrostatic pressure. Internal recirculation also can influence the creation of stresses in the external continuous liquid phase that can be comparable in magnitude to those mentioned previously. Internal circulation may also reduce flattening of droplets (Foote, 1969; Petrov, 1988) which may also lead to increases in drop-side transport rates (Walcek and Pruppacher, 1984). In summary, internal circulation effects within droplets, the flow profiles in the continuous liquid phase suspending liquid, and changes in the interfacial tension are due to drop deformation, and these are major factors that determine mass transfer and are therefore highly relevant to intensification.

In the vast majority of practical applications of liquid–liquid systems the dispersed phase comprises a multitude of moving drops in close proximity. So another important consideration is the degree of interaction that occurs between drops in such situations. For example, Xhang, Davis, and Ruth (1993) demonstrated significant interactions between drops moving in close proximity, which affect the relative motion between the drops. Such interactions are significant factors that determine the performance of industrial-scale liquid–liquid extraction equipment. It follows that data from single drops in unhindered motion must be treated cautiously if accurate prediction of liquid–liquid behavior at industrial scale is to be achieved (Ghalehchian and Slater, 1999).

In spite of the large and long-standing literature dealing with mass transfer to and from moving droplets, there are few that use anything other than a pseudo-steady state approach. The problem of interfacial mass transfer for droplets that are in a state of continuous deformation has only recently been addressed quantitatively and few solutions existed in the literature of the equations governing simultaneous transport of momentum and mass for deformable drops. Haywood, Renksizbulut, and Raithby (1994) considered the gas–liquid case for evaporation from deforming droplets. Another relevant example is that of Saboni, Gourdon, and Chesters (1993). These papers did not fully present a general approach to quantitative description of transient mass transfer in a liquid–liquid system where significant drop deformation is occurring. More recent work of Albernaz et al. (2017) studied the dynamics of a droplet surrounded by its own vapor in isotropic turbulence, considering the impact of drop deformation and surface tension on heat and mass transfer The primary focus was to analyze the coupling between turbulence parameters, thermodynamic variables, and droplet dynamics.

Another complication is if the partition coefficient is significantly different from unity, then the concentration fields in both phases can be discontinuous, which not only influences concentration driving force but also influences local physical properties. These include density, viscosity, diffusion coefficient, and surface tension.

The work of Petera and Weatherley (2001) showed that it is possible to deal with the complex and changing geometries that are encountered in liquid–liquid contacting in the case of drops of deformable shape and hence variable interfacial area and varying velocity characteristics. The phenomenon of drop deformation is of some importance since it impacts both the internal hydrodynamics of the drops and the flow patterns external to the drops, which in turn affects mass transfer resistances in the continuous phase. In order to obtain a full description of the transport processes occurring in and around moving unstable drops, quantitative determination of the distribution of all species present is conducted as an integral part of hydrodynamic analysis.

The method presented by Petera and Weatherley (2001) used the Lagrangian framework for formulation and solution of the continuity equations, the energy equations, and the equations controlling solute transport with automatic remeshing. The framework provides a set of tools for a range of problems in liquid–liquid systems. In particular, as is discussed later in the chapter (section 3.4.1), this has formed the basis of subsequent work that accounts for Marangoni phenomena as part of the overall treatment of interfacial deformation and instabilities.

Their framework and method will now be summarized. The starting point is consideration of a single moving drop of mean radius R_d (drop volume is $V_d = 4/3\ \pi\ R_d^3$) in axisymmetric flow moving under gravity through an immiscible continuous phase of density ρ_c in a vertical column of radius R_T. The two liquids are assumed to be Newtonian and incompressible, having viscosities μ_d and μ_c for the drop and for the continuous phase, respectively. The interfacial tension σ between the liquids is assumed constant. Subscripts "d" and "c" denote quantities associated with the droplet and the continuous phase, respectively. Figure 3.9 shows the general layout of the problem, with the left-hand figure showing discrete drops moving from the formation nozzle at the top, moving through the continuous phase toward the bottom of the column. The right-hand diagram in Figure 3.9 shows the computational domain, focusing on a single droplet, assuming axial symmetry. The boundary conditions are also included.

In order conduct an efficient computation the entire continuum is not considered but rather an imaginary tube of radius R_t ($R_t \ll R_T$) which was defined as containing the droplet and its surrounding continuous phase liquid. R_t is the radius of the imaginary tube, and R_T is the radius of the actual column. It is further assumed that the continuous phase moves relative to the droplet (rather than the opposite) and thus the velocity assumed is the same as the actual fall velocity of the droplet. This also assumes that the

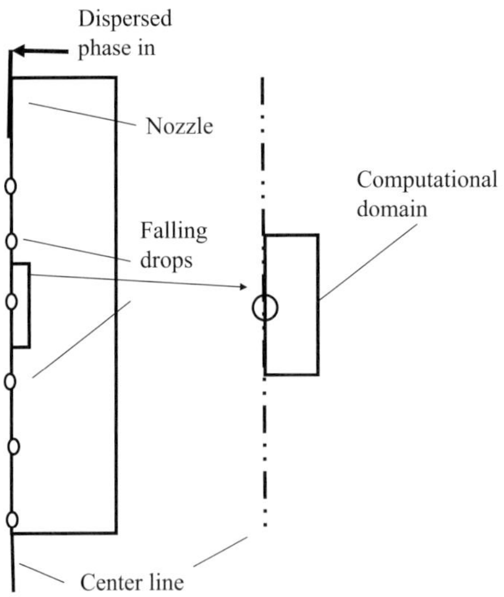

Figure 3.9 Diagram showing the computational domain for calculation of the droplet behavior. Reprinted (adapted) with permission from Petera, J. and Weatherley, L. R. (2001). Modeling of mass transfer from falling droplets. *Chemical Engineering Science*, **56**, 4929–4947. Copyright (2001) Elsevier B.V. All Rights Reserved

tube radius is sufficiently large. Under these circumstances the no-penetration boundary condition is satisfied (see later in equation 3.45).

The controlling equations that determine motion of the drop, the internal and external hydrodynamics, and the inter-phase transport are now summarized from the analysis of Petera and Weatherley (2001).

3.4.1 Fluid Transport

The incompressible Navier–Stokes equations are written firstly for the internal flow inside the drop (equation 3.34) and secondly for the continuous phase (equation 3.36). Continuity equations in each case are written in equations 3.35 and 3.37:

For the drop phase:

$$\rho_{\mathrm{d}}\left(\frac{\partial \mathbf{u}_{\mathrm{d}}}{\partial t}+\mathbf{u}_{\mathrm{d}}\cdot\nabla\mathbf{u}_{\mathrm{d}}\right)=\nabla\cdot\mathbf{T}_{\mathrm{d}} \tag{3.34}$$

$$\nabla\cdot\mathbf{u}_{\mathrm{d}}=0 \tag{3.35}$$

For the continuous phase:

$$\rho_{\mathrm{c}}\left(\frac{\partial \mathbf{u}_{\mathrm{c}}}{\partial t}+\mathbf{u}_{\mathrm{c}}\cdot\nabla\mathbf{u}_{\mathrm{c}}\right)=\nabla\cdot\mathbf{T}_{\mathrm{c}} \tag{3.36}$$

$$\nabla\cdot\mathbf{u}_{\mathrm{c}}=0 \tag{3.37}$$

$\mathbf{u}_{\mathrm{d}}$, p_{d} and $\mathbf{u}_{\mathrm{c}}$, p_{c} denote the velocity vector field and the scalar pressure field in the dispersed phase and continuous phase, respectively.

The stress tensors $\mathbf{T}_{\mathrm{d}}$ and $\mathbf{T}_{\mathrm{c}}$ are written in the following form:

For the drop phase:

$$\mathbf{T}_{\mathrm{d}}=-p_{\mathrm{d}}\mathbf{I}+2\mu_{\mathrm{d}}\mathbf{D}_{\mathrm{d}},\quad\text{where}\quad\mathbf{D}_{\mathrm{d}}=1/2\ [\nabla\mathbf{u}_{\mathrm{d}}+(\nabla\mathbf{u}_{\mathrm{d}})^{T}] \tag{3.38}$$

For the continuous phase:

$$\mathbf{T}_{\mathrm{c}}=-p_{\mathrm{c}}\mathbf{I}+2\mu_{\mathrm{c}}\mathbf{D}_{\mathrm{c}},\quad\text{where}\quad\mathbf{D}_{\mathrm{c}}=1/2\ [\nabla\mathbf{u}_{\mathrm{c}}+(\nabla\mathbf{u}_{\mathrm{c}})^{T}] \tag{3.39}$$

$\mathbf{D}_{\mathrm{d}}$ and $\mathbf{D}_{\mathrm{c}}$ are the rate of deformation tensors for the drop phase and continuous phase, respectively.

$\mathbf{u}_{\mathrm{d}}$ and $\mathbf{u}_{\mathrm{c}}$ are the velocity vectors for the drop phase and continuous phase, respectively.

In cylindrical coordinates, the axisymmetric velocity vectors have components (u_r, u_z) which means that for each phase:

$$\mathbf{u}=u_r\,\mathbf{e}_r+u_z\,\mathbf{e}_z+0\,\mathbf{e}_\theta \tag{3.40}$$

where $\mathbf{e}_r$, $\mathbf{e}_z$, $\mathbf{e}_\theta$ is the unit vector in the r, z, and θ directions, respectively.

At the liquid–liquid interface, conservation of the momentum is expressed by the traction boundary condition on the droplet surface Γ_d,

$$(\mathbf{T}_d - \mathbf{T}_c) \cdot \mathbf{n} = 2H\ \sigma \mathbf{n} \tag{3.41}$$

$\mathbf{n}$ is the outward pointing normal vector, H is the curvature, and σ is the interface tension. The velocity is assumed to be continuous at the phase boundary (equation 3.42).

$$\mathbf{u}_d = \mathbf{u}_c \tag{3.42}$$

$$\mathbf{u}_c = u^0\ \mathbf{e}_z \tag{3.43}$$

The velocity field asymptotically approaches a uniform flow:

$$\mathbf{u}_c \rightarrow u^0\ \mathbf{e}_z \tag{3.44}$$

as $|z| \rightarrow \infty$.

In this coordinate system, the average drop velocity is assumed to be zero, thus:

$$\int_{\Omega_d} \mathbf{u}_d\ d\Omega = \mathbf{0} \tag{3.45}$$

where the integral is calculated over the drop volume ($|\Omega_d| = V_d$).

3.4.2 Mass Transport

In each phase the mass transport of a solute, for example a component "A" moving from the dispersed to continuous phase, is governed by the mass balance differential equations expressed using Fick's law (equation 3.46). The mass flux of component A in each phase is given by:

$$\mathbf{N}_A = c_A(\mathbf{u} + \bar{\mathbf{u}}) - D_{AB}\nabla c_A \tag{3.46}$$

where

$$\mathbf{u} = \begin{cases} \mathbf{u}_d & \text{in dispersed phase} \\ \mathbf{u}_c & \text{in continuous phase} \end{cases} \tag{3.47}$$

The nontransferring component, which can be named "B" in each phase, is assumed to be "inert" as far as the diffusion process is concerned. The flux of B may be formally written also by Fick's law (equation 3.48):

$$\mathbf{N}_B - c_B\mathbf{u} = \mathbf{0} = c_B\bar{\mathbf{u}} - D_{AB}\nabla c_B \tag{3.48}$$

From equations 3.46 and 3.48 the relationship for $\bar{\mathbf{u}}$, the "drift velocity" for this system, can be thus obtained:

$$\mathbf{N}_A - c_A\mathbf{u} = \rho\bar{\mathbf{u}} \tag{3.49}$$

The velocity $\bar{\mathbf{u}}$, however, can be eliminated after using the relationship between density, concentration, and velocity:

$$c_A = a_A \rho \tag{3.50}$$

a_A is the mass fraction of component A. Substitution into equation 3.46 and combining with equation 3.49 yields:

$$\rho \bar{\mathbf{u}}(1 - a_A) = -D_{AB} \nabla c_A \tag{3.51}$$

Replacement of ρ with c_A/a_A and combination of equations 3.46 and 3.49 above yields:

$$\mathbf{N}_A - c_A \mathbf{u} = -\frac{1}{1 - a_A} D_{AB} \nabla c_A \tag{3.52}$$

Thus, the differential mass balance equations may be written, firstly for the internal flow inside the drop:

$$\frac{\partial c_A}{\partial t} + \mathbf{u}_d \cdot \nabla c_A = \nabla \cdot \left(\frac{D_d}{1 - a_A} \nabla c_A \right) \tag{3.53}$$

and then for the continuous phase:

$$\frac{\partial y_A}{\partial t} + \mathbf{u}_c \cdot \nabla y_A = \nabla \cdot \left(\frac{D_c}{1 - a_A} \nabla y_A \right) \tag{3.54}$$

c_A, D_d and y_A, D_c are the mass concentration and diffusion coefficient of the component A in dispersed phase and continuous phase, respectively.

The drop surface (interface) may be assumed to be in local equilibrium with the continuous phase, according to the thermodynamic relationship (equation 3.55):

$$y_A = f(c_A) \tag{3.55}$$

Application of equation 3.55 at the interface yields:

$$y_{Ai} = f(c_{Ai}) \tag{3.56}$$

and

$$y_A = f\left(c_A^*\right) \tag{3.57}$$

c_A^* is a hypothetical concentration in the dispersed phase at equilibrium with the bulk concentration in the continuous phase as assumed in the classical Whitman two film theory of interfacial mass transfer.

The concentration field is at the interface is noncontinuous and therefore imposition of the equilibrium boundary condition on the surface Γ_d is undertaken with care. In principle, equations 3.53 and 3.54 may be solved at each part of the domain and the boundary conditions on surface Γ_d iterated until equation 3.56 is fulfilled. The original approach applied by Petera et al. involved transformation of equation 3.54 using the equilibrium relationship 3.49. Instead of solving for the real concentration field, y_A, in the continuous phase, a hypothetical concentration c_A in this phase was sought, using equation 3.55.

Therefore the equation is equation 3.54 modified by replacement of the term y_A by the function $f(c_A)$. Differentiation of this revised equation leads to equation 3.58:

$$f'\left[\frac{\partial c_A}{\partial t}+\mathbf{u}_c\cdot\nabla c_A\right]=\nabla\cdot\left(\frac{D_c f'}{1-a_A}\nabla c_A\right) \tag{3.58}$$

where

$$f'=\frac{df}{dc_A}$$

Comparing equation 3.58 with the mass balance equation 3.53 in the dispersed phase, one can see that a continuous concentration field c_A can be defined over the whole domain, resulting in a single common equation of the form analogous to equation 3.58 (equation 3.59).

$$f'\left[\frac{\partial c_A}{\partial t}+\mathbf{u}\cdot\nabla c_A\right]=\nabla\cdot\left(\frac{D_{AB} f'}{1-a_A}\nabla c_A\right) \tag{3.59}$$

$\dot{f}$ and D_{AB} in each phase are defined differently, as follows:

$$f'=\begin{cases}1 & \text{in dispersed phase}\\ \dfrac{df}{dc_A} & \text{in continuous phase}\end{cases}$$

$$D_{AB}=\begin{cases}D_d & \text{in dispersed phase}\\ D_c & \text{in continuous phase}\end{cases}$$

The velocity field **u** is defined as previously in equation 3.47.

In order to solve the overall problem of simultaneous momentum and mass transport between the droplet and surrounding liquid, Petera et al. developed a modified Lagrange–Galerkin finite element method that allowed efficient solutions to predict drop shape deformation, velocity fields in and around the droplet, and vitally, mass transfer rates.

The novel aspects that are incorporated in the Petera method is that the dynamic changes in droplet shape were accommodated in the calculation procedure. This was achieved by a process of remeshing of the finite element mesh in order to obtain better predictions of terminal velocity and also accurate calculation of both internal and external mass transfer conditions. It was found that a moving element method for droplet shape tracking causes deformation of the mesh, which after some time led to numerical instability problems. Thus a more efficient, automatic remeshing routine was implemented in the algorithm. This was achieved by iterative adjustment of the aspect ratios of the finite elements comprising the mesh with remeshing, conducted as necessary in order to maintain stability of the program execution.

The outcome of this approach was several-fold. Firstly, it was shown that accurate prediction of droplet terminal velocity is possible from first principles based on the Navier–Stokes equations and continuity equations for oscillating droplets. Prediction of drop deformation allowed accurate calculation of drop velocity in and around the moving drop. In particular it was shown that the time-dependent deformation of the drop can be predicted. The predictions started with a spherical drop that then experiences significant changes in shape as it moves through the continuous phase. Secondly, the velocity profiles near to the interface in the continuous phase were also predicted during the complex and unsteady motion and distortion of the drop shape. Thirdly, the velocity profiles inside the drop were also calculated and the circulation patterns demonstrated. These calculations then led to the accurate prediction of terminal velocity, validated by extensive experimental measurements. Fourthly, it was shown that accurate prediction of mass transfer fluxes and concentration profiles in both phases in the vicinity of the liquid–liquid interface and throughout the interior of the droplet is possible. In light of accurate predictions of drop velocity and oscillation, it was then possible to complete prediction of mass transfer, as shown in Figure 3.10. The composition maps inside and outside the drops were especially interesting since they show the time dependence of the mass transfer and the unsteady state development of the concentration gradients inside a moving drop.

A notable development on the theme of prediction of mass transfer from fundamentals is that of Chen et al. (2015) who address quantitatively the problem of drop deformation and its effect on interfacial mass transfer, particularly at higher Reynolds number, using a semi-Lagrangian advection scheme. The development of the mathematical models by Chen are presented in section 3.5 on Marangoni instability. This scheme was successfully used not only to address the influence of drop deformation but also to describe the effect of solute-induced Marangoni instabilities and the resulting convection close to the liquid–liquid interface. Their predictions of the flow field indicated that Marangoni convection can trigger velocity fields perpendicular to the interfacial boundary, accelerate the interface renewal, and strengthen mixing inside the droplet. The algorithms that were developed successfully predicted the mass transfer coefficients for unsteady mass transfer

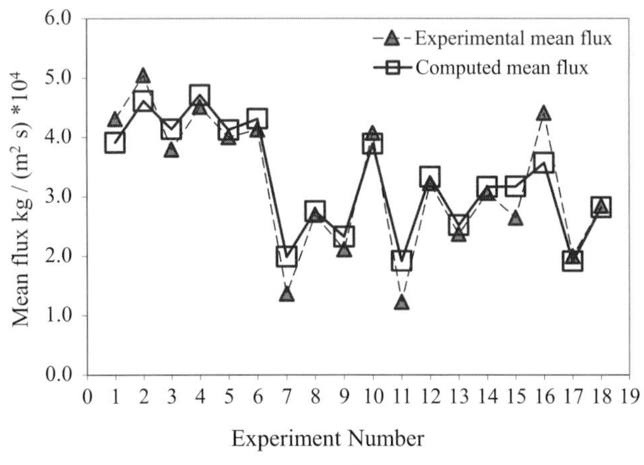

Figure 3.10 Mass transfer flux calculations for discrete drops of ethanol/water mixtures falling through a continuous phase of n-decanol. Comparison between experimental and predicted. Reprinted (adapted) with permission from Petera, J. and Weatherley, L. R. (2001). Modeling of mass transfer from falling droplets. *Chemical Engineering Science*, **56**, 4929–4947. Copyright (2001) Elsevier B.V. All Rights Reserved

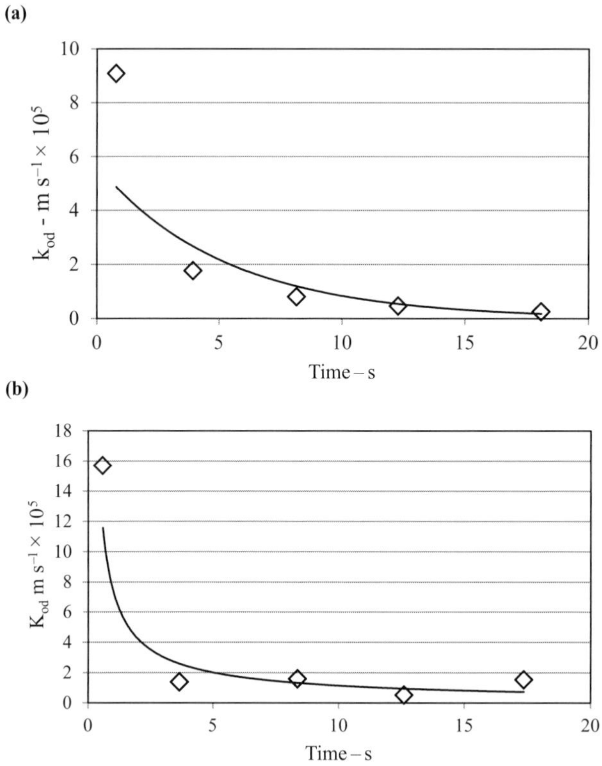

Figure 3.11 Predicted (solid line) and experimental (points) overall mass transfer coefficients as a function of time for the mass transfer of acetic acid from drop phase (MIBK) to continuous phase (water). The conditions as follows: $\rho_d = 841.1$ kg m^{-3}, $\rho_c = 988.2$ kg m^{-3}, $\mu_c = 1.44 \times 10^{-3}$ Pa s, $\mu_d = 3.34 \times 10^{-3}$ Pa s, $\sigma_o = 1.00 \times 10^{-3}$ N m^{-1}, $D_c = 5.5 \times 10^{-10}$ m^2 s^{-1}, $D_d = 2.1 \times 10^{-10}$ m^2 s^{-1}. (a) Drop diameter = 0.98×10^{-3} m. (b) Drop diameter = 1.69×10^{-3} m.

Reprinted (adapted) with permission from Chen, J., Wang, Z., Yang, C., and Mao, Z. (2015). Numerical simulation of the solute-induced Marangoni effect with the semi-Lagrangian advection scheme. *Chemical Engineering and Technology*, **38**(1), 155–163.

to a single deformable drop in the methyl isobutyl ketone (MIBK)/acetic acid/water system. Figure 3.11 shows excellent agreement between the experimentally determined overall mass transfer coefficient and the predicted values based on the advection scheme, which is described in detail in the paper.

3.5 Marangoni Instabilities

Marangoni instabilities are an important phenomenon that have a positive influence on the overall interfacial mass transfer rate. The instabilities arise when interfacial tension gradients occur due to concentration gradients close to the interface. In simple terms, interfacial tension is partially determined by solute concentration. Variations in concentration occur because of diffusional resistance in the regions close to the liquid–liquid interface. These create local differences in interfacial tension, giving rise to hydrodynamic instabilities and interfacial flows. Other properties related to composition include viscosity and density, which can also contribute to interfacial instability. Observations of interfacial instabilities go back a long way in the literature. Sternling and Scriven (1959) in an early study of interfacial instability at planar interfaces claimed that the ratio of kinematic viscosities and the diffusivity ratio play a significant role in the occurrence of interfacial instability effects in liquid–liquid systems. The case for curved interfaces may be different, as stated in recent predictive work by Wegener et al. (2009a), who found no stable configuration for a range of kinematic viscosity ratios. Indeed, Wegener et al. state categorically that the observation of Marangoni instability was always accompanied with an increase in mass transfer.

Interfacial turbulence generally is an important phenomenon in mass transfer processes in liquid–liquid systems and deserves important consideration in exploring intensification of liquid–liquid mass transfer. For example, Golovin (1991) cites numerous experimental studies of interfacial mass transfer which report large increases in mass transfer coefficients attributable to interfacial instabilities. These range from two-fold to several tens-fold with no increase in concentration driving force or interfacial area.

One of the major interfacial instability phenomena is Marangoni instability, which is involved in many cases of interfacial turbulence. Spontaneous interfacial turbulence was first described in the 19th century by Thomson (1855). Interfacial turbulence accompanying the mass transfer of a solute across a flat interface was first reported by Quincke (1888). Detailed investigation was undertaken by Lewis and Pratt (1953), who observed some unusual disturbances at the surface of a droplet when measuring the interfacial tension between two liquids by the drop weight method. Disturbances at a liquid–liquid interface may be induced through the occurrence of local interfacial tension gradients and/or the presence of unstable density gradients often associated with mass transfer and high concentration gradients. Interfacial tension gradient-driven convection is known as Marangoni convection, named by Carlo Marangoni (1869, 1871). Marangoni-type interfacial flows modify the hydrodynamic behavior of the fluid layers near to the interface, thus affecting drop coalescence, jet breakup, and droplet drag coefficients during motion.

The photographs in Figure 3.12 show an example of Marangoni instabilities close to the interface of a pendant drop of water and ethanol suspended in an extracting solvent, n-decanol.

Marangoni convection is driven by interfacial tension gradients that most commonly occur due to local differences in composition of the liquid phases associated with the interface. This occurs because interfacial tension itself is a function of composition close to the interface. In addition, variations in temperature, due to heats of solution or heats of reaction, influence interfacial tension. Electrostatic charge at the interface also can influence variations in interfacial tension, as will be analyzed in Chapter 6.

Sawistowski (1971, 1973, 1981) classified interfacial disturbances according to their origin and in relation to the relative depth of penetration into the liquid bulk phase as follows: (i) Marangoni instabilities; (ii) Marangoni disturbances, and (iii) thin-film phenomena (Sawistowski, 1971, 1973). In the cases of (i) and (ii), the depth of

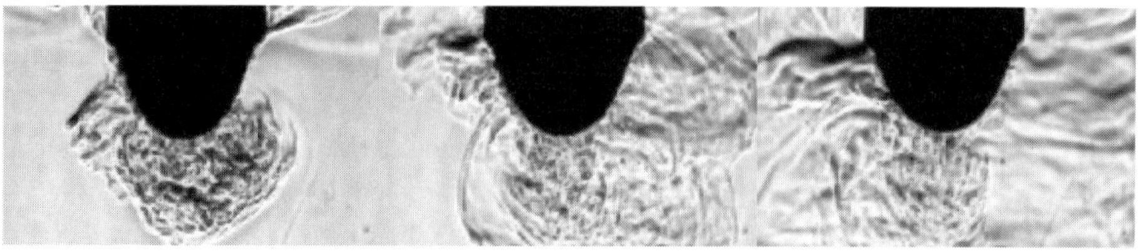

(a) (b) (c)

Figure 3.12 Marangoni instabilities close to the interface of a pendant drop of water and ethanol suspended in an extracting solvent, n-decanol. Time is increasing time (a) – (c) (Qiu, 2010).

penetration of the interfacial convection into the two phases is small with respect to the thickness of the phases. Sawistowski distinguished between Marangoni instabilities and Marangoni disturbances using the magnitude of the mass transfer driving force required to produce interfacial turbulence. For example, in the case of thin films, such as those trapped between drops before coalescence, the depth of the interfacial flow caused by local changes in interfacial tension may be significant to the extent that change in the thickness of the film results. In both cases, Marangoni instabilities and Marangoni disturbances are often associated with increases in mass transfer fluxes across the liquid–liquid interface. These are explained by enhancement of convective forces close to the interface and thus mass transfer rate. Another less obvious phenomenon is the influence of this thin film behavior on the drop shape and the specific interfacial area, with consequences for overall mass transfer rates.

Changes in local physical properties other than surface tension also impact interfacial disturbances. These include the diffusivity of the solute and the viscosities of both phases. The direction of mass transfer may also be an important factor (Golovin, 1990), since this can affect the shape of the interface and hence the impact of locally generated instabilities. For single-solute mass transfer, Golovin also identified there being two critical values of the ratio of the diffusion coefficients in each of the respective phases. These critical ratio values define an envelope within which interfacial turbulence is likely to occur. The critical ratios are independent of the direction of mass transfer. The effects of diffusion coefficient values and viscosity notwithstanding, the intensity of interfacial turbulence mainly depends on the magnitude of the interfacial tension gradient and the relative rate of transfer of mass and momentum between the two liquid phases.

The presence of surfactants can also exert a major effect on the intensity of interfacial turbulence by reducing the interfacial tension and changing the structure of the interface (Agble and Mendes-Tatsis, 2000, 2001; Mendes-Tatsis and Agble, 2000). It is reported that addition of surfactant enhanced Marangoni disturbances in a number of liquid systems. For example, Mendes-Tatsis and Agble (2000) reported for the system isobutanol/water, that interfacial disturbances were enhanced upon the addition of surfactants dodecyl triammonium bromide (DTAB) and sodium decyl sulfate (SDS). In contrast, for the same system, addition of a different surfactant, ATLAS G1300, which is a hydroxylated nonionic surfactant, interfacial disturbances were significantly suppressed.

3.6 Stability Criteria

The establishment of criteria for the conditions under which interfacial instability occurs occupies a significant literature. This is a particularly important consideration when the optimum conditions for high rates of mass transfer are sought. It is clear that interfacial tension gradients and time-dependent changes in interfacial tension are only part of the picture when analyzing instability. Instabilities are by definition time dependent and this adds to the difficulty in arriving at reliable criteria. One of the most significant early attempts to define stability criteria was by Sternling and Scriven (1959), who applied a linearized stability analysis to ternary liquid–liquid systems and proposed the stability criteria for ordered Marangoni instabilities. It is important to note that the widely cited criteria of Sternling and Scriven were defined for quiescent liquid–liquid systems and are not necessarily applicable to flowing bulk phases. The analysis as already briefly discussed, showed that instabilities depend on the interplay between concentration dependence of interfacial tension, solute diffusivities, and viscosities of both phases. This stability criterion has been tested experimentally by many authors and gives reasonable results in many cases (Bakker, van Buytenen, and Beek, 1966; Linde and Schwarz, 1961, 1964; Linde and Shert, 1965; Orell and Westwater, 1961, 1962).

However, there are still some experimental results that cannot be explained by Sternling and Scriven's theory. Since then, some more general models extended the Sternling and Scriven analysis. Perez De Ortiz and Sawistowski (1973a) conducted a stability analysis of binary systems based on the mechanism of thermally induced Marangoni instabilities and developed stability criteria. If the system is binary (one of the phases is also the solute) and the interfacial equilibrium is assumed to be attained instantaneously, under isothermal conditions, the phase rule precludes changes in interfacial concentrations and consequently the occurrence of interfacial tension gradients. Binary systems should therefore be Marangoni stable. However, interfacial turbulence has been observed in these systems (Perez De Ortiz and Sawistowski, 1973b). It was explained in terms of either local variations in interfacial tension due to disturbances to the temperature gradients caused by heats of solution or by dynamic changes of interfacial tension.

Another approach to stability criteria was proposed by Modigell (1981) which built on the Sternling and Scriven criteria, by defining a critical interfacial tension gradient. The critical gradient of interfacial tension must be exceeded if instabilities are to occur and this was demonstrated experimentally but with no quantitative theoretical explanation for the disordered instabilities that were observed. Relative velocity of the two liquid phases close to the interface is also a factor affecting the onset of instability (see Wolf and Stichlmair, 1996). This situation therefore is more realistic when process and equipment impacts are considered. In flowing systems where disordered convection is present, Schott and Pfennig (2004) developed a stability criterion allowing the prediction of the onset of disordered convection using Monte-Carlo simulations. Simulations showed that under certain conditions mass transfer across interfaces induces the formation of nano-droplets in the close vicinity of the interface. Nano-drop behavior close to the interface resulting from both forces of attraction and repulsion was proposed as a mechanism for interfacial instability.

There are some models developed for describing stability criteria, specifically for cases involving surfactants in liquid–liquid systems. Nakache, Duperyrat, and Vignes-Adler (1983) proposed a semiempirical stability criteria for "surfactant" transfer across a liquid–liquid interface that attempts to characterize the system by a ratio of convective flux to diffusive flux at the interface. The nature of the oscillatory behavior at the liquid–liquid interface was shown to be strongly influenced by the composition of the system, the distribution coefficients, and the diffusional properties of the transferring components. The criteria developed were based on the large-scale oscillations that were observed at the interface between an aqueous solution of a long-chain alkyltrimethylammonium halogenide and solution of picric acid in nitrobenzene at compositions very different from equilibrium values. The oscillations were attributed to several interacting effects. These included small-scale interfacial turbulence generated on account of the transfer of surface-active material and slow transfer of solutes by convective diffusion. The study by Nakache et al. also showed the potential importance of dissociation reactions that occur at the interface, which cause obvious compositional changes and consequently changes in the local boundary conditions that influence interfacial flow stability. The significance of this effect is highly dependent on the kinetics of the dissociation. Local reactions can also result in formation of solutes that exhibit significantly lower surface activity which result in changes in interfacial hydrodynamics. Another significant finding reported by Nakache in the case of the model system used in their experiments was the narrow compositional window within which instability was observed. The concentration of surfactant and transferring solutes, together with the dielectric constants of the two phases, were shown to be important parameters, defining the window of instability.

Figure 3.13 shows the time-dependent oscillatory behavior observed for the mass transfer of picric acid from aqueous solution into nitrobenzene in the presence of a surfactant, hexadecyltrimethylammonium chloride (Nakache et al., 1983). The very noticeable aspect

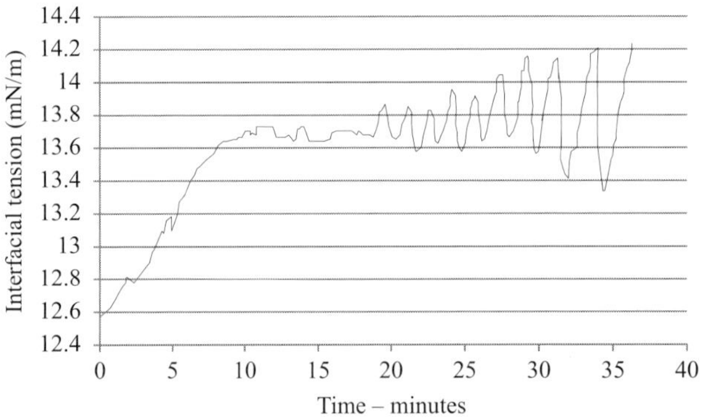

Figure 3.13 Time-dependent variation of interfacial tension for the liquid–liquid system (picric acid/water) (hexadecyltrimethylammonium chloride ($C_{16}C_1$)/nitrobenzene).

Reprinted (adapted) with permission from Nakache, E., Duperyrat, M., and Vignes-Adler, M. (1983). Experimental and theoretical study of an interfacial instability at some oil-water interfaces involving a surfaceactive agent: I. Physicochemcial description and outlines for a theoretical approach. *Journal of Colloid and Interface Science*, **94**(1), 187–200. Copyright Elsevier B.V. All Rights Reserved.

of these observations is the time scale. The data illustrated show the oscillations in interfacial tension developing and expanding over a period of many minutes. The interfacial tension increases and then plateaus to a mean value but with major growing oscillations.

Slavtchev et al. (1998) developed a linear stability theory to predict the onset of Marangoni convection in liquid–liquid systems with added surface-active solutes. They proposed a Marangoni coefficient as an empirical parameter, and its application to some binary systems has been shown to be very successful. They found that the ionic surfactants present in the system increased Marangoni convection and enhanced initial mass transfer fluxes. However, the addition of nonionic surfactants into the system suppressed the interfacial turbulence and reduced the mass transfer fluxes (Agble and Mendes-Tatsis 2000, 2001). It is very difficult to predict accurately interfacial stability theoretically, but the models will be improved greatly with in-depth understanding of interfacial convection and the use of simulation techniques.

3.7 Theoretical Modeling of Marangoni Disturbances

Theoretical analysis of Marangoni disturbances has been undertaken by many researchers. Earlier theoretical analysis by Sawistowski (1973) applied Maxwell's relaxation law to interfacial conditions by regarding interfacial tension as a two-dimensional stress tensor, and they thus derived possible conditions for the appearance of Marangoni convection, according to equation 3.60.

$$\frac{\partial \gamma_{yy}}{\partial y} = \frac{\partial \gamma_\infty}{\partial C}\frac{\partial C}{\partial y} + \frac{\partial \gamma_\infty}{\partial \theta}\frac{\partial \theta}{\partial y} + \frac{\partial \gamma_\infty}{\partial \sigma}\frac{\partial \sigma}{\partial y} + \mu_c \frac{\partial^2 v}{\partial y^2} - \tau_c \frac{\partial^2 (\gamma_{yy} - \gamma_\infty)}{\partial y \partial t} \tag{3.60}$$

where γ_{yy} is the normal component of the dynamic stress tensor or dynamic interfacial tension, γ_∞ is the static interfacial tension, μ_c is the compressional surface viscosity, τ_c is the corresponding surface modulus, v is the y component of the velocity vector, and t denotes time.

The Marangoni effect can therefore result from the appearance in the interface of a gradient of concentration, temperature, or surface density of electric charges, as a result of compression or dilation of a surface film (dynamic effect) and owing to the non-Newtonian behavior of the surface film, which results in the appearance of a time delay between stress and strain. If the interface is Newtonian and dynamic effects of compression and dilation can be neglected, interfacial tension gradients may be created by local changes of solute concentration, temperature, and interfacial electrical potential in liquid–liquid systems. The interfacial solute concentration and temperature can be changed due to mass transfer from the bulk into the interface and the heats of solution.

The starting point for quantitative description of the interfacial flows is based on the Navier–Stokes equations applied to each side of the interface. Wegener (2014) firstly defined a Marangoni number Ma (equation 3.61), where Ma is given by:

$$\mathrm{Ma} = \frac{\partial \sigma}{\partial c}\frac{\Delta c L}{\mu D} \tag{3.61}$$

The term $\partial\sigma/\partial c$ is the interfacial tension gradient, Δc is the concentration gradient, L is a characteristic length, and μ and D are viscosity and diffusivity, respectively. L may be a film thickness or a droplet radius, for example.

The governing equations for a drop in a domain defined as a cylinder of both phases are as follows, similar to the earlier analysis for mass transfer from a single drop as outlined in the previous section in the absence of significant Marangoni convection:

For the flow field we may write the Navier–Stokes equation, the continuity equation, and mass balance equation respectively as shown in equations 3.62–3.64

$$\rho\frac{\partial u}{\partial t}+\rho u\cdot\nabla u+\nabla p-\mu\nabla^2 u=0 \tag{3.62}$$

$$\nabla\cdot u=0 \tag{3.63}$$

$$\frac{\partial c}{\partial t s}+u\cdot\nabla c_S-D_S\nabla^2 c_S=0 \tag{3.64}$$

The momentum balance equation may be transformed for a spherical drop thus:

$$\rho\left(\frac{\partial \underline{u}}{\partial t}+\underline{u}\cdot\underline{\nabla}\,\underline{u}\right)=-\underline{\nabla}p+\underline{\nabla}\cdot\left(2\mu\underline{\underline{D}}\right) \tag{3.65}$$

where $\underline{\underline{D}}$ may be written

$$\underline{\underline{D}}=\frac{1}{2}\left[\underline{\nabla}\,\underline{u}+\left(\underline{\nabla}\,\underline{u}\right)^T\right] \tag{3.66}$$

The boundary conditions defined in detail by Wegener et al. (2009b) include consideration of the boundary condition at the wall of the cylindrical domain, determination of the initial (inlet) velocity condition, and the surface pressure. The concentrations at the liquid–liquid interface are related by the partition coefficient, m as follows:

$$m=\left.\frac{C_{\mathrm{Ad}}}{C_{\mathrm{Ac}}}\right|_{r=R} \tag{3.67}$$

The flux of solute across the interface is expressed as follows:

$$D_{\mathrm{Ad}}\frac{\partial C_{\mathrm{Ad}}}{\partial r}\Big|_{r=R}=D_{\mathrm{Ac}}\frac{\partial C_{\mathrm{Ac}}}{\partial r}\Big|_{r=R} \tag{3.68}$$

The solution algorithm described by Wegener et al. (2009b) assumes that drop shape and shape are constant; also it is assumed that velocity in the radial direction is zero.

Therefore:

$$u_{\mathrm{rd}}\big|_{r=R}=u_{\mathrm{rc}}\big|_{r=R}=0 \tag{3.69}$$

For the tangential velocities, continuity of velocity at the interface results in the following equation:

$$u_{\mathrm{id}}\big|_{r=R}=u_{\mathrm{ic}}\big|_{r=R}\quad (i=\varphi,\Theta) \tag{3.70}$$

In the φ direction:

$$\mu_{\rm c}\left(\frac{\partial \mu_{\varphi,\rm c}}{\partial r} - \frac{u_{\varphi,\rm c}}{R}\right) - \mu_{\rm d}\left(\frac{\partial \mu_{\varphi,\rm d}}{\partial r} - \frac{u_{\varphi,\rm d}}{R}\right) + \frac{1}{R}\frac{\partial \sigma}{\partial \varphi} = 0 \tag{3.71}$$

In the Θ direction:

$$\mu_{\rm c}\left(\frac{\partial \mu_{\Theta,\rm c}}{\partial r} - \frac{u_{\Theta,\rm c}}{R}\right) - \mu_{\rm d}\left(\frac{\partial \mu_{\Theta,\rm d}}{\partial r} - \frac{u_{\Theta,\rm d}}{R}\right) + \frac{1}{R\sin\varphi}\frac{\partial \sigma}{\partial \Theta} = 0 \tag{3.72}$$

The terms in $\partial\sigma$(the last term on the left-hand side of each equation) represent the Maragoni stress resulting from the gradient of interfacial tension.

The latter is conveniently expressed as follows:

$$\frac{\partial \sigma(c_{\rm A})}{\partial i} = \frac{\partial \sigma}{\partial c_{\rm A}} \cdot \frac{\partial c_{\rm A}}{\partial i} \quad (i = \varphi, \Theta) \tag{3.73}$$

The overall outcome of the solution of the above array of equations is the prediction of the time-dependent composition profiles both for the inside of the droplet and for the continuous phase surrounding the droplet. The mass transfer results presented by Wegener et al. (2009a) in Figures 3.14 and 3.15 are for the transfer of acetone from dispersed drops of toluene into a continuous phase of water.

The profiles shown in Figure 3.14 show the simulated internal composition fields inside the drop for three different starting concentrations of acetone in the toluene droplet. Two

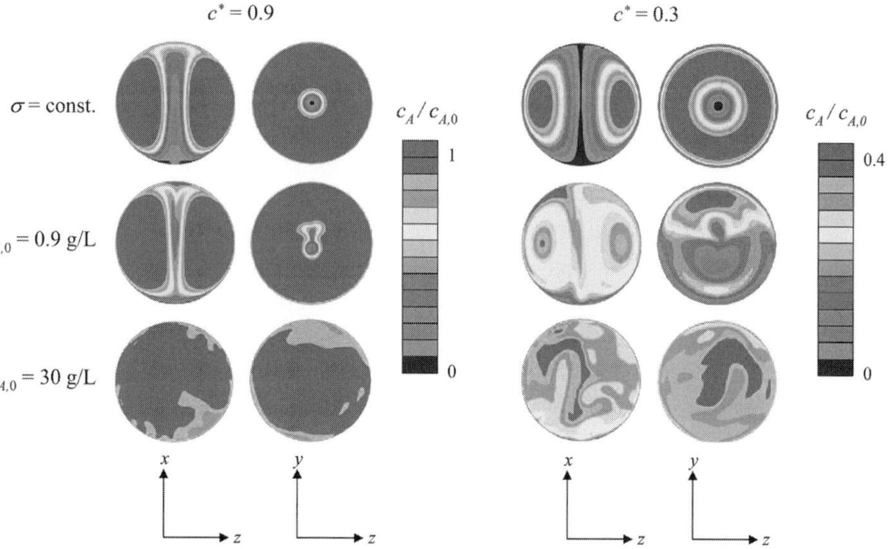

Figure 3.14 Simulations of the internal droplet concentration profiles during the mass transfer of acetone from toluene drops into water for three different initial concentrations at $C^* = 0.9$ (left) and at $C^* = 0.3$ (right) in two different views: in the x–z plane the view is parallel to the flow; in the y–z plane the view is perpendicular to the flow.

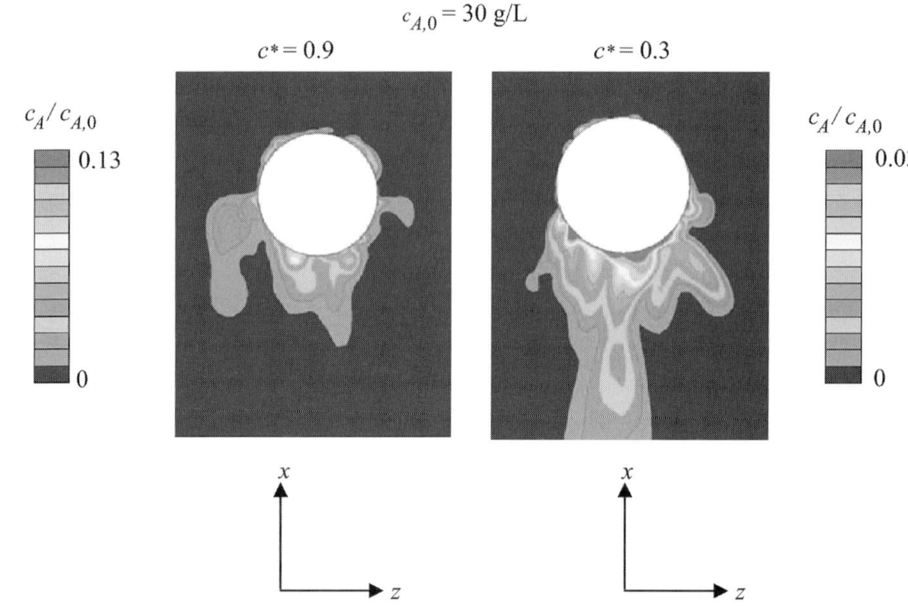

Figure 3.15 Simulations of the continuous phase concentration profiles during the mass transfer of acetone from toluene drops into water. $C_{AO} = 30$ g L^{-1} at $C* = 0.9$ (left) and at $C* = 0.3$ (right) in the x–z plane (view is parallel to the flow).

Reprinted (adapted) with permission from Wegener, M., Eppinger, T., Baumler, K., et al. (2009). Transient rise velocity and mass transfer of a single drop with interfacial instabilities – Numerical investigations. *Chemical Engineering Science*, **64**, 4835–4845. Copyright Elsevier B.V.

mean concentrations are compared (left-hand column 0.9 g L^{-1}; right-hand column 0.3 g L^{-1}). The control is the case of constant interfacial tension (σ = constant) and the topmost image shows the least profile disturbance. The middle image shows the case of a low starting droplet concentration (0.9 g L^{-1}) and in the case of the left-hand image the concentration driving force is low and there is little change compared with the control case, indicating minimal gradient of interfacial tension. In the case of the larger driving forces and higher starting concentrations there is strong evidence of Marangoni-induced instabilities. For example, the bottom drop image, with a starting concentration of 30 g L^{-1}, shows a chaotic composition map, indicating very significant impact of Marangoni disturbances on the internal convection within the droplet.

The profiles of concentration in the surrounding continuous phase shown in Figure 3.15 also reflect the strong impact of concentration driving force-induced interfacial tension gradients on mass transfer.

Simulations for the composition profiles in the continuous phase are shown in Figure 3.15.

Figure 3.16 shows data from the same work, with very good agreement between experimentally observed mean drop concentration variation with time and the simulations for the toluene/acetone/water system. The simulations are valid for two quite disparate starting acetone concentrations, 1.8 g L^{-1} and 30g L^{-1}.

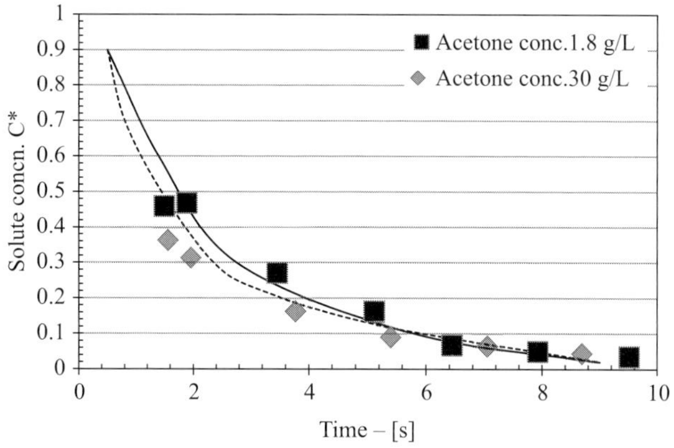

Figure 3.16 Simulated dimensionless mean solute concentration of acetone in a single drop of toluene transferring into water, as a function of time; a comparison between simulations and experimental observations.
Reprinted (adapted) with permission from Wegener, M., Eppinger, T., Baumler, K., et al. (2009). Transient rise velocity and mass transfer of a single drop with interfacial instabilities – Numerical investigations. *Chemical Engineering Science*, **64**, 4835–4845. Copyright Elsevier B.V. All Rights Reserved

These profiles qualitatively compare favorably with experimental visualizations of Marangoni instabilities, for example from Qiu (2010) as shown in Figure 3.12.

Chen et al. (2015) adopted a straightforward approach using the Navier–Stokes equation and the transport equations to determine predicted mass transfer coefficients based on fundamental continuum mechanics but incorporating interfacial disturbance effects. The principle difference in approach compared with that of Petera and Weatherley (2001) is the incorporation of a source term **S** into the Navier–Stokes equation, as seen in equation 3.74. The term **S** represents the extra volume force generated by surface tension gradients, resulting in potential Marangoni disturbances which this analysis takes into account.

$$\rho\left(\frac{\partial u}{\partial t}+u\cdot\nabla u\right)=-\nabla p+\rho g+\nabla\cdot\tau+S \tag{3.74}$$

where τ is the bulk stress tensor:

$$\tau=\mu(\nabla u+(\nabla u)^{\mathrm{T}}) \tag{3.75}$$

Source term $\boldsymbol{S}$ is expressed using the continuum surface force model (Brackbill, Kothe, and Zemach, 1992) in equation 3.74. This is based on the weighted integration method that enables calculation of the local surface tension. Chen suggest that the Dirac delta function, $\delta(\phi)$, may be used to express $\boldsymbol{S}$ as function of the surface tension, σ, concentration, c, and the curvature of the interface, κ, as written in equation 3.76:

$$S=\left(\sigma\kappa n-\frac{\partial\sigma}{\partial c}(I-\mathrm{nn})\cdot\nabla c\right)\delta(\phi) \tag{3.76}$$

where I is the unit tensor, and n is the unit vector perpendicular to the interface. κ represents the mean curvature of the interface, according to equation 3.77:

$$\kappa = -\nabla \cdot \mathrm{n} = \nabla \cdot \left(\frac{\nabla \phi}{|\nabla \phi|} \right) \tag{3.77}$$

In equation 3.77, ϕ is a level set function that Chen defines in terms of the algebraic normal distance to the interface and is expressed by a Hamilton–Jacobi evolution equation:

$$\frac{\partial \phi}{\partial t} + u \cdot \nabla \phi = 0 \tag{3.78}$$

With the level set function, the two-phase flow can be formulated as a single phase with the continuity and Navier–Stokes equations. Chen successfully adopted this approach to simulate Marangoni disturbances, incorporating the concentration dependence of the interfacial tension by a simple polynomial expression of the form:

$$\sigma = \sigma_o(1 + ac + bc^2) \tag{3.79}$$

Figure 3.17(a) shows frames from Chen et al. (2015) for a droplet undergoing mass transfer in the system MIBK/acetic acid/water presented at 0.007 s from the start of the mass transfer (a), and at 0.688 s (i). The convection gradients that the simulations yield show how the convection in and around the drop develops very quickly on account of the strong interactions between the flow field and the concentration field. Figure 3.17(a) clearly shows the development of internal eddy flows within the droplet and these contribute to levelling of concentration gradients within the drop, hence intensifying the mass transfer driving force between the two phases. The development of significant eddy flows in the wake of the drop is also strongly in evidence in image (i).

The very high mass transfer coefficients in the early stages of the mass transfer shown in Figures 3.11(a) and (b) are entirely consistent with the growth in Marangoni disturbances predicted in the first few tenths of a second illustrated in Figure 3.17. Eventually, solute depletion assumes significance, leading to a steep reduction not only in mass transfer rate due to lower driving force but also a reduction in mass transfer coefficient, explainable by reduced Marangoni convection.

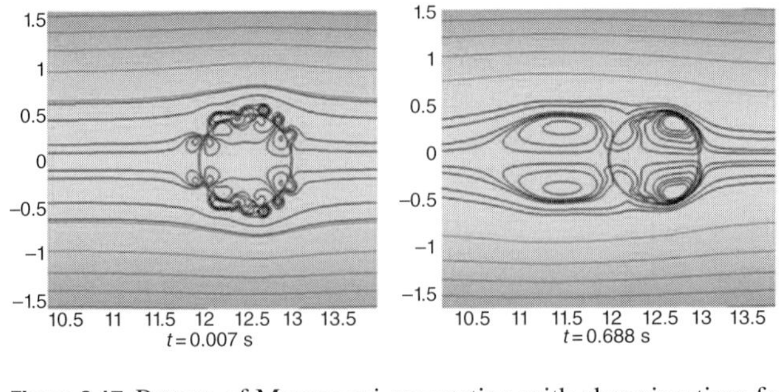

Figure 3.17 Pattern of Marangoni convection with changing time for a moving drop in the system methyl iso-butyl ketone/acetic acid/water.

Reprinted with permission from Chen, J., Wang, Z., Yang, C., and Mao, Z. (2015). Numerical simulation of the solute-induced Marangoni effect with the semi-Lagrangian advection scheme. Chemical Engineering and Technology, **38**(1), 155–163

A further phenomenon is the effect of strong Marangoni flows on the dynamics of drop motion (see Wegener, 2014). It is stated that such flows increase the drag coefficient for rising or falling droplets and thus decrease the droplet velocity with negative influences on the drop Reynolds number and Peclet number and hence the continuous phase mass transfer coefficient.

In further numerical investigations on droplets (curved interfaces), Wegener (2014) claims that no stable configuration may be found for different kinematic viscosity ratios, which is in contrast to Sternling and Scriven's prediction (Sternling and Scriven, 1959). In other words, in the investigated range of kinematic viscosity ratios, Marangoni convection always occurred and mass transfer was always enhanced compared to the case without Marangoni convection, claiming that even a reduction of the interfacial tension gradient by a factor of 100 still showed mass transfer enhancement factors >2 with the smallest concentration gradients having the potential to trigger Marangoni effects and contribute significantly to a better transfer of solute across the interface. They were unable to comment on the effect of diffusivity ratio. The influence of the partition coefficient was also analyzed, with a claim that virtually no mass transfer enhancement is observed at high values of partition coefficient.

Wegener and Paschedag (2011) stated that the interfacial motion due to differences in interfacial tension occur on account of the relationship between local surface energy and interfacial tension differences, with the liquid–liquid interface expanding regions of lower interfacial tension toward regions of higher interfacial tension. The shape of the interface is also a significant factor due to the tangential shear stress that occurs on curved surfaces. The studies by Wegener (2014) also reinforce the findings of Chen et al. (2015) stating that Marangoni effects are more pronounced during the early stages of mass transfer when the concentration gradients are at a maximum (see above). The analysis of Wegener (2014) also summarizes other related phenomena, including the formation of vortex-type structures that enhance mass transfer inside droplets (Figure 3.18).

The analysis describes the main stages of Marangoni convection associated with droplets experiencing mass transfer. The system depicted is based on numerical simulation of 2 mm droplets of toluene containing acetone as solute, extracting into an aqueous continuous phase. Figure 3.18(a) shows the early stages of the mass transfer process when there are large concentration gradients present that are associated with strong Marangoni convection currents. Figure 3.18(b) shows an intermediate stage with decreasing Marangoni convection and onset of internal circulation. Figure 3.18(c) shows a later stage where there is no Marangoni convection but clear evidence of a toroidal flow field inside the droplet.

The significance of Marangoni instabilities in the intensification of mass transfer should not be understated. The toroidal circulatory flow patterns shown in Figure 3.18(c), in the absence of significant Marangoni effects, while promoting some mixing in the tangential axial direction is assumed to promote minimal convection in the direction perpendicular to the streamlines. Further analysis of the promotion of Marangoni instabilities using external fields is covered in Chapter 6.

It is concluded that in the context of intensification of liquid–liquid processes, an understanding and analysis of Marangoni phenomena is important.

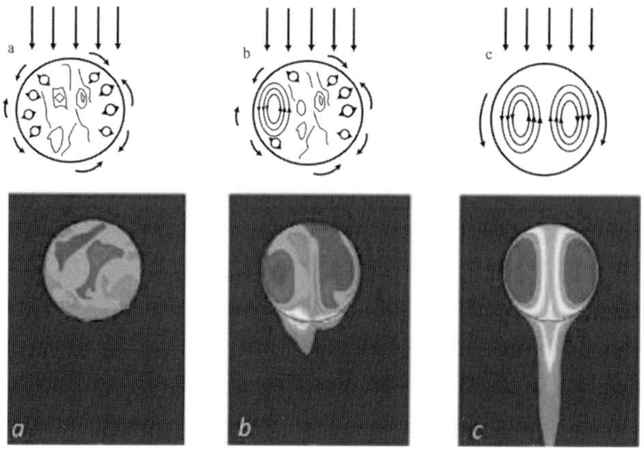

Figure 3.18 Simulation of the development and decay of Marangoni instability for a single drop of toluene with acetone solute transferring into an aqueous continuous phase. Reprinted from Wegener, M. A. (2014). Numerical parameter study on the impact of Marangoni convection on the mass transfer at buoyancy-driven single droplets. *International Journal of Heat and Mass Transfer*, **71**, 769–778 with permission from Copyright Elsevier B.V.

3.8 Swarming Droplet Systems

In industrial liquid–liquid contacting equipment, such as column contactors, phase contact seldom involves the dispersion of discrete single drops and in most cases the dispersed phase progresses through columns in swarms of drops. Therefore, practical extrapolation of correlations and other predictive techniques developed and validated from observation of discrete droplet mass transfer may be unreliable. In the case of dispersion involving swarms of moving droplets, the hydrodynamic situation is altogether more complex. Understanding the underlying mechanisms of mass transfer and mixing in swarming drop liquid–liquid systems is an essential requirement in order to effect intensification. Historically there is a large body of literature describing many innovations in liquid–liquid contactor design, especially in the case of continuous columns, where effective promotion of high rates of interfacial mass transfer, and control of mixing have resulted in much improved performance. Classic examples, which are referred to in more detail in Chapter 1, include packed columns, pulsed plate columns, reciprocating tray columns, rotating disk contactors, and other variants that involve the application of mechanical energy to promote faster and more efficient performance. In virtually all cases the motion of the dispersed phase is complex, with cyclic coalescence and redispersion occurring during passage through the column. This leads to dynamic changes in drop size and size distribution, together with changes in residence time distribution. Forward-mixing and back-mixing of both phases resulting from the complex flow behavior are further complications in accurate description of mass transfer.

There are a number of basic phenomena that impact mass transfer and hydraulics in the flow of swarms of drops compared with the situation with single discrete drops.

Firstly, droplets will interact and collide. This may result in either coalescence or at least a transfer of momentum that will influence the local trajectories of the droplet involved in the encounter. A further possible consequence is the breakup of one or both of the drops. Indeed, coalescence and drop breakup on a cyclic basis during the passage of the dispersed phase through the contactor is highly desirable and promotes mass transfer. The intensification of mass transfer inside each drop occurs on account of the significant hydrodynamic disturbances which occur during breakup and coalescence. Developments in column design over many years have endeavored to maximize the cyclic coalescence and breakup of dispersed drops. Another important effect associated with swarming droplet systems is the momentum transfer between moving drops and the continuous phase that influence the local velocity field in which the droplets are moving. This can impact both the residence time distribution of the droplet phase and of the continuous phase. Variations in residence time distribution can have a major impact on extraction efficiency and in general terms wider residence time distributions are associated with reduction in efficiency, all other factors being equal. Axial back-mixing caused by second-order flows can lead to reduction in overall concentration driving force and hence loss of efficiency.

The prediction of overall mass transfer in swarming droplet systems has been a topic of extensive research over many years. The complexity of swarming droplet systems has challenged development of fundamental analysis for system description. One useful approach uses the concept of enhancement factor, which is applied to the diffusion coefficient values in equations that control the rate of interphase mass transfer between the drops and the continuous phase. At the most basic level, starting equations based on unsteady diffusion in a stagnant drop are cited by Torab-Mostaedi and Safdari (2009) and later by Hemmati et al. (2016) as follows:

$$K_{\mathrm{OD}} = -\frac{d}{6t}\ln\left[6\sum_{n=1}^{\infty} B_n \exp\left(-\frac{4\lambda_n^2 D_{\mathrm{d}} t}{d^2}\right)\right] \tag{3.80}$$

(After Grober, 1925)

$$K_{\mathrm{OD}} = -\frac{d}{6t}\ln\left[\frac{3}{8}\sum_{n=1}^{\infty} B_n^2 \exp\left(-\frac{64\lambda_n D_{\mathrm{d}} t}{d^2}\right)\right] \tag{3.81}$$

(After Kronig and Brink, 1950)

$$K_{\mathrm{OD}} = -\frac{d}{6t}\ln\left[6\sum_{n=1}^{\infty} B_n^2 \exp\left(-\frac{\lambda_n V_{\mathrm{t}} t}{128d(1+k)}\right)\right] \tag{3.82}$$

(After Handlos and Baron, 1957)

Johnson and Hamielec (1960) were among the first to address the prediction of mass transfer coefficients for multidrop systems based on the introduction of an enhancement factor, R, into the expression for stagnant diffusion, as shown in equation 3.83, based in the expression developed by Grober (equation 3.81).

$$K_{OD} = -\frac{d}{6t} \ln\left[6\sum_{n=1}^{\infty} B_n \exp\left(-\frac{4\lambda_n^2 R D_d t}{d^2}\right)\right] \tag{3.83}$$

The use of the R factor to define an enhanced or effective diffusivity value to describe swarming drop systems has been pursued by numerous researchers.

The spray liquid–liquid extraction column is perhaps the earliest and simplest example of continuous countercurrent column industrial contacting equipment. Although a spray column would not normally be regarded as a modern example of intensified extraction equipment it was a good starting point for the study of mass transfer modeling in swarming drop systems. Steiner (1986) provided an early example of predictive correlation of mass transfer in a multidrop system having a degree of hydrodynamic complexity. The study presented by Steiner (1986) developed a correlation based for enhancement factor R, based on mass transfer data collected from spray columns for at least 19 different three-component systems. Given the complexity of the hydrodynamic conditions and the large number of systems used in the comparisions, the results were impressive, as shown for example in Figure 3.19, where the measured values of effective diffusvity ($R/(1 + e)$) versus Re are compared with those predicted by equation 3.84.

$$R = 5.56 \times 10^{-5} \times \left[\mathrm{Re}\frac{2\mu_c}{\mu_c + \mu_{c=d}}\right]^{1.42} (1 - \varepsilon)\mathrm{Sc}_d^{0.67}\mathrm{Bo}^{0.12} \tag{3.84}$$

The validity of the approach was further demonstrated when experimental composition profiles along the length of the column were compared with predictions as shown in Figure 3.20 (from Steiner, 1986). This shows a concentration profile measured in a spray column during the extraction acetone into xylene from water, comparing the experimentally observed values with those calculated using equation 3.84.

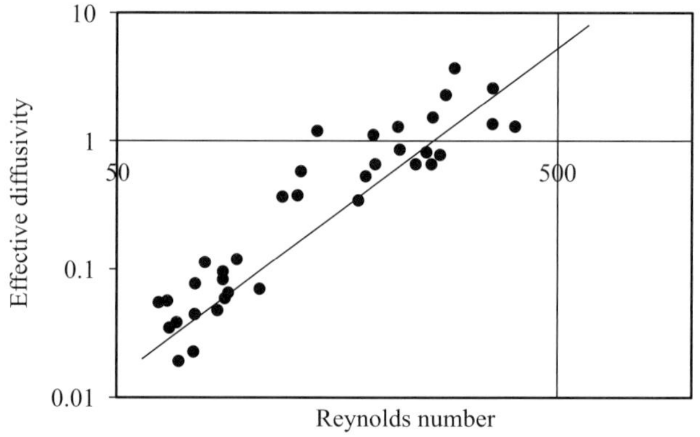

Figure 3.19 Effective diffiusivity versus Re for the system xylene/acetone/water. Comparison of experimental measurements and calculated based on equation 3.84. Reprinted (adapted) with permission from Steiner, L. (1986). Mass-transfer rates from single drops and drop swarms. *Chemical Engineering Science*, **41**(8), 1979–1986. Copyright (1986) Elsevier B.V. All Rights Reserved

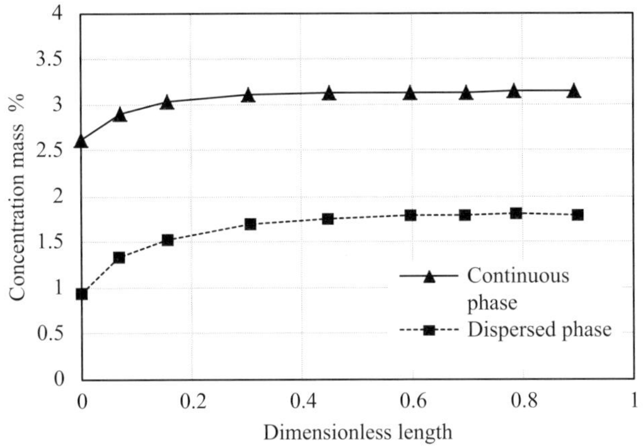

Figure 3.20 Comparison of concentration profiles for dense dispersion of drops in a spray column during mass transfer in the systen xylene/acetone/water. Experimental versus calculated based on equation 3.84.

Reprinted (adapted) with permission from Steiner, L. (1986). Mass-transfer rates from single drops and drop swarms. *Chemical Engineering Science*, **41**(8), 1979–1986. Copyright (1986) Elsevier B.V. All Rights Reserved

A useful summary of the application of the R factor approach to modeling of multidrop mass transfer in enhanced liquid–liquid contacting equipment is listed by Hemmati et al. (2016) as follows.

For a Kuhni column, Hemmati, Torab-Mostaedi, and Asadollahzadeh (2015) proposed:

$$R = 0.51 + 213.35\mathrm{Re}^{-0.257}(1 - x_\mathrm{d})E\ddot{o}^{1.42} \tag{3.85}$$

$$E_{\ddot{o}} = \frac{g\Delta\rho d_{32}}{\sigma}$$

This expression was also successfully tested in a Kuhni column using two systems: toluene/acetone/water, and n-butyl acetate/acetone/water. The comparison between experimental and calculated is shown in Figure 3.21 and indicates good agreement.

Similar equations from a range of sources are presented by Torab-Mostaedi and Safdari (2009) and Torab-Mostaedi et al. (2009, 2010) and for a number of enhanced extraction column contactors.

For the Hanson mixer settler column Torab-Mostaedi et al. (2009) used stage efficiency as a measure for comparison. This was defined in the classical manner according to equation 3.86:

$$E_{\mathrm{Oy}} = \frac{(y_n - y_{n-1})}{(y_n^* - y_{n-1})} \tag{3.86}$$

The application of equation 3.86 and comparison with experimental results used stage efficiency values using two correlations respectively, applicable in two ranges of drop Reynolds number, according to equations 3.87 and 3.88.

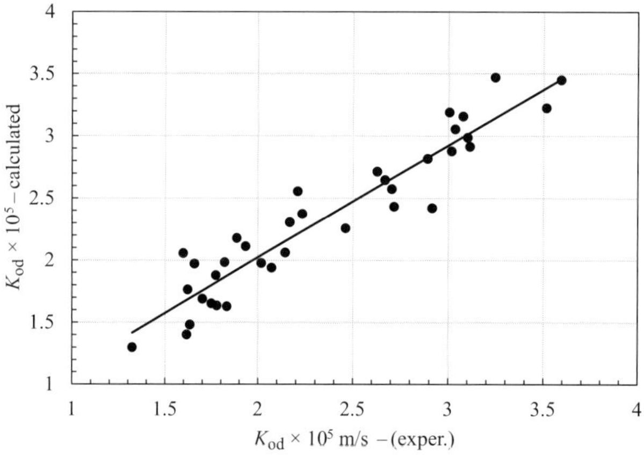

Figure 3.21 Comparison of experimental values of K_{od} with calculated based on equation 3.85 for the Kuhni column.

Reprinted (adapted) with permission from Hemmati, A. R., Torab-Mostaedi, M., and Asadollahzadeh, M. (2015). Mass transfer coefficients in a Kühni extraction column. *Chemical Engineering Research and Design*, **93**, 747–754. Copyright (2015) Elsevier B.V. All Rights Reserved

$$E_{Oy} = 1.417\mathrm{Re}^{0.101}\mathrm{Fr}^{-0.822} \quad \mathrm{Re} > 10 \tag{3.87}$$

$$E_{Oy} = 0.068\mathrm{Re}^{0.874}\mathrm{Fr}^{-0.222} \quad \mathrm{Re} < 10 \tag{3.88}$$

where:

$$\mathrm{Re} = \frac{d_{32} V_{\mathrm{slip}} \rho_c}{\mu_c} \tag{3.89}$$

$$\mathrm{Fr} = \frac{g}{d_R N^2} \tag{3.90}$$

Figure 3.22 shows the comparison of experimental efficiency results against the predictions of equations 3.87 and 3.88 for the system acetone/water/toluene in a seven-stage mixer settler Hansen mixer settle unit.

Torab-Mostaedi and Safdari (2009) also studied the enhancement factor approach for correlation of volumetric mass transfer coefficient mass transfer in a pulsed packed column. This is another early example of an intensified technique for liquid–liquid contacting by the introduction of external mechanical energy into the column by pulsation of the entire liquid volume in the column during contrinuous flow operation. Enhancement factor values were determined experimentally and an empirical correlation derived for prediction of effective diffusivity as a function of Reynolds number, Schmidt number, and viscosity ratio. Good agreement between predictions based on the correlations and experiments was found for all the operating conditions that were investigated.

Equation 3.91 is the correlation presented in that paper with the Reynolds number defined using slip velocity and Sauter mean drop size d_{32}.

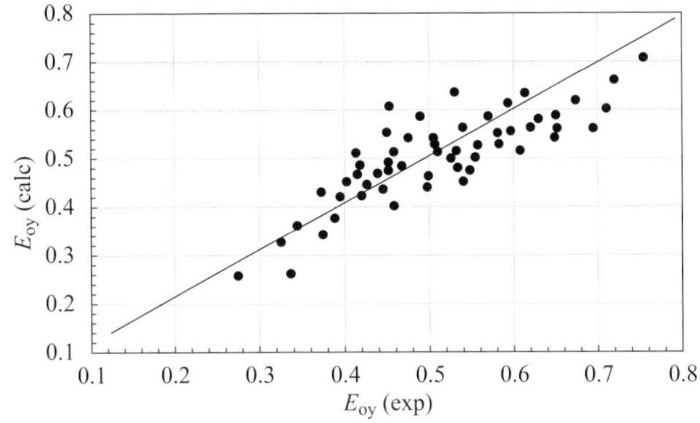

Figure 3.22 Comparison of experimental efficiency results against the calculated values for the Hanson mixer settler.
Reprinted (adapted) with permission from Torab-Mostaedi, M., Safdari, J., Moosavian, M. A., and Maragheh, M. G. (2009). Stage efficiency of Hanson mixer-settler extraction column. *Chemical Engineering and Processing*, **48**, 224–228. Elsevier, All Rights Reserved

$$R = 2.57 + 1326.07\text{Re}^{0.5}\text{Sc}^{-0.94}(1+\kappa)^{-0.80} \tag{3.91}$$

where Re is given by

$$\text{Re} = \frac{d_{32}V_{\text{slip}}\rho_{\text{c}}}{\mu_{\text{c}}} \tag{3.92}$$

As shown in the Figure 3.23, the agreement between the experimental values of K_{od} in the pulsed packed column and the calculated values based on equation 3.91 is very good. Two systems are represented here: toluene/acetone/water and n-butyl acetate/acetone/water.

Torab-Mosteadi, Safdari, and Ghaemi (2010) adopted a similar approach to modeling the mass transfer performance of a pulsed perforated plate extraction column. Mass transfer coefficients were interpreted using a classical axial diffusion model approach, taking account of pulse frequency and amplitude, phase velocities, and operating regimes. In the pulsed plate column these included the mixer settler regime, the emulsion regime, and the transitional regime – which occurs between mixer settler regime and the emulsion regime. These regimes are well known for pulsed plate columns, with an extensive literature showing the existence of a particular regime being strongly dependent on the energy input to the system.

Prediction of dispersed phase overall mass transfer coefficients was based on the development of an expression for enhancement factor, R. A single correlation for R was derived in terms of Reynolds number, Eötvös number, and dispersed phase volumetric holdup, as written in equation 3.93. Study of this correlation showed good agreement with experimental results, as shown in Figure 3.24 for the systems toleune/acetone/water and n-butyl acetate/acetone/water.

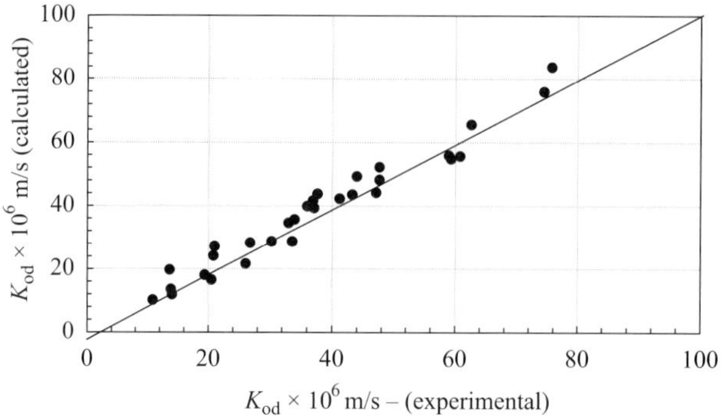

Figure 3.23 For the pulsed packed column, comparison between the experimental values of K_{od} in the pulsed packed column and the calculated values based on equation 3.91.

Reprinted (adapted) under the terms of Creative Commons from Torab-Mostaedi, M. and Safdari, J. (2009). Prediction of mass transfer coefficients in a pulsed packed extraction column using effective diffusivity. *Brazilian Journal of Chemical Engineering*, **26**(4), 685–694

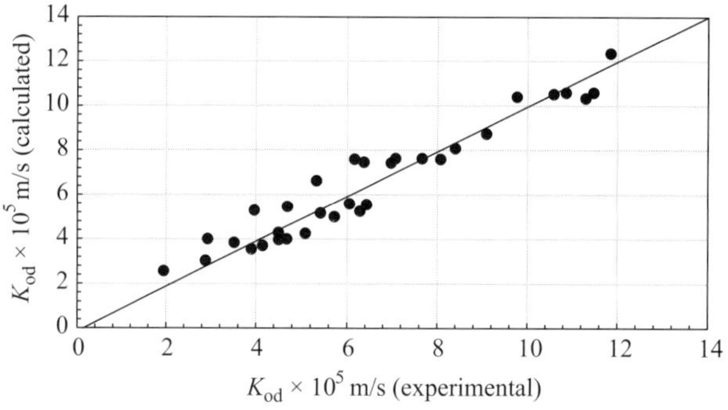

Figure 3.24 For the pulsed plate column: toluene/acetone/water; n-butyl acetate/acetone/water. Comparison between the experimental values of K_{od} in the pulsed column and the calculated values.

Reprinted (adapted) under the terms of Creative Commons from Torab-Mostaedi, M., Safdari, J., and Ghaemi, A. (2010). Mass transfer coefficients in pulsed perforated-plate extraction columns. *Brazilian Journal of Chemical Engineering*, **27**(2), 243–251

For the pulsed sieve plate column, Torab-Mosteadi et al. (2010) present:

$$R = 5.33 + 97.97\text{Re}^{-0.13}(1 - x_d)E_{\ddot{o}}^{0.92} \tag{3.93}$$

where x_d = the dispersed phase fractional volumetric holdup given by equation 3.94:

$$V_{slip} = \frac{V_d}{x_d} + \frac{V_c}{(1 - x_d)} \tag{3.94}$$

$$E_{\ddot{o}} = \text{Eötvös number}$$

$$E_{\ddot{o}} = \frac{g\Delta\rho d_{32}}{\sigma} \tag{3.95}$$

The Reynolds number used in equation 3.93 is defined as in equation 3.92.

The predictive comparisons shown in Figures 3.23 and 3.24 from the work of Torab-Mosteadi and Safdari (2009) and Torab-Mosteadi et al. (2010) for the pulsed packed column and for the pulsed plate column show significant improvements over previous published correlations and are among the best to date based on the empirical approach for intensified contact based on pulsation.

Correlations for the R factor are presented for other continuous liquid–liquid contacting equipment as follows.

For the spray column, Steiner (1986) presents:

$$R = 5.56 \times 10^{-5}\left(\frac{2\text{Re}}{1+\kappa}\right)^{1.42}\left(\frac{g\Delta\rho d_{32}^2}{\sigma}\right)^{0.12}\text{Sc}^{0.67}(1 - x_{\text{d}}) \tag{3.96}$$

For the rotating disk contactor Amanabadi et al. (2009) propose an expression for overall mass transfer coeeficient, based on the classical Newman model (Newman, 1931) and using an expression for effective diffusivity expressed as follows:

$$D_{\text{eff}} = 0.4755 \times 10^{-9}\text{Re}^{0.65} \tag{3.97}$$

The Reynolds number was calculated using slip velocity between dispersed and continuous phases for the velocity term in Re. The calculated value of D_{eff} based on equation 3.97 was used in equations 3.98 and 3.99 in order to study the relationship between K_{od} and mean residence time, t, (or height) in the rotating disk column.

$$\ln(1 - E) = -\frac{6tK_{\text{od}}}{d} \tag{3.98}$$

$$E = \frac{C - C_{\text{o}}}{C^* - C_{\text{o}}} \tag{3.99}$$

$$\frac{C^* - C}{C^* - C_{\text{o}}} = \frac{6}{\pi^2}\sum_{n=1}^{\infty} C_n \exp\left(\frac{-4\lambda_n^2 D_{\text{eff}} t}{d^2}\right) \tag{3.100}$$

Comparison of the calculated values of K_{od} with the experimental values determined for the extraction of acetic acid from a dispersed phase of carbon tertrachloride into an aqueous continuous phase based on equations 3.97–3.100 is shown in Figure 3.25.

The degree of agreement shown in this work was shown to be significantly superior to that in many other studies of mass transfer in rotating disk contactors.

Kumar and Hartland (1999) published a comperhensive study of mass transfer in column contactors, developing expressions for Sherwood number for continuous phase mass transfer, dispersed phase and for the corresponding overall mass transfer coefficients. The Kumar and Hartland analysis for mass transfer in column contactors is neatly summarized by Hemmati et al. (2016). The starting point for the analysis is a set of correlations for Sherwood number for rigid drops.

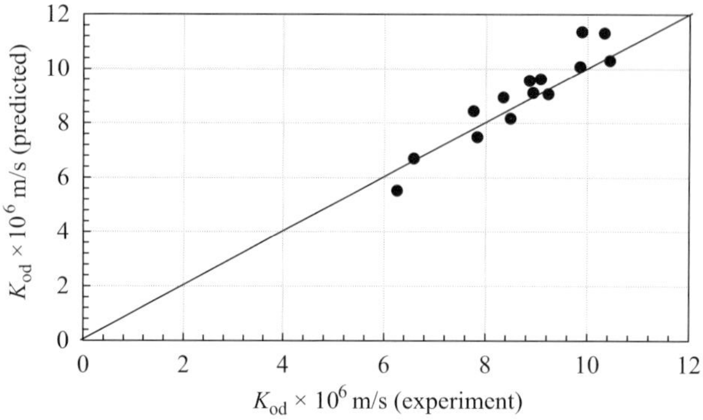

Figure 3.25 Comparison of experimental values of overall mass transfer coeffiicent in a rotating disk contactor for the extraction of acetic acid from a dispersed phase of carbon tertrachloride into an aqueous continuous phase based on equations 3.97–3.100.
Reprinted (adapted) with permission from Amanabadi, M., Bahmanyar, H., Zarkeshan, Z., and Mousavian, M. A. (2009). Prediction of effective diffusion coefficient in rotating disc columns and application in design. *Chinese Journal of Chemical Engineering*, **17**(3), 366–372. Copyright (2009) Elsevier B.V. All Rights Reserved

For example, using continuous phase units:

$$\mathrm{Sh_{c,rigid}} = 2.43 + 0.775\mathrm{Re}^{1/2}\mathrm{Sc_c}^{1/3} + 0.0103\mathrm{Re\,Sc}^{1/3} \tag{3.101}$$

A second expression was defined as follows:

$$\mathrm{Sh_{c,\infty}} = 50 + \frac{2}{\sqrt{\pi}}(\mathrm{Pe_c})^{1/2} \tag{3.102}$$

where $\mathrm{Pe_c} = \frac{HV_c}{E_c}$; $\mathrm{Pe_d} = \frac{HV_d}{E_d}$

where H = height of column; V_d, V_c = linear velocity of dispersed and continuous phases, respectively, and E_d, E_c = axial mixing coefficients of dispersed and continuous phases, respectively.

The full expression for the Sherwood number in continuous phase units is then written:

$$\frac{\mathrm{Sh_c}/(1-x_d) - \mathrm{Sh_{c,rigid}}}{\mathrm{Sh_{c,\infty}} - \mathrm{Sh_c}/(1-x_d)} = 5.26 \times 10^{-2}\mathrm{Re}^{1/3+6.59\times10^{-2}\mathrm{Re}^{0.25}} \times \mathrm{Sc_c}^{1/3}\left(\frac{V_s\eta_c}{\sigma}\right)^{1/3}\frac{1}{1+k^{1.1}} \tag{3.103}$$

In overall dispersed phase unit, the expression for Sherwood number is expressed:

$$\mathrm{Sh_d} = 17.7 + \frac{3.19\times10^{-3}\left(\mathrm{Re\,Sc_d^{1/3}}\right)^{1.7}}{1+1.43\times10^{-2}\left(\mathrm{Re\,Sc_d^{1/3}}\right)^{0.7}}\left(\frac{\rho_d}{\rho_c}\right)^{2/3} \times \frac{1}{1+k^{2/3}} \tag{3.104}$$

Mass transfer coefficients are then available from the definition of Sherwood numbers as follows:

$$\mathrm{Sh_d} = \frac{K_{\mathrm{od}} d_{32}}{D_\mathrm{d}}; \ \mathrm{Sh_c} = \frac{K_{\mathrm{oc}} d_{32}}{D_\mathrm{c}} \tag{3.105}$$

where d_{32} = Sauter mean drop diameter and D_d, D_c = diffusion coefficents in the dispersed and continuous phases, repectively.

A more recent study by Hemmati et al. (2016) examined mass transfer correlations for a perforated plate rotating disk contactor (PRDC) and developed a new correlation for overall mass transfer coefficient, again using an enhancement factor approach. This recent approach was compared with the results of the earlier analysis of Kumar and Hartland.

The enhancement factor R is defined by equation 3.106 as follows:

$$R = -0.51 + 1.74\mathrm{Re}^{0.313}(1 - x_\mathrm{d})E\ddot{o}^{0.165} \tag{3.106}$$

$$E\ddot{o} = \frac{g\Delta\rho d_{32}^2}{\sigma} \tag{3.107}$$

Application of equations 3.106 and 3.107, with equation 3.80 then leads to the expression for K_{od} as written in equation 3.108:

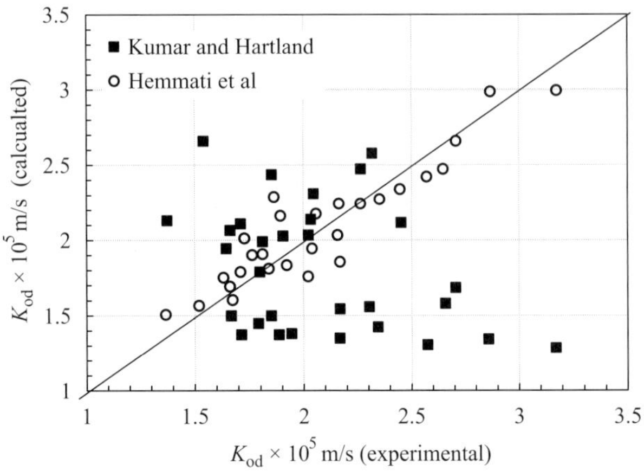

Figure 3.26 Comparison of the predictions of K_{od} using the equations developed by Hartland and Kumar (equations 3.104 and 3.105) with those of Hemmati et al. (equation 3.108) for mass transfer in a perforated plate rotating disc contactor.

$$K_{\text{od}} = -\frac{d}{6t}\ln\left[6\sum_{n=1}^{\infty}\beta_n \exp\left(-\frac{4\lambda_n^2 D_{\text{d}}t}{d^2}\left(-0.51 + 1.74\text{Re}^{0.313}(1 - x_{\text{d}})E\ddot{o}^{0.165}\right)\right)\right] \tag{3.108}$$

Re is defined in equation 3.92 and slip velocity V_s is defined in equation 3.94.

Comparison of the predictions of K_{od} using the equations developed by Hartland and Kumar (equations 3.104 and 3.105) with those of Hemmati et al. (equation 3.108) is shown in Figure 3.26. The comparison is based on experimental mass transfer coefficient data measured in a perforated plate rotating disk contactor. Data for two different extraction systems were collected and used in the comparison. Mass transfer data in a high interfacial tension system, toluene/acetone/water and a medium interfacial tension system butyl acetate/acetone/water (medium interfacial tension were determined over a range of conditions). The continuous phase in all cases was water. The correlations of Kumar and Hartland were developed for a "standard" rotating disk column, but the degree of scatter of the comparisons of predicted versus calculated is much lower in the case of the Hemmati equations.

References

Agble, D. and Mendes-Tatsis, M. A.(2000). The effect of surfactants on interfacial mass transfer in binary liquid–liquid systems. *International Journal of Heat and Mass Transfer*, **43**(6), 1025–1034.

Agble, D. and Mendes-Tatsis, M. A. (2001). The prediction of Marangoni convection in binary liquid–liquid systems with added surfactants. *International Journal of Heat and Mass Transfer*, **44**, 1439–1449.

Albernaz, D. L., Do-Quang, M., Hermanson, J. C., and Amberg, G. (2017). Droplet deformation and heat transfer in isotropic turbulence. *Journal of Fluid Mechanics*, **820**, 6185.

Amanabadi, M., Bahmanyar, H., Zarkeshan, Z., and Mousavian, M. A. (2009). Prediction of effective diffusion coefficient in rotating disc columns and application in design. *Chinese Journal of Chemical Engineering*, **17**(3), 366–372.

Angelo, J. B., Lightfoot, E. N., and Howard, D. W. (1966). Generalisation of the penetration theory for surface stretch: application to forming and oscillating drops. *American Institute of Chemical Engineers Journal*, **12**(4), 751–760.

Bakker, C. A., van Buytenen, B., and Beek, W. J. (1966). Interfacial phenomena and mass transfer. *Chemical Engineering Science*, **21**, 1039–1046.

Bozzi, L. A., Feng, J. Q., Scott, T. C., and Pearlstein, A. J. (1997). Steady axisymmetric motion of deformable drops falling or rising through a homoviscous fluid in a tube at intermediate Reynolds number. *Journal of Fluid Mechanics*, **336**, 1–32.

Brackbill, J. U., Kothe, D. B., Zemach, C., (1992). A continuum method for modeling surface tension. *Journal of Computational Physics* **100**(2), 335–354.

Brauer, H. (1971). *Stoffaustausch einschliesslich chemischer Reaktionen*. Aarau, Switzerland: Verlag Sauerlander.

Brodkorb M.J., Bosse D., von Reden C., Gorak A., and Slater M.J. (2003). Single drop mass transfer in ternary and quaternary liquid/liquid extraction systems. *Chemical Engineering and Processing*, **42**, 825-840.

Chen,J., Wang, Z., Yang, C. and Mao, Z. (2015). Numerical simulation of the solute-induced Marangoni effect with the semi-Lagrangian advection scheme. *Chemical Engineering and Technology,* **38**(1), 155–163.

Clift, R., Grace, J. R. and Weber, M. E., (1978). *Bubbles, Drops and Particles.* New York: Academic Press.

Dandy, D. S. and Leal, L. G. (1989). Buoyancy-driven motion of a deformable drop through a quiescent liquid at intermediate Reynolds numbers. *Journal of Fluid Mechanics*, **208**(1), 161–192.

Elzinga, E. R. and Banchero, J. T. (1959). Film coefficients for heat transfer to liquid drops. *Chemical Engineering Progress Symposium Series*, **55**(29), 149–161.

Foote, G. B. (1969). On the internal circulation and shape of large rain drops. *Journal of Atmospheric Science*, **26**, 179–181.

Gangu, A. S. (2013). Towards in situ extraction of fine chemicals and bio-renewable fuels from fermentation broths using Ionic liquids and the intensification of contacting by the application of electric fields. Ph.D. thesis, University of Kansas.

Garner, F. H. and Skelland, A. H. P. (1955). Some factors affecting droplet behavior in liquid–liquid systems. *Chemical Engineering Science*, **4**(4), 149–157.

Garner, F. H. and Tayeban, M. (1960). The importance of the wake in mass transfer from both continuous and dispersed phase systems, I. *Anales de la Real Sociedad Española de Química*, **56**, 479–490.

Garner, F. H., Foord, A., and Tayeban, M. (1959). Mass transfer from circulating drops. *Journal of Applied Chemistry*, **9**, 315–323.

Ghalehchian, J. S. and Slater, M. J. (1999). A possible approach to improving rotating disc contactor design, accounting for drop breakage and mass transfer with contamination. *The Chemical Engineering Journal*, **75**(2), 131–144.

Golovin, A. A.(1990). The onset of the interracial instability of drops during extraction. *Institution of Chemical Engineers Symposium Series*, **119**, 327.

Golovin, A. A. (1991). Models of mass transfer in the presence of inter-phase mass transfer. *Theoretical Foundations of Chemical Engineering*, **24**(5), 388–403.

Grace, J. R., Wairegi, T., and Nguyen, T. H. (1976). Shapes and velocities of single drops and bubbles moving freely through immiscible liquids. *Transactions of the Institution of Chemical Engineers*, **54**, 167.

Grober, H. (1925). Die Erwarmung und Abkuhlung einfacher geometrischer Korper (The heating and cooling of simpler, geometric matters). *Zeitschrift Des Vereines Deutscher Ingenieure,* **69**, 705–711.

Hadamard, J. (1911). Mouvement permanent lent d'une sphe`re liquide et visqueuse dans un liquide visqueux. *Comptes Rendus de l'Académie des Sciences*, **152**, 1735–1738.

Handlos, A. E. and Baron, T. (1957). Mass and heat transfer from drops in liquid–liquid extraction. *American Institute of Chemical Engineers Journal*, **3**, 127–136.

Haywood, R. J., Renksizbulut, M., and Raithby, G. D. (1994). Transient deformation and evaporation of droplets at intermediate Reynolds numbers. *International Journal of Heat and Mass Transfer*, **37**, 1401–1409.

Heertjes, P. M., Holve, W. A., and Talsma, H. (1954). Mass transfer between isobutanol and water in a spray column. *Chemical Engineering Science*, **3**, 122–142.

Hemmati, A. R., Torab-Mostaedi, M., and Asadollahzadeh, M. (2015). Mass transfer coefficients in a Kühni extraction column. *Chemical Engineering Research and Design*, **93**, 747–754.

Hemmati, A. R., Shirvani, M., Torab-Mostaedi, M., and Ghaemi, A. (2016). Mass transfer coefficients in a perforated rotating disc contactor (PRDC). *Chemical Engineering and Processing*, **100**, 19–25.

Hu, S. and Kintner, R. C. (1955). The fall of single drops through water. *American Institute of Chemical Engineers Journal*, **1**(1), 42–48.

Ihme, F., Schmidtt, H., and Brauer, H. (1972). Theoretical investigation of the flow around and mass transfer to spheres. *Chemie Ingenieur Technik*, **44**(5), 306–313.

Johnson, A. I. and Hamielec, A. E. (1960). Mass transfer inside drops. *American Institute of Chemical Engineers Journal*, **6**(1), 145–149.

Johnson, A. I., Hamielec, A. E., Ward, D., and Golding, A. (1958). End effect corrections in heat and mass transfer studies. *The Canadian Journal of Chemical Engineering*, **October**, 221–227.

Klee, A. J. and Treybal, R. E. (1956). Rate of rise or fall of liquid droplets. *American Institute of Chemical Engineers Journal*, **2**(4), 444–447.

Kronig, R. and Brink, J. C. (1950). On the theory of extraction from falling drops. *Applied Scientific Research Section A – Mechanics, Heat, Chemical Engineering, Mathematical Methods*, **2**(2), 142–154.

Kumar, A. and Hartland, S. (1999). Correlations for prediction of mass transfer coefficients in single drop systems and liquid–liquid extraction columns. *Transactions of the Institution of Chemical Engineers*, **77**(A), 372–384.

Lewis, J. B. and Pratt, H. R. C. (1953). Oscillating droplets. *Nature*, **171**(4365), 1155–1156.

Linde, H. and Schwarz, E. (1961). Untersuchungen zur Charakteristick der freien Grenzflächenkonvektion beim Stoffübergang an fluiden Grenzflächen. *Mber. Deutschen Akademie Wissenschaften Berlin*, **3**, 554.

Linde, H. and Schwarz, E. (1964). Über großräumige Rollzellen der freien Grenzflächenkonvektion. *Mber. Deutschen Akademie Wissenschaften Berlin*, **7**, 330.

Linde, H. and Shert, B. (1965). Schlierenoptischer Nachweis der Marangoni-Instabilität bei der Tropfenbildung. *Mber. Deutschen Akademie Wissenschaften Berlin*, 7, 341–348.

Lochiel, A. C. and Calderbank, P. H. (1964). Mass transfer in the continuous phase around axisymmetric bodies of revolution. *Chemical Engineering Science*, **19**, 471–484.

Marangoni, C. (1869). Sull'espansione delle goccie d'un liquido galleggianti sulla superficie di altro liquido (On the expansion of a droplet of a liquid floating on the surface of another liquid). *Tipographia dei fratelli Fusi*, Pavia.

Marangoni, C. (1871). Über die Ausbreitung der Tropfen einer Flüssigkeit auf der Oberfläche einer anderenAnn. *Annalen der Physik (Leipzig)*, **143**, 337–354.

Mendes-Tatsis, M. A. and Agble, D. (2000). The effect of surfactants on Marangoni convection in the isobutanol/water system. *Journal of Non-Equilibrium Thermodynamics*, **25**, 239–249.

Modigell, M. (1981). Untersuchung der Stoffübertragung zwischen zwei Flussigkeiten unter Berüksichtigung von Gernzflächenphänomener, Dissertation, RWTH Aachen.

Nakache, E., Duperyrat, M., and Vignes-Adler, M. (1983). Experimental and theoretical study of an interfacial instability at some oil-water interfaces involving a surface-active agent: I. Physicochemcial description and outlines for a theoretical approach. *Journal of Colloid and Interface Science*, **94**(1), 187–200.

Newman, A. B. (1931). The drying of porous solids: diffusion and surface emission equations. *Transactions of the American Institute of Chemical Engineers*, **27**, 203–220.

Orell, A. and Westwater, J. W. (1961). Natural convection cells accompanying liquid–liquid extraction. *Chemical Engineering Science*, **16**, 127.

Orell, A. and Westwater, J. W. (1962). Spontaneous interfacial cellular convection accompanying mass transfer: ethylene glycol-acetic acid-ethyl acetate. *American Institute of Chemical Engineers Journal*, **8**, 350–356.

Perez De Ortiz, E. S. and Sawistowski, H. (1973a). Interfacial stability of binary liquid–liquid systems I. Stability analysis. *Chemical Engineering Science*, **28**, 2051–2062.

Perez De Ortiz, E. S. and Sawistowski, H. (1973b). Interfacial stability of binary liquid–liquid systems II. Stability behaviour of selected systems. *Chemical Engineering Science*, **28**, 2063–2069.

Petera, J. and Weatherley, L. R. (2001). Modeling of mass transfer from falling droplets. *Chemical Engineering Science*, **56**, 4929–4947.

Petera, J., Weatherley, L. R., Rooney D., and Kaminski K. (2009). A finite element model of enzymatically catalyzed hydrolysis in an electrostatic spray reactor. *Computers and Chemical Engineering*, **33**, 144–161.

Petrov, A. G. (1988). Internal circulation and deformation of viscous drops. *Vestnik Moskovskowo Universiteta Mekhanika*, **43**, 85–88.

Pratt, H. R. C. and Stevens, G. W. (1992). Selection, design, pilot-testing and scale-up of extraction equipment. In J. D. Thornton, ed., *Science & Practice of Liquid–Liquid Extraction*, Oxford: Clarendon Press, pp. 492–591.

Qiu, J. (2010). Intensification of liquid–liquid contacting processes. Ph.D. thesis. University of Kansas.

Quincke, G. H. (1888). *Annalen der Physik und Chemie*, **7**, 35.

Rose, P. M. and Kintner, R. C.(1966). Mass transfer from large oscillating drops. *American Institute of Chemical Engineers Journal*, **12**(3), 530–534.

Rybczynski, W. (1911). Uber die fortschreitende Bewegung einer ussigen Kugel in einem za¨hen Medium. *Bulletin International de l'Académie des Sciences de Cracovie, Series A*, 40–46.

Saboni, A., Gourdon, C., and Chesters, A. (1993). The influence of mass transfer on coalescence in liquid–liquid systems. In: D. H. Logsdail and M. J. Slater, eds., *Solvent Extraction in the Process Industries*, Vol. 1. Proceedings of ISEC'93, London: SCI. 506–513.

Sawistowski, H. (1971). In C. Hanson, ed., *Recent Advances in Liquid–Liquid Extraction*, Vol. 1. Oxford: Pergamon Press, pp. 293–365.

Sawistowski, H. (1973). Surface-tension-induced interfacial convection and its effect on rates of mass-transfer. *Chemie Ingenieur Technik*, **45**(18), 1093–1098.

Sawistowski, H. (1981). Interfacial convection. *Berichte der Bunsengesellschaft für Physikalische Chemie.*, **85**, 905–909.

Scheele, G. F. and Meister, B. J. (1968). Drop formation at low velocities in liquid–liquid systems: prediction of drop volume. *American Institute of Chemical Engineers Journal*, **14** (1), 9–15.

Schott, R. and Pfennig, A. (2004). Modelling of mass-transfer induced instabilities at liquid–liquid interfaces based on molecular simulations. *Molecular Physics*, **102**, 331–339.

Schroeder, R. R. and Kintner, R. C. (1965). Oscillations of drops falling in a liquid field. *American Institute of Chemical Engineers Journal*, **11**, 5–8.

Skelland, A. H. P. and Wellek, R. M. (1964). Resistance to mass transfer inside droplets. *AIChE Journal*, **10**, 491–496.

Slater, M. J. (1994). Rate coefficients in liquid–liquid extraction systems. In J. C. Godfrey and M. J. Slater, eds., *Liquid–Liquid Extraction Equipment*. Chichester, UK: Wiley, 45–94.

Slavtchev, S., Hennenberg, M., Legros, J.-C., and Lebon, G. (1998). Stationary solutal Marangoni instability in a two-layer system. *Journal of Colloid and Interface Science,* **203**, 354–368.

Steiner, L. (1986). Mass-transfer rates from single drops and drop swarms. *Chemical Engineering Science*, **41**(8), 1979–1986.

Sternling, C. V. and Scriven, L. E. (1959). Interfacial turbulence: hydrodynamic instability and the Marangoni effect. *American Institute of Chemical Engineers Journal*, **5**(4), 514–523.

Thomson, J. (1855). On certain curious motions at the surfaces of wine and other alcoholic liquors. *Philosophical Magazine (London, Edinburgh, Dublin)*, **10**(67), 330–333.

Thorsen, G. and Terjesen, S. G. (1962). On the mechanism of mass transfer in liquid–liquid extraction. *Chemical Engineering Science*, **17**, 137–148.

Torab-Mostaedi, M. and Safdari, J. (2009). Prediction of mass transfer coefficients in a pulsed packed extraction column using effective diffusivity. *Brazilian Journal of Chemical Engineering*, **26**(4), 685–694.

Torab-Mostaedi, M., Safdari, J., Moosavian, M. A., and Maragheh, M. G. (2009). Stage efficiency of Hanson mixer-settler extraction column. *Chemical Engineering and Processing*, **48**, 224–228.

Torab-Mostaedi, M., Safdari, J., and Ghaemi, A. (2010). Mass transfer coefficients in pulsed perforated-plate extraction columns. *Brazilian Journal of Chemical Engineering*, **27**(2), 243–251.

Tsukada, T., Hozawa, M., Imaishi, N., and Fujinawa, K. (1984). Computer simulations of deformation of moving drops and bubbles by use of the finite element method. *Journal of Chemical Engineering of Japan*, 17(3), 246–251.

Tsukada, T., Mikami, H., Hozawa, M., and Imaishi, N. (1990). Theoretical and experimental studies of the deformation of bubbles moving in quiescent Newtonian and non-Newtonian liquids. *Journal of Chemical Engineering of Japan*, **23**(2), 192–198.

Walcek, C. J. and Pruppacher, H. R. (1984). On the scavenging of SO_2 by cloud and raindrops. *Journal of Atmospheric Chemistry*, **1**, 269–289.

Wegener, M. (2014). Numerical parameter study on the impact of Marangoni convection on the mass transfer at buoyancy-driven single droplets. *International Journal of Heat and Mass Transfer*, **71**, 769–778.

Wegener, M. and Paschedag, A. R. (2011). Mass transfer enhancement at deformable droplets due to Marangoni convection. *International Journal of Multiphase Flow*, **37**, 76–83.

Wegener, M., Fevre, M., Paschedag, A. R., and Kraume, M. (2009a). Impact of Marangoni instabilities on the fluid dynamic behaviour of organic droplets. *International Journal of Heat and Mass Transfer*, **52**, 2543–2551.

Wegener, M., Eppinger, T., Baumler, K., et al. (2009b). Transient rise velocity and mass transfer of a single drop with interfacial instabilities. Numerical investigations. *Chemical Engineering Science*, **64**, 4835–4845.

Wolf, S. and Stichlmair, J. (1996). The influence of the Marangoni-effect on mass transfer, Proceedings of the International Solvent Extraction Conference, Melbourne, Australia. In D. C. Shallcross, R. Paimin, and L. M. Prvcic, eds., *Value Adding Through Solvent Extraction*. Australia: The University of Melbourne, pp. 51–56.

Yamaguchi, M. and Katayama, T. (1977). Experimental studies on mass-transfer in continuous and dispersed phases for single drops moving through liquid field at intermediate Reynolds numbers. *Journal of Chemical Engineering of Japan*, **10**(4), 280–286.

Yamaguchi, M., Fujimoto, T., and Katayama, T. (1975). Experimental studies on mass transfer rate in the dispersed phase and moving behavior for single oscillating drops in liquid–liquid systems. *Journal of Chemical Engineering of Japan,* **8**, 361–366.
Zhang, X., Davis, R. H., and Ruth, M. F. (1993). Experimental study of two interacting drops in an immiscible fluid. *Journal of Fluid Mechanics*, **249**, 227–239.
Zheng, H., Ren, W., Chen, K., et al. (2014). Influence of Marangoni convection on mass transfer in the n-propyl acetate/acetic acid/water system. *Chemical Engineering Science*, **111**, 278–285.

4 Membrane-Based and Emulsion-Based Intensifications

4.1 General Introduction

In general terms, membrane separation technology is based on a membrane acting as a barrier between two phases that allows selective transport of one or more components through the barrier. The fluid phase remaining on the feed side of the barrier is referred to as the retentate phase and the fluid phase on the other side of the barrier is referred to as the permeate phase. The driving force for the separation may be based on pressure differences across the membrane, concentration differences, solubility differences, and in some cases the driving force may be an electrical field.

The development of membrane technology over the last 20 years has been rapid, which has been greatly facilitated by advances in material science that have led to the ability to synthesize asymmetric membranes. These developments have led to a much greater range of polymers being applicable for selective separations at both the molecular and macromolecular levels. The ability to apply thin-film technology for the construction of robust membrane–support assemblies has significantly enabled the tailoring of polymer selectivity properties to applications, at the same time ensuring mechanical strength to cope with high pressure differentials necessary for applications such as ultrafiltration and reverse osmosis. Costs have been reduced, allowing membrane separation techniques to be applied to high volume processes such as water treatment, dairy processing, and desalination.

There are many industrial applications involving membrane separations, including production of ultra-pure water, decolorization of supply water, protein separation in the dairy industry, purification of enzymes, kidney dialysis, purification of low molecular weight pharmaceutical products, and separation of whole cell biomass. Most membrane separations are conducted at ambient temperature and therefore tend to be energy efficient. This makes membrane-based techniques particularly attractive for separation of food products such as flavor compounds, and for applications involving viable biomass. They also provide a relatively benign environment for the processing of other sensitive products.

In the context of liquid–liquid systems, the advancement of membrane technology and of polymer chemistry has facilitated design of surface properties that have enabled control of surface wettability. This is highly relevant to phase separation, as described in Chapter 7 discussing intensification of liquid–liquid coalescence. Further developments have facilitated incorporation of highly functionalized liquid solvents into membrane structure, thereby combining the process advantages of membranes with those of versatile extraction properties.

A review of membrane technology is not the main topic of this book but there are areas of overlap between liquid–liquid systems, membrane technology, and process intensification. There are three main areas: (i) liquid membrane systems; (ii) supported liquid membranes, which also includes pertraction; and (iii) membranes for liquid–liquid coalescence.

The deployment of solid membranes for liquid–liquid phase separation is described in Chapter 7, which covers coalescence. In the current chapter, therefore, we will focus on emulsion liquid membranes and related systems, on supported liquid membranes, and on impregnates.

Firstly, some of the fundamentals of liquid–liquid emulsification will be reviewed.

4.2 Emulsions

Emulsions form a large and important category of liquid–liquid systems with applications in separation technology, in reaction engineering, in the formulation of pharmaceutical products, and in the food industry. An emulsion may be described as a liquid–liquid mixture comprising very small drops, dispersed in a second continuous immiscible liquid. Emulsions may be differentiated from other liquid–liquid mixtures on account of their high stability. In many applications, control of stability is an important issue. For example, in the food and pharmaceutical industry there may be a requirement for an emulsion to be stable for a long period of time. On the other hand, in process applications there may be a requirement for the emulsion to be readily "breakable" in order to allow product recovery. Creation and control of emulsions is possible using both physical and chemical methods. For example, Takhistov and Paul (2006) provide a list of possible methods used for the creation of emulsions in the food industry. These include high-speed mixers, colloid mills, high-pressure valve homogenizers, ultrasonic homogenizers, microfluidizers, and membrane and microchannel homogenizers. Intensive physical mixing of two liquids can result in coarse emulsions, but for fine and highly unstable emulsions other techniques are more effective. These include high pressure homogenization, high shear mixing, and the use of ultrasonics. Another approach is by the addition of surfactants, including macromolecules such as proteins that may also enhance emulsion stability (Urbina-Villalba, 2004; Wilde et al., 2004).

Fryde and Mason (2012) provide some useful definitions in the context of liquid membranes and emulsions, as follows:

> Emulsification may be defined as the process by which an emulsion is formed, typically involving flow-induced rupturing of the dispersed phase of larger droplets into smaller droplets within the continuous phase in the presence of a surfactant that inhibits subsequent coalescence.
>
> A microemulsion is a thermodynamically stable lyotropic phase that can be composed of materials similar to those used for emulsions; greater mutual solubility of components enables spontaneous self-assembly (not a nano emulsion).

Ostwald ripening may be defined as the coarsening of an emulsion through migration of molecules of the dispersed phase through the continuous phase; molecules are driven from smaller droplets to larger droplets through differences in Laplace pressures.

Understanding the behavior of emulsions is important in separations for a number of reasons. Perhaps the best known and most important is the need to avoid formation of stable emulsions in liquid–liquid extraction operations. Clean phase separation, either in mixer settling or in continuous contact equipment is a key requirement for efficient product recovery and solvent recycling. Emulsions can also be used in separations, for example as micellar systems or as liquid membranes, though despite much research, industrial application of such systems is yet to be widely available. Emulsions are also very important in the food and pharmaceutical industry and in the production of certain polymers. Control of emulsion quality, expressed in terms of stability, drop size distribution, and specific interfacial area, is especially important. A number of emulsion systems are used in drug formulation for the purpose of controlled release in drug delivery. Emulsions are also generated prior to the production of microparticles and nanoparticles by techniques that include coacervation and interfacial polymerization, again for drug delivery systems (Anton and Vandamme, 2011; Hathout et al., 2010).

Figure 4.1 shows a micrograph of an emulsion of water drops dispersed in gas oil. The drops appear to be almost perfectly spherical and show little evidence of interfacial distension or instability, and are consistent with the existence of a stable emulsion. The droplets may coexist with no significant coalescence due to repulsive surface forces that inhibit drop–drop attraction and associated coalescence. The presence of electrical charge and the electrical properties of the two liquids constitute a significant controlling influence over emulsion stability. As is discussed in Chapter 7 (on coalescence), electrically attractive forces may be harnessed to intensify drop–drop interactions and coalescence. In the case of stable emulsions, the inverse is true and local electrical forces produce drop–drop repulsion, thereby imparting stability.

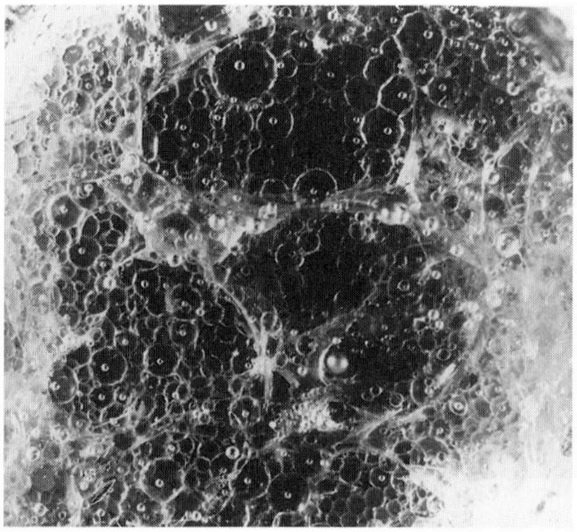

Figure 4.1 Micrograph of an emulsion of water in gas oil in the presence of surfactants.

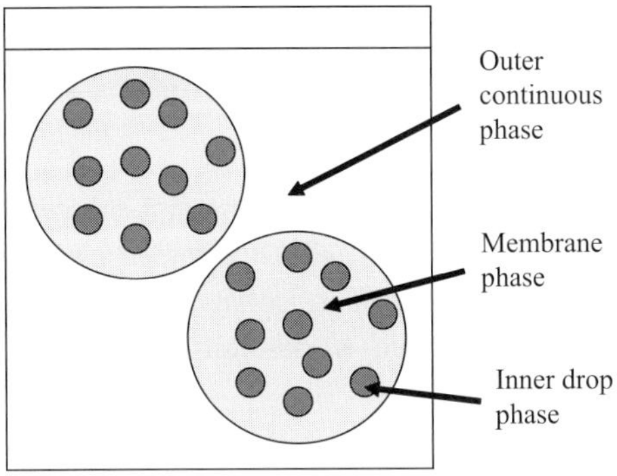

Figure 4.2 Basic principle of the emulsion liquid membrane.

Figure 4.2 shows the principle of the stable emulsion dispersed as larger drops into an external continuous phase.

Some consideration of the thermodynamics of emulsification is relevant to the creation of emulsion liquid membrane systems. A straightforward summary of the thermodynamics is presented by Tauer (2005). In the case of a two-phase, totally or partially miscible, liquid–liquid system, the creation of an emulsion requires the input of energy in order to disperse one phase in the second phase. A good starting point is to consider the Gibbs free energy of a two-phase liquid–liquid system and the change in that free energy involved in the creation of an emulsion. Change in Gibbs free energy in classical thermodynamics is written according to equation 4.1 as follows:

$$\Delta G = \Delta H - T \cdot \Delta S \tag{4.1}$$

where ΔG = change in free energy at constant composition and pressure, ΔH = change in enthalpy, and ΔS = change in entropy.

In the context of emulsion formation, the enthalpy change ΔH may be viewed as the energy input required to achieve a defined emulsion drop size, sometimes referred to as the "binding energy" of the bulk phase. The entropy change ΔS is a measure of the "disorder" resulting from the emulsification process. This may be reflected in the reduction in drop size and the increase in the drop number per unit volume. ΔS would be positive the greater the disorder and can be considered as some measure of emulsion stability – the higher the stability, the greater the entropy change.

In the absence of significant volume change, the enthalpy change ΔH comprises two parts. One contribution to enthalpy change is the energy dissipated to create the dispersion by breaking up the liquid–liquid interface into droplets. The second contribution is the accompanying heat dissipation during the emulsification. If the heat dissipation term is defined as q, then we may express the component of enthalpy change resulting in liquid breakup by the term ΔW, according to equation 4.2:

$$\Delta W = \Delta H - q \tag{4.2}$$

ΔW represents the increase in the energy of the emulsion relative to the state of the two starting unmixed liquids. We may now relate this energy increase to the interfacial tension and the change in overall interfacial area and define the following expression (equation 4.3):

$$\Delta W = \sigma \Delta A \tag{4.3}$$

where ΔW = increase in energy of the system (excluding thermal energy and assuming negligible volume change), σ = interfacial tension, and ΔA = increase in interfacial area.

Another way of defining ΔW, considered by Tauer, is that: "ΔW is the free energy of the interface and corresponds to the reversible work brought permanently into the system during the emulsification process."

Another major factor affecting emulsion stability is the presence of surfactants, and this will be discussed in more detail in the next section.

4.3 Surfactants and Emulsion Stability

As stated in the previous section, the stability of a liquid–liquid emulsion is an important property, the control of which is of significance either when a stable non-coalescing emulsion is required or if emulsion breakage is required as a step in a separation process. Noncoalescing emulsions may be required in many biomedical applications, for example when controlled release of a drug in vivo is required. In that context, the interaction of the chemical environment into which the emulsion is introduced will likely impact the stability of the emulsion. Parameters such as pH, ionic strength, temperature, and presence of surface-active components each could impact the stability of the emulsion. In some controlled release applications, controlled breakdown of the emulsion in vivo and those factors that impact the rate of breakdown require quantification.

In the context of process intensification and liquid–liquid processes, interest in emulsion properties and stability focuses on optimization of emulsion drop size for the enhancement of interfacial area versus ease of phase disengagement. In most industrial liquid–liquid extraction processes, the presence of surface active agents is to be minimized if possible. In some applications, such as in the solvent extraction of products from fermentation media, the presence of surfactants is unavoidable. In other applications, addition of surfactants may actually be desirable in order to promote coalescence and phase separation, and therefore may be considered as having a role in process intensification. On the other hand, if surfactants are added, it may be essential to remove them later in the process as a requirement for product purity. In all cases, understanding of the fundamentals behavior of surfactants in liquid–liquid systems and their impact on the stability of liquid–liquid mixtures is of significant merit.

The role of surfactants in determining emulsion stability is complex and not totally understood. Incorporation of surfactants into a dispersed liquid–liquid system is a dynamic process, which is influenced by rate of surfactant adsorption, and by the transport of surfactant along the interface in creating a gradient of interfacial tension. A brief overview of the relevant fundamentals is now presented.

In principle, the action of surfactants at a liquid–liquid interface is by adsorption with the polar end of the surfactant molecule attached to the hydrophilic part of the interface, and with the nonpolar end attached to the nonpolar, or hydrophobic, side of the interface. The adsorption of the surfactant results in a lowering of the interfacial tension, with a strong relationship between the surfactant concentration and the value of interfacial tension. The relationship is classically expressed according to the Gibbs isotherm (equation 4.4):

$$\Gamma = -\frac{1}{RT}\frac{\partial \gamma}{\partial \ln C} \tag{4.4}$$

where: Γ = excess interfacial concentration of the surfactant (mmol m^{-2}), γ = interfacial tension (mN m^{-1}), C = bulk surfactant concentration (mmol m^{-3}), R = gas constant (J mmol^{-1} K^{-1}), and T = absolute temperature (K).

Inclusion of surfactants into a liquid–liquid system is depicted in Figure 4.3. Here the gradient of interfacial tension generated in the presence of a curved interface is important. The mechanical stresses generated by the distension of the interface create the deviation from a flat interface, resulting in undulations in the interface and creating gradients of interfacial tension. The stresses that are created are strongly associated with emulsion stability. Such gradients are also strongly associated with Marangoni flows, which are usually in the direction of higher surfactant concentration to lower surfactant concentration.

Surfactant molecules congregate close to the liquid interface either as monolayer films (Figure 4.4a), as liquid crystalline films (Figure 4.4b), or as polymers (Figure 4.4c).

The accumulation of the surfactant molecules at the drop interface in the various configurations may lead to stabilization of the drop. The stabilization may be as a result of the inhibitory effect of the surfactant on coalescence. The surfactant may act as a physical barrier to coalescence, preventing adjacent drops from merging. Another possible mechanism is depicted in Figure 4.5, showing the adsorption of polymer on to the surface of a droplet. In this case the presence of the adsorbed polymer results in an electrostatic charge on the surface of the drop. This in turn results in repulsive forces between two or more drops that are in close proximity, thus further inhibiting coalescence.

Emulsified liquid–liquid systems may also be divided into three physical types according to degree of hydrophobicity (Figure 4.6).

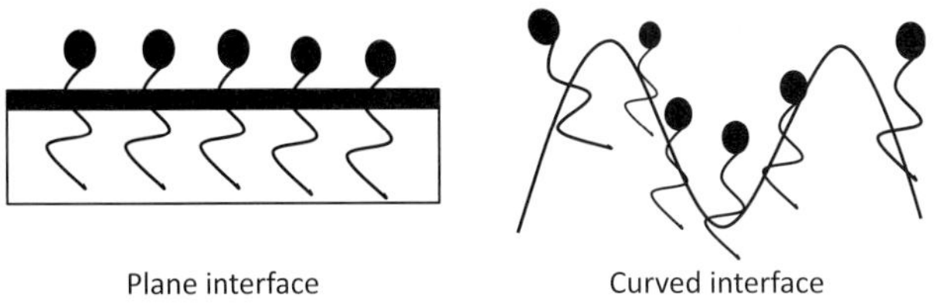

Figure 4.3 Schematic of surfactant at a liquid–liquid interface; comparison between flat interface and a curved interface (Tauer, 2005).

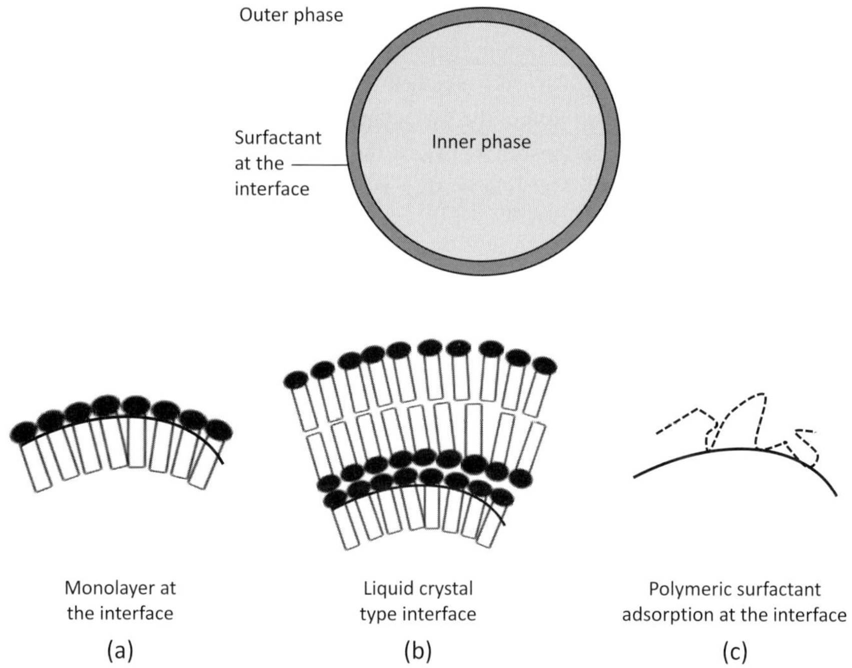

Figure 4.4 Schemes for surfactant attachment at a liquid–liquid interface.

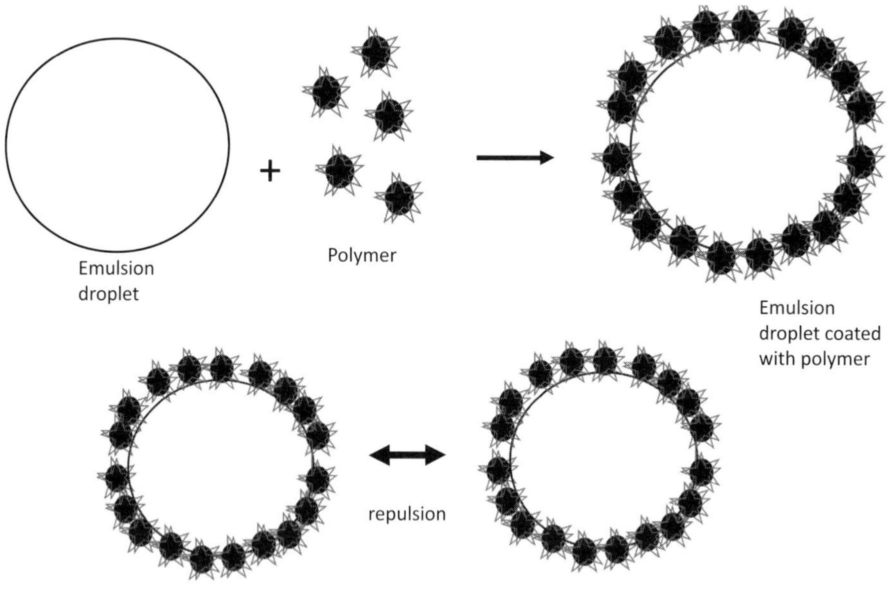

Figure 4.5 Stabilization of an aqueous drop by surface adsorption of a polymer. Adapted from Urrutia (2006)

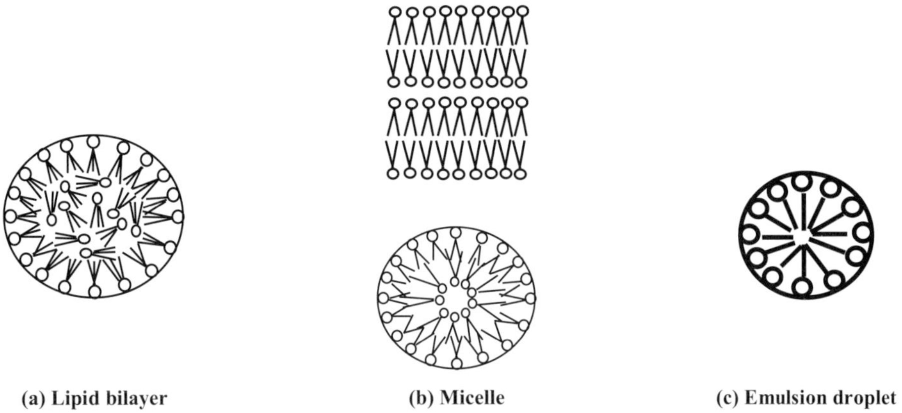

Figure 4.6 Types of emulsified liquid–liquid systems.

1 Emulsions: systems in which hydrophobicity is significantly stronger than hydrophilicity (Figure 4.6c). Such systems tend to form into dispersed single liquid drops suspended or moving a single liquid continuous phase. Such systems may also result in monolayers.
2 Bilayers: on increasing the hydrophilic part of the molecule with the addition of charged groups, water becomes soluble in the polar headgroups, resulting in a bilayer structure (Figure 4.6a).
3 Micelles: a further increase in the hydrophobic/hydrophilic balance results in the formation of micelles (Figure 4.6b). These comprise of spherical or rod-shaped entities, Solutions that are micellar in nature tend to be optically clear, with hydrophobic molecules in rapid equilibrium with the molecules in free solution. Using the known size of the micelle or vesicle it is possible to calculate the interfacial area, which may be directly proportional to the lipid concentration.

Adsorption of surfactant that influences drop–drop interactions is a dynamic process and may be controlled by diffusion rates of the surfactant to the liquid–liquid interface. Urbina-Villalba (2004) provides useful background on the surface adsorption of surfactant. The rate of surfactant adsorption when added to a liquid–liquid system is reflected in the rate of change of the interfacial tension. The relationship between surfactant adsorption kinetics and charge in interfacial tension is, however, not straightforward. The creation of an electrostatic barrier around the surface of drops that are close to each other may be prevalent when high concentrations of an ionic surfactant are present. The presence of an electrostatic barrier not only creates repulsive forces between drops, but also slows down the rate of adsorption of the surfactant. The surfactant may continue to build up until eventually the electrostatic barrier may be overcome, allowing faster adsorption. Addition of electrolyte is a method for reducing the role of the electrostatic barrier and there are many documented studies showing the rapid reduction in interfacial tension that occurs when electrolyte is added to the system.

According to Urbina-Villalba, another important factor affecting emulsion stability and drop size distribution is competition between the surfactant and the average mean free path between drops. The mean free path is a function of the volume fraction of the droplet phase and the spatial distribution of the droplets resulting from the emulsification process.

The mean free path depends on the volume fraction of internal phase, the spatial distribution of the drops resulting from emulsification, and the effective value of the Hamaker constant between the drops. Drop size in emulsion systems may be predicted according to the Kolmogorov relationship, expressed in the correlation presented in equation 4.5 by Mlynek and Resnick (1972).

$$d_{32} = 0.058\left[\frac{\gamma}{\rho V_{\text{imp}} {D_{\text{imp}}}^{3}}\right]^{0.6}(1 + 5.4\phi)D_{\text{imp}} \tag{4.5}$$

where ρ is the density of the continuous phase, γ is the interfacial tension, ϕ is the volume holdup of the dispersion, V_{imp} is the impeller speed, and D_{imp} is the impeller diameter.

Surfactant adsorption also may result in repulsive forces between two or more adjacent drops and this may be quantified as defined in equation 4.6.

$$\Gamma(t) = 2[D_s/\pi]^{1/2}Ct^{1/2} \tag{4.6}$$

where D_s is the surfactant diffusivity, C is the surfactant concentration in the bulk phase, t is the time, and $\Gamma(t)$ is the surfactant excess at time t.

While there is a significant literature on the chemistry and interfacial science associated with emulsion formation and stability, research on the prediction of drop size characteristics in emulsions is much less well developed. Kiss et al. (2011) emphasize that generation of emulsions by static mixers operating in a laminar flow regime is rarely discussed in the literature. Generally static mixers provide a viable alternative to other controlled release particle production methods such as coacervation and spray drying. The production of an emulsion of controlled drop size followed by particle production is shown to be an effective method for controlling ultimate particle size (Moinard-Chécot et al., 2008).

Kiss et al. (2011) present nondimensional correlations for predicting Sauter mean oil drop size for oil-in-water emulsions produced in the laminar flow regime as a function of static mixer operational parameters and liquid properties. Validation of the correlation (equation 4.7) was demonstrated for a wide range of variables. Their approach is based on application of the Ohnesorge number to the emulsification process, as defined in equation 4.8:

$$\frac{\eta_d}{\sqrt{d_{32}\sigma\rho_d}} = 8.2084\left(\frac{\eta_d}{\sqrt{d\sigma\rho_d}}\right)^{0.8338}\left(\frac{u\eta_d}{\sigma}\right)^{0.2501}\left(\frac{\eta_c}{\eta_d}\right)^{0.1328} \tag{4.7}$$

where d is the diameter of static mixer, d_{32} is the Sauter mean drop diameter, η is the viscosity.

$$\text{Ohnesorge number}: \text{Oh} = \frac{\text{We}^{0.5}}{\text{Re}} \tag{4.8}$$

$$\text{Weber number}: \text{We} = \frac{du^2\rho}{\eta} \tag{4.9}$$

$$\text{Reynolds number}: \text{Re} = \frac{du\rho}{\eta} \tag{4.10}$$

Kiss et al. also provide a useful review of earlier drop size correlations for emulsions produced in static mixers. As already stated, many of these are derived for the case of turbulent mixing and based on Kolmorogrov's theory of isotropic turbulence. One of the earliest is that of Middleman (1974), which is as follows:

$$\frac{d_{32}}{d} = C_1 \text{We}^{-3/5} f^{2/5} \tag{4.11}$$

where C_1 is a constant dependent upon the design of mixer, and f is a "friction" factor of the type used typically in pipeflow calculations in turbulent flow, and is written as follows:

$$f = \frac{d\Delta P}{2\rho u^2 L} \tag{4.12}$$

Then using a correlation for the friction factor itself $f = \text{Re}^{-1/4}$, Middleman presented this correlation as:

$$\frac{d_{32}}{d} = C_1 \text{We}^{-3/5} \text{Re}^{-1/10} \tag{4.13}$$

This correlation was validated in later work and further refined by Berkman and Calabrese (1988) who incorporated a term to account for viscosity ratio of the two liquids, thus:

$$\frac{d_{32}}{d} = C_1 \text{We}^{-3/5} \left[1 + 1.38 V_i \left(\frac{d_{32}}{d}\right)^{1/3}\right]^{3/5} \tag{4.14}$$

where

$$V_i = \frac{\eta_d u}{\sigma} \left(\frac{\rho_c}{\rho_d}\right)^{0.5} \tag{4.15}$$

Returning to the recent work of Kiss et al. (2011), Figure 4.7 shows the accuracy of their correlation for prediction of Sauter mean drop size for emulsions comprising water as the continuous phase and solutions of ethyl acetate and benzyl alcohol as the dispersed phase. The liquid properties were adjusted by the addition of polymer PLGA to the dispersed phase and PVA to the aqueous phase. The correlation shown in equation 4.7 was shown to be valid for a range of dispersed phase viscosities (in the range

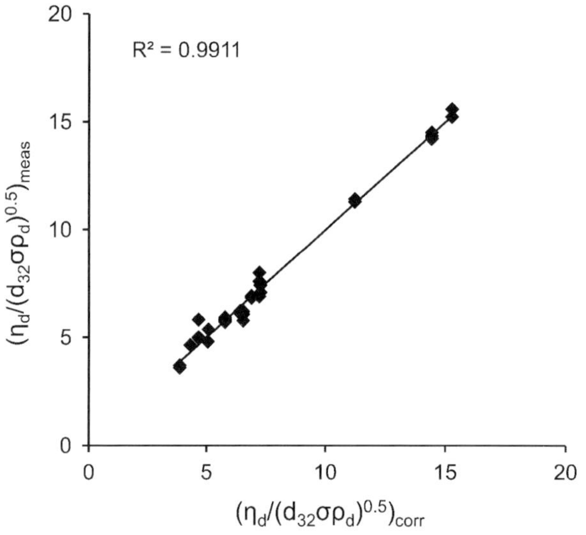

Figure 4.7 Predicted versus experimental Sauter mean drop size in emulsions of water/ethyl acetate, benzyl alcohol/poly-vinyl alcohol.

Reprinted (adapted) with permission from Kiss, N., Brenn, G., Pucher, H., et al. (2011). Formation of O/W emulsions by static mixers for pharmaceutical applications. *Chemical Engineering Science*, **66**, 5084–5094. Copyright (2011) Elsevier B.V. All Rights Reserved

0.0257–0.4105 Pa s), dispersed phase holdups in the range 9–40 volume %, and interfacial tension in the range 0.0019–0.00855 kg s^{-2}.

4.4 Hollow Fiber Technology and Pertraction

Pabby and Sastre (2013) reviewed the state-of-the-art of hollow fiber-based technology as applied to enhancement of extractive separations and reactions. The driving forces for the development of the technology include stricter environmental requirements and legislation. The paper highlights the advantages of membrane-based contactors. These include high turn-down ratios, avoidance of phase separation, reduced equipment and inventory volumes, and feasible operation at high effective phase ratios. There are many examples of promising applications, including extraction of cobalt (Verbeken, Pinoy, and Verhaege, 2009), zinc (Fouad and Bart, 2007), and copper (Ren et al., 2007, 2010). Pabby and Sastre (2013) also highlight application of hollow fiber membrane contactors to gas–liquid processes in addition to liquid–liquid contacting operations. Examples of gas–liquid demonstrations include separations of ammonia, CO_2, SO_2, and H_2S.

Incorporation of ionic liquids into hollow-fiber membranes is another promising area of research and development with potential advantages of stable and consistent performance. This approach is especially interesting given the enormous range of structures available with ionic liquids, making them the ultimate designer solvents

(Chapter 8). The high costs of the liquids developed to date can be mitigated by the low inventories involved in their incorporation into hollow fiber systems.

Wang et al. (2016) provide a comprehensive review of ionic liquid membranes. In describing incorporation of ionic liquids into membrane supports, Wang draws a distinction between supported ionic liquid membranes (SILM) and quasi-solidified ionic liquid membranes (QSILM). The supported ionic liquid membrane comprises a three-phase system in which the ionic liquid is held in place through capillary forces within the pore structure of the membrane. Wang et al. described the preparation methods, which include direct immersion, application of vacuum, or application of pressure. The latter driving forces are often necessary in light of the high viscosity exhibited by many ionic liquids.

Quasi-solidified ionic liquid membranes are prepared using an alternative technique based on incorporation of the ionic liquid during the preparation of the membrane. Preparation by this method reduces the likelihood of leakage of the ionic liquid from the membrane, thus providing a more robust support. It is also reported that ionic liquid membranes prepared in this way can offer very high degrees of selectivity.

The integration of solid membranes into liquid–liquid systems takes a number of different forms. One of the simplest is shown in Figure 4.8, which shows a three-phase system. This comprises phase 1, which is a liquid phase and this may be defined as the feed phase containing the component to be separated. Phase 2 is also a liquid phase, and may be regarded as a stripping solution into which the product component will be recovered. The two liquid phases are separated by a solid membrane. The transport of the solute through the membrane can be by at least two mechanisms. One mechanism may be based on selective solubility of the solute into the membrane itself, with molecular diffusion through the membrane structure as the mode of transport. In this situation, the target component would require a membrane exhibiting a selective

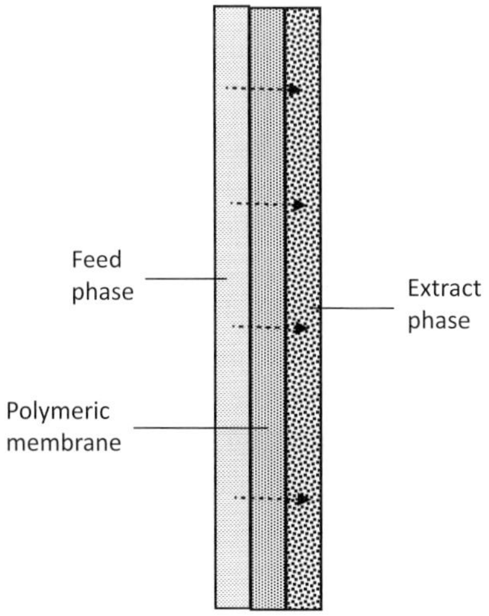

Figure 4.8 Simple configuration of liquid–liquid-membrane.

solubility for the target molecule. Such a system would be very similar to a dialysis system in which the phases 1 and 2 are both aqueous phases.

An alternative arrangement would involve, for example, an aqueous solution as phase 1, and an organic phase as phase 2. Transfer of the solute through the membrane could be in the same mode as described for the wholly aqueous system, as summarized in the previous paragraph. A more common arrangement would involve choice of a membrane that preferentially hosts the organic phase 2. This may be achieved, for example, using a hydrophobic polymer for the membrane, and a nonpolar organic liquid as phase 2. The organic phase is incorporated into the membrane and the target solute enters and is transported through the membrane on account of its affinity for the organic phase that is immobilized within the membrane. This system is referred to as a supported liquid membrane system, and is sometimes referred to as a pertraction system. The membrane polymer itself may be inert to the solute and may not exhibit any selectivity toward the target but acts to host the extracting solvent.

The extracting solvent, as in conventional liquid–liquid extraction, is selected on the basis of its selectivity for the component or components in the feed phase. The chemistry of extraction in most cases is very similar to that in liquid–liquid extraction (Parhi, 2013). Extractants are chosen based on chemical formation, solvation, and ion-pair formation and many familiar extracting agents are potential candidates for incorporation into a solid membrane support. Parhi lists examples that include the LIX family of extractants, Cyanex 272, and D2EHPA, which perform based on their chelating properties. Also highlighted are amine-based extractants, which operate through ion-pair formation. Examples include Alamine-336, Aliquat-336, and Alamine-304 used on supported liquid membranes for the extraction of chromium, vanadium, and molybdenum.

Often the membrane will be of the microfiltration type and on account of their hydrophobicity and structure these provide a large contact area between the feed phase 1, and the organic extractant. The membrane polymer is chosen to provide a stable platform for the organic solvent, preventing significant leakage of the organic phase into the feed stream. The organic solvent is usually held in place by capillary forces that exist within the pores.

The advantage of this type of system is that the need for liquid–liquid phase separation is avoided on account of the membrane barrier between the two participating liquids. The control of hydraulic flows on either side of the membrane is much easier to achieve compared to that in a conventional liquid–liquid contactor. Optimization of the process is thus improved. An additional advantage is the high feed : solvent ratios that are achievable, resulting in overall intensification of the process and reduction in equipment size. Another variation is where the membrane is impregnated with a highly selective solvent component, differing from either the feed phase or the stripping phase. As referred to in Section 4.12, of current interest is the development of the supported liquid membrane concept for exploiting ionic liquids as excellent designer solvents but which come with the limitation of high cost.

In all cases, the solid membrane supports generally comprise hydrophobic microporous supports but which must be carefully matched with the surface tension properties of the solvent. This is necessary in order for the solvent to effectively wet the internal surfaces of the membrane pore structure. Parhi (2013) reviewed the general principles of supported liquid membrane extraction and states that the solid membrane support generally is polymeric in nature with either hydrophobic/hydrophilic properties. The degree of control now available for membrane synthesis provides opportunities for the application of both heterogeneous and homogeneous structures. Anisotropic membranes, symmetric structures, asymmetric structures may also be considered, as may different types of polymers incorporating charged and polar sites. Chemical and thermal stability are also important requirements. Commonly available polymers reviewed as suitable for supported liquid membrane extraction include PTFE, polysulfones, and polypropylene. In the case of polysulfones, flexible functionality is possible by the variation of the functional groups attached to the polymer backbone. These may include monophenyl and bisphenyl or phenoxy groups. The physical properties of the membranes may also be tailored to the required application by development of composite membranes through processes such as lamination, extrusion, stretching, phase inversion, and annealing.

Supported liquid membranes may also be applied where the extraction mechanism is based on facilitated transport, as detailed below. There are established protocols, see for example Wang et al. (2016), for the impregnation of membrane polymers with solvent, with application of pressure as a means of intensifying impregnation. Successful operation of supported liquid membranes involving solvent impregnation requires close control of the transmembrane pressure to prevent loss of the solvent from the membrane due to high transmembrane pressure. The choice of impregnating liquid should be based on selectivity for the target molecule relative to both the feed phase and on the stripping phase, so that not only can efficient transport into the membrane phase be achieved but also that the target molecule can be effectively recovered into the stripping phase. Favorable transport properties are also very important.

In summary, the arrangement shown in Figure 4.8 is known as pertraction and combines the advantages of solvent extraction while at the same time avoiding direct contact of the two liquid phases. There are a number of advantages that accrue from this arrangement. In particular, the formation of stable emulsions is avoided, as is the need for separate arrangements for phase disengagement and coalescence. Liquid transport through the membrane may also be regulated by imposition of a controlled pressure gradient across the membrane. Isolation of the two liquid phases from each other may also be especially advantageous for separation of biological products from fermentation liquor when biomass viability requires preservation with minimal exposure of living microorganisms to a potentially toxic environment. A related issue that a pertraction system can address is product inhibition, which can be minimized by isolation of biomass from the product components on account of the membrane. Another advantage is the avoidance of liquid–liquid phase separation by coalescence, as in conventional solvent extraction. An extension of pertraction involving live biomass being immobilized on one side of the membrane is also a possibility.

A hollow fiber bioreactor concept is also possible, in which biomass is immobilized on the "shell side" side of the module, with extracting solvent fed through the tube side.

The hollow fiber system also has the advantage of separate control of each of the liquid streams, thus providing greater flexibility than would be the case in conventional countercurrent liquid–liquid process equipment where consideration of liquid entrainment, flooding, and back-mixing constrains the range of operating conditions.

Membrane-based techniques require careful consideration of the properties of the membrane, both in terms of transport properties and surface characteristics. Hydrophobicity and hydrophilicity are important properties, especially when biological materials are processed. For example, a hydrophobic surface will tend to exclude microorganisms, which may be desirable for biological product separation.

There are a number of demonstrations of the application of membrane and hollow fiber-based pertraction separations, especially those requiring mild conditions, which include biological product processing. Early demonstrations include ethanol extraction from yeast-based fermentation solutions (Matsumura and Markl, 1986). A study from the same period demonstrated a multi-membrane system for ethanol recovery, using tri-n-butyl phosphate (TBP) as extracting agent (Cho and Shuler, 1986). TBP is a versatile solvent, though it is toxic to many microorganisms so its application to biological product separations involving direct contact with live biomass or enzymes is limited. The specific microorganisms also can have a profound effect on the success of application in membrane-based systems. The work of Matsumura and Markl (1986) showed that deployment of pertraction membrane separation for systems involving filamentous organisms may be problematical on account of mass transfer limitations associated with film formation on the membrane surfaces. Use of hydrophobic membranes may provide a solution in such cases. Comparison of ethanol extractive separations from fermentation solutions based on *Zymomonas mobilis* (a filamentous organism) with those based on *Saccharomyces cerevisiae* (a flocculating yeast) showed this to be the case. Other studies using oleyl alcohol and dibutyl phthalate (Kang, Shukla, and Sirkar, 1990) successfully exploited the high distribution coefficient for ethanol exhibited by these solvents in a hollow fiber-based pertraction system.

Interest in acetone/butanol/ethanol (ABE) production by fermentation has recently received further stimulus through the concerns at CO_2 emissions from the use of fossil fuels. The production of platform chemicals from renewables is of significant interest for the chemical industry and ABE-based technology is one of the leading contenders. The development of hollow fiber-based pertraction technology is a promising option for benign product separations and purification. Use of nonporous tubular silicone membranes has been demonstrated for the in situ extraction of ABE using oleyl alcohol and propylene glycol as solvents (Grobben et al., 1993; Shukla, Kang, and Sirkar, 1989).

This early work, both experimental and theoretical, showed that the total productivity during the ABE fermentation could increase by over 40% in a hydrophobic hollow fiber-based tubular fermenter extractor with 2-ethyl-1-hexanol as the extractant. Other documented examples include lactic acid production (Hano et al., 1996) using both hydrophilic and hydrophobic membranes, and recovery of 1-butanol using oleyl alcohol as the extracting agent (Groot, Vanderlans, and Luyben, 1992).

4.5 Hybrid Liquid Membrane Systems

Yet another variant of liquid membrane systems is the arrangement shown in Figure 4.9. This system comprises two solid membranes separated by an inner solvent phase, with the feed and stripping liquids flowing on the outsides of the membranes. The inner solvent, present as a liquid, provides the selectivity for the target species. The reverse situation can also be considered, in which the feed liquid flows through the center channel of the fiber with the extracting solvent on the other side of the polymeric membrane.

It is also possible to further vary this arrangement, for example using a hollow fiber membrane system in which fibers are configured to perform the functions of both extraction and stripping, as shown in Figure 4.10. In this configuration, sometimes referred to as the hollow fiber contained liquid membrane, the feed flows through the tube side of a proportion of the hollow fibers, while the stripping solution flows through the tube side of the fraction of fibers. The fibers are surrounded by the extracting solvent, which remains on the shell side.

The main advantage of the supported liquid membrane system and the hybrid liquid membrane systems is that the dispersion and coalescence cycle of conventional liquid–liquid extraction is avoided. The need to achieve controlled dispersion is avoided, as is the risk of creation of undesirable stable emulsions. The latter phenomenon is a particular issue in the case of certain separations in biological systems where surfactant contamination and the presence of solids and colloidal species can prevent successful application of solvent extractions. Another significant advantage that is especially relevant to intensification is that the inventory of the highly selective species that are impregnated or incorporated into the membrane system can be very small. This is a major economic and environmental consideration.

As is the case with many membrane-based systems, the stability of supported liquid membranes needs to be considered when considering the feasibility of application of

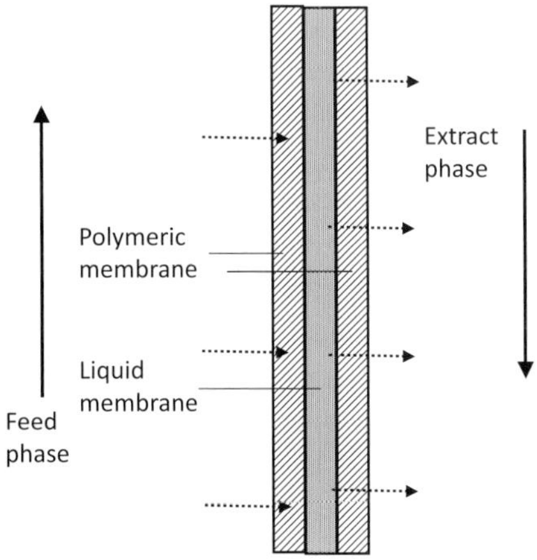

Figure 4.9 Hybrid liquid membrane.

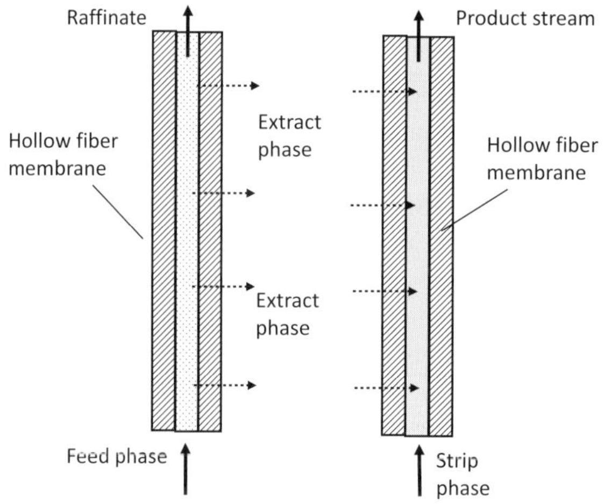

Figure 4.10 Principle of hollow fiber contained liquid membrane.

these systems. Parameters such as osmotic pressure gradient, water solubility, interfacial tension, and polymer wettability are all potentially important factors in determining membrane stability. In the case of hybrid liquid membranes (Figure 4.9), strong adhesion between extractant and membrane is a requirement and this generally leads to increased stability. A specific issue sometimes encountered in bio separations is affinity between whole cells having negative surface charge and the membrane itself. Anionic exchange membranes are especially susceptible to this and it may lead to fouling of the membrane. In some cases this problem can be avoided by coating the external surface of the membrane with a negatively charged layer.

There are important differences between types of liquid membrane systems and some comparisons and contrasts are worthy of consideration. In the case of the simplest emulsion liquid membrane system (ELM), in which drops of the emulsion are dispersed into a continuous liquid phase, breakdown of a small number of the internal drops should not significantly affect the efficiency of the separation. In contrast, the failure of a small number of pores in the supported liquid membrane system would rapidly result in contamination and substantial loss of performance. Hybrid liquid membranes (HLM) (Figure 4.9) are generally acknowledged to be less prone to cross-contamination. Supported liquid membranes (Figure 4.8) are often associated with larger phase volumes compared with, for example, emulsion liquid membranes that are characterized by very high interfacial areas and thus intensification of mass transfer. Emulsion liquid membrane systems are generally highly efficient in terms of solvent inventory. Another significant advantage of emulsion liquid membrane systems is that the stripping phase (i.e. the internal drop phase as shown in Figure 4.2) is of significantly smaller volume compared with the feed phase and thus the recovered product may be at much higher concentration. This is dependent on the nature of the stripping mechanism, which should ensure that the stripped form of the target product has low solubility in the membrane phase. The ability to achieve concentration in addition to separation represents an advantage of the ELM system compared with either the hybrid system or the supported liquid membrane system.

A drawback to supported liquid membrane technology is that incorporation of scrubbing operations is not feasible. In contrast, the hybrid liquid membrane system involves the extractant as a continuous stream, which allows scrubbing, stripping of several extracted compounds in a sequence, reducing the build-up of impurities in the extractant.

From a process engineering angle, the development of ELM-based technology in principle should be straightforward since the principle of dispersion, transport, stripping, and product recovery is essentially the same as for a conventional liquid–liquid extraction process. Commercial application of supported liquid membrane technology and hybrid liquid membrane technology is more challenging, requiring significant investment and development of new hardware and materials.

4.6 Liquid Membrane Applications in Bioprocessing

The separation and purification of biological products often requires process conditions that are relatively benign, both physically and chemically, in order to maintain the functionality of the final product. Operations at close to ambient temperature and atmospheric pressure are normal requirements. Another constraint on biological product separation is that target compounds are often present in primary feed solutions at low concentration and with many other components present. This, combined with high purity specifications, requires highly selective separation tools. Liquid membrane systems offer significant scope for bio separations. Products with potential for recovery by liquid membrane-based separation include molecules having ionizable functional groups such as amino acids, phospholipids, and organic acids such as lactic acid and citric acid.

Boey, Delcerro, and Pyle(1987) discuss the application of liquid membrane systems for in situ extractive fermentation of citric acid. Application to extractive fermentations can be particularly useful since there is an essential requirement that the separation technique not only recovers the product in functional form, but also that enzymes or viable biomass that are to be recycled are not denatured in the separation. Interest in extractive fermentations and extractive biotransformations is stimulated by the ability to reduce the effects of product inhibition on both live biomass and active enzymes, by continuous product removal (Chaudhari and Pyle, 1987). The results of the latter, recently further confirmed by Konzen et al. (2014), show effective separation of citric acid but also highlight the swelling of the internal drop phase in an ELM system as the extraction process takes place. Swelling is explained by the osmotic gradient which exists between the internal and external aqueous phases. Swelling and ultimate rupture of the internal drop phase would lead to significant reduction in total capacity of the ELM, and possible premature breakdown of the emulsion droplet and membrane phases. Another challenge in these systems is water permeation across the membrane phase. This can be reduced by the use of nonpolar solvents. This was demonstrated by

Chaudhari and Pyle (1987) and by Boey et al. (1987), who showed that tertiary amines performed well as extractants for both citric and lactic acid using a predominantly aliphatic solvent such as heptane and other paraffinic solvents. The stripping phase used for organic acid recovery was aqueous sodium carbonate. Extraction efficiency improved with decreasing external phase pH, showing that the undissociated acid is taken up. Konzen et al. conducted similar studies using emulsion liquid membranes comprising aqueous solutions of sodium acetate as the inner phase, mixtures of Alamine 336 and ECA 4360 dissolved in Exxsol D240/280 as the membrane phase, and aqueous citric acid as the feed phase.

Similar studies show the potential for ELM applications for lactic acid separation, including those of Chaudhari and Pyle (1987), Patnaik (1995), and Lazarova and Peeva (1994). Separation efficiencies of approximately 90% are possible for the recovery of lactic acid from fermentation broths with high degrees of purity and with concentration factors of three-fold. In the case of amine-based extractants, the pK_{HA} value of the acid species is an important determinant for the selectivity of the separation. Other studies using ELM systems for lactic acid separation include those of Scholler, Chaudhuri, and Pyle (1993), which successfully demonstrated the extraction of lactic acid from clarified fermentation broth using a tertiary amine carrier, although extraction efficiency was lower compared with experiments conducted using pure lactic acid. The reduction was explained by interference by the presence of other organic acids and similar coproducts.

Other applications in bioprocessing include the extraction of beta-lactam antibiotics such as penicillins and cephalosporins. The sensitivity of these molecules to chemical and physical conditions make the use of both emulsion liquid membranes and supported liquid membrane potential techniques for recovery and purification.

The application of supported liquid membranes was reported by Friesen, Babcock, and Chambers (1986) for the recovery of citric acid from fermentation liquor using mixtures of tertiary amines (primary extractant), long-chain alcohols (in the role of modifier), and hydrocarbons (as diluent). The liquids were supported and immobilized on a porous substrate and high purity recovery and separation of the citric acid was reported. Other related work by Rockman, Kehat, and Lavie (1996) demonstrated the possibility of further enhancing citric acid recovery in a supported liquid membrane system by the imposition of temperature gradients across the membrane.

A further extension of hollow fiber-based extraction was reported by Yang et al. (2007). Removal of a thiol from a simulated naphtha using a reactive extraction technique was demonstrated in an ultrafiltration contactor using cellulosic hollow fiber membranes. The use of reactive extraction resulted in very high partition coefficients approaching 10^3 for the distribution of the thiol between an alkaline solution and the oil phase.

Another interesting area of development involving membrane systems for extraction is in the exploitation of near critical liquid solvents, based on dense-phase gases as extractant on one side of the membrane (Gabelman, Hwang, and Krantz, 2005). While this chapter is not specifically concerned with near critical fluid extraction, the benefits of these fluids include high mass transfer rate (on account of low density), tunable properties (that are available by variation of temperature and pressure or by the addition of entrainers), and easy product recovery (by pressure let down, heating, or a combination of both).

4.7 Membrane Emulsification

A promising technique for the production of emulsions having controlled size distribution is by the use of porous membranes. Santos et al. (2015) present a comprehensive study of the technique. The basis of the technique is that the phase to be dispersed is pumped under pressure through a porous membrane into an agitated continuous phase that is situated in the "permeate" side of the membrane (Figure 4.11).

There are two methods of membrane-based emulsification that may be used. The first method is based on direct permeation of the pure liquid to be dispersed, through the membrane. The passage of the liquid through the fine pore structure of the microporous membrane results in exposure of the liquid to high shear forces that serve to break up the liquid into fine drops that emerge on the permeate side of the membrane, whence they are detached from the membrane surface on account of the agitated continuous phase present in that region. The second approach is to pump a premixed mixed phase comprising a relatively coarse dispersion of drops in the continuous phase liquid through the membrane. The use of a premixed phase has several advantages, including higher fluxes of the dispersed phase and significantly smaller droplet to pore size ratios. The ratio of final drop size to pore size is a measure of the efficiency of the breakup, with lower values associated with increased reduction in drop size. Santos et al. (2015) quote typical droplet to pore size ratios for direct membrane emulsification in the range 2–50. This compares with quoted range of 0.6–10 for premixed membrane emulsification. It is also stated that process conditions are easier to control in the case of the premixed membrane emulsification. One disadvantage referred to is higher polydispersity, which may be observed in the case of premixed membrane emulsification.

Another positive aspect of this technology is energy efficiency. The energy requirements of membrane emulsification are significantly lower when compared with other

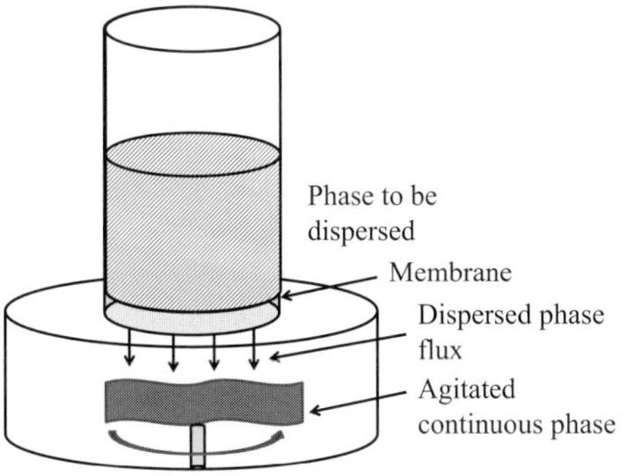

Figure 4.11 Principle of membrane emulsification.

mechanically based techniques such homogenizers. Santos reports energy densities required to achieve a mean droplet size of 1–10 μm using premixed membrane emulsification typically in the range 10^4–10^6 J m^{-3}, compared with values reported by Karbstein and Schubert (1995) for rotor stator devices and high pressure homogenizers in the range 10^6–10^8 J m^{-3}. A range of membrane materials are listed by Santos et al. (2015) for liquid–liquid emulsification, including porous glass, nitrocellulose-based membranes, polycarbonate polymeric membranes, nylon, and nickel microsieves.

4.8 Membrane-Based Extraction Processes/Liquid Membrane Processes

As already stated, a significant variation from conventional two-phase liquid–liquid extraction is to be found in liquid membrane systems. Emulsion liquid membrane systems (ELM) are now considered in further detail. The systems are liquid-based systems usually comprising three distinct liquids, a minimum of two of which are essentially immiscible.Liquid membrane systems are referred to as emulsion liquid membrane systems since two of the liquids are dispersed in such a way as to create a relatively stable emulsion. Such systems came to prominence in the 1970s in light of the work of Li (1968). Understanding of the formation and breakage behavior of liquid–liquid emulsions is central to the useful development of emulsion liquid membrane systems. Liquid membranes that contain only liquid phases are known as (double-) emulsion liquid membrane (ELM) systems. The basic principle of a liquid membrane system is shown in Figures 4.12–4.17. The first step is the creation of an emulsion, which could typically be a water-in-oil emulsion, comprising drops of aqueous phase encapsulated in larger drops of the oil phase (Figure 4.12). In some cases this may require the addition of a surfactant. The oil phase thus forms a membrane layer surrounding a number of smaller aqueous drops. The membrane phase consists of three major components : the carrier, a water-in-oil emulsion stability surfactant, and the membrane solvent. The oil/water entity of drops is now dispersed into an aqueous feed phase that is not dispersed and forms a new mixed phase (Figure 4.13). In practice, this

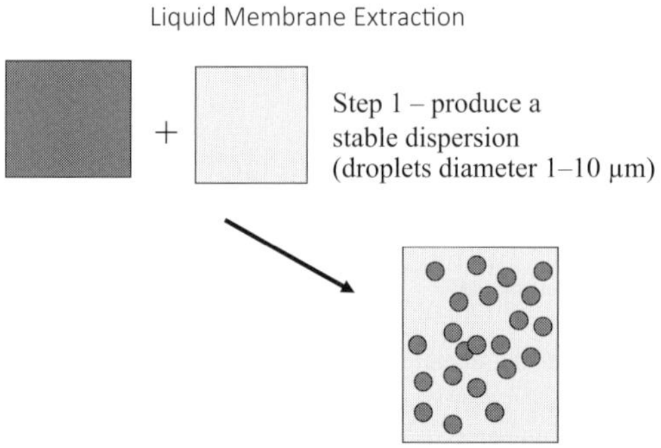

Figure 4.12 Step 1: produce a stable dispersion (droplets diameter 1–10 μm).

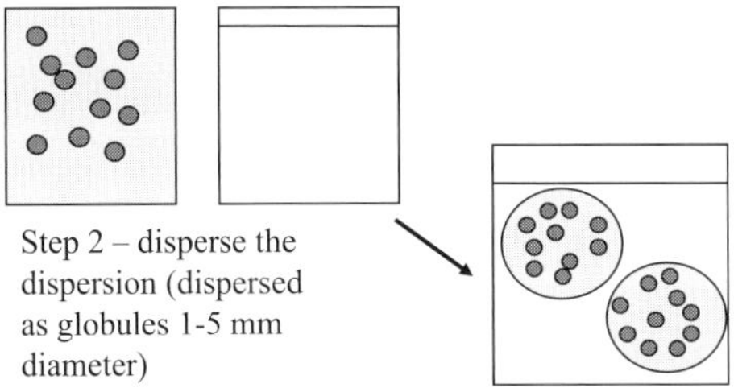
Step 2 – disperse the dispersion (dispersed as globules 1-5 mm diameter)

Figure 4.13 Step 2: disperse the dispersion, dispersed as globules 1–5 mm diameter.

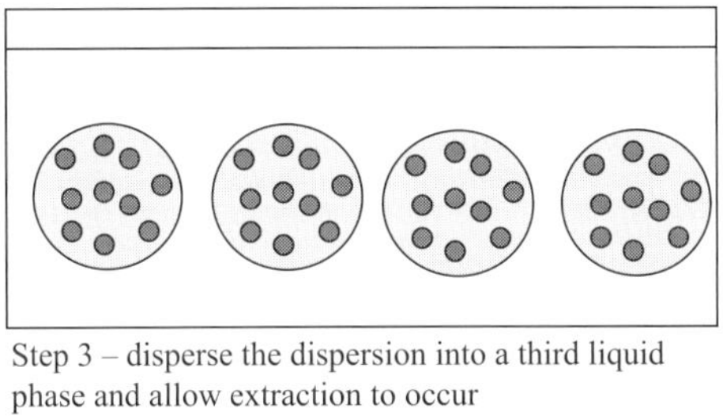
Step 3 – disperse the dispersion into a third liquid phase and allow extraction to occur

Figure 4.14 Step 3: disperse the dispersion into a third liquid phase and allow transport to occur.

new mixed phase is an unstable emulsion but one that may be regarded as a single liquid.

Figure 4.14 shows the third step in the sequence of operations, in which the emulsion is dispersed as droplets into a third liquid phase, which would normally be a solution containing the component to be separated (extracted) and impurities.

There are also examples of liquid membrane systems in which a carrier molecule forms a complex with the desired target molecule that is to be transported into the internal drop phase. The complex is soluble in the membrane phase, thus facilitating diffusion through the membrane and to the interface with the inner drop phase. At this point, decomplexation occurs and the target transfers into the inner drop phase and is essentially recovered there. The complexing agent back-diffuses through the membrane phase and becomes available back at the interface between the membrane phase and the outer "feed" phase. This is detailed later in this section.

With reference to Figure 4.15, in the simplest configuration of an emulsion liquid membrane separation, a solute A, initially present in an external aqueous phase for example, is transferred across the first phase boundary into the organic membrane film phase. The solute A diffuses through the membrane and then is transferred across the second liquid–liquid interface into the internal drop phase. Back diffusion is prevented by a chemical reaction of A as it enters the internal drop phase. The reaction is designed to result in a product that, while retaining the basic characteristics of A, is insoluble in the organic membrane phase. As in conventional solvent extraction, the membrane phase may be a pure organic liquid, or a mixture of extractant and diluent. The overall selectivity for a particular solute depends on its solubility in the membrane phase, and

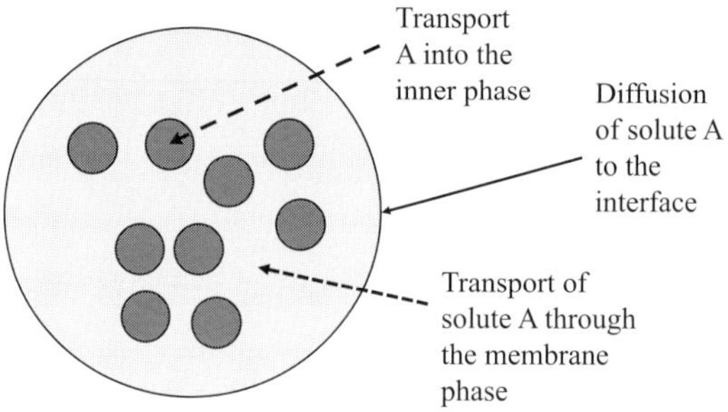

Figure 4.15 Schematic showing solute transport through the liquid membrane droplet entity.

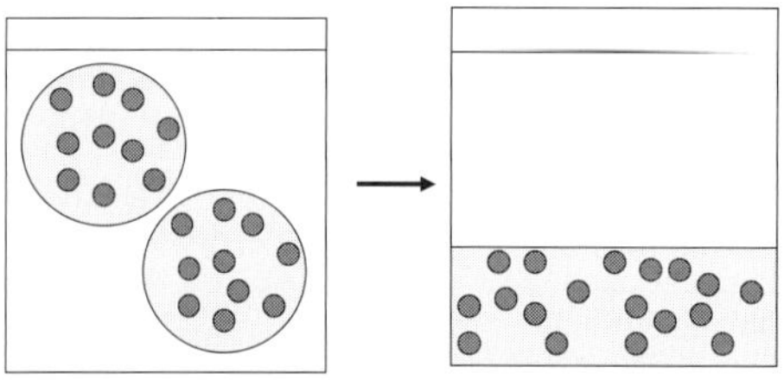

Figure 4.16 Step 4: disengage the membrane emulsion phase from the main continuous phase.

on its transport properties when present in that phase. In cases of mixtures of low molecular weight solutes requiring selective separation, the diffusional properties may be similar and therefore the efficiency of separation is highly dependent on the selectivity of the membrane phase for the solute. As in other liquid–liquid processes, the rates of separation are dependent on driving force, availability of interface, and convection. Upon contact with the inner drop phase uptake may occur followed by reaction depending on the chemical composition of the inner drop phase. The internal phase, phase 3, is emulsified at high shear in the membrane phase, to form a dispersion of droplets (diameter 1–10 μm, typically). The solvent is chosen to provide maximum solvation of the carrier and the solute–carrier complex, so neither component will be dissipated into the aqueous phases. It should have a low solubility in water and a reasonably low viscosity to maximize solute diffusivity while maintaining globule stability. The emulsion is then itself dispersed as globules 1–5 mm diameter in the continuous external aqueous phase. The contact may be performed either in a stirred tank or in a continuous column contactor. The membrane is the continuous phase within the droplet.

The next step in the sequence is shown in Figure 4.16, where the transport of product to the inner drop phase is complete and the emulsion phase, now enriched with the desired product, is separated from the continuous feed phase. This step is analogous with the phase disengagement step in a conventional solvent extraction.

The final step is the breakage of the emulsion (Figure 4.17).

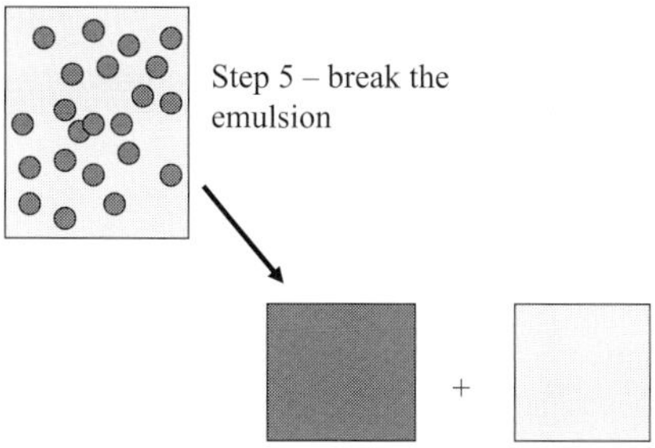

Figure 4.17 Step 5: emulsion breakage and product recovery.

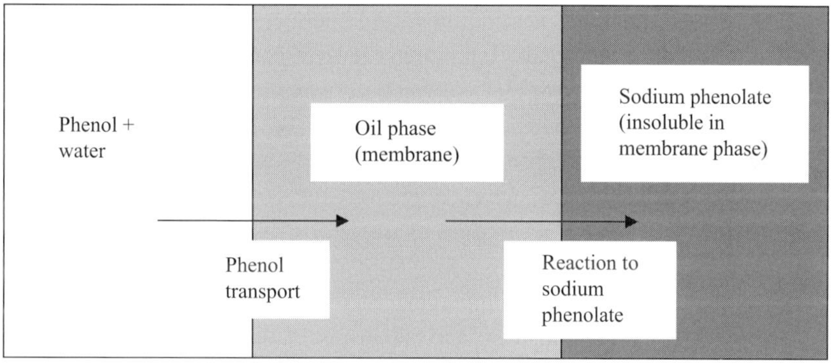

Figure 4.18 Summary of the principle of liquid membrane extraction; phenol removal from water.

Emulsion liquid membrane systems of the type described in general terms here may be useful for the intensification of nonequilibrium separations and reactive separation processes. The membrane phase itself acts usually as an organic carrier with a high preference for a specific component in a mixture.

The internal drop phase or stripping phase is designed to prevent back-extraction of the solute. This may be achieved by ensuring that the solute is rendered insoluble in the membrane phase by including a reagent that either ionizes the solute or facilitates another chemical reaction to produce an insoluble derivative. Extraction of an organic acid neutralized by a base to form an insoluble salt would be an example of such a reaction. Upon completion of the extraction, the emulsion and the aqueous phase may be separated, for example by gravity settling and coalescence. Then the emulsion is broken to release the stripping phase and hence recover the extracted solute. The solvent and carrier may be recovered and recycled.

A common example of emulsion liquid membrane extraction is the separation of phenol from dilute aqueous solution see below (Cahn and Li, 1974) (Figure 4.18).

phenol + sodium hydroxide → sodium phenolate

A significant difference from conventional liquid–liquid extraction is that the cycle of extraction and stripping is occurring almost simultaneously in the same continuum of liquid phases. The extraction step is occurring at the first interface between the external feed phase and the macro emulsion droplet. The stripping is occurring at the inner or second interface between the membrane phase and the inner microdroplets.

4.9 Facilitated Transport

Figure 4.19 shows a different mechanism, facilitated transport. Here the species to be separated, shown here as ion A^+ in the left-hand region of the figure, complexes with a selective carrier or extractant, shown here as species R. At the left-hand interface, the ion A^+ forms a reversible complex with R. The complex A^+–R is selectively soluble in the liquid membrane phase (shown in the center of the figure) and diffuses through the membrane to the right-hand liquid–liquid interface. At this point the complex is broken since the inner liquid phase (shown in the right-hand section of the figure) contains (in this case) a proton donor that preferentially complexes with R to form R–H^+, releasing species A^+ into the inner liquid phase. Species A^+ in this example remains in the inner phase until the system is de-emulsified. The net overall effect of the process in this case is an ion exchange of A^+ for H^+.

Figure 4.20 shows an outline example for the recovery of copper ions from an aqueous feed phase (Sengupta, Sengupta, and Subrahmanyam, 2006). The copper ions form a complex with the complexing agent, shown here as R_2. The CuR_2 complex is mobile in the membrane phase and is transported down the concentration gradient to the interface with the inner drop phase that comprised sulfuric acid. The copper ion is decomplexed from R_2 through the formation of copper sulfate and is retained in the inner drop phase. The R_2 complex agent binds with the released hydrogen ion and is transported back to the interface with the feed phase. At this point the hydrogen ion associated with the R_2 complex exchanges with more copper and the process is repeated. The overall net effect is analogous to ion exchange of copper ions for hydrogen ions.

A number of complexing agents (R) have been successfully reported for copper extraction using emulsion liquid membranes. These include the LIX family of solvents (Frankenfeld, Cahn, and Li, 1981; Marr and Kopp, 1982; Volkel, Halwachs, and

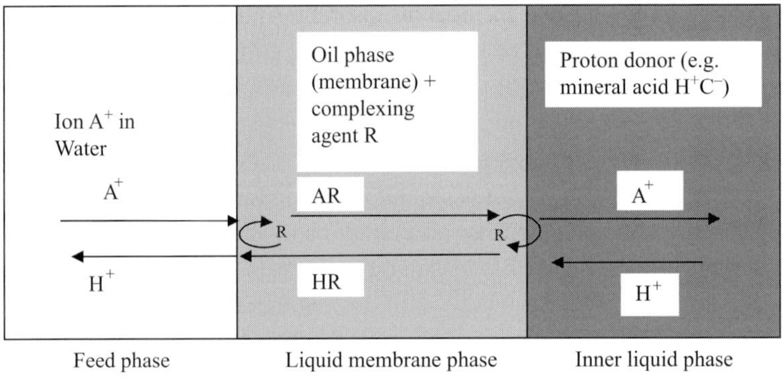

Figure 4.19 Example of facilitated transport in a liquid membrane system.

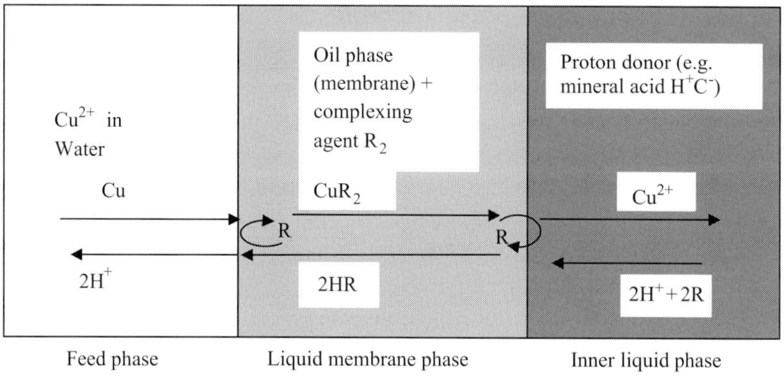

Figure 4.20 Example of facilitated transport of copper ions in a liquid membrane system.

Schugerl, 1980), D2HEPA (Strzelbicki and Charewicz, 1980), and SME 529 (Teramoto et al., 1983).

Facilitated transport in an emulsion liquid membrane system implies that the choice of the carrier species is critical as the carrier molecule and the complex must be exclusively soluble in the membrane phase but also must be capable of dissociation as it contacts the internal drop phase. This is controlled by the chemical composition of the inner drop phase. In many cases, the complex formed between the target molecule or ion and the complexing agent will exhibit smaller diffusivity compared with the target itself, so a high solubility in the membrane phase should favor a high driving force for mass transfer.

Another variation of facilitated transport is "co-transport." This involves two solutes in the initial feed phase. One of the solutes, for example solute A, is the target, the other, B, is added to facilitate the formation of a labile complex in the membrane phase. The complex diffuses across the membrane to the stripping phase, when the complexing agent is released and recycled.

Emulsion liquid membrane extraction in principle has several advantages over conventional solvent extraction. Transport kinetics can be very fast, in view of the very thin membrane film and very small drop sizes that may be achieved in the emulsification process. Secondly, the interfacial areas can be very high, consistent with high rates of interfacial mass transfer. The chemical solvent selectivity required is incorporated in the choice of membrane phase and in extraction operations such solvents tend to be high cost. In the case of ELM systems, since the membrane phases may represent a small fraction of the total liquid present, the inventory costs and inventory hazard can be significantly reduced. Another positive feature is the integrated nature of the system, whereby extraction and stripping are achievable within the same droplet, with a very small volume of strip phase. The high efficiency of the membrane transport may also allow the use of high cost reagents since the inventory volumes per unit throughput are much smaller than in a conventional extraction.

Possible drawbacks to commercial application are the requirement of emulsion formation and breakage at large scale. At large scale it may be difficult to achieve an emulsion of uniform properties necessary for efficient transport. Another potential challenge is membrane instability. This can result in leakage of the inner drop phase and loss of the stripping agent. A further aspect of behavior not properly understood is volume changes in the internal drop phase that occur during the uptake process and also can be exacerbated by transport of the external solvent phase (water for example) into the membrane phase and the internal drop phase. Transported water may also carry impurities resulting in loss of selectivity.

4.10 Colloidal Liquid Aphrons

Another relevant subset of liquid membrane systems is the colloidal liquid aphron (CLA) system. Referring to Figure 4.21, a colloidal liquid aphron comprises a small liquid nonaqueous droplet that is surrounded by a thin layer of emulsion. The thin layer typically comprises an aqueous film that is stabilized by the inclusion of a mixture of nonionic and ionic surfactants. The entire droplet essentially forms the dispersed phase in the surrounding continuum, typically an aqueous phase. A comprehensive text on these system is by Sebba (1987). As is the case with emulsion liquid membrane systems, the application of colloidal liquid aphrons require suitable stability of the drop emulsion layer system and favorable partition properties for the target molecule. The size and stability of the aphron is controllable by variation of the surfactant composition and the amount used in the preparation of the aphron. One advantage of colloidal liquid aphrons is that they can incorporate polar solvents such as lower molecular weight esters (for example, ethyl acetate) that have significant solubility in water. This provides significant flexibility when considering choice of extractant, with fewer constraints compared with a conventional solvent extraction system.

Features of CLA systems include low power requirement since surfactant addition is essential, leading to lower power requirements for dispersion. Another positive observation is the high area to volume ratios that are available, which is important for intensification of mass transfer. Reduction in solvent inventory and highly efficient recycle capacity are also reported advantages. Stuckey (1996) discussed the application of CLA systems to bioprocessing and reported novel ways of breaking the CLA for

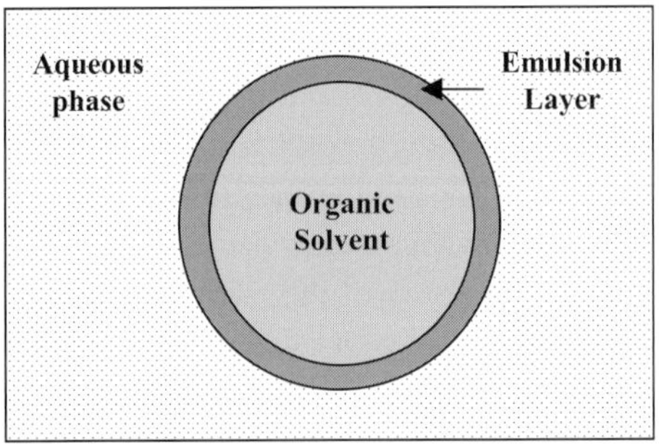

Figure 4.21 Colloidal liquid aphron schematic (Carolan, 1997).

product recovery. Flotation and separation by microfiltration are possible techniques for separation necessary for stripping and recycling of the solvent.

Interest in the application of CLA systems to bioprocessing stems from their ability to reduce emulsion formation associated with solvent extraction when applied to product separation from fermentation broths. Another significant potential advantage is the reduction in biomass accumulation at the interface on account of the ability to deploy anionic surfactants in the CLA system as these show a tendency to repel negatively charged biomass and proteinaceous materials (Matsushita et al., 1992). This effect also may reduce toxicity of the extractant to live biomass. Applications that are reported include extraction of low-solubility antibiotic compounds such as Erythromycin (Lye and Stuckey, 1994, 1996), chiral epoxides such as phenyl glycidil ether (Carolan, 1997), and carboxylic acids (Stuckey, 1996).

4.11 Microextraction

Microextraction is another related intensification technique involving liquid–liquid systems that has seen an increase in research activity. The main application on which recent literature has focused is in the field of analytical chemistry. In particular, the technique is of interest for the determination of compounds that are present in the environment in very low concentration. Such compounds include micropollutants, such as phthalic acid esters used as plasticizers and as solvents in many personal care products (PCPs). Another group of micropollutants comprise those based on polycyclic chlorinated hydrocarbons such as gamma hexachlorocyclohexane (γ-HCH), commonly known as lindane, which have been used extensively as pesticides in a broad range of environments, including agriculture, forestry, and horticulture (Regueiro et al., 2008). These and many similar compounds end up at very low concentration in wastewater, in swimming pool water, in lakes, in domestic tap water, and even bottled water.

Gonzalez et al. (2016) reviewed recent developments of the microextraction technique as a means of accurate determination of such micropollutants. The principle of microextraction is based on conventional solvent extraction involving extraction of the target analyte from an aqueous phase into a single microdroplet of organic extractant. This technique has attracted increasing attention on account of the simplicity of the experimental setup, short analysis time, and that very small amounts of organic solvent are required for each analysis. In the early development of microextraction, instability of the microdrop and relatively low precision were often encountered. An improved microextraction technique is described by Regueiro et al. (2008) based on dispersion of the organic extractant as very small droplets in the presence of a dispersant, to produce almost a homogeneous system known as cloud-point extraction. This approach was shown to improve the speed of extraction, reduce the cost of the analysis, and improve the ability to enrich the extract phase with the target analyte, thus facilitating more accurate analysis. A significant negative aspect of this technique is that the addition of the dispersant can depress the partition coefficient for the distribution of the analyte into the organic drop phase. Requieiro cites the work of Luque de Castro and Priego-Capote (2006, 2007), who effectively demonstrated the use of ultrasonic intensification to produce emulsions of appropriate quality for microextraction, thus avoiding the addition of dispersant agents. Later work by Delgado-Povedano and Luque de Castro (2013) discussed the effectiveness of ultrasound not only in microextraction for analytical purposes but also for nonanalytical applications. The application of ultrasonic techniques for enhanced microextraction exploits several phenomena resulting from exposure of liquid–liquid mixtures to ultrasonic fields.

Apart from the obvious benefits of high interfacial areas associated with sub-micron emulsions created by the phase disintegration due to ultrasonic energy dissipation, there are other effects that enhance microextraction. These include the generation of acoustic flows in the continuous phase, highly localized increases in temperature, and cavitational collapse, which results in steep pressure and temperature gradients close to the liquid–liquid interface.

4.12 Recent Developments in Membrane Engineering

The further recent development of membrane-based techniques relevant to liquid–liquid systems focuses on the application of ionic liquid systems combined with membranes. Another important area is the development of more effective solvent-based complexing agents that are immobilized into membrane supports. A third area is the development of in loco biosurfactant synthesis techniques for improving the economics of emulsification and the generation of a liquid membrane extractant. A fourth area of emerging importance is the development of polymer inclusion membranes. There are a number of recent reviews that summarize the latest developments in combined membrane extraction processes, including those of Pabby, Swain, and Sastre (2017) and Rynkowska, Fatyeyeva, and Kujawski (2018). In particular, Pabby defines a category of novel extractive membrane techniques by the term "smart integrated membrane-assisted liquid extraction" or SIMALE. These are proposed as new effective separation techniques that lend themselves to industrial chemical separations. A major feature of the innovation is the stability of the membrane system. Pabby's review concludes that the combination of liquid membranes and membrane contactors can achieve significant process intensification due to the large increases in interfacial mass transfer area that this technology facilitates. Reductions in equipment size mentioned include a claimed 30-fold reduction compared to extraction in conventional packed towers and a 10-fold reduction compared with stirred tanks for solvent extraction.

As quoted by Pabby, this is based on "a modified supported liquid membrane using pseudo emulsion based hollow fiber strip dispersion or nondispersive solvent extraction techniques." In the pseudo emulsion based hollow fiber strip dispersion technique, which is described, the configuration consists of an emulsion phase comprising the extracting agent and a back-extraction agent, flowing through the shell side of a single hollow fiber module. The aqueous feed phase, containing the target molecule, flows through the tube side of microporous fibers.

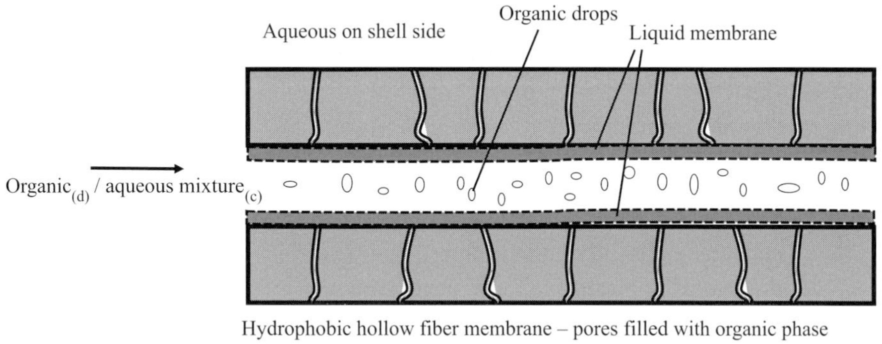

Figure 4.22 Principle of the hollow fiber liquid renewal membrane
Reprinted (adapted) with permission from Pabby, A. K., Swain, B., and Sastre, A. M. (2017). Recent advances in smart integrated membrane assisted liquid extraction technology. *Chemical Engineering & Processing: Process Intensification*, **120**, 27–56. Copyright (2017) Elsevier B.V. All Rights Reserved

In another configuration, Pabby et al. (2017) describe a new technique, based on a hollow fiber renewal liquid membrane concept (Figure 4.22). The surface renewal theory is the basis of this concept that, according to Pabby et al., combines the advantages of hollow fiber membrane extraction with liquid film permeation. The hollow fiber membrane, which is made of hydrophobic polymer, is conditioned by wetting with the organic phase so that the pores in the membrane are saturated with organic solvent. The mode of operation involves metering the aqueous feed phase through the shell side of the hollow fiber assembly. The mixed organic/aqueous feed is metered through the tube (lumen) side. This mixed organic/aqueous feed comprises the extract phase (organic) and the stripping phase (aqueous). On account of the hydrophobicity of the hollow fiber wall, the organic phase preferentially wets the inner surface of the lumen, forming a film, as shown in Figure 4.22. It is postulated that the film of organic phase periodically peels off, due to local shear forces, into small droplets that are mixed with the flowing mixed phase flowing through the interior of the lumen. The film is reestablished due to hydrophobic interaction of the organic drops in the mixed phase, thus creating a cycle of surface renewal. It is claimed that this surface renewal, together with the large interfacial areas generated in the mixed phase, results in acceleration of mass transfer rate and reduction in mass transfer resistance inside the lumen. The review cites successful demonstration of the hollow fiber renewal liquid membrane system by Ren et al. (2009) for the simultaneous removal and recovery of copper(II) from aqueous solutions in the system $CuSO_4$/D2EHPA/kerosene/hydrochloric acid.

Other variants reviewed include the incorporation of ionic liquids into combined membrane/liquid emulsion membrane, and the combination of the liquid membrane concept with dense gas phase/supercritical fluid systems.

The incorporation of ionic liquids into supported liquid membranes offers the possibility of broadening the deployment of ionic liquids for industrial separations involving liquid–liquid concept separations and gas liquid. Typically in a hollow fiber membrane system the ionic liquid is essentially immobilized in the pores structure of the membrane by capillary action. The cost of ionic liquids, together with the limitations associated with the high viscosity of some ionic liquids, mean that large-scale application as free liquids is likely to be limited. Incorporation into a membrane may reduce the inventory of ionic liquid required. This will lower the risk of solvent losses, and will allow processing with two liquid phases comprising the same solvent, for example two aqueous phases. For example, in the latter case aqueous feed and strip solutions would be circulated on the lumen side and on the shell side of a hollow fiber system, with the membrane with the hosted ionic liquid providing a barrier between the two streams. Successful application of the membrane-immobilized ionic liquid concept has been amply demonstrated in numerous works. For example, Ferreira et al. (2014) demonstrated successful removal of thiols from jet fuels using ionic liquid membrane extraction. Venkatesan, Meera, and Begum (2009) demonstrated emulsion liquid membrane pertraction of benzimidazole using a room temperature ionic liquid as carrier.

Turget et al. (2017) examined the performance of several alkyl imidazolium bromide ionic liquids incorporated with PVDF-co-HFP-based ionic polymer inclusion membranes for the separation of Cr(VI) ions. Butyl, hexyl, octyl, and decyl substituted versions of alkyl imidazolium bromide ionic liquids were evaluated. The conclusion was that there is a direct relationship between transport of the Cr(VI) ion and alkyl chain length. The ionic liquids studied were 1,3-didecyl-1*H*-imidazol-3-ium bromide, 1,3-dioctyl-1*H*-imidazol-3-ium bromide, 1,3-dihexyl-1*H*-imidazol-3-ium bromide, and 1,3-dibutyl-1*H*-imidazol-3-ium bromide. The mechanism is anion exchange of the $HCrO_4^-$ ion with the bromide ion associated with the ionic liquid. The 1,3-didecyl-1*H*-imidazol-3-ium bromide appeared to provide the best transport of the Cr(VI) through the membrane. Other studies of Cr(VI) extraction using ionic liquid-supported liquid membranes include those of Alguacil, Garcia-Diaz, and Lopez (2012) and Guo et al. (2012). Nosrati et al. (2013) evaluated transport of vanadium(IV) through supported liquid membranes.

Polymer inclusion membranes (PIM) are a further development in membrane engineering. They are claimed to provide metal ion transport with high selectivity and are straightforward to construct and utilize for metal separations. Polymer inclusion membranes are based on a novel structure in which the liquid extractant and a plasticizer are immobilized through entanglement with the base polymer of the membrane support (Yoshida et al., 2019). Kaya, Alpoguz, and Yilmaz (2013) describe the synthesis of a polymer inclusion membrane demonstrated for the effective removal of Cr(VI). Typically the synthesis of polymer PIMs as described by Kaya et al. (2013) and others, including Nghiem et al. (2006) and Ulewicz et al. (2007), is achieved through casting a solution containing the extractant, plasticizer, and base polymer to provide the structural integrity of the membrane. Plasticizers quoted include *o*-nitrophenyl alkyl ethers. Base polymers include cellulose tri-acetate and poly vinyl chloride. The membrane is prepared by combination of these three components into a thin, flexible, and stable film.

In many cases the extracting solvent serves both purposes of carrier and of plasticizer (Almeida, Cattrall, and Kolev, 2012). The interest in polymer inclusion membranes has partly stemmed from the fact that their preparation is relatively straightforward and, according to a review by Yoshida et al. (2019), they provide better stability than conventional supported liquid membranes. They may also be regarded as providing a green alternative to the use of organic solvents since extraction and stripping may be achieved using just aqueous phases on either side of the membrane, and lower solvent inventories may be considered on account of the effective immobilization of the extracting agent, as mentioned previously. Yoshida et al. (2019) also list several published applications of potential commercial interest. In addition to chromium extraction, documented examples of polymer inclusion membrane applications include the precious metals gold and platinum (Nunez et al., 2006; Yoshida et al., 2017), cobalt (Baba et al., 2016), copper, and zinc (Pospiech, 2014). Kaya et al. (2013) also lists other examples, including zinc(II), strontium(II), lead(II), cadmium(II), and some rare earth metals.

Another important recent development in polymer inclusion membrane technology is the deployment of calixarenes in the structure. Calixarenes have been studied by Kaya et al. (2013, 2016), and Onac et al. (2019) and are shown to have the ability to form complexes with a range of ions and non-ionized compounds and thus may be regarded as effective ion carriers.

Recent developments in the application of supported liquid membranes for multicomponent metal ion separation are also significant. For example, separation of copper, nickel, and cobalt by conventional solvent extraction can be challenging on account of the poor selectivities exhibited by available solvents. Duan et al. (2017) describe a successful technique for the separation of these metal ions by the application of supported liquid membranes impregnated with extracting solvent. Figure 4.23 shows the three-compartment arrangement used by Duan et al. in which the extraction was conducted. The mixed feed of copper(II) ions, cobalt(III) ions, and nickel(II) ions is fed to the left-hand compartment, which is dosed with a mixture of ammonia and ammonium chloride solutions. The compartments are separated by a membrane comprising polyvinylidene difluoride (PVDF) membranes loaded with 20 vol.% Acorga M5640 in kerosene. The second (middle) compartment is conditioned with sulfuric acid (5 g L^{-1}), which is favorable for the transport of the copper(II) ions, and the nickel(II) ions from

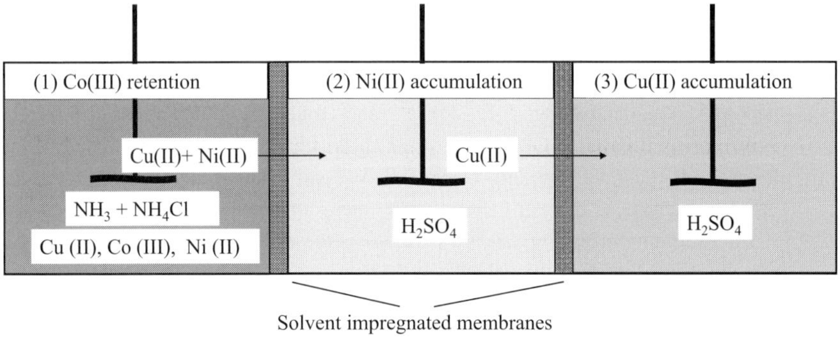

Figure 4.23 A novel sandwich supported liquid membrane system for simultaneous separation of copper, nickel, and cobalt.

Reprinted (adapted) with permission from Duan, H., Wang, Z., Yuan, X., et al. (2017). A novel sandwich supported liquid membrane system for simultaneous separation of copper, nickel and cobalt in ammoniacal solution. *Separation and Purification Technology*, **173**, 323–329. Copyright (2017) Elsevier B.V. All Rights Reserved the first compartment but is unfavorable for the transport of the cobalt(III) ions. Thus a high degree of separation of the cobalt is achieved. The third compartment, to the right of the second, is also conditioned with sulfuric acid but at a higher concentration of 50 g L^{-1}. The conditions thus created in the third compartment are favorable for the transport of copper but unfavorable for nickel. Thus nickel is retained in the center compartment, while the separated copper ions appear in the third compartment. It was claimed that excellent separation of the three ions was achieved overall.

Many of the emulsion liquid membrane systems require the addition of chemical surfactants in order to create the emulsion. In some cases the chemical surfactant itself may be polluting. In a significant recent development, Ferreira et al. (2019) describe the preparation of an in-situ biosurfactant and its direct application for the production of a useful ELM system. These researchers highlighted the problem that while ELM are an up and coming technique, especially for the removal of metal ions from polluted water, the chemical surfactants can create collateral environmental damage. Based on this concern, a substitute synthesized biologically is offered as an alternative. The biological production of ramnolipids at low cost from renewable sources is discussed. These are surfactants that promote the formation of microemulsions and nanoemulsions. Ferreira et al. (2019) describe the biological production of surfactant from glycerol in the presence of *Pseudomonas aeruginosa*. The emulsion liquid membrane comprised biosurfactant, sulfuric acid, EDTA or D_2HPA as carrier, and was successfully demonstrated for the removal of manganese(II) from wastewater. A cost comparison with the chemical surfactant Span 80 showed the biosurfactant to be about one fifth of the cost.

References

Alguacil, F. J., Garcia-Diaz, I., and Lopez, F. A. (2012). Transport of Cr(VI) from HCl media using (PJMTH(+)Cl(−)) ionic liquid as carrier by advanced membrane extraction Processing. *Separation Science and Technology*, **47**, 555–561.

Almeida, M. I. G. S., Cattrall, R. W., and Kolev, S. D. (2012). Recent trends in extraction and transport of metal ions using polymer inclusion membranes (PIMs). *Journal of Membrane Science*, **415–416**, 9–23.

Anton, N. and Vandamme, T. F. (2011). Nano-emulsions and micro-emulsions: Clarifications of the critical differences. *Pharmaceutical Research*, **28**, 978–985.

Baba, Y., Kubota, F., Goto, M., Cattrall, R. W., and Kolev, S. D. (2016). Separation of cobalt(II) from manganese(II) using a polymer inclusion membrane with *N*-[*N*,*N*-di(2-ethylhexyl) aminocarbonylmethyl]glycine (D2EHAG) as the extractant/carrier. *Journal of Chemical Technology and Biotechnology*, **91**, 1320–1326.

Berkman, P. and Calabrese, R. (1988). Dispersion of viscous liquids by turbulent flow in a static mixer. *AICHE Journal*, **34**(4), 602–608.

Boey, S. C., Delcerro, M. C. G., and Pyle, D. L. (1987). Extraction of citric acid by liquid membrane extraction. *Chemical Engineering Research and Design*, **65**, 218–223.

Bremond, N., Thiam, A. R., and Bibette, J. (2008). Decompressing emulsion droplets favors coalescence. *Physical Review Letters*, **100**, 024501.

Cahn, R. P. and Li, N. N (1974). Separation of phenol from waste-water by liquid membrane technique. *Separation Science*, **9**(6), 505–519.

Carolan, N. (1997). Partition studies in whole broth extraction. Unpublished Ph.D. thesis, The Queens University of Belfast.

Chaudhari, J. and Pyle, D. L. (1987). Liquid membrane extraction. In M. S. Verrall and M. J. Hudson, eds., *Separations for Biotechnology*. Ellis Horwood, Vol. 1, pp. 241–259.

Cho, T. and Shuler, M. L. (1986). Multi-membrane bioreactor for extractive fermentation. *Biotechnology Progress*, **2**, 53–60.

Delgado-Povedano, M. M. and Luque de Castro, M. D. (2013). Trends in ultrasound-assisted analytical emulsification-extraction. *Analytical Chemistry*, **45**, 1–13.

Duan, H., Wang, Z., Yuan, X., et al. (2017). A novel sandwich supported liquid membrane system for simultaneous separation of copper, nickel and cobalt in ammoniacal solution. *Separation and Purification Technology*, **173**, 323–329.

Ferreira, A. R., Neves, L. A., Ribeiro, J. C., et al. (2014). Removal of thiols from model jet-fuel streams assisted by ionic liquid membrane extraction. *Chemical Engineering Journal*, **256**, 144–154.

Ferreira, L. C., Ferreira, L. C., Cardoso, V. C., and Filho, U. C. (2019). Mn(II) removal from water using emulsion liquid membrane composed of chelating agents and biosurfactant produced in loco. *Journal of Water Process Engineering*, **29**, 00792.

Fouad, E. A. and Bart, H.-J. (2007). Separation of zinc by a non-dispersion solvent extraction process in a hollow fiber contactor. *Solvent Extraction and Ion Exchange*, **25**, 857–877.

Frankenfeld, J. W., Cahn, R. P., and Li, N. N. (1981). Extraction of copper by liquid membranes. *Separation Science and Technology*, **16**(4), 385–402.

Friesen, D. T., Babcock, W. C., and Chambers, A. R. (1986). Separation of citric acid from fermentation beer using supported-liquid membranes. *Abstracts of Papers of the American Chemical Society*, **191**, 194–195.

Fryde, M. M. and Mason, T. G. (2012). Advanced nanoemulsions. *Annual Reviews in Physical Chemistry*, **63**, 493–518.

Gabelman, A., Hwang, S.-T., and Krantz, W. B. (2005). Dense gas extraction using a hollow fiber membrane contactor: experimental results versus model predictions. *Journal of Membrane Science*, **257**, 11–36.

Gonzalez, J. A. O., Fernandez-Torres, R., Bello-Lopez, M. A., and Ramos-Pay, M. (2016). New developments in microextraction techniques in bioanalysis – A review. *Analytica Chimica Acta*, **905**, 8–23.

Grobben, N. G., Eggink, G., Cuperus, F. P., and Huizing, H. J. (1993). Production of acetone, butanol and ethanol (ABE) from potato wastes – Fermentation with integrated membrane extraction. *Applied Microbiology and Biotechnology*, **39**, 494–498.

Groot, W. J., Vanderlans, R. G. J. M., and Luyben, K. C. A. M. (1992). Technologies for butanol recovery integrated with fermentations. *Process Biochemistry*, **27**, 61–75.

Guo, L., Zhang, J., Zhang, D., et al. (2012). Preparation of poly(vinylidene fluoride-co-tetrafluoroethylene)-based polymer inclusion membrane using bifunctional ionic liquid extractant for Cr (VI) transport. *Industrial and Engineering Chemistry Research*, **51**, 2714–2722.

Hano, T., Matsumoto, M., Hirata, M., et al. (1996). Extraction of fermentation organic acids with hollow fiber membranes. *Proceedings of the International Solvent Extraction Conference*, **2**, 1387–1392.

Hathout, R. M., Woodman, T. J., Mansour, S., et al. (2010). Microemulsion formulations for the transdermal delivery of testosterone. *European Journal of Pharmaceutical Sciences*, 40, 188–196.

Kang, W. K., Shukla, R., and Sirkar, K. K. (1990). Ethanol production in a microporous hollow-fiber-based extractive fermenter with immobilized yeast. *Biotechnology and Bioengineering*, **36**, 826–833.

Karbstein, H. and Schubert, H. (1995). Developments in the continuous mechanical production of oil-in-water macro-emulsions. *Chemical Engineering and Processing*, **34**(3), 205–211.

Kaya, A., Alpoguz, H. K., and Yilmaz, A. (2013). Application of Cr(VI) transport through the polymer membrane with a new synthesized calix[4] arene derivative. *Industrial and Engineering Chemistry Research*, **52**, 5428–5436.

Kaya, A., Onac, C., Alpoguz, H. K., Yilmaz A., and Atar, A. (2016). Removal of Cr(VI) through calixarene based polymer inclusion membrane from chrome plating bath water. *Chemical Engineering Journal*, **283**, 141–149.

Kiss, N., Brenn, G., Pucher, H., et al. (2011). Formation of O/W emulsions by static mixers for pharmaceutical applications. *Chemical Engineering Science*, **66**, 5084–5094.

Konzen, C., Araújo, E. M. R., Balarini, J. C., Miranda, T. L. S., and Salum, A. (2014). Extraction of citric acid by liquid surfactant membranes: bench experiments in single and multistage operation. *Chemical and Biochemical Engineering Quarterly*, **28**(3), 289–299.

Lazarova, Z., and Peeva, L., (1994) Facilitated transport of lactic acid in a stirred transfer cell. *Biotechnology and Bioengineering*, **43**(10), 907–912.

Li, N. N. (1968). Separating hydrocarbons with liquid membranes. US Patent 3,410,794.

Luque de Castro, M. D., and Priego-Capote, F. (2006). *Analytical Applications of Ultrasound*. Amsterdam: Elsevier.

Luque de Castro, M. D. and Priego-Capote, F. (2007). Ultrasound-assisted preparation of liquid samples. *Talanta*, 72(2), 321–334.

Lye, G. J. and Stuckey, D. C.(1994). Extraction of macrolide antibiotics using colloidal liquid aphrons (CLAs). In *Separations for Biotechnology – Volume 3*, D.L. Pyle, ed. SCI Publishing, 539–545.

Lye, G.J. and Stuckey, D.C. (1996). Predispersed solvent extraction of erythromycin using colloidal liquid aphrons. In G. Stevens, ed., *Proceedings of the International Solvent Extraction Conference*. Melbourne: University of Melbourne, pp. 1399–1404.

Marr, R. and Kopp, A. (1982). Liquid membrane technology – a survey of phenomena, mechanisms, and models. *International Chemical Engineering*, **22**(1), 44–60.

Matsumura, M. and Markl, H. (1986). Elimination of ethanol inhibition by pertraction. *Biotechnology and Bioengineering*, **28**, 534–541.

Matsushita, K., Mollah, A. H., Stuckey, D. C., Delcerro, C., and Bailey, A. I. (1992). Predispersed solvent extraction of dilute products using colloidal gas aphrons – Aphron preparation, stability and size. *Colloids and Surfaces*, **69**, 65–72.

Middleman, S. (1974). Drop size distributions produced by turbulent pipe flow of immiscible fluids through a static mixer. *Industrial and Engineering Chemistry Process Design and Development*, **13**(1), 78–83.

Mlynek, Y., and Resnick, R. (1972). Drop sizes in an agitated liquid–liquid system. *AIChE Journal*, **18**, 122–127.

Moinard-Chécot, D., Chevalier, Y., Briançon, S., Beney, L., and Fessi, H. (2008). Mechanism of nanocapsules formation by the emulsion-diffusion process. *Journal of Colloid and Interface Science*, **317**(2), 458–468.

Nghiem, L. D., Mornane, P., Potter, I. D., et al. (2006). Extraction and transport of metal ions and small organic compounds using polymer inclusion membranes (PIMs). *Journal of Membrane Science*, **281**(1–2), 7–41.

Nosrati, S., Jayakumar, N. S., Hashim, M. A., and Mukhopadhyay, S. (2013). Performance evaluation of vanadium (IV) transport through supported ionic liquid membrane. *Journal of the Taiwan Institute of Chemical Engineers*, **44**, 337–342.

Nunez, M. E., de San Miguel, R., Mercader-Trejo, F., Aguilar, J.C., and de Gyves, J. (2006). Gold (III) transport through polymer inclusion membranes: efficiency factors and pertraction mechanism using Kelex 100 as carrier. *Separation and Purification Technology*, **51**, 57–63.

Onac, C., Kaya, A., Ataman, D., Gunduz, N. A. and Alpoguz, H. K. (2019). The removal of Cr (VI) through polymeric supported liquid membrane by using calix[4]arene as a carrier. *Chinese Journal of Chemical Engineering*, **27**, 85–91.

Pabby, A. K. and Sastre, A. M. (2013). State-of-the-art review on hollow fibre contactor technology and membrane-based extraction processes. *Journal of Membrane Science*, **430**, 263–303.

Pabby, A. K., Swain, B., and Sastre, A. M. (2017). Recent advances in smart integrated membrane assisted liquid extraction technology. *Chemical Engineering & Processing: Process Intensification*, **120**, 27–56.

Parhi, P. K. (2013). Supported liquid membrane principle and its practices: a short review. *Journal of Chemistry*, 2013, ID 618236, https://doi.org/10.1155/2013/618236.

Patnaik, P. R. (1995) Liquid emulsion membranes – principles, problems and applications in fermentation processes. *Biotechnology Advances*, **13**(2), 175–208.

Pospiech, B. (2014). Synergistic solvent extraction and transport of Zn and Cu across polymer inclusion membrane with a mixture of TOPO and Aliquat 336. *Separation Science and Technology*, **49**, 1706–1712.

Regueiro, J., Llompart, M., Garcia-Jares, C., Garcia-Monteagudo, J. C., and Cela, R. (2008). Ultrasound-assisted emulsification-microextraction of emergent contaminants and pesticides in environmental waters. *Journal of Chromatography A*, **1190**, 27–38.

Ren, Y. Z. Q., Zhang, W. D., Liu, Y.M., Dai, Y., Cui, C. H. (2007). New liquid membrane technology for simultaneous extraction and stripping of copper(II) from wastewater. *Chemical Engineering Science*, **62**, 6090–6101.

Ren, Z., Meng, H., Zhang, W., Liu, J., and Cui, C. (2009). The transport of copper(II) through hollow fiber renewal liquid membrane and hollow fiber supported liquid membrane. *Separation Science and Technology*, **44**, 1181–1197.

Ren, W. Z., Zhang, M. J., and Shuguang, L. (2010). Extraction separation of Cu(II) and Co(II) from sulfuric solutions by hollow fiber renewal liquid membrane. *Journal of Membrane Science*, **365**, 260–268.

Rockman, J. T., Kehat, E., and Lavie, R. (1996). Thermally enhanced liquid–liquid extraction of citric acid using supported liquid membranes. In *Proceedings of the International Solvent Extraction Conference*, Melbourne, Vol. 2, pp. 857–862.

Rynkowska, E., Fatyeyeva, K., and Kujawski, W. (2018). Application of polymer-based membranes containing ionic liquids in membrane separation processes: A critical review. *Reviews in Chemical Engineering*, **34**(3), 341–363.

Santos, J., Vladisavljevic, G. T., Holdich, R. G., Dragosavac, M. M., and José Munoz, J. (2015). Controlled production of eco-friendly emulsions using direct and premix membrane emulsification. *Chemical Engineering Research and Design*, **98**, 59–69.

Scholler, C., Chaudhuri, J. B., and Pyle, D. L. (1993). Emulsion liquid membrane extraction of lactic-acid from aqueous-solutions and fermentation broth. *Biotechnology and Bioengineering*, **42**(1), 50–58.

Sebba, F. (1987). *Foams and Biliquid Foams – Aphrons*. New York: Wiley and Sons.

Sengupta, B., Sengupta, R., and Subrahmanyam, N. (2006). Process intensification of copper extraction using emulsion liquid membranes: Experimental search for optimal conditions. *Hydrometallurgy*, **84**, 43–53.

Shukla, R., Kang, W., and Sirkar, K. K. (1989). Acetone butanol ethanol (abe) production in a novel hollow fiber fermenter extractor. *Biotechnology and Bioengineering*, **34**, 1158–1166.

Strzelbicki, J. and Charewicz, W. (1980). The liquid surfactant membrane separation of copper, cobalt and nickel from multi-component aqueous solutions. *Hydrometallurgy*, **5**, 243–254.

Stuckey, D. C. (1996). Solvent extraction in biotechnology. In G. Stevens, ed., *Proceedings of the International Solvent Extraction Conference*. University of Melbourne, pp. 25–34.

Takhistov, P. and Paul, S. (2006). Formation of oil/water emulsions due to electrochemical instability at the liquid/liquid interface. *Food Biophysics*, **1**, 57–73.

Tauer, K. (2005). Emulsions – Part 2: A little (theory): Emulsion stability www.mpikg.mpg.de/886743/Emulsions_-2.pdf.

Teramoto, M., Sakai, T., Yanagawa, K., Oshuga, M., and Miyake, Y. (1983). Modeling of the permeation of copper through liquid surfactant membranes. *Separation Science and Technology*, **18**(8), 735–764.

Turgut, H. I., Eyupoglu, V., Kumbasar, R.A., and Sisman, I. (2017). Alkyl chain length dependent Cr(VI) transport by polymer inclusion membrane using room temperature ionic liquids as carrier and PVDF-co-HFP as polymer matrix. *Separation and Purification Technology*, **175**, 406–417.

Ulewicz, M., Lesinska, U., Bochenska, M., and Walkowiak, W. (2007). Facilitated transport of Zn(II), Cd(II) and Pb(II) ions through polymer inclusion membranes with calix[4]-crown-6-derivatives. *Separation and Purification Technology*, **54**, 299–306.

Urbina-Villalba, G. (2004). Effect of dynamic surfactant adsorption on emulsion stability. *Langmuir*, **20**, 3872–3881.

Urrutia, P. I (2006). Predicting water-in-oil emulsion coalescence from surface pressure isotherm. MSc Thesis, University of Calgary.

Venkatesan, S., Meera, K. M., and Begum, M. S. (2009). Emulsion liquid membrane pertraction of benzimidazole using a room temperature ionic liquid (RTIL) carrier. *Chemical Engineering Journal*, **148**, 254–262.

Verbeken, B. K., Pinoy, L. V., and Verhaege, M. (2009). Cobalt removal from waste-water by means of supported liquid membranes. *Journal of Chemical Technology and Biotechnology*, **84**, 711–715.

Volkel, W., Halwachs, W., and Schugerl, K. (1980). Copper extraction by means of a liquid surfactant membrane process. *Journal of Membrane Science*, **6**, 19–31.

Wang, J., Luo, J., Feng, S., Li, H., and Zhang, X. (2016). Recent development of ionic liquid membranes. *Green Energy & Environment,* **1**, 43–61.

Wilde, P., Mackie, A., Husband, F., Gunning, P., and Morris, V. (2004). Proteins and emulsifiers at liquid interfaces. *Advances in Colloid and Interface Science,* **108–109**, 63–71.

Yang, X., Caoa, Y.-M., Wang, R., and Yuan, Q. (2007). Study on highly hydrophilic cellulose hollow fiber membrane contactors for thiol sulfur removal. *Journal of Membrane Science,* **305**, 247–256.

Yoshida, W., Baba, Y., Kubota, F., Kamiya, N., and Goto, M. (2017). Extraction and stripping behavior of platinum group metals using an amic-acid-type extractant. *Journal of Chemical Engineering of Japan*, **50**, 521–526.

Yoshida, W., Baba, Y., Kubota, F., Kolev, S. D., and Goto, M. (2019). Selective transport of scandium(III) across polymer inclusion membranes with improved stability which contain an amic acid carrier. *Journal of Membrane Science*, **572**, 291–299.

5 High Gravity Fields

5.1 Introduction

The employment of high gravity fields in multiphase processing has a long history and a wide range of industrial application, ranging from the hydrocyclone technology (Figure 5.1) through to the use of centrifugal separators for liquid–liquid separations. The underlying principle in conventional high gravity process equipment is based on differences in density, the respective differences in accelerative forces in each phase, and the resulting effect on motion and position in the continuum. The term "high gravity field" is used synonymously with the term "centrifugal field" in this chapter, since in the majority of cases in liquid processing, centrifugal forces generated by rotational flows are exploited to achieve process intensification. High gravity fields are exploited in a wide range of processes involving multiphase mixtures, which include separation of fine particulates from liquids and gases, gas bubbles from liquids, and for the separation of liquid–liquid dispersions. In some cases, use of high gravity fields may be the only means of achieving a separation, for example in the case of some stable emulsions. In all cases, the accelerative forces achieved acting on the particles or drops can be several thousand-fold that of gravity. The benefits of the presence of the high gravity field alone include shorter residence times, more efficient separations, less material holdup and inventory, and reduction of equipment size.

There are many major reviews and textbooks dealing with hydrocyclone technology. For example, Svarovsky (1991) deals with the topic in considerable detail, particularly for solid–liquid phase separations. Judd et al. (2014) discuss the application of hydrocyclone technologies for offshore oil industry applications where there is a great incentive for process intensification. An example is the treatment of produced water, requiring oil–water separation as part of treatment. Another important intensification phenomenon frequently associated with processing in the presence of high gravity fields is the promotion of high shear forces, which result in high heat and mass transfer rates, for example between the phases being processed but also, in the case of heat transfer, between the fluid phase and heat transfer surfaces. In liquid–liquid systems, the occurrence of high shear rates at the liquid–liquid interface gives rise to high rates of mass transfer. In addition, although detailed knowledge is uncertain due to difficulties in obtaining live physical measurements, it is also postulated that the behavior of liquid–liquid dispersions in high gravity fields may also involve significant increases in cyclic drop breakup and coalescence, which further enhances diffusion rates.

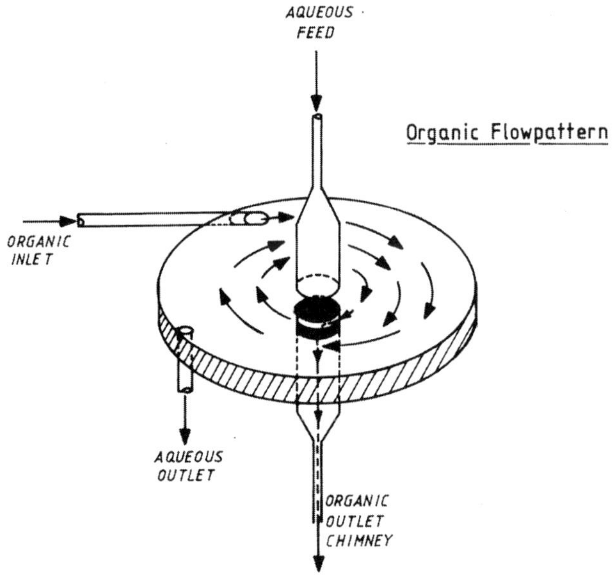

Figure 5.1 Schematic of a hydrocyclone contactor / separator.

As is the case with other types of intensification technologies described in this book, there are often trade-offs for industrial applications between increased capital cost of equipment, energy costs, space saving, maintenance costs, safety, and reliability. This is particularly the case with high speed centrifugal technologies, which involve significant power consumptions to drive rotating machinery. An important requirement is for reliable seals and bearings on rotating shafts. Another factor is the requirement of special materials for components performing in high torque environments. Protection against overheating bearings, which may present a fire/explosion risk, is also an important consideration.

There are several examples of technology based on centrifugal fields specifically for processing of liquid–liquid systems and which are well established in industrial applications. A review of the various types of contactors is presented elsewhere. Examples include the Podbielnak family of centrifugal contactors, developed in the 1950s for the separation of antibiotics from fermentation media. The Robatel contactor, the decanter extractor developed by Westafalia, and rotary contactors by Flottweg are further examples.

5.2 Spinning Disk Technology

In this section we consider a family of liquid–liquid contactors that is relatively novel and that represents some of the more recent developments in the exploitation of high gravity fields to intensify liquid–liquid contact. As seen in Chapter 1, high gravity fields can be successfully applied in rotating contactors, such as the Podbielniak and the GEA decanter extractor, which enable processing of feed solutions containing solids and the successful rapid recovery of unstable compounds whose recovery is not viable using conventional column contactors or mixer settlers.

Spinning disc contactors represent one of a number of process intensification technologies employing high gravity fields in which there is significant interest. Research and development has been aimed at systems involving fast liquid/liquid and gas/liquid reactions involving large heat effects, such as nitrations, sulfonations, and polymerizations. Figures 5.2 and 5.3 show a typical spinning disc reactor arrangement described by Oxley et al. (2000) and further developed by others (Boodhoo and Jachuck, 2000a, 2000b; Boodhoo et al., 2006). The design shown in these figures comprises an outer cylindrical vessel, the walls of which may be equipped with internal heating or cooling coils. Inside the vessel a flat disk is located on the top of a rotating shaft that spins the disk at a preset rotational speed. The upper lid of the outer vessel is equipped with inlet ports for the introduction of one or more liquid streams. For gas/liquid applications, the outer vessel may also be equipped with separate inlet and outlet ports for the continuous introduction of gas flow. A high gravity field involving centrifugal forces is created by rotation of the flat disc mounted in a horizontal plane onto a rotating vertical shaft, all contained within a cylindrical vessel. When a liquid is introduced onto the disc surface at or adjacent to the spin axis, the liquid flows radially outward under the centrifugal force in the form of a thin film (Figures 5.4 and 5.5). At rotational speeds of

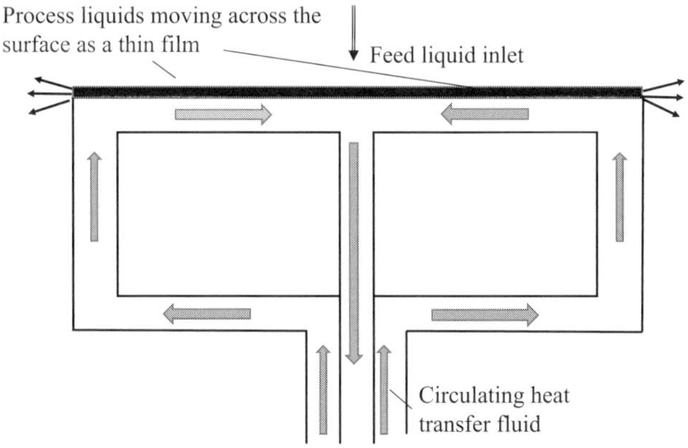

Figure 5.2 Spinning disk reactor. (Oxley et al., 2000)

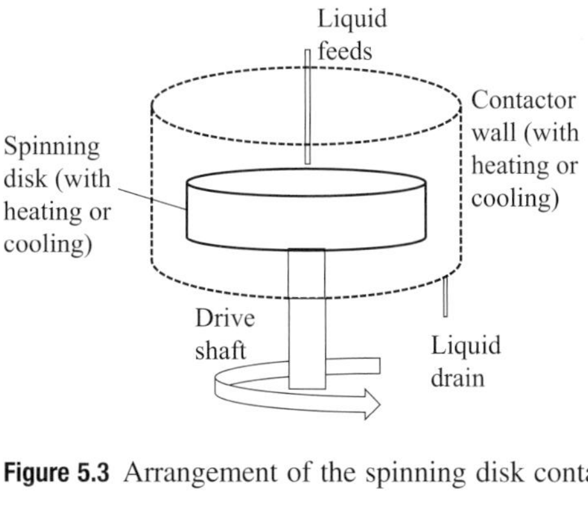

Figure 5.3 Arrangement of the spinning disk contactor. (Oxley et al., 2000)

Figure 5.4 Internal view of a spinning disk reactor developed for styrene polymerization. Reprinted from Boodhoo, K. V. K. and Jachuck, R. J. (2000a). Process intensification: spinning disk reactor for styrene pol. *Applied Thermal Engineering*, **20**, 1127–1146. Copyright (2000), with permission from Elsevier

approximately 1000 rpm, the films created are less than 100 μm thick and thus offer a short diffusion path length. In a variation of the basic design, a shroud plate may be included above the rotating disk. The shroud plate enhances flow conditions in the immediate vicinity of the rotating disk surface. Under some conditions the gas velocity and mixing in the space between the surface of the rotating disk and the lower surface of the shroud plate may be enhanced, providing enhanced conditions for gas–liquid mass transfer, and heat transfer. At the outer edge of the spinning disk, the film of liquid detaches from the disk surface and is ejected under the influence of centrifugal force. The ejected liquid collides with the inner wall of the outer containment vessel, coalesces, and flows to an outlet incorporated into the base of the containment vessel.

The liquid film formed on the surface of the rotating disk can exhibit unstable flow with significant wave formation on the free surface at the gas–liquid interface (Figure 5.5). Unsteady film surface waves on the disc surface, coupled with the shearing action of the rotating surface, ensure that micromixing and excellent mass and heat transfer are achieved. Extensive mass and heat transfer studies (Ramshaw, 2004) show that convective heat transfer coefficients as high as 14 kW m^{-2} K^{-1} and mass transfer coefficients K_L values as high as 30 × 10^{-5} m s^{-1} and K_G values as high as 12 × 10^{-8} m s^{-1}, can be achieved. Micromixing and an appropriate fluid dynamic environment for achieving faster reaction kinetics are features of spinning disk contactors. Residence times on the spinning disc range from a few seconds down to fractions of a second, and it is therefore well suited to fast processes where the inherent reaction kinetics are of the same order or faster than the mixing and diffusional kinetics.

The spinning disk contactor has the following characteristics:

- intense mixing in the thin liquid film;
- high heat and mass transfer coefficients;
- short residence time;
- plug flow characteristics.

Another feature shown in Figures 5.2 and 5.3 is that the disk itself may be hollow. This allows heat transfer media to be circulated up the drive shaft and through the

(a) (b) (c)

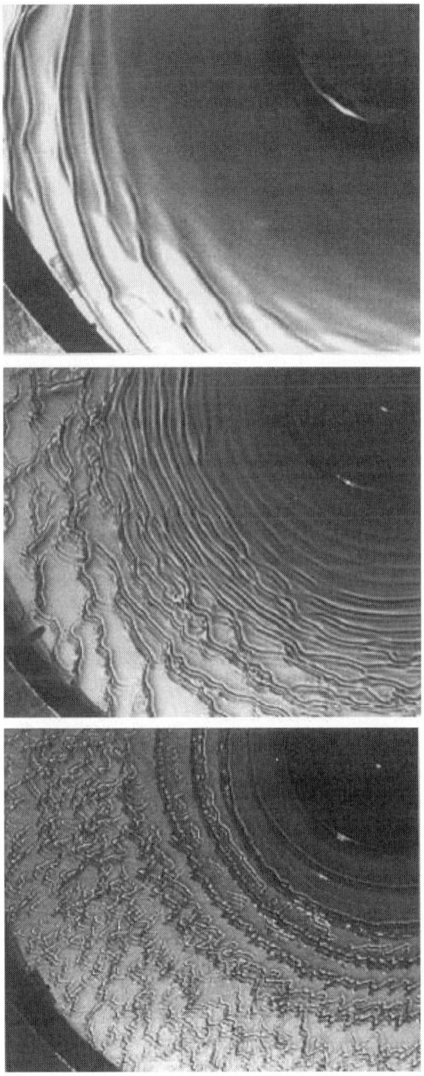

Figure 5.5 Wavy flow variation with disc speed observed on the surface of a spinning disk during a polymerization reaction involving a liquid-liquid feed. (a) 100 rpm (b) 200 rpm (c) 600 rpm. Reprinted from Matar, O. K., Grigori, M., Sisoev, G. M., and Lawrence, C. J. (2006). The flow of thin liquid films over spinning discs. *The Canadian Journal of Chemical Engineering*, **84**, 625–642 with permission from John Wiley and Sons (2006)

interior of the disk. On account of the local high shear fields close to the surfaces of the disk, very high heat transfer coefficients are available to promote efficient heat removal or addition from or to the liquid film on the outer surface of the disk. Additional temperature control is also available by the circulation of heat transfer fluids through heating or cooling coils incorporated in the external walls of the outer containment vessel.

A range of applications of spinning disk contactors has been explored, including for polymerization reactions that involve condensation, free radical, and ionic mechanisms

(Boodhoo et al., 2006). These have been successful, even with polymers of high viscosity. Significant enhancements in polymerization rates have been obtained within the thin films generated on the rotating reaction surface, compared to classical stirred tank reactors. Another feature of the thin film of liquid reaction mixture that is created on the surface of the spinning disk is that polymerizations which are photo initiated may be successfully conducted in the reactor. The light source, which may be visible or UV, can be external to the actual contactor. The thinness of the film of reacting liquid present on the disk results in minimal attenuation of the light, thereby allowing more uniform initiation across the polymerizing continuum.

Another example where the thin film occurring on the free surface of a spinning disk reactor is exploited was demonstrated by Boiarkina, Pedron, and Patterson (2011) for photocatalytic degradation of methylene blue in the presence of ultraviolet light. In this case, the spinning disk was coated with an anatase TiO_2 photocatalyst. The unique hydrodynamic features of the spinning disk contactor gave rise to potential intensification of both solid/liquid transport between the lower boundary of the liquid film and the solid photocatalyst, and gas/liquid transport between the upper surface of the liquid film and surrounding gas.

Table 5.1 shows data from Leveson et al. (2003), comparing the rate constants required to achieve a fast polymerization of styrene to polystyrene in a 5-second residence time, comparing conventional batch reaction with a continuous spinning disk reactor. The three rate constants k_t, k_p, and k_d refer to rate constants for initiation, polymerization, and termination respectively. Significantly, distribution of molecular weight properties was improved. Another application is the continuous production of nanosize and microsize particles via reactive crystallization (Jachuck and Scalley, 2003; Oxley et al., 2000; Tai, Tai, and Liu, 2006). Due to short residence times, micromixing time is less than the induction time for clustering of particles, encouraging the formation of the very smallest particles.

Published applications of spinning disk contactors are largely limited to fast reactions with large heat effects or to those requiring short residence times. In light of the characteristics of spinning disc contactors, they can potentially be used to intensify both liquid–liquid and gas–liquid reactions. This is due to the achievement of high mass transfer rates, uniform micromixing, and high heat transfer rates between liquid films and the disk surface. In order to attain high reaction rates within a very limited time of several seconds, high speed, intense, and forced molecular interdiffusion of the reactant molecules is required. The typical design of an open surface spinning disk reactor is shown in Figures 5.2–5.4. Designs have been further developed to achieve enhanced transport and reaction rates, with several novel developments based on Ramshaw's initial concept.

Table 5.1 Kinetic parameters for styrene polymerization comparing batch reactor with a spinning disk system. Kinetic parameters to achieve 10% conversion in 5 s (Leveson, Dunk, and Jachuck, 2003).

Variable	Batch reactor (363 K)	Required for spinning disk reactor (SDR) conversion (363 K)	Factor
k_t	1.23×10^8	1.8×10^3	7×10^4
k_p	9.27×10^8	2.3×10^5	4×10^3
k_d	8.18×10^{-5}	n/a	n/a

Application of the design of contactor initially described by Ramshaw and co-workers to liquid–liquid systems had not been directly studied and demonstrated. The application of the concept to intensification of liquid–liquid contact requires several modifications. An upper shroud plate may be located very close to the surface of the spinning disk, leaving a small gap between the lower surface of the shroud plate and the upper surface of the rotating disk. Secondly, the phases may be introduced in opposing directions, the phase of greater density being introduced through an orifice in the center of the shroud plate, and the lighter phase being introduced in an upward direction through a small orifice located in the center of the lower rotating disk. This led to the development of intensified liquid–liquid contacting by incorporating impinging jets into a high gravity device.

5.3 Impinging Jets

The impinging jet liquid–liquid contactor may be regarded as a development of the spinning disk contactor. The principle of the technique is illustrated in Figure 5.6 (Dehkordi, 2002a). The two immiscible liquid phases are fed coaxially but in opposite directions from the center of the lower disk and the center of the upper disk. The upper disk is rotated while the lower disk remains stationary. This arrangement allows intensive mixing in the region of impingement followed by exposure of the resulting two phase mixture to high shear rate forces within the interdisk space.

The hydrodynamic conditions in the small space between the static lower disk plate and the spinning disk promote high rates of shear and very intense mixing of the two phases. The net flow of the phases is radially outward, and, as in the original Ramshaw

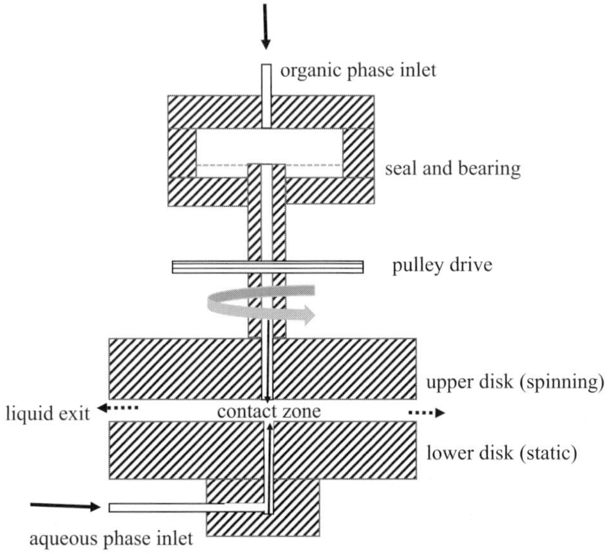

Figure 5.6 Schematic of impinging jet spinning disk contactor.
Reprinted (adapted) with permission from Dehkordi, A. M. (2002). Liquid–liquid extraction with chemical reaction in an impinging jets reactor. *American Institute of Chemical Engineers Journal*, **48**(10), 2230–2239. Copyright (2002) John Wiley and Sons – All Rights Reserved

design, the mixed phases are ejected from the outer edge of the rotating disk, hitting the inner wall of the containment vessel where phase disengagement may occur. In gas–liquid systems, if gas–liquid separation occurs efficiently, the gas phase product is removed via a separate exit port fitted to the outer vessel. In the case of liquid–liquid systems the phase separation may be achieved in a separate coalescence region or in an external separator. This concept has been applied to both gas/liquid and liquid/liquid systems but without quantitative insights into the fluid mechanics that control reaction, heat, and mass transport phenomena.

Dehkordi analyzed the performance of the impinging jet contactor for the extraction of n-butyl formate into water, measuring the rate of extraction at different disk speeds, phase rates, and disk separation distances. In these devices the interdisk spaces are very small and therefore physical measurement of the flow conditions and interfacial area are challenging.

The mass transfer rates that were measured by Dehkordi et al. are displayed in Figure 5.7 shown as overall mass transfer coefficient as a function of the rotation speed of the upper disk at three different aqueous flow rates. There is significant intensification of mass transfer as the disk speed is increased. The mass transfer is expressed using a volumetric mass transfer coefficient that incorporates the effects of specific area and other influences such as shear and turbulence. Dehkordi describes a novel method of estimating the interfacial area by repeating the experiments using a dilute sodium hydroxide solution to saponify the formate under conditions of very fast reaction kinetics. The "Danckwerts plot" method was used to extract specific intefacial area values for each set of conditions of disk speed, interdisk space, and phase flow rate. Large increases in interfacial area were shown to be achieved as disk speed increases, as shown in Figure 5.8. The best case shows that a doubling of disk speed results in approximately a 50% increase in intefacial area. Examination of the mass transfer coefficient data in Figure 5.7 shows a corresponding increase with disk speed, suggesting that an increase in interfacial area plays a major role in increase of the overall rate.

Another variation of the spinning disk contactor described by Dehkordi is illustrated in Figure 5.9, which is titled the "Non-impinging jet spinning disk" contactor. In this

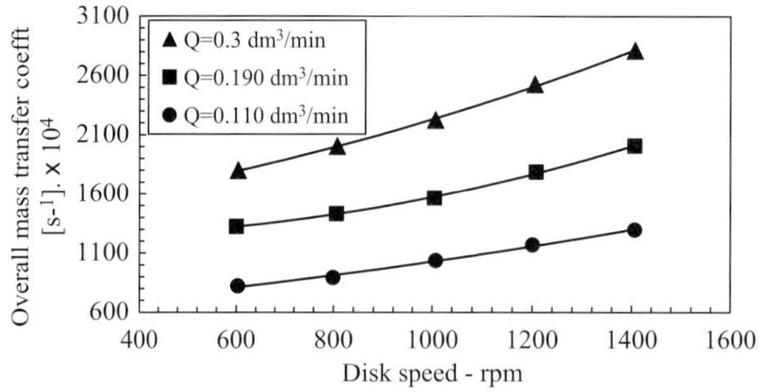

Figure 5.7 Overall volumetric mass transfer coefficient versus disk speed for the extraction of n-butyl formate into water (disk diameter 0.2 m, interdisk space 0.0015 m).

Reprinted (adapted) with permission from Dehkordi, A. M. (2002). Liquid–liquid extraction with chemical reaction in an impinging jets reactor. *American Institute of Chemical Engineers Journal*, **48**(10), 2230–2239. Copyright (2002) John Wiley and Sons – All Rights Reserved

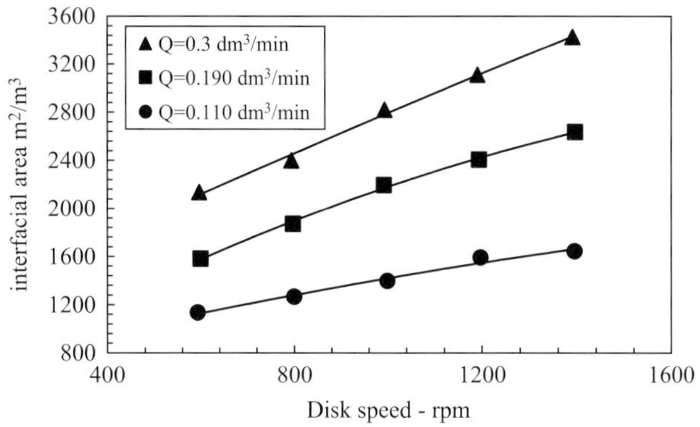

Figure 5.8 Specific interfacial area versus disk speed for the extraction of n-butyl formate into water in an impinging jet rotary liquid–liquid contactor (disk diameter 0.2 m, interdisk space 0.0015 m). Reprinted (adapted) with permission from Dehkordi, A. M. (2002). Liquid-liquid extraction with chemical reaction in an impinging jets reactor. *American Institute of Chemical Engineers Journal*, **48**(10), 2230–2239.

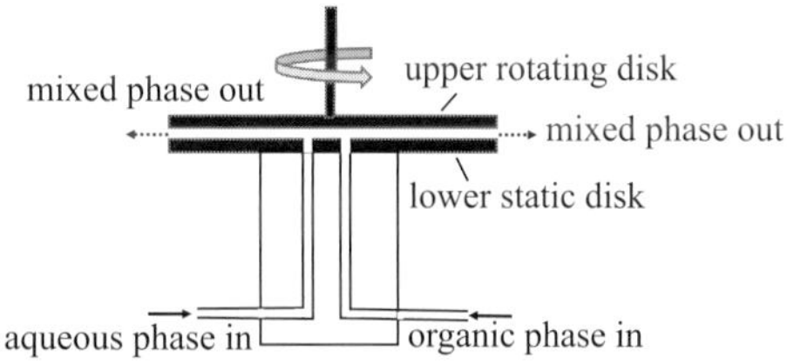

Figure 5.9 Non-impinging jet contactor, designed by Dehkordi (2002b).
Reprinted (adapted) with permission from Dehkordi, A. M. (2002). A novel two-impinging-jets reactor for copper extraction and stripping processes. *Chemical Engineering Journal*, **87**(2), 227–238.

arrangement the two feed liquids are introduced through twin orifices located in the lower static disk. The upper disk rotates and the two liquids are subject to intense mixing and high shear forces in the narrow interdisk space, typically of the order of 5 mm.

Holl (2006) describes another variation of the spinning disc contactor that employs a small spinning cylinder to facilitate high reaction rate. A feed system suitable for two or more immiscible reactants is provided with an inlet located above the disc surface at the center of disc. The second phase is fed into the film developed by the first reactant via an annular Venturi nozzle located at some distance from the first entry. The reactor also utilizes a retaining surface, coextensive with the disc surface. The thickness of the gap between the disc surface and retaining surface can be adjusted to less than 1.00 mm. The retaining surface is used as a means of heat exchange to control the reactants' temperatures passing in the gap. The reactor can effect uniform mixing of two or more liquids within several milliseconds with a mixing length of less than 5 mm. The forced interdiffusion from the high shear rates can allow a very rapid encounter of all reactant molecules as immediately as possible, thus accelerating the observed reaction rate.

5.4 Variants of the Spinning Disk Contactor

A further variation of the spinning disk contactor is described by Meeuwse, van der Schaaf, and Schouten (2010) in which a gas–liquid mixture is co-fed into the spinning disk reactor, reporting a six-fold increase in mass transfer in the space between the rotor and stator due to the very high shear environment. This design was further developed by Visscher et al. (2012) for the intensification of liquid–liquid contacting, as shown in Figure 5.10. One of the key questions to be addressed is: "Can spinning disk technology be scaled to meet the throughput needs of industrial separations?"

Taking into account the studies conducted on spinning disk contactors, and on the different variants, there is a consistent picture showing the following key advantages:

- order of magnitude increases in heat and mass transfer rates;
- ability to greatly enhance multiphase contact involving viscous liquids;
- scaleability;
- very short residence time contact and very low liquid volumetric holdup;
- the ability to integrate the absorption/desorption process in a compact process unit that is readily scalable.

In light of the high transport fluxes available in spinning disk contactors, their application is well suited to systems that exhibit slow mass transfer, such as those involving high viscosity liquids. Since spinning disk contactors generally exhibit short residence times and very low volumetric holdup, they may be considered for application involving high-cost solvent and products where there is an economic requirement to minimize inventory. This would also be important in cases where expensive catalysts are deployed in liquid–liquid systems; for example, for reactions involving homogeneous phase transfer catalysis. Spinning disk contactors may also have a role in the application of heterogeneous catalysts, either in liquid suspension or as part of the spinning disk itself.

There are several examples of spinning disk technology for enhanced liquid–liquid extraction described in recent literature, and these are briefly reviewed.

Visscher et al. (2012) present data for a rotor–stator spinning disk contactor in which the extraction of benzoic acid from n-heptane into water was studied. A schematic of this contactor is shown in Figure 5.10. The contactor consists of a rotating disc driven by a rotor that is enclosed by two stationary discs (the stators) and a stationary cylindrical housing. The velocity gradient created in the space between the rotor and the stator creates intense shear forces on the liquids inside the contactor. The resulting turbulence causes a high degree of dispersion and thus a high interfacial area. Two very significant findings arose from this work.

A spectacular increase in mass transfer was observed with respect to disc rotational velocity, as shown in Figure 5.11. The volumetric mass transfer coefficient value $ka_L\varepsilon_{org}$ for the benzoic acid transfer ranged from a value of 0.17 $m^3\ m^{-3}\ s^{-1}$ at 100 rpm (water rate of $2.5 \times 10^{-6}\ m^3\ s^{-1}$) to a value of 51.47 $m^3\ m^{-3}\ s^{-1}$ at 1600 rpm and at a water rate of $12.5 \times 10^{-6}\ m^3\ s^{-1}$. Comparison with the same extraction conducted in a packed column indicated an increase in the mass transfer rate of approximately 25-fold.

Experimental determination of the flow regimes inside the spinning disk contactors is difficult and much reliance is placed on simulation studies in order to gain information

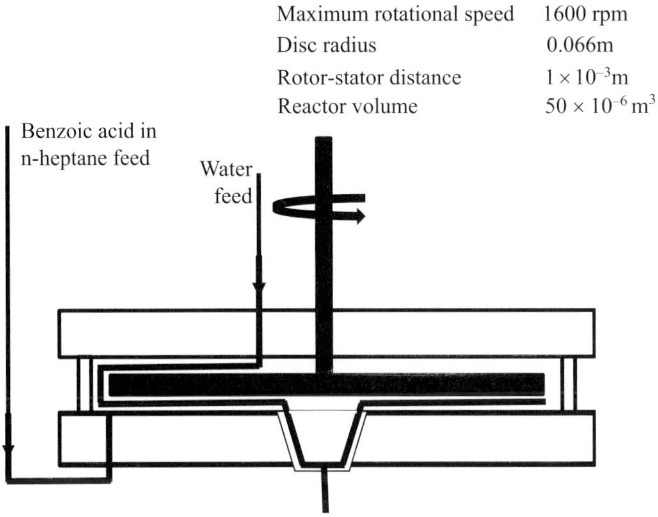

Figure 5.10 Schematic drawing of the rotor–stator spinning disc contactor.
Reprinted (Adapted) from Visscher, F., van der Schaaf, J., de Croon, M. H. J. M., and Schouten, J. C. (2012). Liquid-liquid mass transfer in a rotor–stator spinning disc reactor. *Chemical Engineering Journal*, **185/186**, 267–273. Copyright (2012), with permission from Elsevier

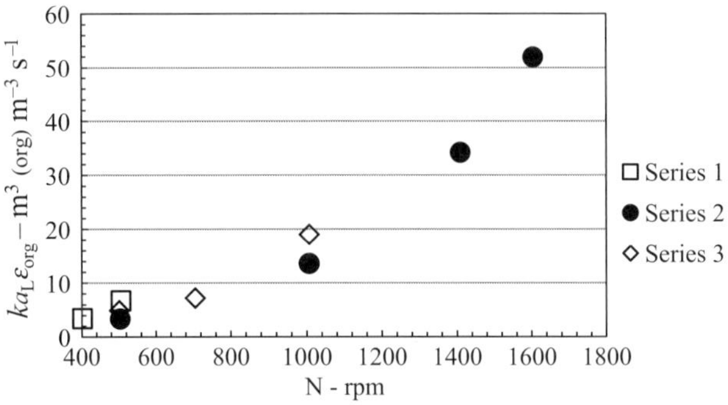

Figure 5.11 Overall mass transfer rate $ka_L\, e_{Lorg}$ as a function of the disc rotational speed, aqueous : organic rate (q), and total liquid rate (f_{TOT})

Series 1: $q = 1.1$; $f_{TOT} = 4.7 \times 10^{-6}\ m^3\ s^{-1}$
Series 2: $q = 5.6$; $f_{TOT} = 14.7 \times 10^{-6}\ m^3\ s^{-1}$
Series 3: $q = 6.8$; $f_{TOT} = 17.2 \text{ x } 10^{-6}\ m^3\ s^{-1}$

Reprinted (Adapted) from Visscher, F., van der Schaaf, J., de Croon, M. H. J. M., and Schouten, J. C. (2012). Liquid-liquid mass transfer in a rotor-stator spinning disc reactor. *Chemical Engineering Journal*, **185/186**, 267–273. Copyright (2012), with permission from Elsevier

for optimizing mass transfer and reaction performance. The work of Visscher et al. (2012, 2013), and Martinez et al. (2017) have made some significant progress in experimental flow interrogation techniques.

In regards to the rotor–stator spinning disc contactor shown in Figure 5.10, an important finding of the work of Visscher et al. (2012) is that the flow pattern of the

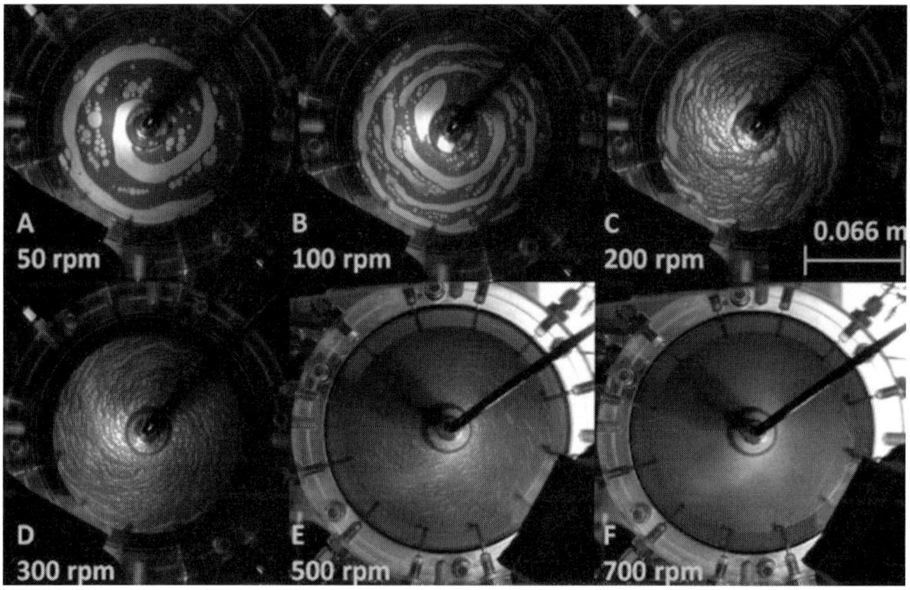

Figure 5.12 Images of flow pattern inside the rotor–stator spinning disc contactor.
Reprinted (Adapted) from Visscher, F., van der Schaaf, J., de Croon, M. H. J. M., and Schouten, J. C. (2012). Liquid–liquid mass transfer in a rotor–stator spinning disc reactor. *Chemical Engineering Journal*, **185/186**, 267–273. Copyright (2012), with permission from Elsevier

dispersed phase, in this case the n-heptane, was shown to be plug flow with uniform fluid residence time. Figure 5.12 shows a sequence of high speed photographs taken from the underside of the contactor at difference rotor speeds. The lighter portion of the image shows the movement of the n-heptane in a spiral, moving toward the center of the contactor with minimal evidence of mixing and thus it is assumed that the residence time for all fluid elements is approximately the same.

Three distinct flow patterns were observed as the rotational disc speed was increased. At speeds below 100 rpm, a radial inward flow of the organic n-heptane phase is observed as a continuous spiral with little evidence of back-mixing. The spiral flow of the organic phase appears entwined with a spiral flow of the aqueous phase in which droplets of organic phase may also be observed. From the experimental observations of the flow patterns, it was deduced that both organic and aqueous phases exhibit plug flow behavior at low rotational speed. At rotational speeds between 100 and 300 rpm, breakup of the continuous organic phase into a dispersion was observed, explained by the high shear forces generated between the rotor and stator. The droplets resulting from the breakup were observed to move still on a spiral trajectory with no evidence of back-mixing, consistent with uniform residence time. Above 300 rpm, significantly greater dispersion of the organic phase was observed, with large reductions in drop diameter to values less than 10^{-3} m being recorded. Significant back-mixing was also seen at higher rotational speeds and this was explained by formation of boundary layer flows on the rotor and stator surfaces (Poncet, Chauve, and Schiestel, 2005). Based on these observations it was assumed that both phases approach ideal mixing as the rotational speed is increased.

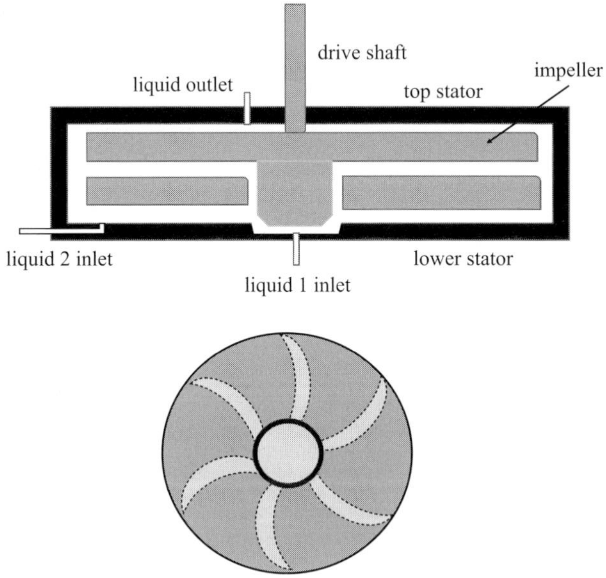

Figure 5.13 Arrangement of the impeller–stator spinning disk contactor.
Reprinted (adapted) from Visscher, F., Nijhuis, R. T. R., de Croon, M. H. J. M., van der Schaaf, J., and Schouten, J. C. (2013). Liquid–liquid flow in an impeller–stator spinning disc reactor. *Chemical Engineering and Processing*, **71**, 107–114. Copyright (2013), with permission from Elsevier B.V. All Rights Reserved

Another significant finding was the reduction in the volumetric holdup of the dispersed organic phase as the rotational speed of the disk was increased. The decrease in volume fraction was explained by the reduction in drop size of the dispersed organic phase due to creation of higher shear rates between stator and rotor. The resulting smaller drops experience enhanced entrainment by the continuous phase and thus lower residence time in the contactor.

These studies were further developed for the spinning disk contactor principle for liquid–liquid contact by replacing the solid rotor in the contactor described above, with an impeller (Visscher et al., 2013). The impeller design studied was comparable to that used in a conventional centrifugal pump. The fluid dynamics in the contactor are very similar to those in a centrifugal pump such that the fluid is forced outwards from the center of the impeller acquiring increased kinetic energy in the process. The design of impeller in the spinning disk contactor shown in Figure 5.13 promotes selective flow-through of the liquid of the lighter phase and increases the holdup of the heavier phase, as shown in Figure 5.14. It was claimed that the flow regime in the impeller–stator spinning disk contactor may be exploited to promote sequential mixing and separation of the phases, which is an important phenomenon for optimization of mass transfer. Using this type of impeller, six distinct flow regimes were identified for the liquid–liquid system n-heptane/water.

The observed flow regimes in this impeller–stator spinning disc reactor were distinguished by the liquid that is present as the continuous phase and by the degree of

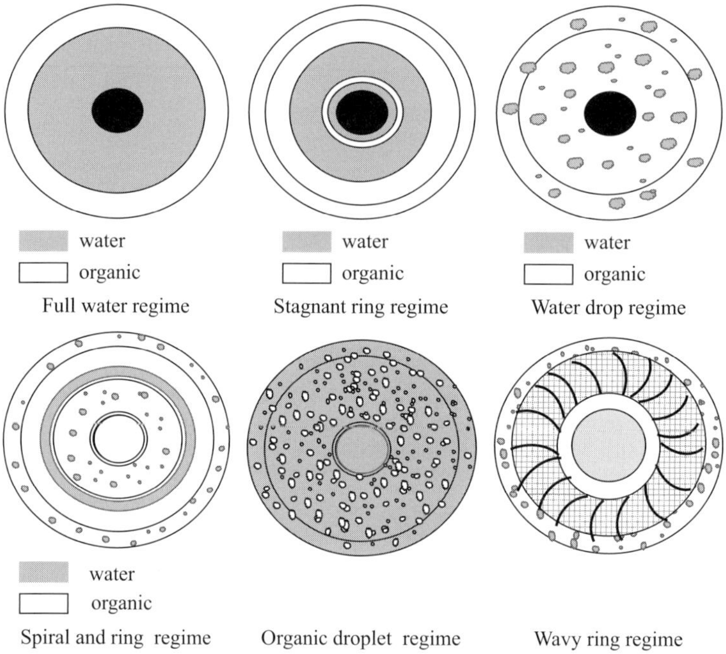

Figure 5.14 Six regimes of flow determined in the impeller–stator spinning disk liquid–liqiud contactor as described by Visscher et al. (2013).

dispersion of the dispersed phase. The representation of the flow regimes as a function of the rotational speed and the aqueous volume fraction is further denoted as the flow map:

- full water regime;
- stagnant ring regime;
- water droplet regime;
- heptane spiral and ring regime;
- heptane droplet regime;
- wavy ring regime.

The existence of a particular regime was detemined by the speed of the impeller and the phase ratio. The relationship is complex and is represented graphically as shown in Figure 5.15 for the system water/n- heptane. Notable is the phase inversion that occurs as the impeller speed is increased at low values of water/organic ratio in the upper left of Figure 5.15.

5.5 Combined Field Contactors

Another variation on the spinning disk contactor for liquid–liquid systems is the combined field contactor, as described by Arnott (1993) and by Arnott and Weatherley (1994a, 1994b). The arrangement is shown in Figure 5.16.

The main differences of this contactor compared with other spinning disk designs are three fold. Firstly, the design was conceived as a development of electrically enhanced

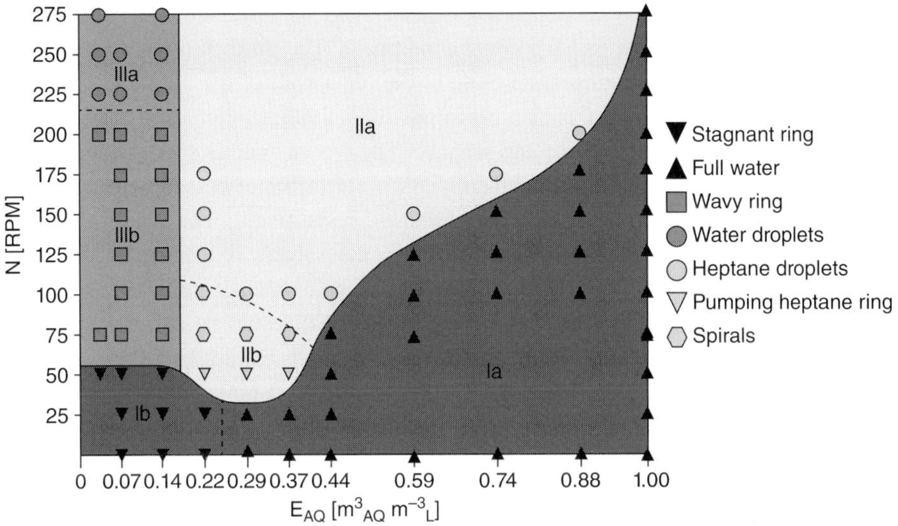

Figure 5.15 Flow regimes for the system n-heptane/water in the impeller–stator spinning disk liquid–liquid contactor.

Reprinted (adapted) from Visscher, F., Nijhuis, R. T. R., de Croon, M. H. J. M., van der Schaaf, J., and Schouten, J. C. (2013). Liquid–liquid flow in an impeller–stator spinning disc reactor. *Chemical Engineering and Processing*, **71**, 107–114. Copyright (2013), with permission from Elsevier B.V. All Rights Reserved

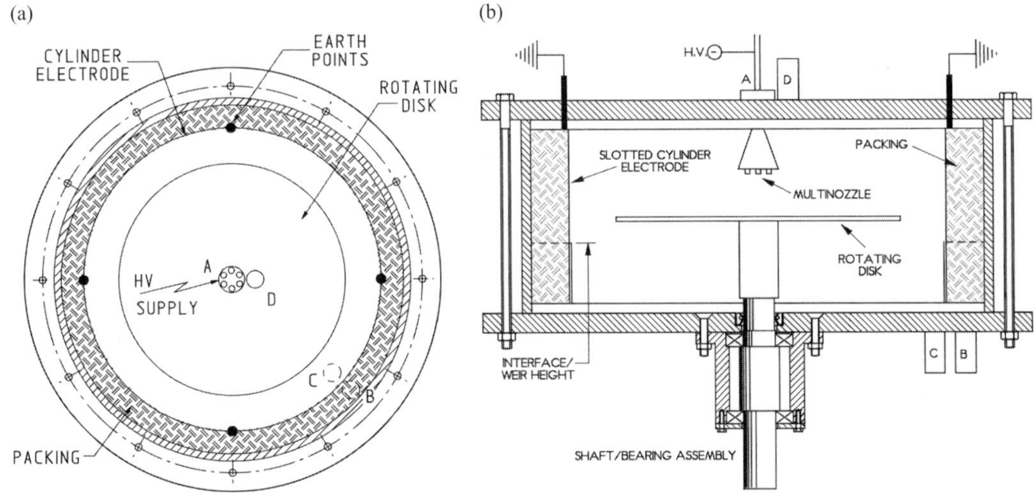

Figure 5.16 Arrangement of a novel spinning disk liquid–liquid contactor, combining intensified dispersion and coalescence.

liquid–liquid contact, and the dispersed liquid is introduced into the second liquid via an electrostatically charged multi nozzle shown, which is fitted to the top lid of the main vessel, as shown in Figures 5.17(a) and (b). The second difference is that the contactor vessel is filled with the continuous phase with arrangements for continuous phase feed and exit labeled in Figure 5.16 at D and B respectively. The rotation of the disk provides the main function of promoting shear and localized mixing between the electrically dispersed phase and the continuous phase. The third difference is the incorporation of an annular packed section close to the outer wall of the cylindrical containment vessel. The function of the packing is to provide a relatively quiescent zone in which the mixed phases can coalescence and separate. There is an annular space between the outer surface of the packed section and the inner wall of the main vessel, with exit ports for separate removal of the light and heavy phases, respectively. The intensification achievable in separation in liquid–liquid systems is the subject of Chapter 7, but the aim of this contactor design was to exploit the mass transfer intensification achievable by electrostatic dispersion, in a small contactor geometry that would be suitable for scale up. Liquid–liquid contact in a device that combines the advantages of large reduction in drop size achieved by electrostatic fields at the liquid feed point with the application with high gravity flow conditions should make this possible. The electrical field is created across the

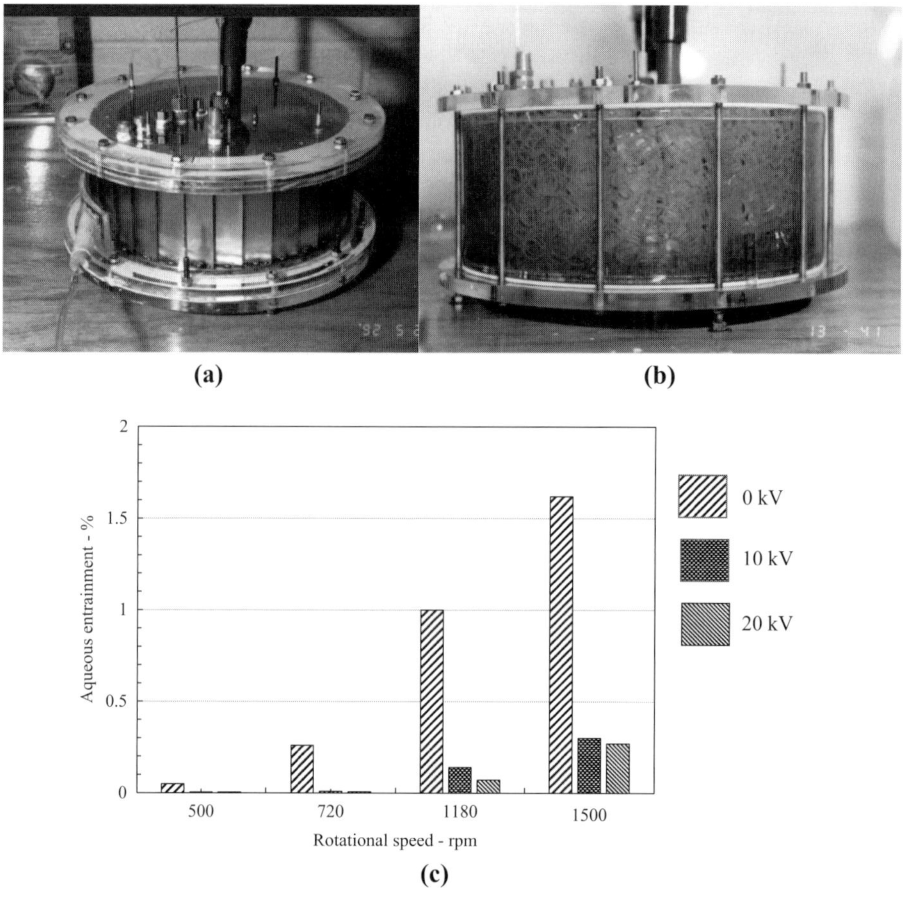

Figure 5.17 Arrangement of a novel spinning disk liquid–liquid contactor, combining intensified dispersion and coalescence. (a) The slotted partition dividing the inner mixing section from the outer annular coalescence section. (b) The packing in place in the outer coalescence section. (c) Entrainment reduction in a combined field intensified liquid–liquid contactor. Heavy distillate/water mixtures with aqueous dispersed ratio of 20:1. (Arnott and Weatherley, 1994a, 1994b)

radius of the contactor, with charge applied the dispersion nozzle located at the top of the main vessel and a corresponding earth electrode that forms the gridded containment of the cylindrical coalescence packed section, as shown in Figures 5.17(a) and (b). The net effect of the electrical field is to further accelerate the movement of droplets toward the outer region of the contactor, thereby augmenting centrifugal force with the electrostatic forces acting on the swarm of droplets moving radially outwards.

Entrainment data for the combined field contactor are shown in Figure 5.17(c). The volumetric entrainment of aqueous phase in the organic phase outlet is shown as a function of rotor speed, and with respect to the magnitude of the DC voltage applied to the nozzle electrode. There are two points of significance in these data. Firstly, a significant increase in entrainment of aqueous phase is observed as the rotor speed is increased. This is not surprising. The most notable observation is the very large decrease in entrainment when a high DC voltage is applied, with greater than 5-fold reduction observed when voltage is present at a rotor speed of 1500 rpm. At the lower rotor speed of 1180 rpm, the reduction in entrainment is even greater, with a 14-fold reduction in aqueous entrainment observed when a DC voltage of 20 kV is applied.

Another conceptual design of a novel intensive liquid–liquid contactor is described by Miller and Weatherley (1989). This was based on a patented design by Bowe, Oruh, and Singh (1985) known as the vortex contactor. The arrangement of the contactor is shown in Figure 5.18(a). The operating principle is that of a two chambered mixer settler with introduction of a liquid–liquid dispersion tangentially into the lower cylindrical chamber of the device (labeled A). In this chamber the dispersion is exposed to high gravity force created by the rotational flow promoted by the tangential feed arrangement. The kinetic energy of the tangential feed associated with the flow promotes rapid and intensive mixing within the chamber. There are conceptual similarities with the cyclonic liquid–liquid mixer discussed by Treybal (1955) and later by Bowe et al. (1985). The dispersion is then discharged through a vertical port (labeled B) leading from the center of the upper face of the lower chamber, discharging into an upper settling section located above the mixing chamber. The mixture exits from the upper end of the discharge port into the upper section of the contactor (labeled C), which provides a quiescent settler zone in which phase separation may occur. The light phase is discharged from the upper chamber through a port located at the upper end of the chamber and the heavy phase from a port located at the lower end of the chamber.

Figures 5.18(b) and (c) show a further development of the vortex contactor, also described by Millar and Weatherley (1989), in which the two feed liquid phases are predispersed by mean of continuous electrostatic spraying. The organic and aqueous phases are fed perpendicular to each other in the cruciform assembly shown on the right-hand side of Figure 5.18(b). The electrical dispersion is generated by means of a high voltage being imposed across the center of the cruciform perpendicular to the direction of flow of the organic phase. The dispersion is then fed into the vortex contactor as shown in Figure 5.18(a) and as described above. Sample results for the solvent extraction of ethanol from an aqueous solution into a hydrocarbon solvent are shown in Figure 5.18(c). The progress of the extraction is expressed as the percent approach to equilibrium during extraction using the contactor in recirculating batch mode. Comparison between extraction rates using the electrically generated dispersion compared with only hydrodynamic

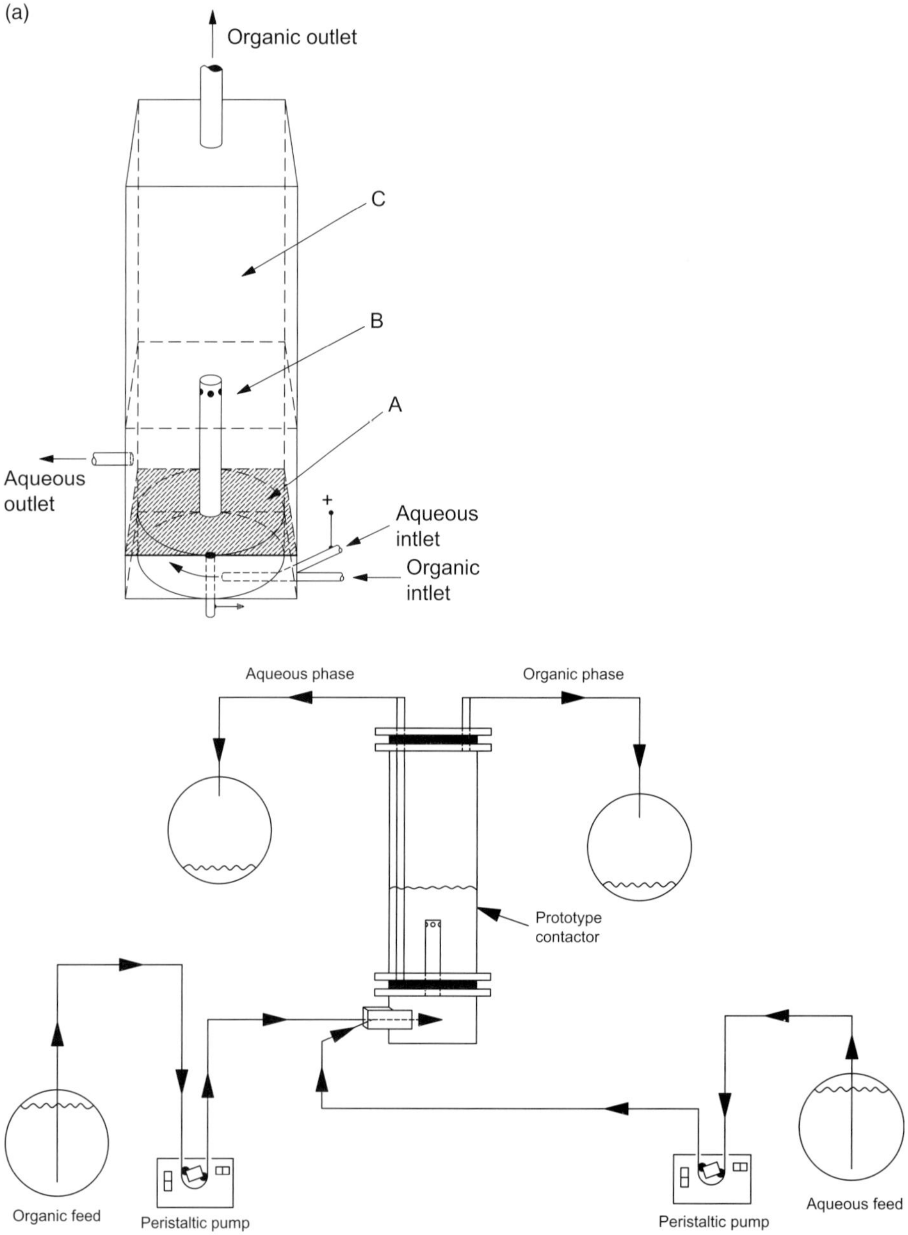

Figure 5.18 (a) Vortex contactor for intensive liquid–liquid contact: general arrangement. (b) Vortex contactor for intensive liquid–liquid contact, combined with electrically intensified predispersion. (c) Ethanol extraction in the vortex contactor showing comparison of hydrodynamic pre-dispersion with combined hydrodynamic and electrostatic pre-dispersion. Millar and Weatherley (1989) – *re-used with permission from Elsevier*

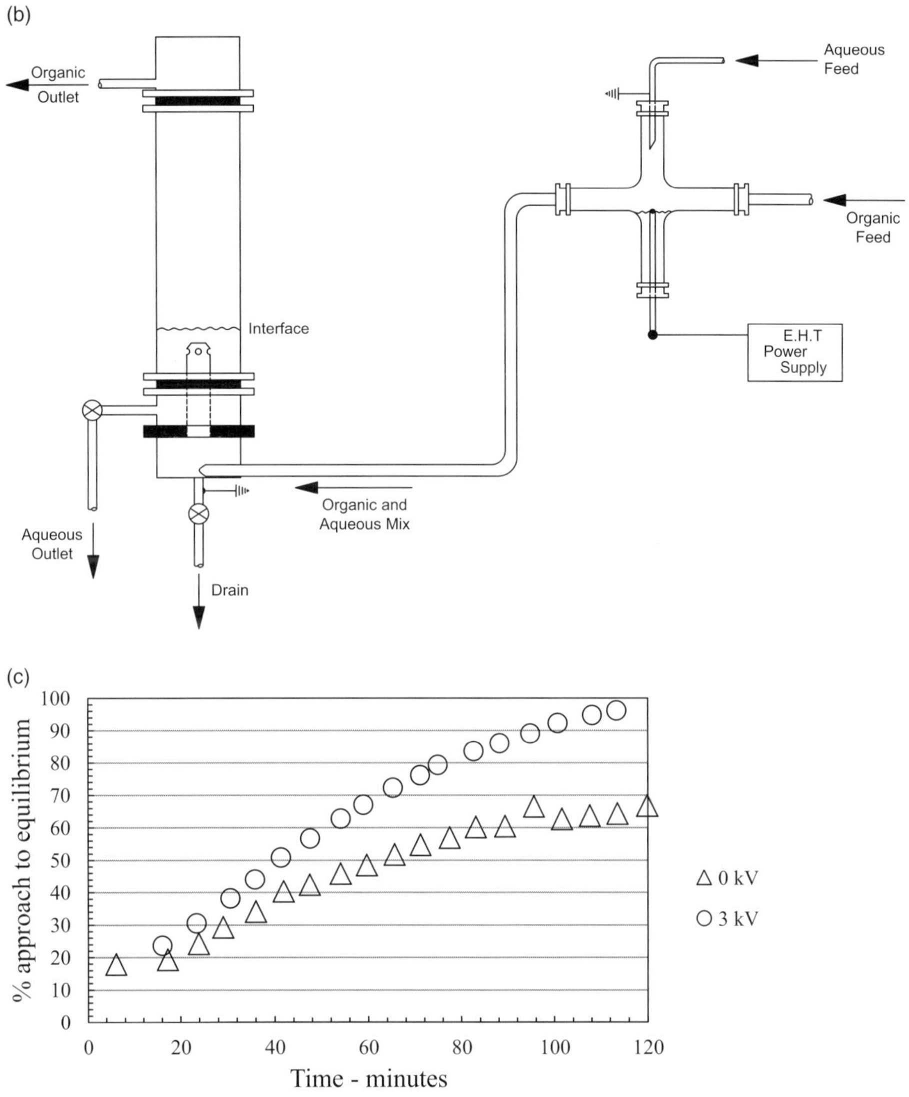

Figure 5.18 (cont.)

dispersion (with no applied electrical field) in Figure 5.18(c) shows favorable intensification of extraction rate with the electrically generated dispersion.

5.6 Modeling of Liquid–Liquid Systems in High Gravity Fields

5.6.1 Fundamental Summary

This section is a fundamental summary of controlling equations for fluid dynamics of liquid–liquid dispersions in a centrifugal field (e.g. conventional liquid–liquid centrifugal separators).

The application of centrifugal forces for the enhanced processing of liquid–liquid mixtures was initially concerned with acceleration of phase separation in order to reduce equipment size, to facilitate the separation of mixtures that experienced slow rates of settling due to small density differences or for the separation of mixtures that are partly stabilized by the presence of surfactants. As a starting point for theoretical analysis of liquid–liquid system behavior in the presence of centrifugal fields, a brief summary of the fundamental relationships is presented.

Jammoal and Lee (2015) present one of the few recent quantitative analyses of the motion of liquid droplets under centrifugal forces in a liquid–liquid system. The starting point is a consideration of a rigid spherical drop moving in a stagnant continuous phase; the governing equation is written as equation 5.1. This shows the net accelerative force acting on the drop as the sum of the gravitational force, accelerative force exerted due to motion of the continuous phase (in this case zero), and the drag force. The final term on the right-hand side of equation 5.1 is known as the acceleration history integral, which is significant in the case of transient flows.

$$m_\mathrm{d}\frac{dU_\mathrm{i}}{dt} = (m_\mathrm{d} - m_\mathrm{c})g_\mathrm{i} - \frac{1}{2}m_\mathrm{c}\frac{dU_\mathrm{i}}{dt} - 6\pi a\mu U_\mathrm{i}(t) - 6\pi a^2\mu\int_0^t\left(\frac{dU_\mathrm{i}(\tau)/d\tau}{[\pi\upsilon(t-\tau)]^{1/2}}\right)d\tau \quad (5.1)$$

Jammaol and Lee develop this further for radial motion in a centrifugal field where the force on the moving drop continually changes. This specifically impacts the description of the drag forces. Equation 5.2 expresses the force balance taking this into account. The net drag force, F_D, acting on the drop in the case of unsteady flow is shown, and comprises three components: steady state drag, added mass term, and the acceleration history integral. Δ_A and Δ_H are empirical coefficients relevant to flows outside the creeping flow regime:

$$F_\mathrm{D} = -\frac{\pi d^2}{8}\rho_\mathrm{c}C_\mathrm{D}U^2 - \Delta_\mathrm{A}\frac{\pi d^3}{12}\rho_\mathrm{c}\frac{dU}{dt} - \frac{3}{2}\Delta_\mathrm{H}d^2\sqrt{\pi\rho_\mathrm{c}\mu}\int_0^t\frac{(dU/dt)_{t=\sigma}}{\sqrt{t-\sigma}}d\sigma \quad (5.2)$$

Equation 5.2 is developed further for the actual case of a drop moving in a centrifugal field. Coriolis forces and Magnus forces (associated with spinning objects) are assumed to be insignificant and thus equation 5.3 is written to include just the additional effect of the centrifugal force induced by the circular motion of the continuum where ω is the angular velocity.

$$\begin{aligned}\frac{\pi d^3}{6}\rho_\mathrm{d}\frac{dU}{dt} &= \frac{\pi d^3}{6}\Delta\rho\omega^2 r - \frac{\pi d^2}{8}\rho_\mathrm{c}C_\mathrm{D}U^2 - \Delta_\mathrm{A}\frac{\pi d^3}{12}\rho_\mathrm{c}\frac{dU}{dt}\\ &\quad - \frac{3}{2}\Delta_\mathrm{H}d^2\sqrt{\pi\rho_\mathrm{c}\mu}\int_0^t\frac{(dU/dt)_{t=\sigma}}{\sqrt{t-\sigma}}d\sigma\end{aligned} \quad (5.3)$$

It is assumed that the initial drop velocity is zero, and Stokes law is assumed to describe drag force coefficient C_D. The velocity term, U, and characteristic time for the drop are expressed in dimensionless form respectively as $U = U/U_\mathrm{o}$, and $\tau = \rho_\mathrm{D}{}^2/18\mu$. Using Laplace transforms, the overall equations of motion (force balance) were rewritten in explicit form, allowing numerical solution to obtain velocity and positional data for individual drops moving through the continuous phase under the influence of the centrifugal field.

The predicted trajectories were accurately validated by experimental measurements of droplets moving in a rotary contactor of diameter 0.4 m, as shown in Figure 5.19, showing comparisons of experimental measured droplet velocities with calculated values based on the solution of the above equations (Jammoal and Lee, 2015). The motion of drops were considered in two different systems: water/butanol, which has a low density difference and low interfacial tension, and the system water/dodecane, which exhibits high density difference and high interfacial tension. In both cases excellent agreement is shown. The drop trajectories are presented as mean velocity against radial position. The drops are introduced close to the center of the rotating assembly and move outwards in both cases. The velocity tends to increase as the drops move outwards from the center. The subsequent rapid drop in velocity reflects the drops moving out of the region of the rotating assembly at its outer edge.

5.6.2 Modeling of Spinning Disc Contactors

Oxley et al. (2000) provide a concise theoretical analysis of thin film flow across the free surface of the spinning disk contactor described earlier (see Figures 5.2 and 5.3). The flow arrangement analyzed is summarized in Figure 5.13, where the spinning disk is shown as a horizontal circular plate with the liquid feeds pumped to a well at the center of the disk. The case study used was the production of pharmaceutical product via a phase-transfer catalysis. The liquids flow at high velocity across the radius of the disk, forming a film of thickness δ. The liquid mixture exits the disk at the periphery and flows to the base of the outer assembly. As summarized earlier, the shear stresses to which the liquids are exposed can be significant, giving rise to high heat and mass transfer coefficients at the interface between the liquid surface and surrounding gas, and between the lower interface of the liquid with the disk surface. The analysis presented assumes that the flow of the liquid film is a rippling type flow that is inherently unstable and disturbed. The mixing intensity and residence time distributions are key factors that may be considered usefully by theoretical analysis, since these factors have major influence on the degree of separation or conversion that may be achieved. Shear stress determines the efficiency of mixing and is based on the assumptions of laminar flow, which is also circumferentially uniform. Oxley et al. express the shear stress according to equation 5.4:

$$\tau = -1.5\eta \frac{v_{\text{rel}}}{\delta} = -1.5\sqrt[3]{\frac{\eta \rho^2 Q_{\text{V}} \omega^4 r}{18\pi}} \tag{5.4}$$

It is noteworthy that the angular velocity, ω, is a dominant term in equation 5.4 and therefore rotational speed is a key operating variable with a major influence on shear stress, where ω is related to disk rotational speed (equation 5.5):

$$\omega = \frac{2\pi}{60} \tag{5.5}$$

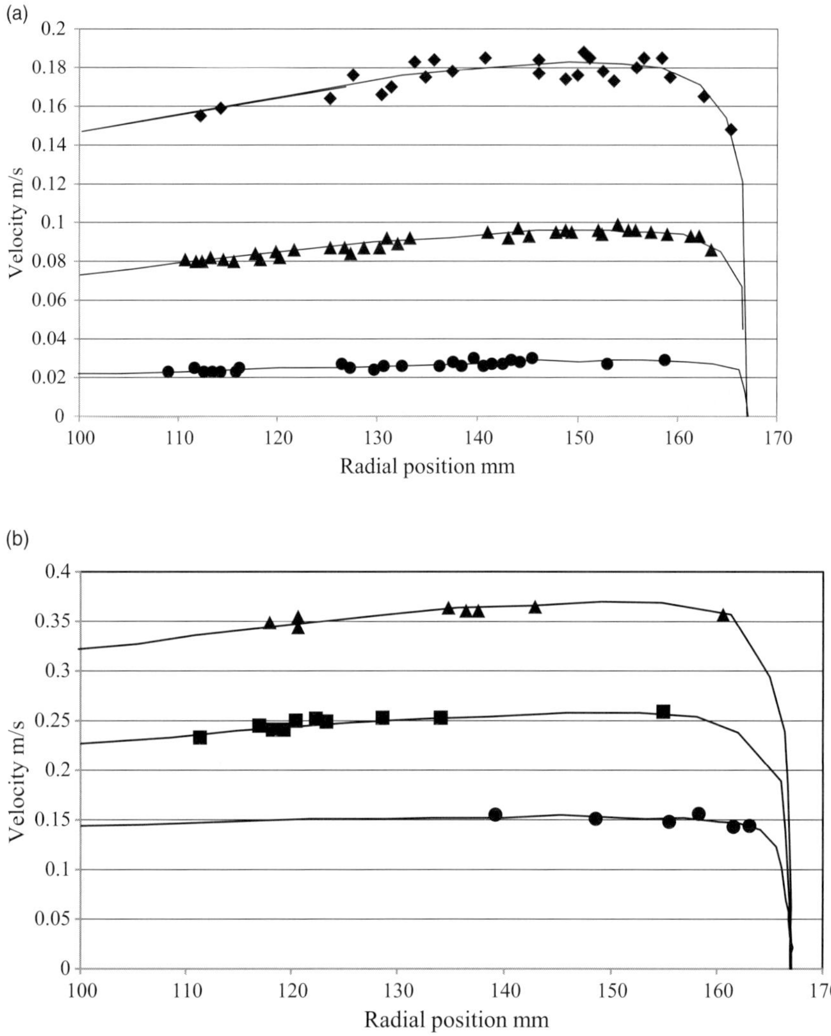

Figure 5.19 (a) The effect of rotational speed on drop motion in the butanol/water system: diamonds are 200 rpm, $d = 0.427$ mm; triangles are 600 rpm, $d = 0.426$ mm; circles are 400 rpm, $d = 0.421$ mm; continuous lines show model fits. (b) The effect of rotational speed on drop motion in the dodecane/water system: diamonds are 200 rpm, $d = 1.43$ mm; triangles are 600 rpm, $d = 1.43$ mm; circles are 400 rpm, $d = 1.417$ mm; continuous lines show model fits. Reprinted (adapted) with permission from Jammoal, Y. and Lee, J. M. G. (2015). Drop velocity in a rotating liquid–liquid system. *Chemical Engineering Research and Design*, **104**, 638–646. Copyright (2015) Elsevier B.V. All Rights Reserved.

Equations 5.4 and 5.5 provide a useful relationship between rotational speed and shear stress. In the case of laminar flow, equation 5.4 would be regarded as quantitative. The equations also provide a qualitative description in the case of turbulent flow. Oxley et al. cite laminar flow in the film to be present when Re < 4 and turbulent at Reynolds numbers greater than 20.

Oxley et al. also present a relationship for residence time on the disk (equation 5.6):

$$t_{\text{res}} = \left(\frac{81\pi^2\eta}{16\omega^2 Q_{\text{V}}^2\rho}\right)^{1/3}\left(r_{\text{o}}^{4/3} - r_{\text{i}}^{4/3}\right) \tag{5.6}$$

This shows that residence time on the disk is inversely proportional to angular velocity $\omega^{0.67}$. The conclusion here, taking account of equations 5.4 and 5.6, is that achievement of high stress is only achievable at low residence times, for example 0.5 s is quoted as a typical figure for high conversions in the case of a reacting system.

5.6.3 Modeling Spinning Disc Contactors: Impinging Jet Systems (Qiu, Petera, and Weatherley, 2012)

The development of a validated model for the spinning disk contactors enables efficient and low-cost exploration of alternative designs, operating conditions, and solvents. A quantitative approach provides valuable insights into the detailed hydrodynamic conditions on the disk surface and in the case of the impinging jet spinning disk contactor, the conditions within the interdisk space in the case of multidisk geometries. On account of the complexity, modeling methods build upon the successful discrete-phase finite element method approach to the simulation of mixing, and single and swarming drop motion in liquid–liquid systems, as described in Chapter 3. The system of equations describing the hydrodynamics close to the surface of the disk is based on classical Navier–Stokes for momentum conservation, equation of continuity, energy balance, and diffusional transport. On the basis of numerical modeling, it can be shown that novel finite element analysis is a feasible method for calculation of the hydrodynamic conditions and mass transfer in the high shear environment of the impinging jet spinning disk reactor. The influence of flow rate, reactor geometry, and liquid properties are the key variables a model is required to accommodate. The key parameters are external hydrodynamic conditions, temperature, and physical properties.

The approach to modeling can be broken down into three parts:

1. modeling of the motion of a single phase liquid to determine predictions of back-mixing, recirculatory flows, and residence time distributions;
2. modeling of the liquid–liquid flow regimes to establish the effect of disk geometry upon a two-phase flow regime, drop size, and residence time distributions of both liquid phases;
3. prediction of overall contactor performance through combining two-phase fluid flow modeling with mass transfer and reaction kinetics to predict the overall performance of the spinning disk reactor (SDR).

The first part has been successfully demonstrated through mathematical modeling of single-fluid, nonhomogeneous, multicomponent flow in an impinging jet SDR. The approach was classed as the so-called mixture model (Petera and Weatherley, 2001), which can be used in different ways. For example, the approach can be adopted for multiphase flows where the phases move at different velocities but assume local equilibrium over short spatial length scales. It can also be used to model homogeneous multiphase flows with very strong coupling and phases moving at the same velocity and to calculate non-Newtonian viscosity. As mentioned, the mixture model uses a *single-fluid* approach, allowing the phases to be interpenetrating. The volume fractions can be equal to any value between 0 and 1, depending on the space occupied by a phase.

The mixture model allows the phases to move at different velocities, using the concept of slip velocities to be deployed, or they can be assumed to move at the same velocity. In the latter case the mixture model is then reduced to a homogeneous multiphase model, as is described below.

The system of equations to be solved is based on classical Navier–Stokes analysis for momentum conservation, the equation of continuity, energy balance equation, and diffusional transport. In the case of a reacting system, reaction kinetics may also be incorporated.

Here we take as an example the case of a parallel plate impinging jet spinning disk contactor, shown in Figure 5.20.

The spinning disk contactor shown in Figure 5.20 comprises a stationary lower disk, which is coaxially spaced adjacent to a rotating parallel disk separated by only a fraction of a millimeter. Rather than using a high speed jet, the two immiscible liquid phases are pumped from the centers of the stationary disk and the rotating disk, respectively. The example shown here from Qiu et al. (2012) shows the intensification that can be achieved in this type of device for a liquid–liquid application. The example presented is biodiesel synthesis through the transesterification of a triglyceride ester and methanol in the presence of an alkaline catalyst, in this case sodium hydroxide. With reference to Figure 5.20, the reactor arrangement comprises a top feed liquid of canola oil, and an

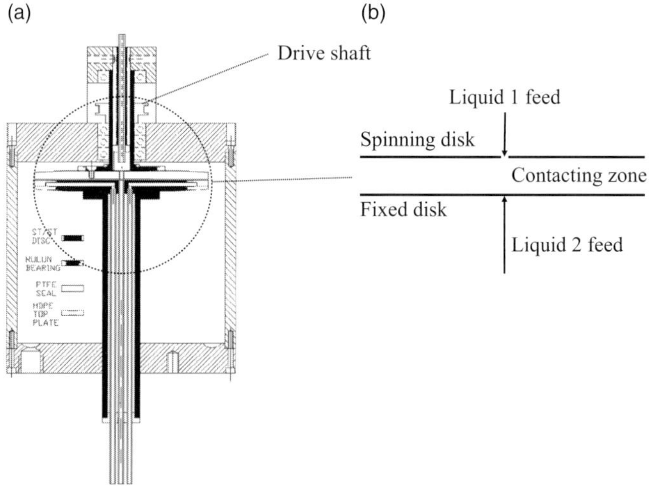

Figure 5.20 Parallel plate spinning disk liquid–liquid contactor.
Reused from Qiu, Z., Petera, J., and Weatherley, L. R. (2012). Biodiesel production using an intensified spinning disc reactor. *Chemical Engineering Journal*, **210**, 597–609. Copyright (2012), with permission from Elsevier B.V. All Rights Reserved

impinging feed of sodium methoxide (sodium hydroxide/methanol) fed through the center of the lower disk. This type of contactor combines a high mixing rate achievable through the impingement process combined with the high shear forces associated with the fluid in contact with the disk surfaces. The high gravity forces achieved through the disk rotation greatly increase the accelerative forces present in the interdisk space. Substantial improvements in interfacial mass transfer rate in a range of systems are demonstrated. This is attributed to higher specific interfacial area and to improved internal circulation rates within droplets and due to enhanced rates of interfacial shear on account of increased slip velocity between dispersed and continuous phases.

However, increase in mass transfer and reaction rate due to increase in rotational speed may be small and the effect of the impinging jet on the conversion, especially when the gap between the two disks is very small, can be significant (Dehkordi, 2002a). Data from Qiu et al. (2012) also confirm this in the case of biodiesel synthesis from canola oil, as shown in Figure 5.21. A major observation is the strong influence on the conversion of the gap width between the two discs. On the one hand the inverse relationship between conversion and gap size can be explained in terms of high shear rate and greatly intensified transport rates. On the other hand, the increase in gap width increases the effective reactor volume and hence increases the residence time, assuming that the flow rates are constant, as in this case. These data show that the intensification of heat transfer, mass transfer, and mixing achieved at the smaller gap widths more than offsets any opposing effect of reduced residence time.

Theoretical analysis of this example is presented, with the major focus on the fluid dynamics present in the interdisk gap (see Figure 5.20). Prediction of the flow conditions within the gap under different disk speeds and gap widths allows the continuum in the gap to be considered as a reactor with volume determined by the

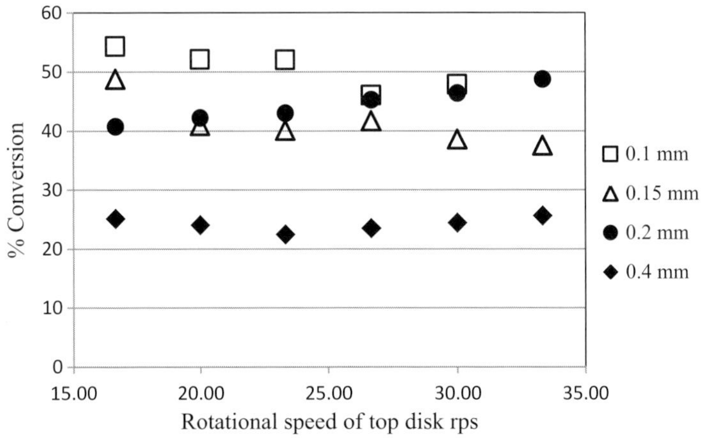

Figure 5.21 Effect of rotational speeds on the conversion of canola oil at 25 °C to biodiesel in a parallel plate impinging jet spinning disk liquid–liquid reactor as different interdisk gap widths. Reprinted/reused from Qiu, Z., Petera, J., and Weatherley, L. R. (2012). Biodiesel production using an intensified spinning disc reactor. *Chemical Engineering Journal*, **210**, 597–609. Copyright (2012), with permission from Elsevier B.V. All Rights Reserved

product of disk area and gap width. Consideration of reaction kinetics, together with heat and mass transfer, enables determination of the conversion achieved in the fluid leaving the outer edge of the gap. With reference to Figure 5.20(b), axial symmetry may be assumed, with the net flow of the liquid–liquid mixture directed from the center, radially outwards, to the edge of the interdisk space.

The following section is an excerpt reprinted with permission from Qiu, Z. Petera, J., and Weatherley L.R. (2012). Biodiesel production using an intensified spinning disc reactor. *Chemical Engineering Journal*, **210**, 597–609. Copyright (2012) Elsevier B.V. All Rights Reserved.

On account of axial symmetry, the computational domain may be assumed to be the radial section of the gap between the plates, and this is depicted in Figure 5.22. The algorithm for solving the controlling equations was developed assuming transient mode, starting with the interdisk space filled with one of the phases, for example the less dense phase, at zero velocity. The simulation is started with the injection of the heavy phase through the central orifice in the lower disk, and with the commencement of rotation, which is assumed to be instantaneous.

The domain of interest is the interdisk space, as shown typically in Figure 5.22.

The equations controlling the fluid dynamic are listed in equations 5.7– 5.16 below.

The equations are written both in general vector form and also in the axisymmetric form thus:

(a) Continuity:

$$\nabla\cdot\ \mathbf{v} = 0;\ \frac{1}{r}\frac{\partial(rv_r)}{\partial r} + \frac{\partial v_z}{\partial z} = 0 \tag{5.7}$$

$\mathbf{v}$ = velocity vector; r = radial coordinate; z = interdisk distance; p = pressure; μ = viscosity; v_r, v_z, = velocity in radial and axial directions, respectively.

(b) Momentum transfer (which is required for the velocity field calculation)

$$\mathbf{D} = \frac{1}{2}\left[\nabla\mathbf{v} + (\nabla\mathbf{v})^T\right] \tag{5.8}$$

$$\rho\frac{d\mathbf{v}}{dt} = \rho\mathbf{g} - \nabla p + \nabla\cdot(2\mu\mathbf{D})$$

Equations 5.8 may be expanded as follows:

In the radial (r) direction:

$$\rho\left(\frac{\partial v_r}{\partial t} + v_r\frac{\partial v_r}{\partial r} + v_z\frac{\partial v_r}{\partial z} - \frac{v_\theta^2}{r}\right) = -\frac{\partial p}{\partial r} + \frac{\partial}{\partial r}\left(\mu\frac{1}{r}\frac{\partial(rv_r)}{\partial r}\right) + \frac{\partial}{\partial z}\left(\mu\frac{\partial v_r}{\partial z}\right) \tag{5.9a}$$

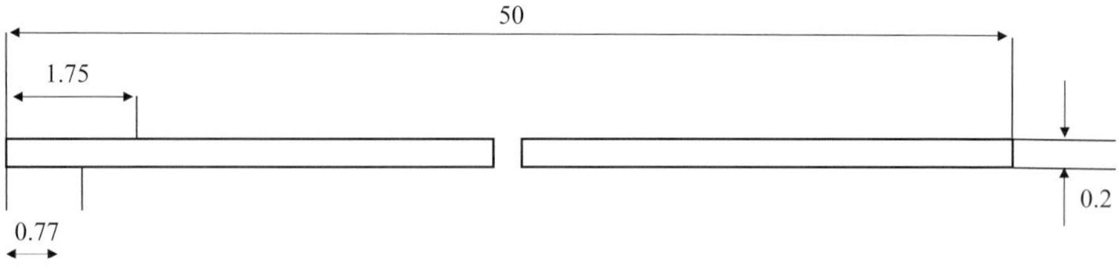

Figure 5.22 Computational domain: the space between spinning disk and fixed disk.

In the tangential *(θ)* direction:

$$\rho\left(\frac{\partial v_\theta}{\partial t}+v_r\frac{\partial v_\theta}{\partial r}+v_z\frac{\partial v_\theta}{\partial z}+\frac{v_r v_\theta}{r}\right)=\frac{\partial}{\partial r}\left(\mu\frac{1}{r}\frac{\partial(rv_\theta)}{\partial r}\right)+\frac{\partial}{\partial z}\left(\mu\frac{\partial v_\theta}{\partial z}\right) \tag{5.9b}$$

In the axial (z) direction:

$$\rho\left(\frac{\partial v_z}{\partial t}+v_r\frac{\partial v_z}{\partial r}+v_z\frac{\partial v_z}{\partial z}\right)=-\frac{\partial p}{\partial r}+\frac{1}{r}\frac{\partial}{\partial r}\left(\mu r\frac{\partial v_z}{\partial r}\right)+\frac{\partial}{\partial z}\left(\mu\frac{\partial v_z}{\partial z}\right)+\rho g \tag{5.9c}$$

The viscosity of the mixture, μ, may be calculated additively on the basis of concentrations and viscosities of individual components.

(c) The energy conservation equations required for calculation of the temperature field, may be written as follows:

$$c_p\rho\,\frac{dT}{dt}=\nabla\cdot(\lambda\ \nabla T)+\mu\ \dot{\gamma}^2 \tag{5.10}$$

$$\frac{\partial T}{\partial t}+v_r\frac{\partial T}{\partial r}+v_z\frac{\partial T}{\partial z}=\frac{1}{r}\frac{\partial}{\partial r}\left(\lambda r\frac{\partial T}{\partial r}\right)+\frac{\partial}{\partial z}\left(\lambda\frac{\partial T}{\partial z}\right)+\mu\dot{\gamma}^2 \tag{5.11}$$

The conductivity of the mixture may be calculated additively on the basis of concentrations and conductivities of individual components.

(d) The equations controlling mass conservation, (equations 5.12 and 5.13) are required for calculation of the concentration field. These include a reaction rate term, r_{xi}, which is derived from the detailed reaction kinetics of the reaction taking place between the two liquid phases.

$$\frac{dx_i}{dt}=\nabla\cdot\left(D_{eff,i}\nabla x_i\right)+r_{xi} \tag{5.12}$$

$$\frac{\partial x_i}{\partial t}+v_r\frac{\partial x_i}{\partial r}+v_z\frac{\partial x_i}{\partial z}=\frac{1}{r}\frac{\partial}{\partial r}\left(D_{i,eff}r\frac{\partial x_i}{\partial r}\right)+\frac{\partial}{\partial z}\left(D_{i,eff}r\frac{\partial x_i}{\partial z}\right)+r_{xi} \tag{5.13}$$

In order to establish a solution to these equations with the goal of calculating velocity and composition profiles in the interdisk space, two additional items of information are required. Firstly, the physical properties of the mixed phases are required and include density and viscosity. The density in many cases may be assumed constant, based on weighted average of the density values of the two feed liquids. The viscosity may exhibit a significant composition dependence in which case predictive methods such as the Wilke–Chang modification of the Stokes–Einstein equation can be built into the calculation algorithm. Secondly, equations are required describing the reaction kinetics that allow quantification of the rate term r_{xi} in equations 5.9 and 5.10. For the example of biodiesel synthesis, Qiu et al. (2012) provide a detailed analysis of the reaction kinetics and their incorporation into the calculation algorithm.

The algorithm described in detail by Qiu et al. is summarized as follows. Starting from the initial condition for velocity, temperature, and concentration at fixed boundary conditions listed below, the evolution of all these variables may be obtained consecutively from the differential conservation equations solution. The initial conditions may be imposed by assuming that at the start the whole domain (the gap) was

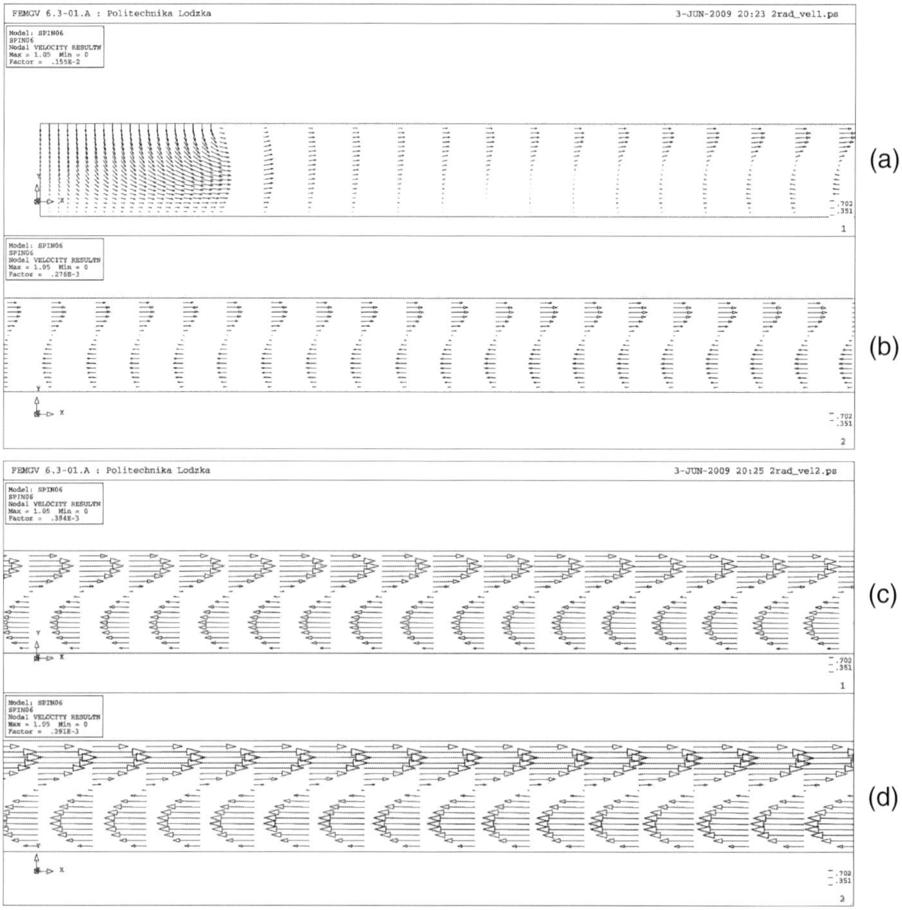

Figure 5.23 Predicted flow maps for the fluid flow through the interdisk space in the presence of rotation and continuous flow in the parallel plate spinning disk contactor (Petera, private communication). (a) Velocity map showing development of flow across the radius of the disk close to the center entrance port. (b)–(d) Velocity maps showing development of single-fluid multicomponent flow across the radius of the disk: (b) closer to center, (d) approaching the outer edge.

filled with fluid at rest and from this point the process started by injecting fresh liquid through the orifice near the central axis of the disks (far left in Figure 5.23). The next stage in the algorithm is the solution of the continuity and momentum conservation equations in the form as shown above for calculation of the velocity and pressure fields.

An example of the flow maps that can be predicted for the fluid flow through the interdisk space in the presence of rotation and continuous flow is shown in Figure 5.23.

The image in Figure 5.23(a) shows an example of the velocity field profile formation close to the inlet of the impinging jet spinning disk contactor. The boundary conditions for this example are as follows:

(i) $t = 0;\ R > r > 0$

$$v_z = v_r = v_\theta = v_{rot} = 0$$

(ii) $t > 0;\ r = 0$

$$z = 0;\ v_z = 7.874 \times 10^{-3}\ \text{m s}^{-1}$$

$$z = 2 \times 10^{-7}\ \text{m};\ v_z = 6.652 \times 10^{-3}\ \text{m s}^{-1}$$

The lower images (Figure 5.23b–d) show the predicted velocity map for development of single-fluid multicomponent flow across the radius of the disk moving outwards from the center. The image on the lower right (Figure 5.23d) shows the predicted velocity map toward the outer edge of the disk. This shows clear evidence of back-mix flow below the centerline. The predicted velocity maps in Figure 5.23 illustrate the complexity of the flow in the interdisk space during rotation. Validation of these predictions using experimentally obtained velocity profiles is yet to be established, but a degree of validation has been obtained through the measurement of reactor performance, as is now discussed briefly.

The concentration field predictions by Qiu et al. (2012) for multicomponent, multiphase flow with reaction are presented in Figure 5.24. As mentioned, the convection–diffusion–reaction (equations 5.4 and 5.5) were solved for individual components. The shading indicates in the form of contours the component concentrations coinciding with the zones of individual phases near the domain axis that do not mix near the inlet. Further toward the outlet the phases mix (according to the assumptions of the mixing model) and

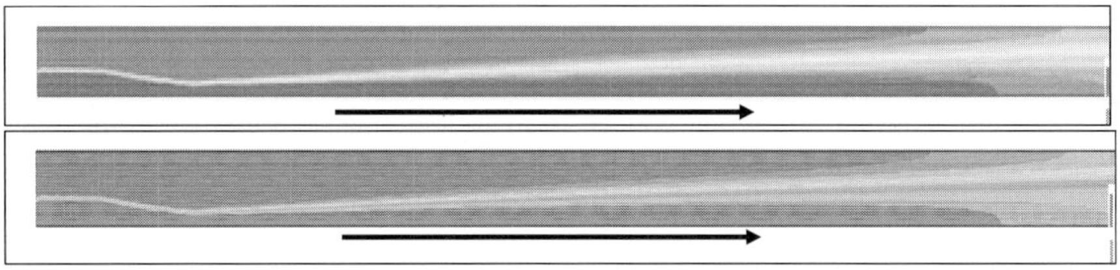

Oil phase concentration profile reducing left to right (darker to lighter)

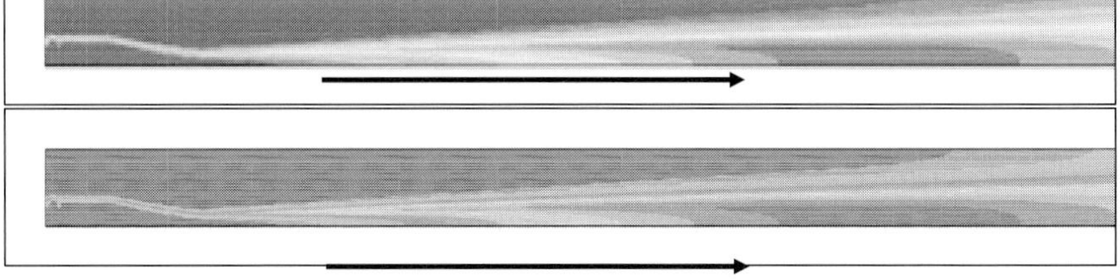

Biodiesel phase concentration profile increasing left to right (darker to lighter)

Figure 5.24 The concentration field with reaction and mass transfer in the gap near the inlet with an interdisk gap of 0.20 mm. Top: max = 1, min = 0; bottom: max = 0.947.

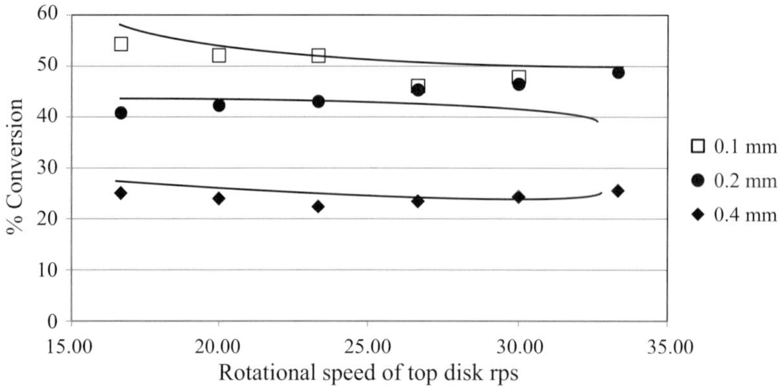

Figure 5.25 Effect of rotating speeds and gap size on the conversion of canola oil at 25 °C; comparison of simulations with experimental conversion for the parallel plate impinging jet spinning disk liquid–liquid reactor. (re-used with permission from Elsevier. Qiu, Z. Petera, J., Weatherley L.R. (2012). Biodiesel production using an intensified spinning disc reactor. *Chemical Engineering Journal*, **210**, 597–609)

there is are more distinct borders between them. The upper chart shows in dark color the upper layer of canola oil "shrinking" in thickness from left to right. The lower chart shows the development of the biodiesel, initially as a very fine light area between the upper and lower phases, expanding as the distance from the center increases from left to right.

Figure 5.25 compares experimental measurements of conversion of the canola oil to biodiesel in the spinning disk reactor with the simulations, showing good agreement.

The principle outcome of simulations such as are summarized here is development of computational tools that can be used for optimization and economic development of better designs of spinning disk systems for use for enhanced liquid–liquid contacting.

5.7 Spinning Tubes

Spinning tube contactors for the intensification of liquid–liquid processing represent another important development. As with the impinging jet liquid–liquid spinning disk contactor, the spinning tube contactor promotes high intensity hydrodynamic conditions by virtue of shear forces promoted in the liquid phases by the proximity of high velocity rotating surfaces. The spinning tube contactor typically comprises a rotor located coaxially to a stator, as shown diagrammatically in Figure 5.26. The rotor and stator are separated by a narrow annulus into which the two immiscible liquids are introduced. The relative angular motion of the rotor to the stator creates a high shear field across the narrow liquid continuum comprising the two liquid phases. Under the appropriate conditions, a Couette flow regime is generated that greatly intensifies transport between the two phases (Holl, 2006).

The basic concept upon which the design of spinning tube contactors is based is the creation of excellent hydrodynamic conditions in a small gap between two concentric cylinders of slightly different diameter (see Figure 5.26).

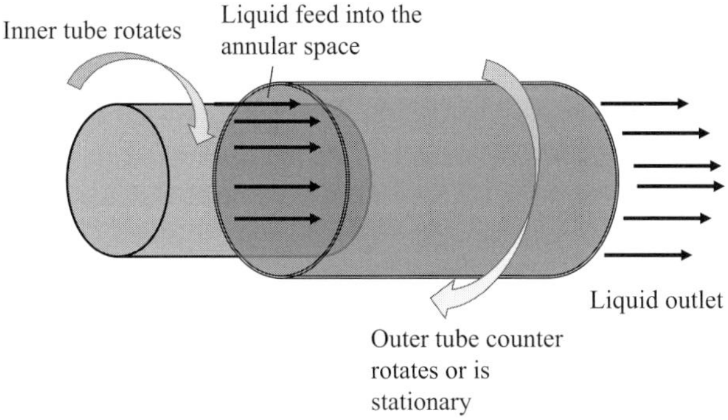

Figure 5.26 Conceptual arrangement of the spinning tube liquid–liquid contactor.

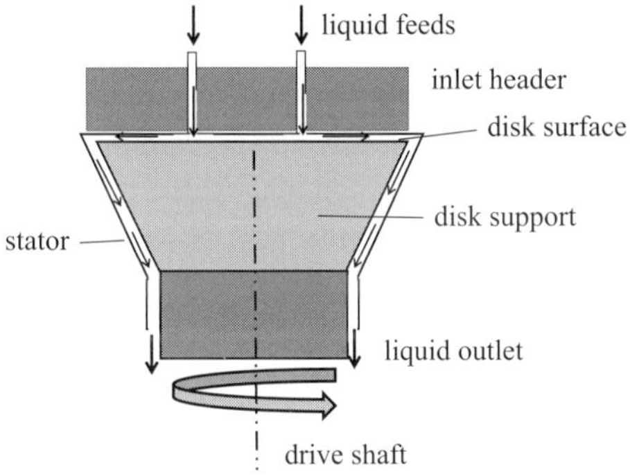

Figure 5.27 Outline flow scheme of a novel spinning cylinder liquid–liquid contactor. Adapted from Holl (2006)

One or both of the cylinders rotate, creating a complex flow field in the annular gap between the two cylinders. In general terms the operational variables for this system are the thickness of the annulus, the flow rates, volumetric holdup, orientation of the tubes from the vertical, rotational speed, and direction of rotation. Studies have shown that a variety of flow regimes can exist within the annular space, depending upon the variables listed above and upon the physical properties of the liquids.

Holl (2006, 2010) presented several variations on the spinning disc contactor concept. The first to be considered here is shown in Figure 5.27. The contactor employs a small spinning conical cylinder to promote intensive liquid–liquid contact. There are two liquid inlets, with the first inlet located above the surface of the cylinder but at the center of it. The second reactant is fed in such a way that it is intimately contacted with the film formed by the entry of the first liquid. The second liquid is thus fed through an annular Venturi nozzle located at some distance from the first feed point. The exterior of the contactor comprises a retaining surface that is coaxial with the disc surface and that allows the creation of a small gap between the outer surface and the rotating conical cylinder. The thickness of the gap between the cylinder surface and retaining surface can be adjusted to less than 1.00 mm. The retaining surface may also act as a heat exchange surface to control the temperature of the liquid mixture flowing through the

Table 5.2 Spinning tube in tube reactor performance (Holl), comparison with conventional CSTRs.

Reaction	Intensification in a spinning tube system (STT) compared with conventional CSTR
Styrene polymerization	1000
Amino acid synthesis	1200
Saponification	260
Free radical polymerization of acrylates	120
Sulfidation	100
Hydrosilyation	100

gap. It is claimed that the contactor can achieve uniform mixing of two or more liquids within several milliseconds. The mixing length is reported to be less than 5 mm.

As with the parallel plate impinging jet spinning disc contactors, the high shear field created in the gap promotes forced inter diffusion between the two liquids and thus very high rates of mass transfer. A limitation of this design may be the difficulty of injecting liquid through a Venturi nozzle into the very narrow gap. High energy consumption may be an issue in some cases. On the other hand, the arrangement can permit processing of more than two liquid feeds.

The impressive increases in reaction and transfer kinetics claimed in the invention give rise to the following possible explanation put forward by the authors of the invention, summarized in the words of the Holl patent as follows: "It is believed that achievement of fast inter diffusion is hampered significantly by the diffusion-retarding preponderance of what may be termed molecular clusters or swarms, inherently occurring in liquids or gases, within which clusters or swarms the molecules are anisotropically ordered from a kinematics point of view. Such ordering impedes rapid, natural inter diffusion due to the oscillation mode of the molecules within the clusters or swarms, consisting of large numbers of molecules oscillating in unison and unidirectionally on a scale <100 nm."

Holl put forward the argument that the molecular clusters present in multiphase systems mentioned above are not broken down in regular mixing equipment and thus mass transfer is still limited. The claim is made that the very high shear field created in the small gap between the disc surfaced and gap between the conical cylinder in the device breaks down molecular clusters, leading to much higher rates of reaction and transport.

A further patent by Holl (2010) describes a novel spinning tube contactor that comprises an outer stator and an inner rotor, separated by a small annular gap. The corresponding increases in kinetics are claimed to be 100–1000 times faster than in stirred tanks, and up to 100 times faster than in microreactors (Table 5.2) (Holl, 2003).

5.8 The Annular Centrifugal Contactor

A contactor design that is similar in concept to the spinning tube system (STT) contactor is the annular centrifugal contactor, described extensively in a series of papers (Meikrantz, Macaluso, and Flim, 2002; Tamhane, Joshi, and Patil, 2014; Vedantam et al.,

2012). The annular centrifugal liquid–liquid contactor (Figure 5.28), like the STT, comprises a pair of concentric cylinders, one of which is the inner rotor and the other, the stator in this case, is a fixed outer cylindrical containment vessel. The contacting of the two liquids occurs in the annular gap between the two cylinders as the inner rotor spins. The hydrodynamic regime in the annular gap is complex and highly dependent on

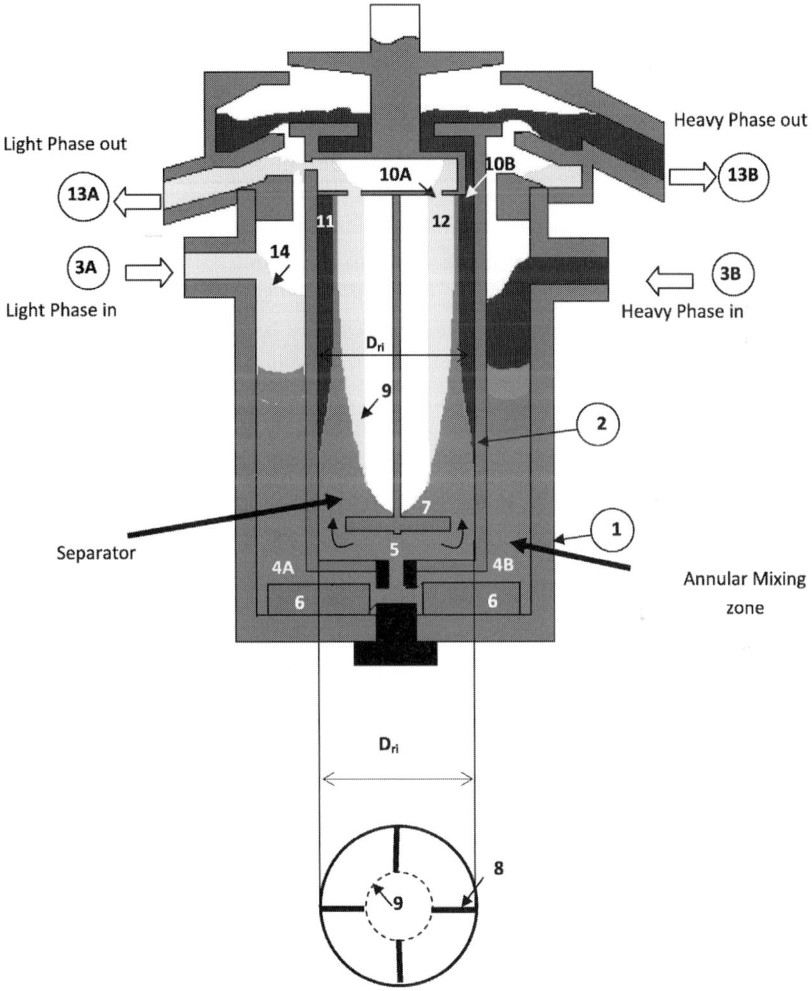

Figure 5.28 Schematic of the annular centrifugal contactor conceptual diagram of annular centrifugal extractor: (1) stationary cylinder, (2) rotating cylinder, (3a) light phase inlet, (3b) heavy phase inlet, (4a) and (4b) region below rotating cylinder, (5) central opening for rotating cylinder, (6) radial baffles on the stationary bottom plate, (7) deflecting baffle in the rotor, (8) vertical baffles in the rotor, (9) interface between air and light phase, (10a) and (10b) overflow weirs for lighter and heavier phase, respectively, (11) clean width for heavy phase, (12) clean width for light phase, (13a) and (13b) outlets for light and heavy phases, respectively, and (14) liquid level in the annulus.

Reprinted from Tamhane, T. V., Jyeshtharaj, B., Joshi, J. B., et al. (2012). Axial mixing in annular centrifugal extractors. *Chemical Engineering Journal*, **207–208**, 462–472.

the speed of rotation, the width of the gap, and the physical properties of the liquid mixture. The spinning of the rotor imparts power, which, according to Tamhane et al. (2012), may be in the range of 20–600 kW m^{-3}. The exposure of the two liquids to this power density results into a very fine dispersion. With reference to Figure 5.28, the liquid–liquid dispersion flows downwards through the annular region and most of the mass transfer occurs in this region. The mixture then flows toward the center, close to the base of the inner spinning rotor, and then into the base of the hollow rotor, as shown at point number 5 in Figure 5.28. The dispersed mixture entering at this point is now deflected toward the inner wall of the cylindrical rotor and this is achieved by the presence of a horizontal baffle (7) that directs the liquid flow accordingly. The heavy phase, shown in dark color in Figure 5.28, migrates to the inner wall of the rotor, and the whole of the lighter phase shown in light color migrates toward the center of the axis of rotation. Weirs are located at the upper end of the contactor that direct the respective separated light and heavy phase flows to the outlet ports. The flow regime promoted inside the rotor achieves highly efficient phase separation due to the high "g" forces within; thus fully separated liquid phases may be discharged at the upper end of the contactor at points 13A and 13B (as shown in Figure 5.28). It is very important that the length of the rotor is such that efficient phase disengagement can occur as the liquid mixture flows upwards through the rotor interior. Another novel feature of the design not shown on the diagram is the fitment of addition baffles inside the rotor that effectively create a series of compartments. Design considerations for efficient performance include drop size and density difference and location of the overflow weirs at the top of the contactor.

Several advantages of this particular contactor design that are highly relevant to the concepts of intensification are claimed. These include low volumetric holdup, short residence time, reduced liquid inventory, and flexible operating conditions (high turndown ratio). High mass transfer rates have been demonstrated on account of the high specific interfacial areas available in the annular region, consequential upon the high specific power dissipation rates that may be achieved (Kadam et al., 2008, 2009).

Kadam et al. (2009) present a comprehensive study of the annular contactor and list validated correlations for volumetric holdup, specific interfacial area, Sauter mean drop diameter, and energy consumption (mean energy dissipation) for the annular centrifugal contactors as listed below:

volumetric holdup, ε_D, is expressed in equation 5.14:

$$\varepsilon_D = 0.583\left(\frac{P}{V}\right)^{0.11}\left(\frac{Q_D}{Q_C + Q_D}\right)^{0.93}\left(\frac{\sigma^3\Delta\rho}{\mu_C^4 g}\right)^{-0.03}\left(\frac{\mu_D}{\mu_C}\right)^{-0.04} \tag{5.14}$$

Specific interfacial area $\underline{a}$, is expressed as follows in equation 5.15:

$$\underline{a} = 510\left(\frac{P}{V}\right)^{0.32}\left(\frac{Q_D}{Q_C + Q_D}\right)^{0.96}\left(\frac{\mu_D}{\mu_C}\right)^{-.15}\sigma^{-0.65}(1 + 8.5\varepsilon_D)^{-1} \tag{5.15}$$

Sauter mean drop diameter, d_{32}, is expressed as follows in equation 5.16:

$$d_{32} = 2.2 \times 10^{-2} \left(\frac{P}{V}\right)^{-0.33} (\rho_{\mathrm{C}})^{-0.19} \sigma^{0.6} \left(\frac{\mu_{\mathrm{D}}}{\mu_{\mathrm{C}}}\right)^{0.1} (1 + 8.5\varepsilon_{\mathrm{D}}) \tag{5.16}$$

Energy dissipation, P, is expressed in equation 5.17:

$$\frac{P}{N^3 D_{\mathrm{iO}}^5 \rho} = 41.55 \left(\frac{N D_{\mathrm{iO}}^2 \rho}{\mu}\right)^{-0.5} \left(\frac{\Delta r}{D_{\mathrm{iO}}}\right)^{-0.066} \left(\frac{h_{\mathrm{C}}}{D_{\mathrm{iO}}}\right)^{0.84} \tag{5.17}$$

where D_{iO} is the outer diameter of the rotor, Δr is the width of the annular gap, and h_{C} is the dispersion height in the annulus.

Applications of annular centrifugal liquid–liquid contactors include specialized areas such as in nuclear fuel separation and purification where containment, low maintenance, and sub-critical geometry are major safety considerations. Other applications where controlled shear fields are necessary, for example when handling the separation of sensitive biological molecules. This includes synthesis of monodisperse silica particles, regeneration of activated carbons used in water treatment, and chemical syntheses such as multiphase esterification and hydrolysis reactions (Vedantam and Joshi, 2006).

The design and simulation of high gravity contactors such as the annular centrifugal contactor present significant challenges. Optimization of the design must ensure that the internal geometries promote conditions for mass transfer rates, at the same time avoiding flooding and reducing the effects of back-mix flow. The important factors in understanding the hydrodynamic behavior in the annular centrifugal contactor are analyzed in some depth by Vedantam et al. (2012) and Tamhane et al. (2012). With reference to Figure 5.28, the key dimensions that influence hydrodynamic performance include the annular space between the inner rotating cylinder and the outer stationary cylinder. It is in this region where mass transfer occurs under intensive conditions. The internal weir heights are important dimensions, and the arrangement of internal baffles in the inner cylinder, which affects axial mixing, is also important. As with other liquid–liquid processing equipment, the optimization of phase separation is also essential and therefore the residence time and hydrodynamic conditions within the inner cylinder must be understood. The challenge for mathematical simulation, even for single phase flows, is formidable, and more so for multiphase liquid–liquid and liquid–liquid–gas flows. With regards to two-phase liquid–liquid flows, it is suggested that a pseudo-homogeneous liquid be assumed and, based on that assumption, useful insights into the effect of the above design parameters on the hydrodynamic performance may be obtained.

Considering initially the conditions in the annular region of this contactor, the development of unstable flows is of critical importance in promoting mass transfer. Much has been written on instability in annular geometry that is subject to external shear. Analysis by Taylor (1923) set criteria for instability phenomena for viscous fluids in a rotating concentric cylinder system by defining the Taylor number:

$$\mathrm{Ta} = \frac{4\Omega_i^2 d^4(\theta - \eta^2)}{\upsilon^2(\eta^2 - 1)} \tag{5.18}$$

$$\mathrm{Re} = \frac{R_i \Omega_i^2 d}{\upsilon^2} \tag{5.19}$$

The Taylor number, as defined in equation 5.18, in principle represents the ratio of centrifugal forces to viscous forces and was used to define the transition from Couette flow to Taylor vortex flow. This occurs at a value of *Ta* defined as the critical Taylor number $\mathrm{Ta_{crit}}$. $\mathrm{Ta_{crit}}$ has an approximate value of 1708.

Given the importance of instabilities for promoting interphase contact in liquid–liquid systems, some consideration of the various regimes is worthwhile. Fruh (2002), Syed and Fruh (2003), and Nemri et al. (2013) provide imagery and analysis to summarize the various regimes that can exist in multiphase flow in through an annulus subject to shear that is induced by rotation of one or both cylinders that comprise the annulus. With reference to Figure 5.29, for low angular velocities, measured by the Reynolds number Re, the flow is steady and purely azimuthal and is known as circular Couette flow. Taylor (1923) showed that when the angular velocity of the inner cylinder is increased above a certain threshold, Couette flow becomes unstable, leading to the onset of axisymmetric toroidal vortices, known as Taylor vortex flow, as shown in the outer part of the annular space (see Moser et al., 2000). Further increase of the angular velocity leads to further generation of instabilities, which lead to flows of additional complexity known as wavy vortex flow. If the two cylinders rotate in opposite directions then spiral vortex flow arises. Beyond a certain Reynolds number there is the onset of turbulence.

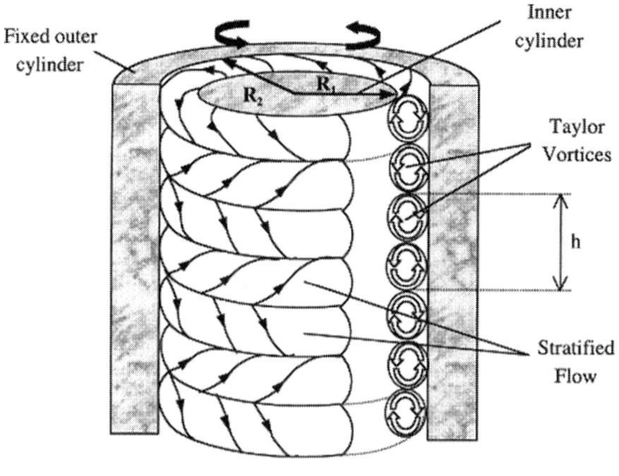

Figure 5.29 Taylor–Couette flow.

Reprinted from Moser, K. W., Raguin, L. G., Harris, A., et al. (2000).Visualization of Taylor–Couette and spiral Poiseuille flows using a snapshot FLASH spatial tagging sequence. *Magnetic Resonance Imaging*, **18**, 199–207.

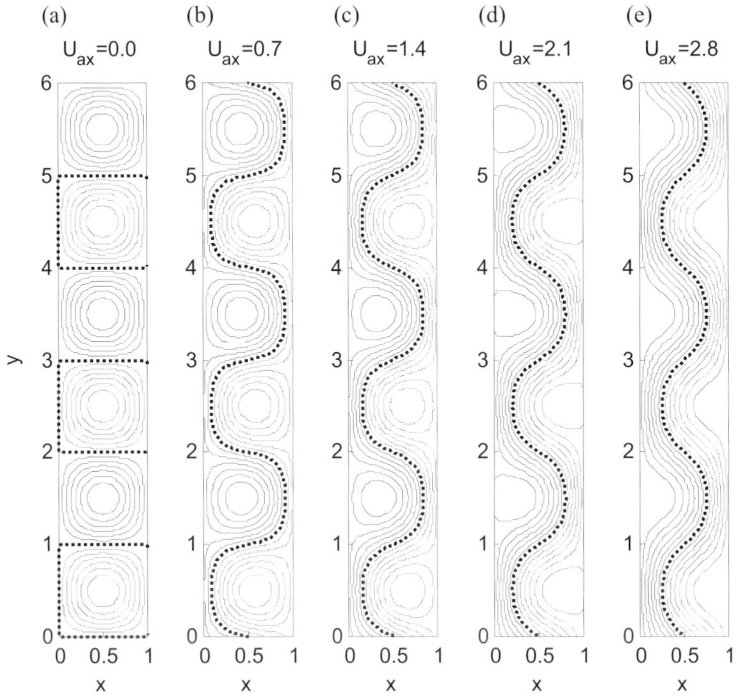

Figure 5.30 Regimes of flow in the annular region of a spinning tube contactor: (a) discrete vortex cells; (b), (c) moving vortices with secondary flows; (d) wavy axial flow; (e) turbulence. Reprinted with permission from Dr W. Fruh

A further illustration of the complexity of the flow regime is provided in Figure 5.30 (Fruh, 2002). This figure shows a two-dimensional flow map for the fluid flowing through the annular space between the coaxial rotating tubes. The degree of mixing is highly sensitive to the linear velocity, with the regime transitioning from a series of discreet vortices at zero net flow, through to moving vortices, to wavy axial flow, and ultimately to turbulent flow.

With reference to the annular centrifugal contactor, Vedantam et al. (2012) depict the transitions from Couette flow to turbulent Taylor vortex flow regimes in the annular region of contactor, as shown in Figure 5.31.

Diagram 1 in Figure 5.31 shows Couette flow, with diagrams 2–5 showing the transitions to turbulent Taylor vortex flow with the highest degrees of instability. Chandrasekhar (1962) studied the effect of an axial flow the transition from Couette flow (CF) to Taylor vortex flow and showed that the introduction of axial flow delays the occurrence of instability, with the critical Taylor number increased by a factor dependent on the square of the Reynolds number based on the flow velocity on the axial direction, as shown below in equation 5.20.

$$\mathrm{Ta}_{\mathrm{Cr}} = 1708 + 27.15\mathrm{Re}_Z^2 \tag{5.20}$$

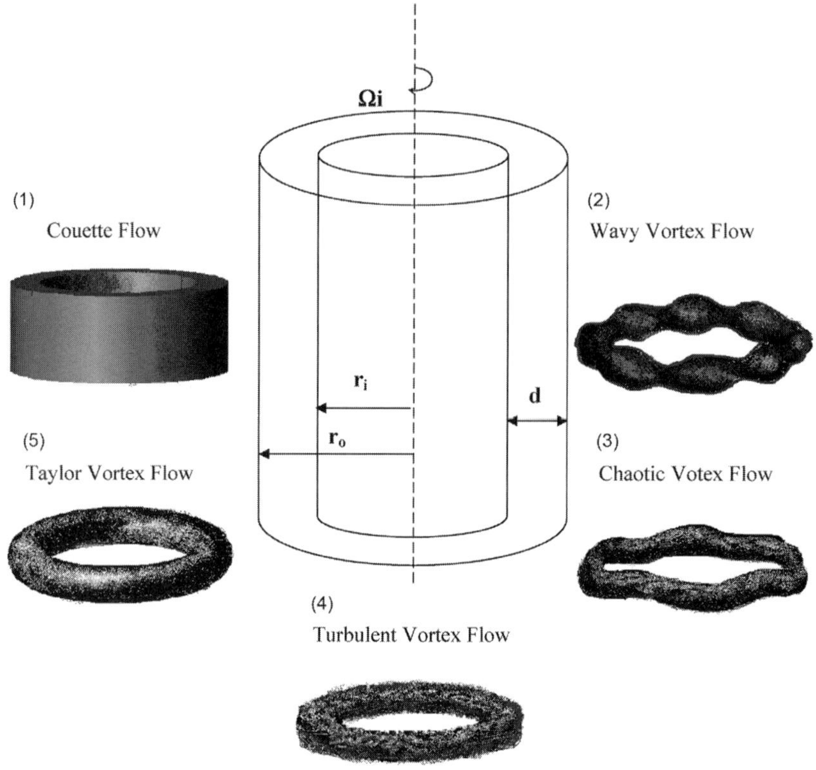

Figure 5.31 Various vortex patterns in the absence of axial flow: (1) Couette flow, (2) wavy vortex flow, (3) chaotic vortex flow, (4) Taylor vortex flow, (5) turbulent Taylor vortex flow. Reproduced from Vedantam, S., Wardle, K. E., Tamhane, T. V., Ranade, V. V., and Joshi J. B. (2012). CFD simulation of annular centrifugal extractors. *International Journal of Chemical Engineering*, Article ID 759397, an open access article distributed under the Creative Commons Attribution License, which permits unrestricted use, distribution, and reproduction in any medium, provided the original work is properly cited. Copyright © 2012

5.9 New Applications of High Gravity Systems

5.9.1 Enantioselective Separations

One particularly interesting new application of high gravity liquid–liquid extraction is presented by Feng et al. (2016) where the successful enantioselective liquid–liquid extraction (ELLE) of 2-phenylpropionic acid (2-PPA) enantiomers from 1,2-dichloroethane in a multistage countercurrent cascade of centrifugal contactor separators (CCSs) using hydroxypropyl-β-cyclodextrin (HP-β-CD) as extractant. This example serves to show the potential for intensification of liquid–liquid contact by application of high gravity forces to the production of high value, high specification compounds that are difficult and expensive to isolate and produce using other technologies.

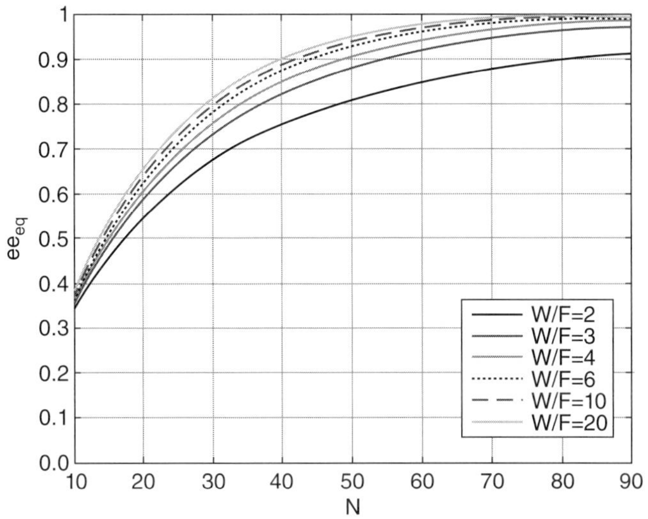

Figure 5.32 Influence of number of stages on ee_{eq} for separation of α-cpma enantiomers at different w/f ratio. Conditions: w/o = (0.001, 2), pH = 2.5, [α-cpma] = 4 mmol L^{-1}, t = 5 s, feed in the middle stage.
Reprinted from Zhang, P., Feng, X., Tang, K., and Weifeng, X. W. (2016). Study on enantioseparation of α-cyclopentyl-mandelic acid enantiomers using continuous liquid–liquid extraction in centrifugal contactor separators: Experiments and modeling. *Chemical Engineering and Processing*, **107**, 168–176 with permission from Elsevier.

2-Phenylpropionic acid is an important 2-arylpropionic acid class of compound widely used to synthesize chiral drugs. 2-Arylpropionic acid enantiomers exhibit their useful pharmacological activity as the (*S*)-enantiomer, while the (*R*)-enantiomer is an inactive species; therefore, separation of racemic mixture is highly desirable. Such chiral syntheses and purification at commercial scale are challenging and historically have relied heavily upon chromatographic separation techniques that are not very amenable to commercial scale-up. In the example studied by Feng et al., *m*-HP-β-CD was used in the aqueous phase as chiral selector, which preferentially recognizes the (*R*)-enantiomer of 2-phenylpropionic acid (Feng et al., 2016). Demonstration of successful separation exploited the main advantages of the centrifugal contactor with its compact geometry, low material inventory, suitability for continuous flow equipment, and possession of excellent mass transfer properties. In this example, reaction and separation are combined in a single device. As would be expected, the overall extraction performance is controlled by extract phase/washing phase ratio, the extractant concentration, the pH value of the aqueous phase, and the number of stages. Successful separation of α-cyclopentyl mandelic acid (cpma) enantiomers was also demonstrated using the same approach (Zhang et al., 2016). The enantiomeric excess (ee) is a measure of the optical purity of the raffinate and the extract. Figure 5.32 (Zhang et al., 2016) shows the strong relationship between the number of centrifugal contactors in series and the enantiomeric excess for the desired (*S*)-enantiomer.

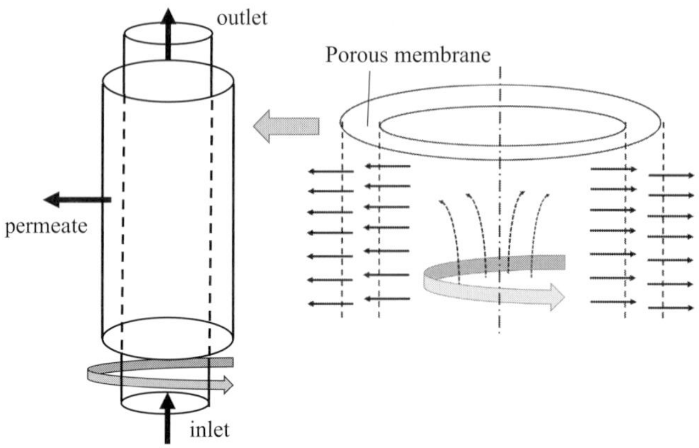

Figure 5.33 Geometry of a cross-flow filtration system in which the ceramic membrane is rotating about the vertical axis.
Reprinted (adapted) with permission Motin, A., Tarabara, V. V., and Bénard, A. (2015). Numerical investigation of the performance and hydrodynamics of a rotating tubular membrane used for liquid–liquid separation. *Journal of Membrane Science*, **473**, 245–255. Copyright (2015) Elsevier B.V. All Rights Reserved.

5.9.2 The Rotating Tubular Membrane

In a recent development, Motin, Tarabara, and Bénard (2015) described a rotating tubular ceramic membrane that exploits centrifugal forces to achieve intensification of liquid–liquid separations. The conceptual system is shown in Figure 5.33, which depicts the tubular ceramic membrane rotating on a vertical axis with the liquid–liquid dispersion feed entering at the base with preferential flow of the permeate (separated liquid phase) through the membrane in a direction perpendicular to the axis of rotation. Although the study is a computer simulation, the results show the potential for greatly intensifying the rate and a degree of liquid–liquid separation due to the additional high gravity forces acting on the dispersed droplets. A quantitative theoretical analysis of the droplet trajectories within the membrane channel showed that the accumulation of drops at the membrane interface is reduced and the separation efficiency is increased. The authors show that the Reynolds number, the Swirl number, and the Stokes number defined respectively in equations 5.21, 5.22, and 5.23, are the controlling parameters that determine pressure and velocity field, shear stresses, droplet cutoff size, and separation efficiency.

$$\text{Swirl number: Sw} = \frac{\omega R}{\overline{U_Z}} \tag{5.21}$$

$$\text{Reynolds number: Re} = \frac{4\dot{m}_W}{\pi \mu_W D} \tag{5.22}$$

$$\text{Stokes number: Stk} = \frac{d_d^2(\rho_c - \rho_d)\bar{U}_Z}{18\mu_d D} \tag{5.23}$$

As is the case with the majority of cross-flow filtration systems, increase in shear stresses closer to the membrane surface was predicted as the major factor in decrease of

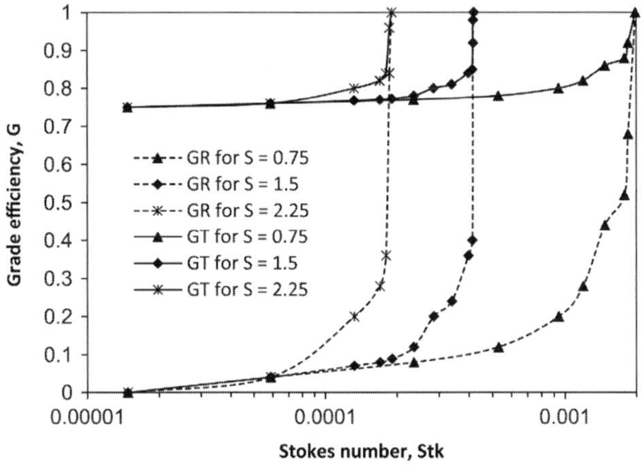

Figure 5.34 Effect of swirl number on grade efficiency for the cross-flow filtration system at Re of 1415.

Reprinted (adapted) with permission Motin, A., Tarabara, V. V., and Bénard, A. (2015). Numerical investigation of the performance and hydrodynamics of a rotating tubular membrane used for liquid–liquid separation. *Journal of Membrane Science*, **473**, 245–255. Copyright (2015) Elsevier B.V. All Rights Reserved.

drop accumulation at the membrane surface. The Swirl number and the Stokes number showed a strong influence on separation efficiency, but less influence of Reynolds number. The possibility of removing very fine drops by increasing Swirl number is also proposed.

Grade efficiency, G_T, is defined as the ratio of total mass of drops removed through the membrane to that in the initial feed.

$$G_T = \frac{\dot{m}_{do}}{\dot{m}_{di}}$$

$$\begin{aligned} \dot{m}_{do} &= \text{mass flow of drops at outlet} \\ \dot{m}_{di} &= \text{mass flow of drops at inlet} \end{aligned} \tag{5.24}$$

$$\begin{aligned} G_R &= \frac{G_T - R_F}{1 - R_F}; \quad R_F = \frac{Q_o}{Q_i} \\ Q_o &= \text{total flowrate out} \\ Q_i &= \text{total flowrate in} \end{aligned} \tag{5.25}$$

Figure 5.34 shows the strong effect that increase in Stokes number (equation 5.23) has on the grade efficiency. The negative influence of Swirl number, Sw (equation 5.21), is also shown, and as the value of Sw is increased the grade efficiency is reduced for simulations at each fixed value of Stokes number. The reduction is explained by the increased centrifugal forces associated with higher Swirl number that in turn increases the radial pressure gradient thus reducing the outward component of the radial velocity. Data for the relationship between grade efficiency, Stokes number, and Swirl number where grade efficiency is expressed as a reduced grade efficiency, G_R, as defined in equation 5.25 are also included in Figure 5.34.

References

Arnott, I. (1993). Solvent extraction of fermentation products using electrostatic and centrifugal fields. Ph.D. thesis, Heriot Watt University.

Arnott, I. A. and Weatherley, L. R. (1994a). Hydraulic studies in a combined field liquid–liquid contactor. Proceeding of EXTRACTION 94, Edinburgh, UK, October 1994. Institution of Chemical Engineers Symposium Series, London.

Arnott, I. A. and Weatherley, L. R. (1994b). Entrainment studies in a combined field liquid–liquid extractor. Proceedings of the Fourth IChemE Irish Research Symposium, Dublin, March 1994, 17-30, The Institution of Chemical Engineers, London.

Boiarkina, I., Pedron, S., and Patterson, D. A. (2011). An experimental and modelling investigation of the effect of the flow regime on the photocatalytic degradation of methylene blue on a thin film coated ultraviolet irradiated spinning disc reactor. *Applied Catalysis B: Environmental*, **110**, 14–24.

Boodhoo, K. V. K. and Jachuck, R. J. (2000a). Process intensification: spinning disk reactor for styrene pol. *Applied Thermal Engineering*, **20**, 1127–1146.

Boodhoo, K. V. K. and Jachuck, R. J. (2000b). Process intensification: spinning disk reactor for condensation polymerization. *Green Chemistry*, **2**, 235–244.

Boodhoo, K. V. K., Dunk, W. A. E., Vicevic, M. et al. (2006). Classical cationic polymerization of styrene in a spinning disc reactor using silica-supported BF_3 catalyst. *Journal of Applied Polymer Science*, **101**, 8–19.

Bowe, M. J., Oruh, S. N., and Singh, J. (1985). UK patent application GB 2,155,803A.

Chandrasekhar, S. (1962). The stability of spiral flow between rotating cylinders. *Proceedings of the Royal Society A*, **265**(1321), 188–197.

Coles, D. J. (1965). Transition in circular Couette flow. *Journal of Fluid Mechanics*, **21**, 385–425.

Dehkordi, A. M. (2002a). Liquid–liquid extraction with chemical reaction in an impinging jets reactor. *AIChE Journal*, **48**(10), 2230–2239.

Dehkordi, A. M. (2002b). A novel two-impinging-jets reactor for copper extraction and stripping processes. *Chemical Engineering Journal*, **87**(2), 227–238.

Feng, X., Tang, K., Zhang, P., and Yin, S. (2016). Experimental and model studies on continuous separation of 2-phenylpropionic acid enantiomers by enantioselective liquid–liquid extraction in centrifugal contactor separators. *Chirality*, **28**, 235–244.

Fruh, W. (2002). www.pinetwork.org/pubs/W_Fruh.pps

Holl, R. (2003). Reactor design: adding some spin. *The Chemical Engineer*, 742, 32–34.

Holl, R. A. (2006). Methods of operating surface reactors and reactors employing such methods, US Patent 7,125,527 B2.

Holl, R.A. (2010). Spinning tube in tube reactors and their methods of operation, US Patent 7,780,927 B2.

Jachuck, R. J. and Scalley, M. J. (2003). Process technology for continuous production of nano-micron size particles. *AICHE Spring National Meeting. Safety and Sustainability: Core Issues Shaping Tomorrow*, New Orleans, Louisiana, USA.

Jammoal, Y. and Lee, J. M. G. (2015). Drop velocity in a rotating liquid–liquid system. *Chemical Engineering Research and Design*, **104**, 638–646.

Judd, S., Qiblawey, H., Al-Marri, M., et al. (2014). The size and performance of offshore produced water oil-removal technologies for reinjection. *Separation and Purification Technology*, **134**, 241–246.

Kadam, B. D., Joshi, J. B., Koganti, S. B., and Patil, R. N. (2008). Hydrodynamic and mass transfer characteristics of annular centrifugal extractors. *Chemical Engineering Research and Design*, **86**(3), 233–244.

Kadam, B. D., Joshi, J. B., Koganti, S. B., and Patil, R. N. (2009). Dispersed phase hold-up, effective interfacial area and Sauter mean drop diameter in annular centrifugal extractors. *Chemical Engineering Research and Design*, **87**(10), 1379–1389.

Leveson, P., Dunk, W. A. E., and Jachuck, R. J. (2003). Numerical investigation of kinetics of free-radical polymerization on spinning disk reactor. *Journal of Applied Polymer Science*, **90**, 693–699.

Martínez, A. N. M., van Eeten, K. M. P., Schouten, J. C., and van der Schaaf, J. (2017). Micromixing in a rotor–stator spinning disc reactor *Industrial and Engineering Chemistry Research*, **56**, 13454–13460.

Meeuwse, M., van der Schaaf, J., and Schouten, J. C. (2010). Mass transfer in a rotor–stator spinning disk reactor with cofeeding of gas and liquid. *Industrial and Engineering Chemistry Research*, **49**, 1605–1610.

Meikrantz, D. H., Macaluso, L. L., and Flim, W. D. (2002). A new annular centrifugal contactor for pharmaceutical processes. *Chemical Engineering Communications*, **189**, 1629–1639.

Millar, M. K. and Weatherley, L. R. (1989) Whole broth extraction in an electrically enhanced liquid/liquid contact system. *Chemical Engineering Research and Design*, **67**, 227–231.

Moser, K. W., Raguin, L. G., Harris, A., et al. (2000).Visualization of Taylor–Couette and spiral Poiseuille flows using a snapshot FLASH spatial tagging sequence. *Magnetic Resonance Imaging,* **18**, 199–207.

Motin, A., Tarabara, V. V., and Bénard, A. (2015). Numerical investigation of the performance and hydrodynamics of a rotating tubular membrane used for liquid–liquid separation. *Journal of Membrane Science*, **473**, 245–255.

Nemri, M., Climent, E., Charton, S., Lanoëa, J., and Ode, D. (2013). Experimental and numerical investigation on mixing and axial dispersion in Taylor–Couette flow patterns. *Chemical Engineering Research and Design,* **91**, 2346–2354.

Oxley, P., Brechtelsbauer, C., Ricard, F., Lewis, N., and Ramshaw, C. (2000). Evaluation of spinning disk reactor technology for the manufacture of pharmaceuticals. *Industrial Engineering Chemistry Research*, **39**, 2175–2182.

Petera, J. and Weatherley, L. R. (2001). Modeling of mass transfer from falling droplets. *Chemical Engineering Science*, **56**, 4929–4947.

Poncet, S., Chauve, M. P., and Schiestel, R. (2005). Batchelor versus Stewartson flow structures in a rotor–stator cavity with throughflow. *Physics of Fluids*, **17**, 075110–075115.

Qiu, Z., Petera, J., and Weatherley, L. R. (2012). Biodiesel production using an intensified spinning disc reactor. *Chemical Engineering Journal*, **210**, 597–609.

Ramshaw, C. (2004). The spinning disc reactor. In A. Stankiewicz and J. A. Moulijn, eds., Re-engineering the Chemical Processing Plant Process Intensification. Marcel Dekker.

Svarovsky, L. (1991). Solid–Liquid Separation, 3rd ed. Oxford: Butterworth Heinemann.

Syed, A. and Fruh, W. (2003). Modelling of mixing in a Taylor-Couette reactor with axial flow. *Journal of Chemical Technology and Biotechnology*, **78**, 227–235.

Tai, C. Y., Tai, C. T., and Liu, H. S. (2006). Synthesis of submicron barium carbonate using a high gravity technique. *Chemical Engineering Science*, **61**, 7479–7486.

Tamhane, T. V., Jyeshtharaj, B., Joshi, J. B., et al. (2012). Axial mixing in annular centrifugal extractors. *Chemical Engineering Journal*, **207–208**, 462–472.

Tamhane, T. V., Joshi, J. B., and Patil, R. N. (2014). Performance of annular centrifugal extractors: CFD simulation of flow pattern, axial mixing and extraction with chemical reaction. *Chemical Engineering Science*, **110**, 134–143.

Taylor, G. I. (1923). Stability of a viscous liquid contained between two rotating cylinders. *Philosophical Transactions of the Royal Society A*, **223**(605–615), 289–343.

Treybal, R. E. (1955). Liquid–Liquid Extraction. New York: McGraw Hill.

Vedantam, S. and Joshi, J. B. (2006). Annular centrifugal contactors: a review. *Chemical Engineering Research and Design*, **84**(7), 522–542.

Vedantam, S., Wardle, K. E., Tamhane, T. V., Ranade, V. V., and Joshi, J. B. (2012). CFD simulation of annular centrifugal extractors. *International Journal of Chemical Engineering*, Article ID 759397.

Visscher, F., van der Schaaf, J., de Croon, M. H. J. M., and Schouten, J. C. (2012). Liquid–liquid mass transfer in a rotor–stator spinning disc reactor. *Chemical Engineering Journal*, **185/186**, 267–273.

Visscher, F., Nijhuis, R. T. R., de Croon, M. H. J. M., van der Schaaf, J., and Schouten, J. C. (2013). Liquid–liquid flow in an impeller–stator spinning disc reactor. *Chemical Engineering and Processing*, **71**, 107–114.

Zhang, P., Feng, X., Tang, K., and Weifeng, X. W. (2016). Study on enantioseparation of α-cyclopentyl-mandelic acid enantiomers using continuous liquid–liquid extraction in centrifugal contactor separators: Experiments and modeling. *Chemical Engineering and Processing*, **107**, 168–176.

6 Electrically Driven Intensification of Liquid–Liquid Processes

6.1 Introduction

The application of external electrical fields to liquid–liquid systems has attracted attention for a number of years. The action of electrical fields appears in different forms, ranging from electrically induced acceleration of ionic and other charged species through a fluid continuum, to electrostatic dispersion and coalescence processes involving immiscible and partially miscible liquids. Potential advantages of electrically induced phenomena in liquid–liquid systems include higher rates of mass transfer, control of drop size and interfacial area in dispersions, higher rates of phase separation, and enhanced mixing.

The employment of electric fields and electrostatics to improve liquid-based processes is well known in a number of manufacturing areas and applications. One significant example is the electrostatic enhancement of crop spraying, demonstrated through the application of electrostatic breakup of solutions of pesticide. Similar applications include those demonstrated for paint spraying and other surface-coating processes (Barletta and Gisario, 2009; Coffee, 1980; Law, 2001; Singh et al., 2013). In these examples, the liquid phase is electrostatically sprayed to effect large reductions in drop size. Another important effect that is exploited in this technique is electrostatic attraction between the sprayed droplet and the target surface. This has resulted in improved efficiency of the process, since the percentage of liquid feed that arrives and adheres to the target surface may be significantly increased.

Another notable application of electrostatic enhancement is to boiling heat transfer. It has been reported that electric fields have been successfully used to control nucleation rates during boiling heat transfer and thereby achieve a continuous rise in heat transfer coefficient with respect to heat flux (McGranaghan and Robinson, 2014; Quan et al., 2015).

Likewise for liquid–liquid extraction, there is substantial published work that illustrates the intensifications that may be attributed to electrostatic effects. These are based largely on achieving reduction in drop size, increases in interfacial shear rate, reduction in drop stability, enhancement of interfacial instability, and increase in coalescence rates of unstable drops. The potentially positive effects of externally applied electrical fields on liquid–liquid processes stem from a number of fundamental physical phenomena associated with electrostatics, and liquid and solid surfaces. These may be listed as follows:

- Enhanced motion of dispersed drops due to the action of electrical forces.
- Enhanced motion of molecular and macromolecular species.
- Enhanced motion of ionic species.
- Interfacial disruptions resulting from imbalance of forces at liquid–liquid interfaces created by the presence of electrical charge and an external electrical field. This ultimately can result in the quantitative break up of a liquid phase into a very fine spray with large specific interfacial areas.
- Electrically induced interfacial flows due to nonuniform distribution of electrical surface charge.

6.2 Summary of Fundamental Equations: Electrostatic Processes

An understanding of the above phenomena requires brief analysis of the relationships between electrostatic charge, electrical field, and the resultant electrical forces acting on particle, drops, and liquid–liquid interfaces. Cross (1987) provides a comprehensive review of electrostatics fundamentals from first principles. Here we confine treatment to the fundamentals pertaining to electrostatic intensification of liquid–liquid processes.

The relevant controlling equations will now be summarized.

6.2.1 Coulomb's Law

Firstly, the force between two charged entities, liquid drops or particles, in a surrounding medium, may be described by the classical Coulomb's law:

$$F = \frac{Cqq'}{d^2} \tag{6.1}$$

where F = force, d = distance of separation, q = electrostatic charge on first entity, in coulombs, q' = electrostatic charge on second entity, in coulombs, C = constant.

In SI units, the constant C may be defined as follows:

$$C = \frac{1}{4\pi\varepsilon_0\varepsilon_\tau} \tag{6.2}$$

where ε_0 = permittivity of free space; ε_τ = relative dielectric constant of the surrounding medium.

In the presence of an electrical field E, the force exerted on an individual charge, q, is expressed thus:

$$F = q{\cdot}E \tag{6.3}$$

Combining equations 6.1–6.3 results in:

$$\frac{qq'}{d^2}\frac{1}{4\pi\varepsilon_0\varepsilon_\tau} = qE \quad \text{or}$$

$$E = \frac{q'}{d^2}\frac{1}{4\pi\varepsilon_0\varepsilon_\tau} \tag{6.4}$$

Electrical potential, V, at a point in space may be considered as the energy required to move unit charge to that point from a point of zero potential. The units of potential are volts, equivalent to energy per unit charge. A key quantity for analysis of electrically intensified processing is the electric field. This may be considered as the rate of change of potential with distance. As will be seen later, these quantities have a critical effect on the dynamics of drop formation and motion.

Alternatively, electric field E may be written as potential gradient as follows, where V is electrical potential and s is distance:

$$E = -\frac{dV}{ds} \tag{6.5}$$

In three dimensions E is more commonly is written as:

$$E = -\text{grad} \quad V = -\nabla V \tag{6.6}$$

Equations 6.4 and 6.5 may be combined to express electrical potential, V, explicitly as:

$$V = \frac{q'}{4\pi\varepsilon_0 d} \tag{6.7}$$

In the case of dielectric liquids, polarization of the liquid can be an important factor that influences the electric field and hence the electrostatic force acting on a charged entity. Under these circumstances, combined equations 6.1 and 6.2 are modified as in equation 6.4, to give the following for the force expression F:

$$F = \frac{qq'}{d^2}\frac{1}{4\pi\varepsilon_0\varepsilon_\tau} \tag{6.8}$$

6.2.2 Gauss's Law

Gauss's law is an important fundamental relationship governing electrical phenomena in liquid systems. It relates electrical flux to electrical field and to surface charge, stating that the net electric flux through any hypothetical closed surface is equal to 1/ε times the net electric charge within that closed surface where ε is the electrical permittivity (Serway, 1996).

The electrical flux through a closed surface $\boldsymbol{S}$, Φ, is written as follows in terms of the surface integral of the electrical field $\boldsymbol{E}$ across the entirety of a surface $\boldsymbol{S}$. The term $d\boldsymbol{A}$ is a vector quantity equivalent to an infinitesimal element of the surface area.

$$\Phi_{\mathrm{E}} = \oiint_S E \cdot dA \tag{6.9}$$

and

$$\Phi_{\mathrm{E}} = \frac{Q}{\varepsilon_0} \tag{6.10}$$

Equations 6.9 and 6.10 are simple expressions of Gauss's law in integral form.

The implications of Gauss's law are important when the electrical charging of liquids is considered. Consistent with Gauss's law, there can be no free charge within a liquid (or solid conductor) and electrical charge must lie at the surface. This is especially significant when we consider the acquisition and behavior of electrical charges on drops and at liquid–liquid interfaces.

The corollary to this is that a conducting entity having surface charge cannot give rise to electrostatic charge inside the entity.

It also follows from Gauss's law that the charge density on the surface of the entity can obtained by writing Gauss's law as follows:

$$\int \mathbf{E}\cdot d\mathbf{S} = \sum \frac{q}{\varepsilon_0 \varepsilon_\tau} \tag{6.11}$$

If charge density is defined as Q_s (in coulombs per m^2) then integration of equation 6.11 results in:

$$E = \frac{Q_s}{\varepsilon_0 \varepsilon_\tau} \tag{6.12}$$

6.2.3 Poisson's Equation

Another fundamental relationship relevant to electrically intensified liquid–liquid operations is Poisson's equation. This relates the electrical potential, V, to the volumetric charge density, defined as Q_V.

As before, but working in spherical coordinates, we may define the electrical potential as the work done when a charge q is moved through a differential distance dr in an electrical field $\boldsymbol{E}$.

This may be expressed in the following equation:

$$V = -q\mathbf{E}\cdot dr \tag{6.13}$$

The actual work done may be calculated for a finite distance, for example from a position x_1 to x_2 by integration of equation 6.13.

Combination of Gauss's equation with equation 6.13 leads to Poisson's equation 6.15:

$$\operatorname{div} \mathbf{E} = \frac{Q_V}{\varepsilon_0 \varepsilon_\tau} \tag{6.14}$$

From equation 6.6:

$$\mathbf{E} = -\text{grad } V = -\nabla V$$

we therefore get:

$$\operatorname{div}(\text{grad } V) = \nabla^2 V = -\frac{Q_V}{\varepsilon_0 \varepsilon_\tau} \tag{6.15}$$

6.2.4 Electrically Charged Drops

In the context of intensification of liquid–liquid processing, a useful starting point for fundamental consideration is the case of induction charging of particle or drops, considered in detail by Cross (1987) and Felici (1984).

Firstly, we may consider the case of a conducting sphere (or liquid drop) in a uniform electrical field E_0 (Figure 6.1).

The field at the surface of the sphere may be expressed as:

$$E = 3E_0 \cos\theta \tag{6.16}$$

The coulombic charge per unit area Qs on the particle due to the presence of the electric field E_0 may be written:

$$Q_s = 3\varepsilon_0 E_0 \cos\theta \tag{6.17}$$

The total charge on the surface of the particle (as a hemisphere) may be obtained by integration of equation 6.17 across the surface of the particle and may be written:

$$q = 4\pi\varepsilon_0(1.64E_0a^2) \tag{6.18}$$

where a = radius of the particle.

For a full sphere of radius a, the total charge due to the field may be expressed as follows:

$$q = 4p\,\varepsilon_0(1.64E_0a^2) \tag{6.19}$$

The corresponding force on the particle is expressed as:

$$F = 4\pi\varepsilon_0(1.37E_0{}^2a^2) \tag{6.20}$$

One of the primary areas of interest in the intensification of liquid–liquid processes using electrical enhancement is the ability to better control the motion of droplets in a dispersion. Acceleration of drops relative to a continuous phase and resulting in enhanced film mass transfer is of importance. A second area of interest is improved control of drop trajectory and residence time in liquid–liquid contactors in order to improve hydraulic capacity and utilization of volume. Fundamental understanding of the role that electrical forces can play in influencing droplet behavior is central to the development of intensification technology based on electrical fields.

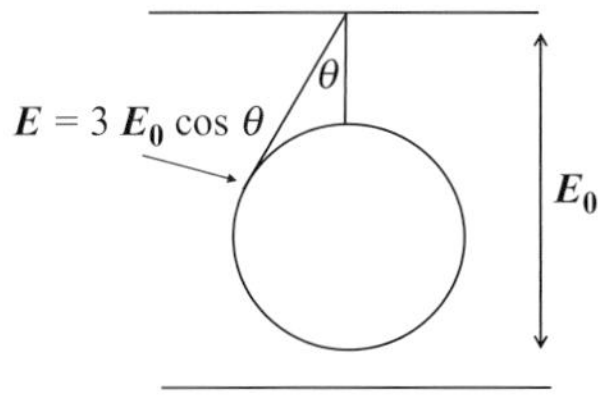

Figure 6.1 A conducting droplet in a uniform electric field.

A second and equally important phenomenon requiring analysis is electrostatic disruption of liquid interfaces. Disruption is a mechanism central to dispersion of liquids into small drops. Control and reduction of drop size is an important component in the intensification of separations and chemical reactions in liquid–liquid systems. Much of the fundamental science dates back to the 19th century and was based on consideration of gas–liquid interfaces, but much of the theory is also relevant to phenomena at liquid–liquid interfaces. The concept of electrically induced interfacial disruption is based on the ability of a liquid surface to hold an electrostatic charge. In so doing, the liquid surface is potentially exposed to an electrical force that is opposed to the interfacial tension force maintaining the integrity of the interface. At a certain magnitude of charge, the interface is destabilized by the electrostatic force, and disruption occurs. The point at which this occurs is known as the Rayleigh limit.

The limit may be quantified by the following equation:

$$q = (64\pi^2\varepsilon_0\gamma a^3)^{1/2} \tag{6.21}$$

where γ = interfacial tension, in N m^{-1}.

Another way of expressing the limit is in terms of the mass to charge ratio:

$$\frac{q}{m} = \left[\frac{36\varepsilon_0\gamma}{a^3\rho}\right]^{1/2} \tag{6.22}$$

Disruption of the liquid–liquid interface is also a central factor in electrostatic dispersion, which is of prime importance in increasing interfacial area per unit volume in liquid–liquid mixtures. Gross increases in interface represent one of the major mechanisms by which intensification of mass transfer and reaction occur. The conditions under which each of these three phenomenon occur – dispersion, interfacial disruption, and drop motion – may be described in terms of the electrical field, the electrical properties of the phases, the geometry of the system, and the composition of the phases. The conditions under which electrostatic charge can be induced at an interface require analysis.

The placement of electrostatic charge onto liquid droplets can occur by five mechanisms, as listed by Cross (1987):

1. disruption of the electrical double layers that are present at a liquid–liquid (or gas–liquid) interface;
2. disruption of the electrical double layers at a solid–liquid interface;
3. charge transfer due to contact with an electrically charged surface that could include a charged nozzle or an electrode;
4. charge separation in one or both liquid phases;
5. charging due to the division of randomly distributed ions within one or both of the liquids.

All of these mechanisms may play a part in the intensification of liquid–liquid processing. For example, disruption of the electrical double layers at a liquid–liquid interface is a primary mechanism in the creation of disturbances at the interface that impact local transport kinetics between the two phases. Charge transfer between a surface and a liquid, for example a forming droplet at a conducting nozzle, is a primary mechanism for imposing surface charge on the drop, which then has influence on the drop size at detachment and ultimately on the drop motion in an electrical field.

6.3 Electrokinetic Phenomena

The effects of the application of electrical fields to liquid-phase environments may involve a number of different mechanisms under the generic heading of electrokinetic phenomena. These include electrophoresis, electroosmosis, and electrohydrodynamic flow. Another phenomenon highly relevant to intensification, which is strongly influenced by the presence of external electrical fields and electrostatic surface charge, is interfacial disturbance and surface breakup.

There are a variety of entities in a liquid continuum that may be responsive to the presence of an electrical field. These include macromolecules, such as proteins and colloids, submicro-sized particles or drops, and discrete droplets of sizes in the macro range. The size of the entities will to some extent determine their response to an electrical field. In most cases there will exist at the interface an electrical double layer that will also determine the nature of the response. In simple terms, the application of an external electrical field will result in the particle or drop moving in the field according to the direction of the field and the polarity of the charge on the particle. The movement of the particle or droplet is accompanied by movement of the surrounding continuum in several directions due to displacement. In the case of a neutral particle, polarization of the particle may occur and be accompanied by alignment of the particle in accordance with the direction of the field. Neutral particles will only move in response to the presence of the field if the field is nonuniform.

Another related phenomenon is the response of a surrounding liquid continuum and this may be considered at the molecular level. Therefore, ions and molecules that are dipolar will also move in response to the electrical field. As will be discussed in Section 6.6, the movement of charged entities at this level is a significant factor in the phenomenon of space charge migration. The resulting force from the motion of charged ions and molecules affects the carrier liquid on account of collisions that result in liquid movement, known as electroconvection or electrohydrodynamic flow.

Electrophoresis may be regarded as the enhanced motion of particles in a fluid under the influence of a uniform electric field. The enhanced motion occurs because of the presence of charge at the interface of the particle that in the presence of the uniform electrical field results in an accelerative force on the particle. It can be important in liquid–liquid systems where externally applied electrical fields may result in the enhanced transport of charged entities in the vicinity of a liquid–liquid interface. This may be important particularly for chemical reactions that occur at liquid–liquid interfaces, such as in phase-transfer catalysis. The main application of electrophoresis relates to separation and characterization of macromolecules that carry surface charge, such as proteins and enzymes.

In electrophoresis, the velocity of the charged entity is related to the magnitude of the applied external electrical field by:

$$\mathbf{v} = \mu_e \cdot \mathbf{E} \tag{6.23}$$

where μ_e = the electrophoretic mobility and $\mathbf{E}$ is the applied field.

$$\mu_e = \frac{\varepsilon_T \varepsilon_0 \zeta}{\eta} \tag{6.24}$$

Equations 6.23 and 6.24 are generally limited to systems in which the electrical double layer around the particle is thin, with the thickness being less than the radius of curvature of the particle. At higher thicknesses, the motion becomes dependent also on the shape and size of the particle.

The electrophoretic forces $\mathbf{F}_{EP}$ acting on a charged particle/droplet due to an external electric field are in direct proportion to the electric field strength; see equation 6.25, which is similar to equation 6.3 but using vector notation:

$$F_{EP} = q\ E \tag{6.25}$$

As before, q is the electric charge on the particle/droplet and has time dependence, behaving in accordance with the permittivity and conductivity of the liquid phase, as expressed in equation 6.26:

$$q = q_0 \exp\left(-t/\lambda\right) \tag{6.26}$$

where λ is the relaxation time, defined as follows:

$$\lambda = \varepsilon_0 \varepsilon_r / \kappa \tag{6.27}$$

$\mathbf{E}$ is the resultant electric field strength from the externally applied electric potential and from contributions of other particles/droplets in the close neighborhood if these are present.

A related phenomenon that is also important when considering electrical intensification of liquid–liquid systems is dielectrophoresis. This may be defined as the motion of an uncharged (electrically neutral) dielectric particle/droplet in a spatially nonuniform electrical field.

Petera et al. (2007) summarize the controlling equations in the context of dispersed, moving, electrically charged drops in a liquid–liquid system. The nonuniformity of the field is critical to the nature of the forces acting on the droplet.

They describe the dielectrophoretic force $\mathbf{F}_{DEP}$ acting on the drop as follows:

$$\mathbf{F}_{DEP} = \alpha\ \mathbf{E} \cdot \underline{\nabla} \mathbf{E} \quad = \quad 1/2\ \alpha\ \underline{\nabla} |\mathbf{E}|^2 \tag{6.28}$$

This expression for the electrophoretic force may be used when a particle/droplet is present as a discrete entity in another dielectric phase, liquid or gas. α is the polarizability, which for a sphere is:

$$\alpha = \frac{1}{2} \pi\ \varepsilon_1 \frac{\varepsilon_2 - \varepsilon_1}{\varepsilon_2 + 2\varepsilon_1} d^3 \tag{6.29}$$

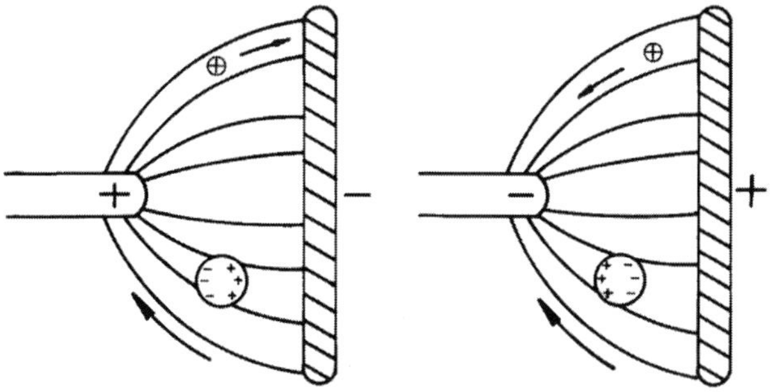

Figure 6.2 Principle of dielectrophoresis: movement of a drop due to induced polairity in a nonlinear electric field.

ε_1, ε_2 are the permittivities of the medium and dispersed phase, respectively, and d the particle diameter.

The principle of dielectrophoresis is illustrated in Figure 6.2.

In the context of electrostatic intensification of liquid–liquid processes, electrophoretic phenomena may be important and can influence the fluxes of substrate toward a liquid–liquid interface, and also influence the migration of reaction products away from the interface. Both phenomena are important, particularly in the case of equilibrium reactions, influencing both conversion and kinetics. Likewise, the occurrence of hydrodynamic flows and related phenomena such as electroosmosis also may influence the composition conditions at a liquid–liquid interface, thus affecting extraction kinetics and in some cases reaction kinetics. The influence of electrical fields is also relevant in phase-transfer catalysis as the transport of highly polar phase-transfer catalysts may be enhanced in the presence of an electrical field due to one or more of the above phenomena. The final destination of used catalysts, both homogeneous and heterogeneous, is a concern in green engineering.

Electroosmosis is a phenomenon that becomes important in processes when the liquid phase is confined in a very small geometry where the thickness of the electrical double layer is of similar magnitude to the characteristic length in the fluid environment. In this situation the fluid motion *per se* can be strongly influenced by an electrical field. Example of such geometries include fiber-based membranes, capillary devices, and microchannel contactors.

Electroosmosis, or electroosmotic flow, results from an electrical force acting on the charged double layer (Levich, 1962) in the vicinity of an interface. In the context of moving drops and interfacial flows it may be regarded as a slippage velocity at the very thin layer (a few nanometers) very close to the phase boundary. Petera et al. (2007) defined the electroosmotic flow velocity according to equation 6.30:

$$\mathbf{u}_{\mathrm{EOS}} = -\frac{\varepsilon_1 \varsigma}{\mu_c}\,\mathbf{E} \tag{6.30}$$

where ζ is the zeta potential, μ_c the viscosity, and ε_l the liquid-phase permittivity. Electroosmosis increases the slip velocity between the dispersed phase and continuous

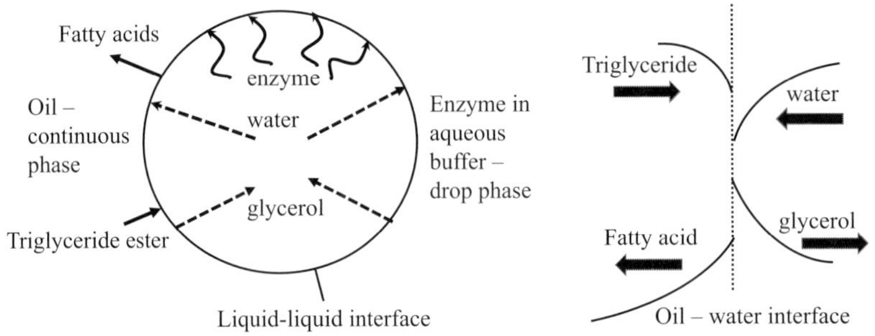

Figure 6.3 Interfacial enzymatic hydrolysis of a triglyceride ester.

phase. This is thought to occur on account of enhanced flow in the thin layer at the phase boundary. The flow is in the opposite direction to the particle's electrophoretic flow since the layer is of opposite charge. The role of electroosmosis in intensified liquid–liquid processing is not well understood at this point. Overall, the movement of drops in an electrical field is controlled by a combination of electrical forces and mechanical forces. Electrokinetic forces coupled with mechanical forces dictate the motion of each individual drop or particle.

The enhancement of liquid–liquid contact by the application of electrical fields is not only on account of improvement in fluid dynamics, but enhancement of diffusion across the liquid–liquid interface is also an important consideration. An example of a well-known reacting liquid–liquid system is used to demonstrate this point.

Figure 6.3 shows hypothetical concentration profiles in the vicinity of a liquid–liquid interface for the case of an enzymatically catalyzed interfacial hydrolysis reaction of a triglyceride ester with water to produce glycerol and free fatty acid. The concentration profiles of each of the four components shown are determined by reaction kinetics and by the transport rates of each. In the presence of an electrical field and electrical charge at the interface, transport rates may not only be governed by purely mass transfer effects. Electrokinetic transport phenomena may play a role in the transport rates close to the interface and hence affect the available substrate concentrations present for reaction, and at the same time influence the transport of reaction products away from the interface. In the case of equilibrium reactions such as enzymatic hydrolysis, the removal of reaction products from the locus of reaction has a major influence on the rate of reaction and final conversion. The role of electrokinetic transport phenomena on interfacial reaction kinetics relative to Fickian diffusion and convective transport is still a matter of some conjecture due to the experimental difficulties involved in the independent measurement of the individual phenomena that are involved.

6.4 Drop Formation

In this section the influence of electrostatic charge and the imposition of external electric fields upon drop behavior is analyzed. Creation of improved conditions for interfacial

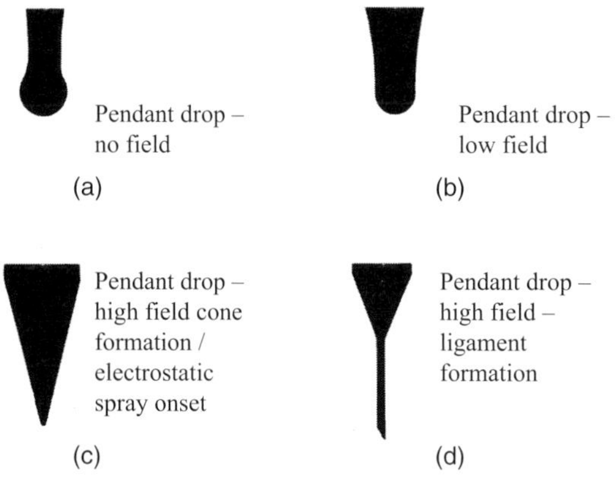

Figure 6.4 Charged pendant drop formation at an electrically charged nozzle.

mass transfer is a critical goal for successful intensification. Electrostatic forces and imposition of electric fields represent major pathways to reduction in drop size, increase in specific interfacial area, and acceleration of dispersed drops through the continuous phase. Each of these phenomena can be deployed to intensify interfacial transport processes.

The experimental basis of electrically enhanced extraction systems is well documented. With reference to Figure 6.4, the method relies upon the ability to charge droplets of the dispersed phase as they are formed at a nozzle from which they disperse as a charged entity into a continuum of immiscible organic phase. The enhancement mechanism is described in detail by Stewart and Thornton (1967a, 1967b) and Bailes (1981).

Four significant phenomena are associated with the production of charged droplets under the influence of a high-voltage electrical field:

i. Generation of a continuous spray of very small electrically charged droplets, having a high interfacial area per unit volume.
ii. Droplet oscillation is enhanced, which in turn enhances internal circulation within the droplet thus improving dispersed-phase mass transfer coefficients.
iii. Enhanced motion of dispersed electrically charged droplets on account of the influence of an externally imposed electrical field. The enhanced motion can result in changes in residence time behavior, enhanced mixing in both phases, and increases in external mass transfer kinetics due to enhanced relative motion between the drop phase and continuous phase.
iv. Inducement of enhanced Marangoni disturbances at the liquid–liquid interface.

Thornton (1976) distinguished two droplet formation regimes under which electrically enhanced extraction occurs, see Figure 6.4.

These may be best described initially by considering a pendant forming drop, as shown in Figure 6.4(a). The droplet is forming in contact with a second immiscible phase. The forming drop may acquire a surface charge if the nozzle through which it is forming is connected to a voltage source, and if the liquids present have suitable

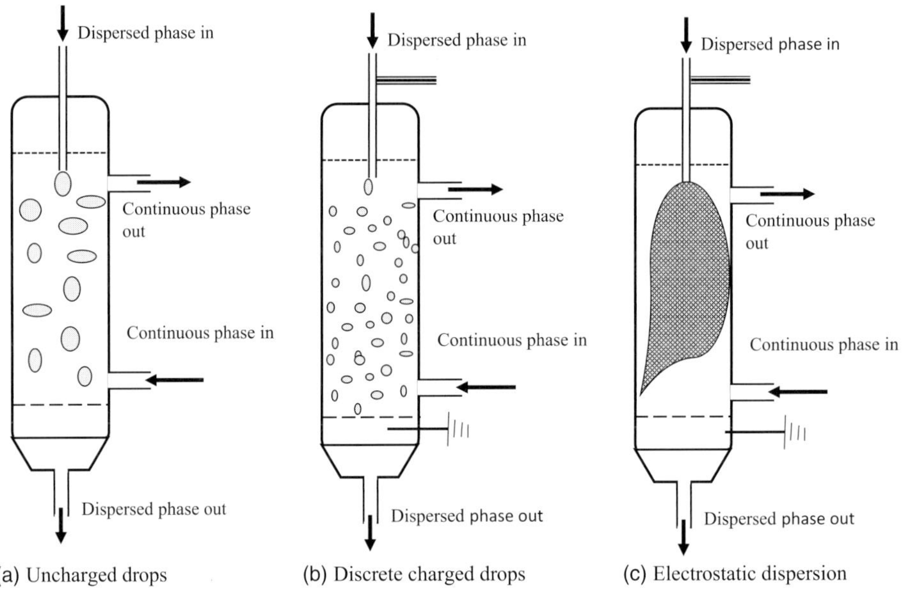

Figure 6.5 Conceptual diagram of spray formation in a continuous countercurrent spray with electrical dispersion: (a) uncharged drop formation, (b) discrete charged drop formation, (c) electrostatic dispersion.

electrical properties. The presence of the surface charge may result in the onset of distortion of the interface as the electrical forces at the interface gain significant influence on determining the shape of the interface. As higher voltage is applied, depicted for example in Figure 6.4(b), the forming drop may continue to distend. Eventually a point maybe reached at which the interface destabilizes, resulting in emission of fine dispersion, as depicted in Figure 6.4(c).

Figure 6.5 shows a similar depiction of the respective droplet dispersion regimes that would occur for example in a countercurrent spray liquid–liquid spray column. Here the surface charge on the dispersed phase is shown via a high-voltage source linked to the single nozzle at the top of the column. An earth electrode is shown near the base of the column in Figure 6.5(b) and (c).

The location of the electrodes allows the creation of an electrical gradient across the entire liquid continuum in the column and the resulting electrical field influences the motion of the electrically charged drops during formation and subsequent to detachment. Stewart and Thornton (1967a, 1967b) suggested boundaries on the values of conductivity and dielectric constant of the two liquid phases in order for electrostatic intensification of drop formation to be effective. It was proposed that electrical enhancement would only be achievable if the dispersed-phase conductivity exceeds $10^{-10}\ \Omega^{-1}\ m^{-1}$and if the dispersed-phase liquid dielectric constant is greater than 10. Similar constraints were proposed for the continuous phase liquid suggesting a required limits of conductivity not exceeding $10^{-12}\ \Omega^{-1}\ m^{-1}$ and dielectric constant not exceeding 4.

Table 6.1 Values of conductivity and dielectric constant (relative permittivity) for a selection of some liquids deployed in liquid–liquid systems.

Liquid	Conductivity Ω^{-1} m^{-1}	Dielectric constant
n-Heptane	1×10^{-15}	1.9
Diethyl ether	4×10^{-11}	4.3
Kerosene	1×10^{-10}	1.74
Decyl alcohol	2.1×10^{-7}	8.1
Isopropyl alcohol	3.5×10^{-6}	18.3
Chloroform	2.5×10^{-6}	4.9
n-Butyl acetate	1.5×10^{-5}	5.1
Water	1×10^{-4}	80

Analysis of the underlying physics can help explain the principle behind the limits on electrical properties. Firstly, consideration of a forming pendant droplet, for example depicted in Figure 6.4(a), shows that as a voltage is applied to the nozzle under suitable conditions, the droplet will acquire surface charge. The rate at which that charge is acquired is governed by equation 6.31, where λ is the "relaxation time" and is directly proportional to the dielectric constant, ε_0, and inversely proportional to the electrical conductivity, K.

$$q = q_0 \quad \exp(-t/\lambda) \tag{6.31}$$

where $\lambda = \varepsilon_0 \varepsilon_r / \kappa$ and is defined as the relaxation time of the liquid, $\varepsilon_r =$ the relative permittivity of the liquid, $\varepsilon_0 =$ the dielectric constant of free space, $\kappa =$ the conductivity of the liquid [Ω^{-1} m^{-1}], and $t =$ time variable [s].

In order for charge growth on the forming drop to occur quickly, according to equation 6.31 and from the definition of λ, κ should be large and ε small. The magnitude of the charge acquired by the drop is also determined by the effective nozzle voltage and the formation time of the drop prior to detachment. Relaxation time continues to be very important after detachment, not only that of the drop phase, but also the relaxation time for the continuous phase. Here inverse criteria apply since the leakage of charge from the detached drop also depends upon the dielectric properties of the continuous phase. Equation 6.31 also applies to the continuous phase and therefore κ must be small and ε large in order for the loss of charge to the continuous phase to be as slow as possible. In principle, equation 6.31 dictates the criteria listed in the prior paragraph. Values of K and ε for some common liquids are shown in the Table 6.1 (Weatherley, 1993).

Figure 6.6 shows a hypothetical calculation of rate of charge acquisition by a drop of water forming into a continuum of n-heptane. The points referring to nozzle to water illustrate the high rate of charge transfer from the charged nozzle to a forming water drop. The calculation is based on a single initial charge on the nozzle (q_o), decaying as it transfers rapidly to the water drop. The high conductivity of the water and its high dielectric constant are reflected in the rapid loss of charge. In contrast, the corresponding calculation for the water drop phase, which should experience some loss of charge to the n-heptane, shows that no change occurs, reflected in the horizontal set of points

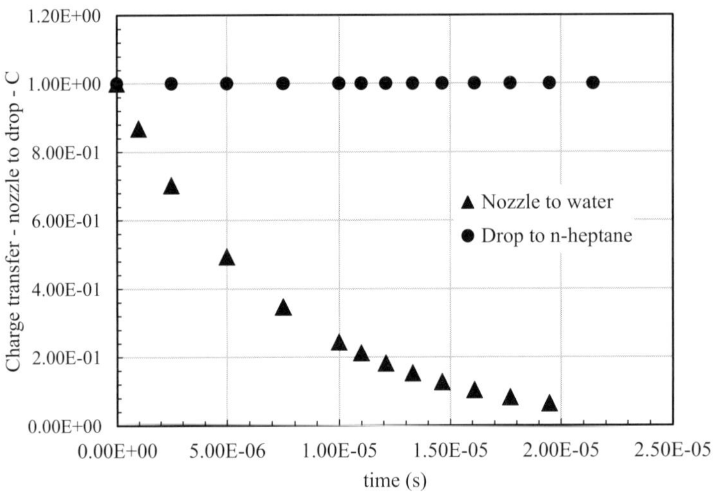

Figure 6.6 Comparison of charge acquisition/loss for water and n-heptane.

labeled "drop to n-heptane." This is consistent with the very low conductivity and low dielectric constant of the n-heptane.

6.5 Discrete Drop Size

In the case of an electrically intensified system, the drop size, motion, and mixing phenomena are strongly influenced by the presence of the charge on the nozzle(s) at which drop formation occurs and on the electrical field to which the detached drops are exposed. The local electrical field may impact the role of interfacial disturbance on mass transfer, and may also promote other related transport phenomena such as electroosmosis, electrophoresis, and dielectrophoresis.

The theoretical description of an electrically intensified liquid–liquid system is complex but in principle and in common with modeling of liquid–liquid systems in general, prediction of mass transfer, reaction kinetics, and hydraulic behavior requires an understanding of the following:

- Drop formation and prediction of drop size. The number and size of drops created and that are present in the liquid–liquid continuum are essential for determination of interfacial area for mass transfer and kinetic calculations.
- Drop motion, including the motion of single (discrete) drops and sprays (or clouds) of drops. Drop motion impacts residence time, mass transfer coefficients, and volumetric holdup. The fluid dynamics of the continuous phase and the impact of the drop size and motion are vitally important factors that influence mass transfer.
- Interfacial phenomena that influence transport processes. These include Marangoni effects, electroosmosis, electrophoresis, and dielectrophoresis.
- The impact of contactor geometry on all of the above.

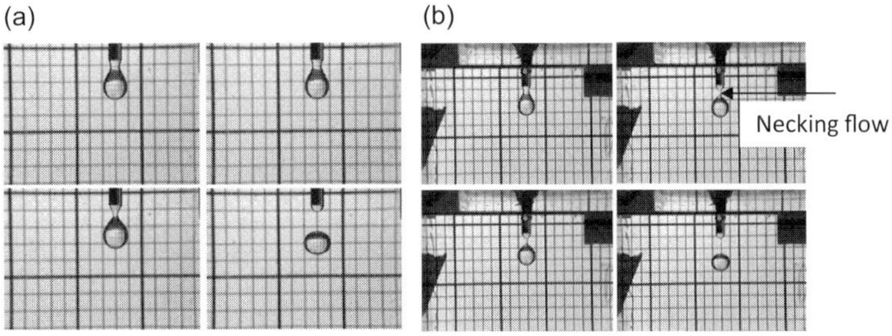

Figure 6.7 Detaching drop sequence for carboxmethyl cellulose (cmc)/water dispersing into mineral oil: (a) no electrical field; (b) −5 kV d.c. applied to formation nozzle. (Gangu, 2013)

There are two underlying differences between electrically intensified liquid–liquid systems and more conventional liquid–liquid systems. Firstly, the creation of liquid drops carrying surface charge and their introduction into an electrical field subjects those drops to additional electrical forces that can profoundly influence motion. Secondly, the phenomenon of electrostatic dispersion can result in clouds of highly compacted streams of fine electrically charged drops whose hydrodynamic behavior may be different from clouds of uncharged droplets. While the overall impacts on mass transfer, reaction, and hydraulic behavior are essentially similar, the theoretical understanding of each case differs significantly. Each case is now considered.

Knowledge of drop size and mechanisms for controlling it in liquid–liquid systems are vitally important. Drop size is a major factor in determining interfacial area for mass transfer or reaction. The size of drops also has major impact on velocity, residence time in a device, and holdup. Drop size taken with density and other properties of both liquids is used to determine drop velocity, which is turn is used to establish equipment size, for example in continuous column contactors.

Drop formation and initial drop size in a liquid–liquid system in the absence of any external electrical field involve a number of forces and stages. The uncharged case is discussed in detailed in Chapter 2. The photographic sequence displayed in Figure 6.7 shows the development of a drop emerging from a single nozzle, illustrating the change in shape that occurs during the formation process. Notable is the formation of a fine filament of liquid that occurs just prior to detachment of the drop. This is known as a "neck" and the flow associated with the flow as "necking flow." The necking flow, as is subsequently illustrated, plays a role in determining drop size after detachment and can also influence the creation of very small satellite drops that accompany the main drop shortly after detachment.

The sequence of photos in Figure 6.7 show the effect of the application of an electrical voltage to the formation nozzle with reduction in drop size clearly shown when the two parts are compared.

Both sets of images in Figure 6.7 show the phenomenon of necking flow that occurs momentarily during drop detachment. Figure 6.7 also shows the effect of an applied voltage on the drop detachment and in particular the various stages of shape

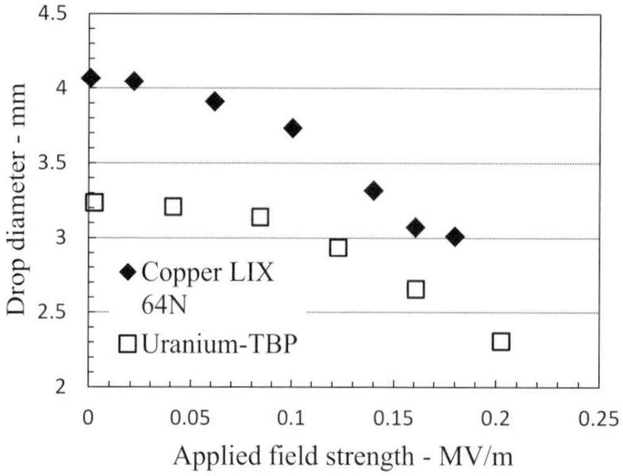

Figure 6.8 Droplet size versus nominal field strength for the system copper–LIX 64N and for the system uranium–TBP. Negatively charged aqueous droplets.
Reprinted (adapted) with permission from Bailes, P. J. (1981). Solvent extraction in an electrostatic field. *Industrial and Engineering Chemistry Research*, **20**, 564–570. Copyright (1981) American Chemical Society

deformation that detaching charged droplets experience. The theoretical analysis of this phenomenon is discussed briefly below. The images in Figure 6.7(a) show drop detachment of a drop of carboxmethyl cellulose in water dispersing into mineral oil with no external electrical charge applied to the nozzle. This may be compared with the set of images in Figure 6.7(b), which shows the effect of an applied voltage of −5 kV to the formation nozzle. The droplet is clearly in the discrete drop regime and of similar drop size but with significant elongation of the necking filament formed prior to detachment.

Specific knowledge of the relative velocity of drops to that of the continuous phase in countercurrent flow is required to avoid reverse flow and specifically, axial backflow of the drop phase. Another major reason for understanding drop motion is the momentum transfer that occurs between moving drops and the continuous phase, which creates secondary flows in the continuous phase, which in turn impacts mass transfer and mixing phenomena in a continuous liquid–liquid contactor. The introduction of an electrical field across the contactor, taken together with the presence of charged drops, raises a question of impact on holdup and flooding limit relationships necessary for design and for informing operational decisions.

Some of the early published data showing the reduction in drop size that may be achieved using electrostatic fields for systems of industrial interest are by Bailes and Thornton (1971). Figure 6.8 shows measured data for drop size as a function of nominal field strength for the extraction systems for copper extraction from aqueous solution into LIX 64N solvent, and for the system uranium from aqueous uranium nitrate solutions into tri-n-butyl phosphate/diluent. In both cases the reduction in mean drop size is significant, representing up to a three-fold increase in specific interfacial area for mass transfer.

The starting point for prediction of drop size in an electrically enhanced system is consideration of the factors that influence the size of a single uncharged drop detaching from a nozzle and then to take into consideration the additional forces acting on the

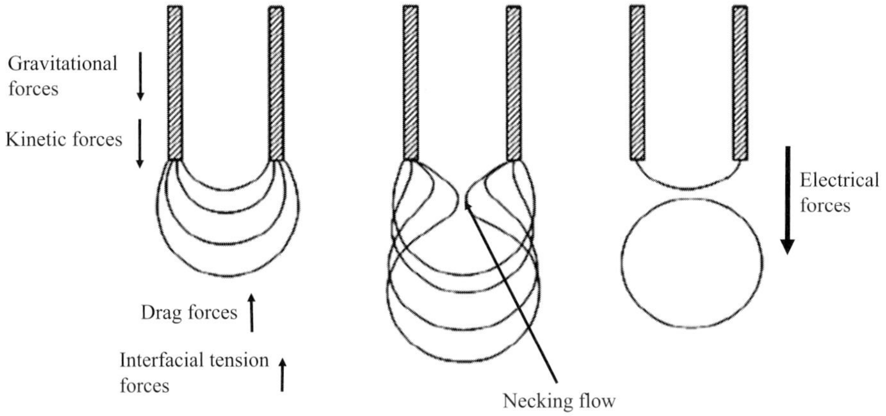

Figure 6.9 Single drop formation and detachment.

forming drop due to the presence of an electrical field. One of the classical analysis of drop formation in a liquid–liquid system is that of Scheele and Meister (1968). Their correlations for the prediction of drop volume correctly predicted drop volumes to within an error of 11% for the liquid–liquid systems studied involving Newtonian liquids and assuming spherical droplets.

In an uncharged system and with reference to Figure 6.9, four forces may be considered to be acting on a drop during formation and that ultimately determine the size of the detached drop. Buoyancy forces are assumed to be a function of gravity and density difference. Kinetic forces, which also act to detach a drop from the nozzle, occur on account of the kinetic energy of the fluid entering the nozzle to form the drop. Opposing forces comprise drag forces and interfacial tension force, which act to maintain the drop on the nozzle prior to detachment, as discussed in Chapter 2. When the buoyancy and kinetic forces exceed the drag and interfacial tension forces the drop detaches from the nozzle. Subsequent motion is assumed to be controlled by balance of drag forces, buoyancy forces, and inertial forces. Other phenomena incorporated into the analysis by Scheele and Meister that also affect the size of the detached drop include "necking flow" that is associated with the very fine filament of drop phase that momentarily forms immediately prior to detachment. A further factor is the dissipation of the kinetic energy associated with the velocity of the drop phase as it is forced through the formation nozzle.

The correlation for the size of a detached drop developed by Scheele and Meister (1968) to calculate the volume of the detached drop for a parabolic velocity profile in the nozzle itself, which takes into account these factors, is as follows.

$$V_{\mathrm{F}} = F\left[\underbrace{\frac{\pi\gamma D_{\mathrm{N}}}{g\Delta\rho}}_{\text{Interfacial tension}} + \underbrace{\frac{20\mu_{\mathrm{c}}Q_{\mathrm{v}}D_{\mathrm{N}}}{D_{\mathrm{F}}^{2}g\Delta\rho}}_{\text{Drag}} - \underbrace{\frac{4\rho_{\mathrm{d}}Q_{\mathrm{v}}U_{\mathrm{N}}}{3g\Delta\rho}}_{\text{Kinetic flow}} + 4.5\underbrace{\left(\frac{Q_{\mathrm{v}}^{2}D_{\mathrm{N}}^{2}\rho_{\mathrm{d}}\gamma}{(g\Delta\rho)^{2}}\right)^{1/3}}_{\text{Necking flow}}\right] \tag{6.32}$$

Equation 6.32 expresses the drop volume, V_F, after detachment, calculated taking account of the roles of interfacial tension forces between forming drop and nozzle, the role of drag force in determining the point of detachment, the role of kinetic flow, which is associated with the velocity of the fluid through the nozzle, and fourthly the role of necking flow.

As stated earlier, the actual volume of the detached drop is less than that predicted under ideal conditions, so the factor F is applied to allow for the small fraction of liquid that is retained at the nozzle or that detaches as a small satellite drop. In the case of detaching drops that are electrostatically charged and that are discharging into an electrical field then equation 6.32 is modified to incorporate the additional electrical force that acts on the detaching drop. This was first examined in detail by Takamatsu et al. (1981, 1982) who used the Scheele and Meister equation as a starting point to develop a new correlation for the volume of drops detaching from an electrically charged nozzle, thus:

$$V_F = F\left[\frac{\pi\gamma D_N}{g\Delta\rho} + \frac{20\mu_c Q_v D_N}{D_F{}^2 g\Delta\rho} - \frac{4\rho_d Q_v U_N}{3g\Delta\rho} - \frac{F_E}{g\Delta\rho} + 1.15\left(\frac{Q_v{}^2 D_N{}^2 \rho_d \gamma}{(g\Delta\rho)^2}\right)^{1/3}\right] \quad (6.33)$$

This correlation includes the term $\frac{F_E}{g\Delta\rho}$ to represent the role of the additional electrical force that influences the volume at detachment. Takamatsu et al. also varied the coefficient on the necking flow term (the 5th term in the expression) to achieve a better fit to experimental drop size data (Figures 6.10 and 6.11).

In equation 6.33 the term F_E presents a challenge for prediction since the electrical force is a function of both the Coulombic charge on the forming drop and of the local electrical field strength, which is many cases is nonlinear and until recently was difficult to calculate in real continua.

Takamatsu presented a semiempirical approach to prediction of F_E that used details of the nozzle geometry and position, assuming a value of electrical field strength based on electrode voltage and interelectrode distance. Thus a nominal value of field strength

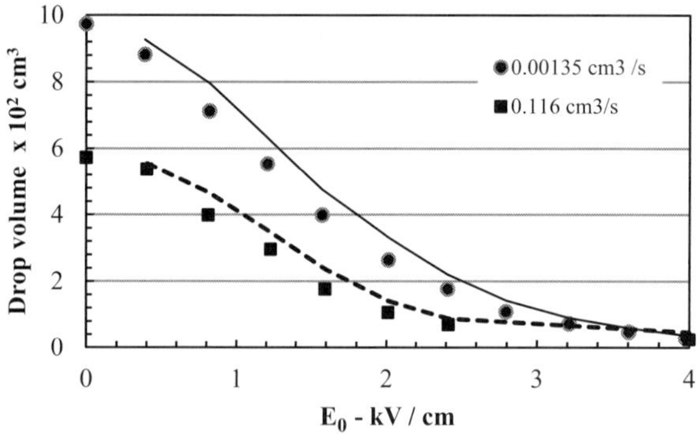

Figure 6.10 Drop volume versus nominal field strength for water drops forming into an insulating oil phase.

Reprinted (adapted) from Takamatsu, T., Yamaguchi, M., and Katayma, T. (1982). Formation of single charged drops in liquid media under a uniform electric field. *Journal of Chemical Engineering of Japan*, **15**(5), 349–355. With permission from the Society of Chemical Engineers of Japan

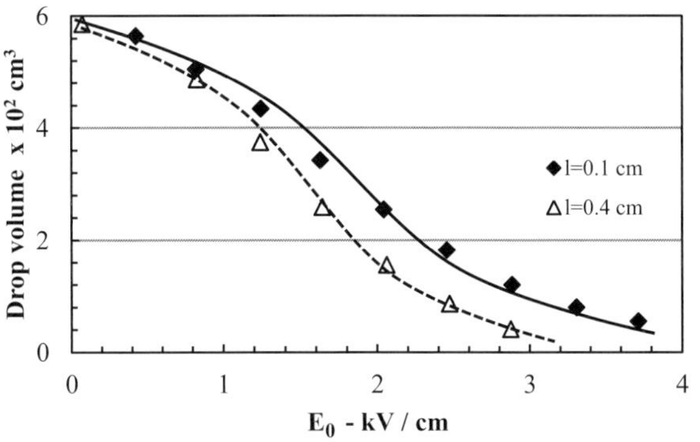

Figure 6.11 Drop volume versus nominal field strength for water drops forming into a cyclohexane phase, showing the effect of nozzle immersion depth on drop size (lines show predicted values). Reprinted (adapted) from Takamatsu, T., Yamaguchi, M., and Katayma, T. (1982). Formation of single charged drops in liquid media under a uniform electric field. *Journal of Chemical Engineering of Japan*, **15**(5), 349–355 with permission from the Society of Chemical Engineers of Japan

E_0 was used and expressed as the voltage difference between the electrode divided by the distance. Their equations are presented as follows.

• Initially a linear electrical field was assumed. The electrical force acting on the forming drop was expressed in terms of E_0, the drop radius R, and a parameter β defined in equation 6.36.

$$F_E = 4\pi\varepsilon\beta R^2 {E_0}^2 \tag{6.34}$$

• Charge Q_E on the forming drop was expressed as:

$$Q_E = 4\pi\varepsilon\beta R^2 E_0 \tag{6.35}$$

where β is given by

$$\beta = \begin{cases} 1.12 & (\alpha = 1) \\ \alpha + 0.34 & (\alpha > 1) \end{cases} \tag{6.36}$$

and

$$\xi = \alpha + 0.63(\alpha \geq 1)$$

• ξ and β are charge coefficients used to model the data.

$$\alpha = (l + R)/R \tag{6.37}$$

• The nozzle protrusion length l is the distance that the nozzle protrudes from the electrode and R is the droplet radius. Data for water drops forming into a continuous phase of cyclohexane produced a good fit even assuming a linear electrical field.

A similar approach applied to the prediction of drop size of water drops forming into silicone oil at an electrically charged nozzle resulted in a poor fit.

The above analysis, which is based on the assumption of uniform electric field strength across the entire continuous phase, is therefore not accurate in all cases. In actuality the field strength can be highly nonlinear, with much higher values close to the

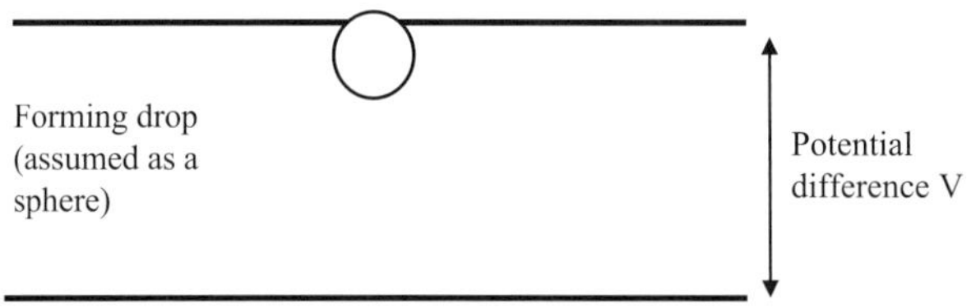

Figure 6.12 Sphere to plane electrode geometry used to model drop formation in a nonlinear electrical field.

electrodes including the nozzle at which drops are forming. In such cases the electrical forces acting at the formation nozzle will be much greater and therefore exert a much greater influence on the drop formation process. Takamatsu et al. (1983) developed an improved analysis for calculation of drop size for the nonlinear electrical field case as follows.

The starting point for modeling of drop formation in the nonlinear electrical field is the sphere to plane depiction shown in Figure 6.12. It was assumed that the field strength in the nonlinear field E' be expressed in terms of the field strength based on a linear field E_0 and the ratio of the maximum field strength at the leading edge of each forming drop E_a and E_B. Using the method of images, Takamatsu developed an improved expression for the electrical force acting in each forming droplet, shown in equations 6.38–6.41 as follows:

$$E' = E_0 \frac{E_a}{E_B} \tag{6.38}$$

$$E' = 0.238\left[1 + \frac{r}{2L}\right]\frac{V}{r} \tag{6.39}$$

$$F_e = 4\pi\varepsilon(0.0634)\left[1 + \frac{r}{2L}\right]^2 V^2 \tag{6.40}$$

$$Q = 4\pi\varepsilon(0.388)\left[1 + \frac{r}{2L}\right] rV \tag{6.41}$$

Use of equation 6.34 for the electrical force term F_e in the force balance equation governing drop detachment gave an accurate prediction of drop volume as a function of the applied voltage (potential difference) V, as shown in Figure 6.13.

The effect of surface charge upon interfacial tension and the consequences for drop size and interface stability are central to understanding how the application of external electric fields can intensify the behavior of liquid–liquid systems. The phenomenon of interfacial tension results from the imbalance of intermolecular forces at a liquid–liquid interface, consistent with the theory of surface tension of liquids. Stewart and Thornton (1967a, 1967b) used the relationship between force, f, interfacial tension, and radii of curvature at an interface, as expressed in equation 6.42, as a starting point for theoretical analysis of interfacial tension at an electrically charged interface:

$$f = \tau\left(\frac{1}{r_1} + \frac{1}{r_2}\right) \tag{6.42}$$

where r_1 and r_2 are the principle radii of curvature at the interface, and τ is the interfacial tension.

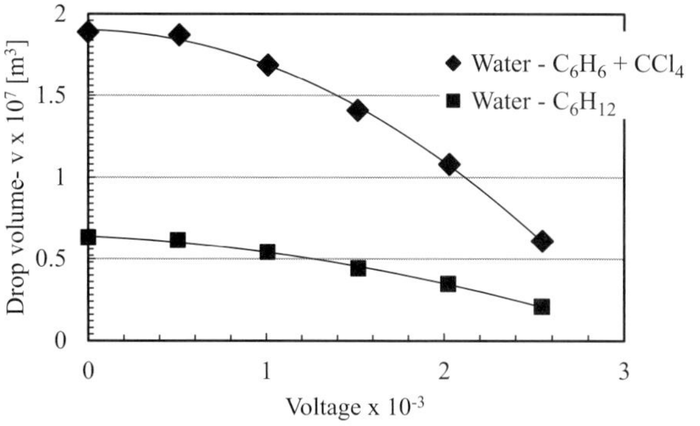

Figure 6.13 Comparison between theoretical and experimental drop volumes for water drops forming into dielectric liquids.
Reprinted from Takamatsu, T., Yamaguchi, M., and Katayma, T. (1983). Formation of single charged drops in liquid media in a non-uniform electric field. *Journal of Chemical Engineering of Japan*, **16**(4), 267–272 with permission from the Society of Chemical Engineers of Japan

The electrical force F_E, at the charged interface is described in equation 6.43 in terms of the electrical charge density at the interface σ and the dielectric properties of the liquid ε, ε_0. It may be assumed that the force acts in the normal but opposite direction to the interfacial tension force, thus lowering interfacial tension.

$$F_E = -\frac{\sigma^2}{2\varepsilon\varepsilon_0} \tag{6.43}$$

ε, ε_0 represent the dielectric constant of the liquid and the permittivity of free space respectively.

The charge density, σ, is assumed for an isolated spherical drop under ideal conditions and is equivalent to the total charge, Q, on the drop per unit area, as in equation 6.44:

$$\sigma = \frac{Q}{\pi d^2} \tag{6.44}$$

The net inward force f_n acting across the interface is obtained by combination of equations 6.42, 6.43, and 6.44 to result in equations 6.45 and 6.46:

$$f_n = f - F_E \tag{6.45}$$

and

$$f_n = \tau\left(\frac{1}{r_1} + \frac{1}{r_2}\right) + \frac{\sigma^2}{2\varepsilon\varepsilon_0} \tag{6.46}$$

If the effective interfacial tension for the electrically charged drop is defined as τ' then we may write:

$$f_n = \tau'\left(\frac{1}{r_1} + \frac{1}{r_2}\right) \tag{6.47}$$

For an interface of spherical curvature $r_1 = r_2 = d/2$, with d = diameter of the spherical drop, we may then write the effective interfacial tension τ' as:

$$\tau' = \tau - \frac{\sigma^2 d}{8\varepsilon\varepsilon_0} \tag{6.48}$$

Finally, the analysis of Stewart and Thornton considered the case of a spherical drop in a uniform electrical field and the resulting charge density of the drop. The charge density is expressed in two parts. The initial charge on the drop, Q, as in equations 6.43 and 6.44 provides a component term contributing to charge density, σ (equation 6.44). In the case of the charged drop located in an external electric field, the actual charge density is affected by the field and varies over the surface of the drop. The classic expression for charge density in this case is written as:

$$\sigma = \frac{Q}{\pi d^2} + 3\varepsilon\varepsilon_0 E \cos\theta \tag{6.49}$$

where E is the nominal field strength and θ is the coaltitude angle of a point on the surface of the drop.

Equation 6.49 indicates that the charge density on the surface of the drop varies with position on the surface and thus by inspection of equation 6.48 the interfacial tension also varies with position on the drop surface. In principle this means that for an electrically charged drop in a uniform external field, gradients in the interfacial tension are likely to lead to highly localized interfacial flows, which may be considered similar to Marangoni-type effects, beneficial to mass transfer. This is considered in depth in Sections 6.9 and 6.10.

Thus use of equation 6.49 to quantitatively calculate the charge distribution on a drop in an external electric field is limited by the difficulty in predicting the local field strength E, which cannot be assumed to be linear. The nonlinearity of E occurs on account of migration of "space charge," which refers to the electrically induced migration of very small but finite conducting species present in the surrounding liquid continuum. This problem is further exacerbated by the presence of species of both negative and positive charge that possess different electrical mobilities. A third factor is the overall geometry of the system, which includes the shape of the containment vessel, and the position of inlet nozzles and electrodes.

Much of the fundamental work on electrical enhancement of droplet formation processes in liquid–liquid systems for separation applications has focused on the dispersion of the conducting liquid (often an aqueous phase) into the nonconducting liquid, usually an organic solvent phase, using d.c. voltages. Gneist and Bart (2002) present a study of charged drop formation for droplets of isododecane and for toluene drops forming into a continuous phase of water in the presence of an a.c. field. This work showed successful dispersion with large reductions in drop size from the order of 2000 μm to less than 300 μm. The drop size data was successfully correlated with the power input (rather than voltage) (Figure 6.14). The work also showed the influence of increasing the conductivity of the continuous phase and the resulting further reduction in drop size with increasing conductivity.

Gniest and Bart correlated drop size data using a semiempirical approach based on balance of gravitational forces, viscous forces, and electrical forces and presented equation 6.50 for mean drop size:

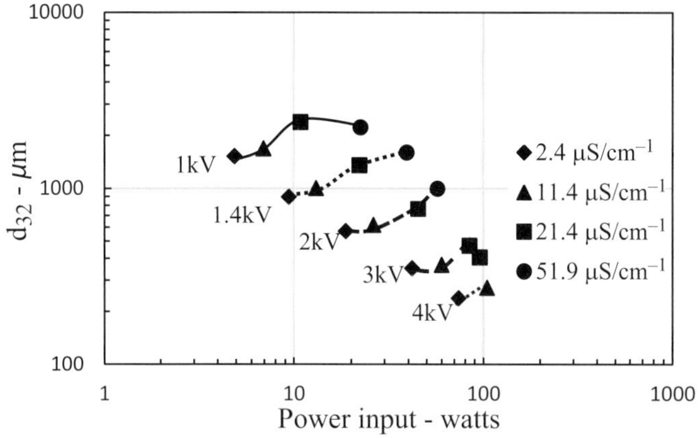

Figure 6.14 Sauter mean drop diameter for electrically charged drops of isododecane in water versus power input for different aqueous phase conductivites.
Reprinted (adapted) with permission Gneist, G. and Bart, H. J. (2002). Electrostatic drop formation in liquid-liquid systems. *Chemical Engineering Technology*, **25**(9), 899–904.

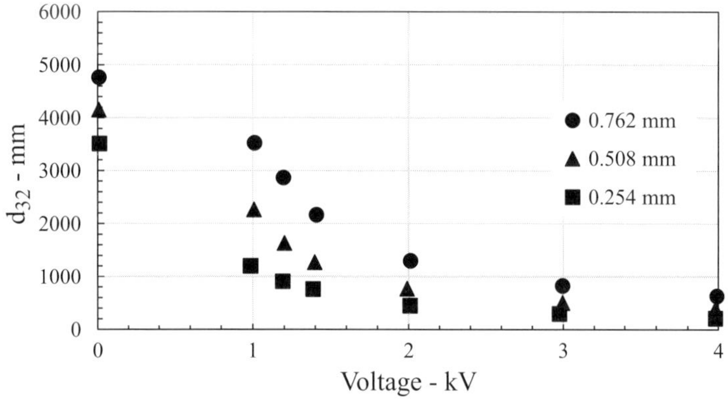

Figure 6.15 Sauter mean diameter for toluene drops in distilled water with respect to a.c. voltage for differing capillary diameters.
Reprinted (adapted) with permission Gneist, G. and Bart, H. J. (2002). Electrostatic drop formation in liquid-liquid systems. *Chemical Engineering Technology*, **25**(9), 899–904.

$$\psi\left(\frac{r_C}{v_D^{1/3}}\right)\pi d_C \gamma = \Delta\rho g \frac{\pi}{6} d_D^3 + \frac{C_1 d_D^{C_2}}{(C_3 + d_D)^{C_4}} U^{C_5} \tag{6.50}$$

where C_1, C_2, C_3, C_4, C_5 = correlation constants, d_C = nozzle diameter, in m, d_D = drop diameter, in m, r_c = nozzle radius, in m, v_D = drop volume, in m^3, U = voltage, in kV, $\Delta\rho$ = density difference between the drop phase and the continuous phase, in kg m^{-3}, γ = interfacial tension, in N m^{-1}, and ψ = Harkins correction factor (see Chapter 2).

Figure 6.15 shows experimental values of drop size versus applied voltage compared with values calculated using equation 6.50 (Gniest and Bart, 2002), with very good agreement.

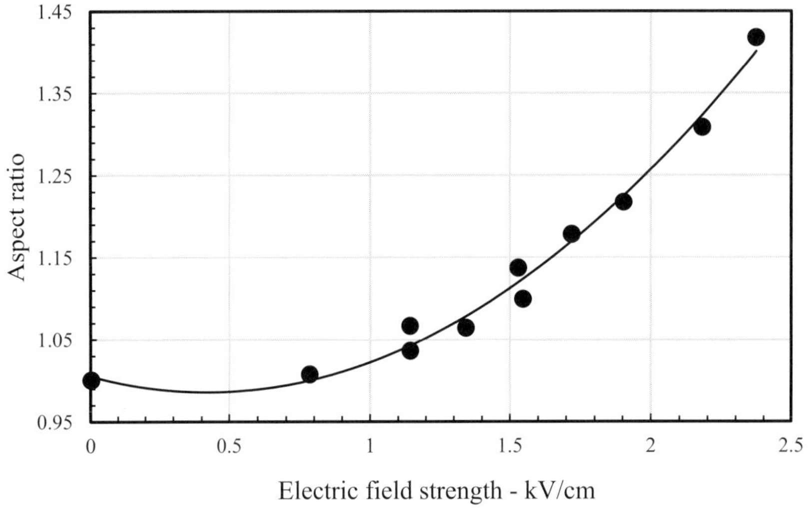

Figure 6.16 Droplet distortion versus electrical field strength.
Reprinted (adapted) with permission Wham, R. M. and Byers, C. H. (1987). Mass transfer from single droplets in imposed electric fields. *Separation Science and Technology*, **22**, 447–466.

Much of the published work on drop size and motion assumes spherical drop shape. As is clearly evident from experimental measurements in the case of electrically charged detaching droplets, this is often not the case, as is shown in Figure 6.7. Wham and Byers (1987) studied the relationship between drop shape and applied electrical field for a single water drop during the mass transfer into 2-ethylhexanol. Their data showed a strong relationship between drop shape and field strength, as shown in Figure 6.16.

Mathematical prediction of the changing shape of the forming drops is a challenge. This is important since the initial shape of the detached drop may have a significant impact on the subsequent motion of the drop and on the interfacial area. Both the motion and the interfacial area impact hydraulics and rate processes. Yang et al. (2013) present a rigorous analysis of the problem of an electrically charged drop forming in a high-voltage field.

A 3D phase field model was developed based on the coupled nonlinear governing equations for the electric field, the fluid flow field, and free surface deformation. The model of Yang et al. compared well with deformation theory developed by G. I. Taylor (Taylor, 1964, 1966) for the deformation of a single dielectric droplet in an electric field. Computed results for the deformation of a leaky dielectric droplet in an electric field that align with visual observations of various stages of deformation prior to attainment of a final oblate shape are shown in Figure 6.17. This approach considered phase field equations that describe the relationship between surface free energy, surface tension, and interface thickness, each of which ultimately influences drop shape. A second set of equations describes the electric field using the well-known Poisson equation, relating free charge density, dielectric properties, and electrical potential to establish the electric field distribution. This will be discussed in Sections 6.6 and 6.7, in the context of modeling of charged drop motion. The electric field calculation also used the free charge conservation

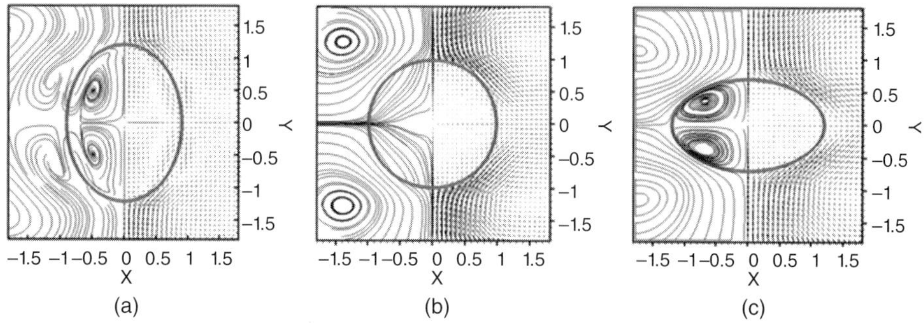

Figure 6.17 Transient development of droplet deformation, where the cross-section at $z = 0$ is shown rather the 3D shape, along with the fluid flow field and phase field. The initial shape is spherical (not shown). The deformation undergoes (a) the prolate shape first, then (b) the spherical shape, and then (c) eventually the oblate shape.

relationship and assumed negligible diffusion effect. The model also utilized a third set of equations based on mass conservation and on the Navier–Stokes equations.

6.6 Drop Motion in an Electrical Field: Discrete Drops

6.6.1 Calculation of the Electrical Field

In terms of theoretical prediction of drop size, drop motion, and interfacial disturbances, taking account of the application of electrical fields to the liquid–liquid system, the key is the accurate modeling of space charge migration and calculation of the electrical field strength as a function of spatial position. This is essential in order to accurately establish the electrical environment to which a moving or forming drop is exposed at any point in time. This problem was addressed in detail by Petera, Rooney, and Weatherley (1998), and Hume, Petera, and Weatherley (2003, 2004) and the following summary is based on their approach.

Petera et al. (1998) initially considered charged spherical particles of known constant size moving in a simple cylindrical column equipped with a single inlet nozzle, through which an aqueous phase containing the particles was dispersed into the continuous organic phase (Figure 6.18). An electrode was located near to the base, as shown, with the electrical field imposed across the continuous phase between the inlet point and the fixed electrode. Thus the mode of operation is based on the creation of electrically charged drops at the nozzle, the subsequent motion of which is strongly influenced by the magnitude of the local electrical field (as already stated). The theoretical development aimed to establish a map of the local field strength taking account of space charge leakage within the continuous phase, charge leakage through the walls of the contactor, the geometry of the nozzle, earth electrode, and contactor shell, and the magnitude of the applied voltage. As will be described, the controlling equations are set up to describe each of the above. The geometry of the system is configured using a finite element mesh

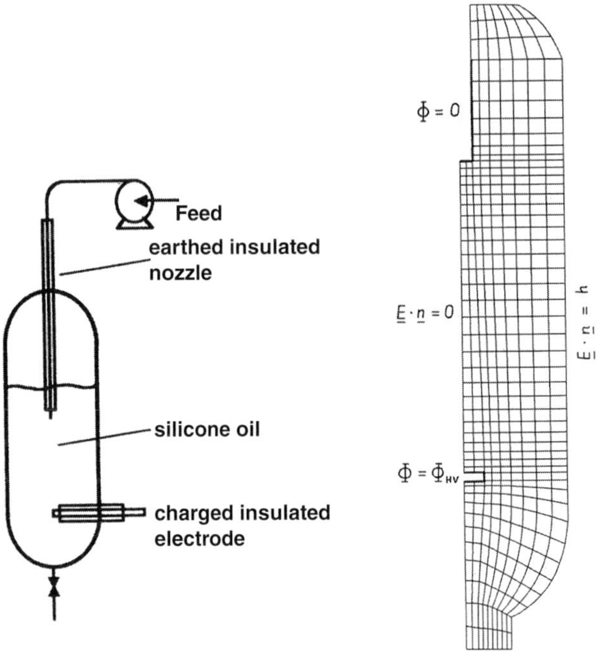

Figure 6.18 Outline of a simple electrically enhanced spray column contactor and the finite element mesh developed for the simulations E is the electric field strength vector; I is the electric potential. Reprinted with permission from Petera, J., Rooney D., and Weatherley, L. R. (1998). Particle and droplet trajectories in a nonlinear electrical field. *Chemical Engineering Science*, **53**(22), 3781–3792. Copyright Elsevier B.V.

(see the right-hand part of Figure 6.18), and the relevant variables represented on each of the finite elements. This approach enabled a quantitative computational solution of the controlling equations to provide description of the electric field in three dimensions, and ultimately to describe the force and flow environment to which individual particles and drops are subject. The finite element method allows quantitative study of many geometries of significant complexity.

The controlling equations and their application to the electrically intensified liquid–liquid system will now be described.

The following section is reprinted (adapted) as an excerpt with permission from Petera, J., Rooney D., and Weatherley, L.R., Particle and droplet trajectories in a non-linear electrical field. *Chemical Engineering Science*, 1998, **53**(22) 3781–3792. Copyright Elsevier B.V. (1998). All Rights Reserved.

It was assumed that the electric field in the domain could be described by the well-known (Panofsky and Phillips, 1969) Poisson differential equation, as defined in equation 6.14 and re-written as equation 6.51.

$$\underline{\nabla}\cdot(\varepsilon_r\Phi) = -\frac{\sigma}{\varepsilon_0} \tag{6.51}$$

where Φ is the electric potential, ε_r is the relative permittivity of the materials (dielectric liquid or air), ε_0 is the dielectric constant of the free space, and σ is the electric space charge density.

The axisymmetric version of equation 6.51 was used for the column geometry illustrated in Figure 6.18, in a cylindrical coordinate system (r, z) in the following form:

$$\frac{1}{r}\frac{\partial}{\partial r}\left(r\varepsilon_{\mathrm{r}}\frac{\partial \Phi}{\partial r}\right) + \frac{\partial}{\partial z}\left(\varepsilon_{\mathrm{r}}\frac{\partial \Phi}{\partial z}\right) = -\frac{\sigma}{\varepsilon_0} \tag{6.52}$$

With reference to Figure 6.18, the boundary conditions used in the solutions to the above equation are:

(a) of the Dirichlet type : Φ prescribed on the part Γ_ϕ of the total boundary Γ, belonging to the electrodes, $\Phi = 0$ on the earthed electrode,

$$\Phi = \Phi_{\mathrm{HV}} \text{ on the charged electrode} \tag{6.53}$$

(b) of the Neuman type on the remaining part Γ_h of the boundary Γ, with

$$\underline{E} \cdot \underline{n} = h \tag{6.54}$$

where $\underline{E}$ is the electric field strength vector,

$$\underline{E} = -\underline{\nabla}\Phi \tag{6.55}$$

$\underline{n}$ is the outward unit vector normal to the boundary, on which a function (of the position) h is given. If no flux is assumed on the axis of symmetry, the function h may be assumed to be zero. The calculated values of the electrical field vectors and forces acting on individual drops are sensitive to the boundary conditions of the Neuman type, shown in equation 6.54. The electric flux at the wall determined by the function h present as a result of the double layer phenomenon. It may be assumed that the function h is proportional to the local potential gradient normal to the wall, consistent with equation 6.54, and the definition of electric potential in equation 6.53. An adjustable constant of proportionality B may be established for individual sets of particles trajectories (equation 6.56):

$$h = B \quad \underline{\nabla}\Phi \cdot \underline{n} \tag{6.56}$$

It is assumed that space charge is present as a result of impurities in the dielectric continuous phase, as referred to above and described by Cross (1987). In this analysis for the purpose of macroscopic modeling it was assumed that some small particles with order of magnitude diameters less than 10^{-5} m were present and acquired a charge caused by corona charging on particles. As a result, a small but significant electric current was assumed to flow between the electrodes, and it was also assumed that the space charge, treated as a continuum field, accumulates near the electrodes. The field strength and the electric potential change significantly in those regions as a result of the behavior of the space charges (Stewart and Thornton, 1967a, 1967b).

There is uncertainty as to the number of microscopic electric charge carriers that may be present. This was addressed by Petera et al. (1998) by calculation of a possible distribution of the concentration of the space charge carriers using numerical solution of the pure transport equation (equation 6.57):

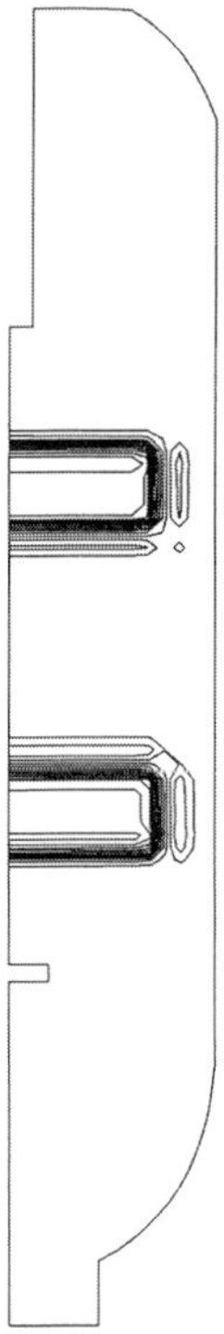

Figure 6.19 Initial assumed arbitrary space charge distribution.
Reprinted with permission from Petera, J., Rooney D., and Weatherley, L. R. (1998). Particle and droplet trajectories in a nonlinear electrical field. *Chemical Engineering Science*, **53**(22), 3781–3792. Copyright Elsevier B.V.

$$\frac{\partial \sigma}{\partial t} + \underline{u}\cdot\bar{\sigma} = 0 \tag{6.57}$$

where t is the time, $\underline{u}$ is a velocity field of the continuous field, σ is the charge density, which is determined by mobility, α, the particles, and the strength of the local electric field strength $\underline{E}$:

$$\underline{u} = \alpha \underline{E} \tag{6.58}$$

With regard to the mobility, α, in this approach there is no need for a real mobility value since the actual timescale of the migration process is not important. Instead, a final result of the migration is sought that is consistent with a plausible space charge distribution determined by the external electrical field. The purpose is to establish the influence of the space charge on the resultant electrical field. A trial and error method is used to determine a constant value of the mobility parameter by using repeating solutions of the simultaneous system of equations 6.51–6.58 using finite element techniques as a transient problem, starting from an arbitrary, artificial concentration field shown in Figure 6.19. The final mobility value used in the main simulation was based on literature values for the time for space charge migration to decay. In the case analyzed by Petera et al., the time value was typically of the order of 20 min. On this basis, a space charge distribution field was calculated, as shown in Figure 6.20.

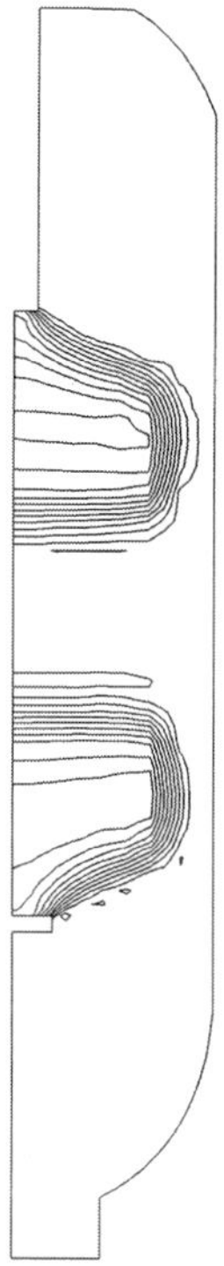

Figure 6.20 Steady-state space charge distribution for electrical field calculations.
Reprinted with permission from Petera, J., Rooney D., and Weatherley, L. R. (1998). Particle and droplet trajectories in a nonlinear electrical field. *Chemical Engineering Science*, **53**(22), 3781–3792. Copyright Elsevier B.V.

Since the space charge distribution is calculated with a degree of confidence, the electrical field for the entire domain may be calculated based on the solution of equations 6.56–6.58. In effect this procedure allowed the superposition of the external electrical field due to the voltage applied to the electrodes and the field arising from the space charge whose distribution is changing with time.

The influence of space charge migration and geometry on the values of electrical field strength across the domain is clearly illustrated in Figure 6.21.

The dotted lines in Figure 6.21(a) show the variation of the electrical potential and the field with respect to interelectrode position for the ideal case of a uniform electrical field. The respective trends for potential and field that take into account the geometry and space charge, are shown in Figure 6.21(b) and a very large deviation from linearity is very evident. The field enhancement is generally greater in the vicinity of the electrodes, with the net effect that the field strength experienced by the drops will be greater than the nominal value (applied voltage/distance) when the drop is near to the electrodes and less in the bulk of the continuum. The nonlinearity has a large influence on the electrical force acting on individual drops that are moving through the continuum. The analysis of charged drop/particle motion on the basis of the calculated nonlinear field will now be presented.

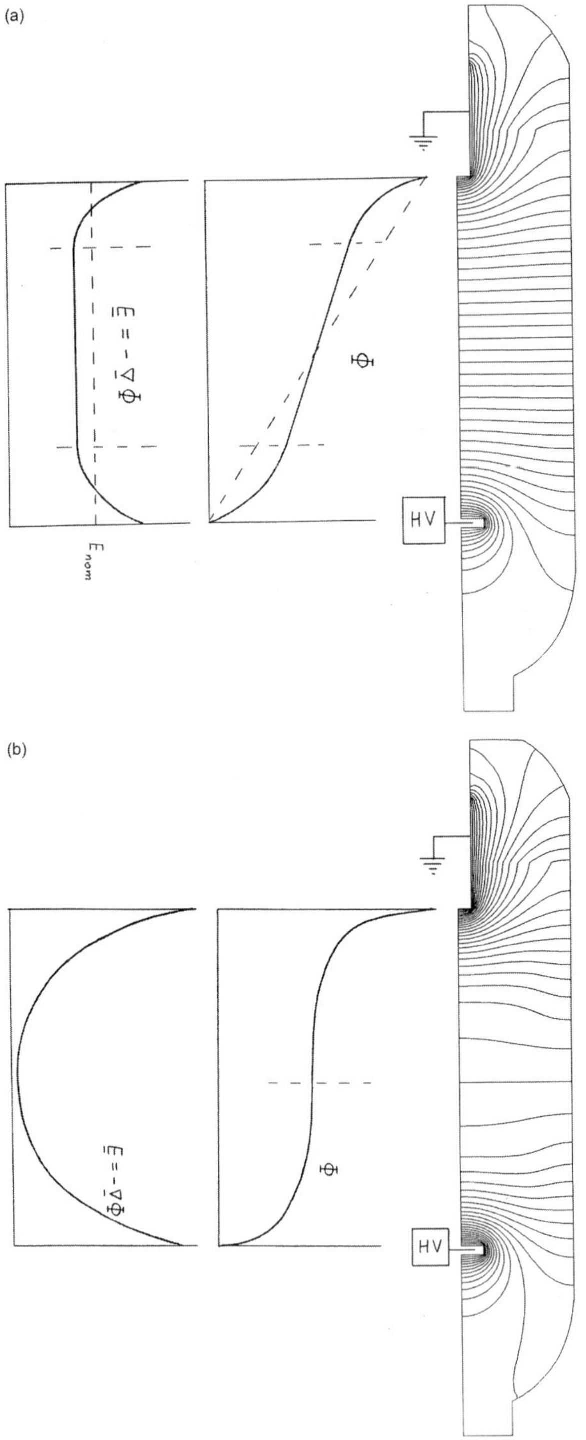

Figure 6.21 Simulation of variation of field strength in an electrically enhanced liquid–liquid contactor: (a) in the absence of space charge migration; (b) in the presence of space charge migration. Reprinted with permission from Petera, J., Rooney D., and Weatherley, L. R. (1998). Particle and droplet trajectories in a nonlinear electrical field. *Chemical Engineering Science*, **53**(22), 3781–3792. Copyright Elsevier B.V.

6.6.2 Prediction of Drop Motion in an Electrical Field

Determination of the electrical field, as described in the previous section, is critical for subsequent calculation of the charged drop motion. As was discussed in the prior section on drop size, the motion of drops following detachment is governed by the sum of the forces acting on each drop. The case of a discrete individual charged drop may be considered initially. It may be assumed that the following forces determine the motion of a charged drop released in a dielectric liquid:

1 The electrical force is expressed as follows:

$$\underline{F}_E = \underline{E}\ q \tag{6.59}$$

where q is the electric charge of the particle, and it may change according to the well-known rule (for constant permittivity ε_r and conductivity κ of the liquid):

$$q = q_0 \exp(-t/\lambda) \tag{6.60}$$

and $\lambda = \varepsilon_0 \varepsilon_r / \kappa$ is known as the relaxation time of the dielectric liquid.

Note that $\underline{E}$is the electrical field vector, as discussed in the previous section.

2 The drag force is expressed according to the Rybczynski Hadamard modification of Stokes law, as shown in equation 6.61:

$$\underline{F}_D = \frac{\pi \mu_c D_F (2\mu_c + 3\mu_d)}{\mu_c + \mu_d} \underline{v} \tag{6.61}$$

where μ_c and μ_d are the viscosities of the continuous and dispersed phases respectively, D_F is the drop diameter, and $\underline{v}$ the velocity.

3 The buoyancy force for the droplet is expressed thus:

$$\underline{F}_B = \frac{\pi d^3}{6} (\rho_S - \rho_L) \underline{g} \tag{6.62}$$

ρ_S and ρ_L are the respective densities of the particle and the liquid, $\underline{g}$ is the gravity vector (0,−9.81) in the coordinate system mentioned above.

4 The inertial force is expressed thus:

$$\frac{\underline{F}_I = \frac{\pi d^3}{6} \rho_S \quad d\underline{v}}{dt} \tag{6.63}$$

Thus the dynamic equation for the acceleration of the drop can now be written as follows:

$$\frac{d\underline{v}}{dt} = A_1\ \underline{E}q\ +\ A_2\ \underline{v}\ +\ A_3\ \underline{g} \tag{6.64}$$

where

$$A_1 = \frac{6}{\pi d^3 \rho_d}$$

$$A_2 = \frac{6\ \mu}{d^2 \rho_d} \left(\frac{2\mu_c + 3\mu_d}{\mu_c + \mu_d} \right) \tag{6.65}$$

$$A_3 = 1 - \frac{\rho_c}{\rho_d}$$

Equation 6.64 may be solved numerically using the simple definition of the particle velocity as defined in equation 6.66:

$$\frac{\underline{v} \ = d\underline{x}}{dt} \tag{6.66}$$

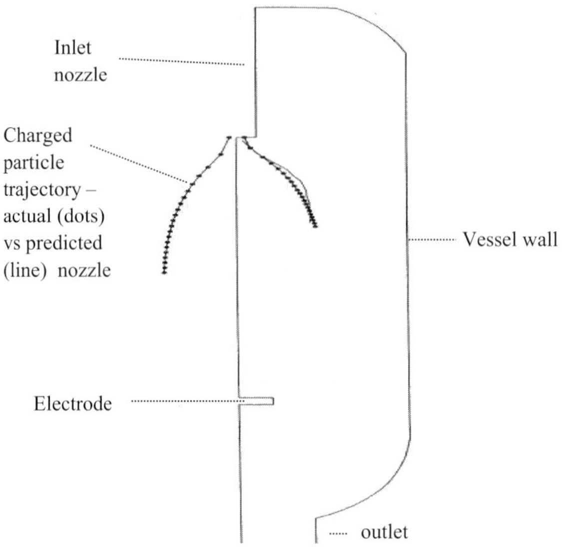

Figure 6.22 The comparison between experimental (continuous line) and simulated (solid circles) trajectories for a charged particle moving in silicone oil at an applied electrode voltage of 15 kV. Reprinted with permission from Petera, J., Rooney D., and Weatherley, L. R. (1998). Particle and droplet trajectories in a nonlinear electrical field. *Chemical Engineering Science*, **53**(22), 3781–3792. Copyright Elsevier B.V.

$\underline{x}$ is the position vector and thus the drop trajectory in the domain calculated in the presence of the electric field, which may be quantitatively determined using the procedure summarized above.

Figure 6.22 shows a sample comparison of an experimentally measured spherical charged particle trajectory in a column of dielectric oil with the calculated trajectory based on the above analysis. Full analysis of the algorithm and calculation procedure are described in Petera et al. (1998).

Hume et al. (2003) present solutions for the case of charged water droplets in a rectangular column containing silicone oil and the comparison with experimental observed trajectories is shown in Figure 6.23. Here the coordinate y refers to the vertical position of the droplet and it can be observed that the droplet moves down the column from a starting point of 5 mm from the top, moves down and after 2 s the trajectory reverses for a period of approximately 1.5 s so the drop moves upwards before recommencing descent. This is consistent with the droplet reflecting from the lower electrode in the column due to charge reversal. Significantly, this is predicted very accurately by the mathematical analysis.

Petera et al. (2007) extended the analysis and validation to a three-dimensional case (Figure 6.24), again with considerable accuracy for water drops dispersing in the discrete drop regime into silicone oil.

These findings were significant as they demonstrated the ability to calculate drop motion in a complex electrohydrodynamic environment from first principles. This makes possible the calculation of the velocity vector for individual drops in light of the physical properties, the geometry of the containment vessel and components (nozzles, electrodes, inlets, and outlets), the electrical field, and flow rates. As will be seen in a later section, this approach enables determination of the hydraulics (mixing profiles in both phases), mass transfer to and from droplets, and ultimately quantitative calculation of volumetric holdup, and flooding condition.

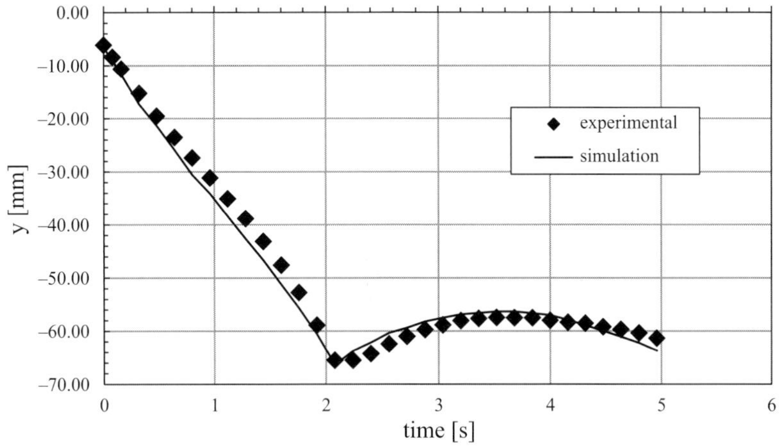

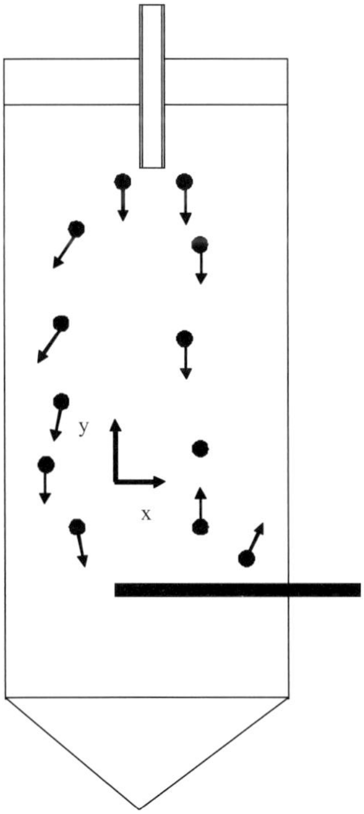

Figure 6.23 One-dimensional droplet trajectory for water drops in silicone oil: external field is −6.00 kV, dispersed phase flow rate = 65 mL h^{-1}.

Reprinted with permission from Hume, A. P., Petera, J., and Weatherley, L. R. (2003). Trajectories of charge drops in a liquid-liquid system. *The Chemical Engineering Journal*, **95**, 171–177. Copyright Elsevier B.V.

6.7 Electrostatic Dispersions (Sprays)

The above analysis is confined to describing the motion of discrete electrically charged drops in an electrical field. The motion behavior of individual drops has a profound influence upon diffusion rates, rates of reaction, system stability, and system

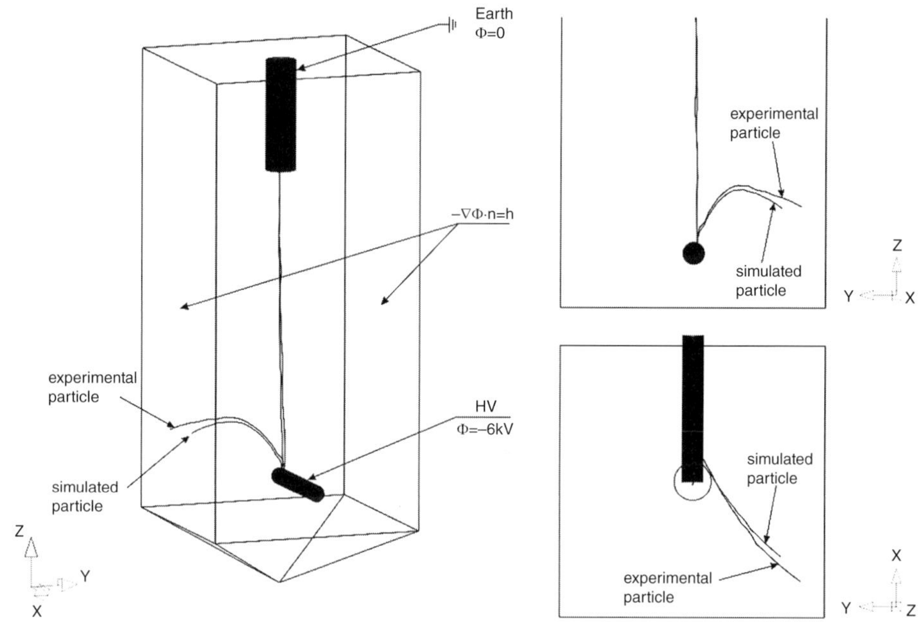

Figure 6.24 Three-dimensional comparison between the experimental and simulated trajectories from droplets in an experimental electrically enhanced contactor in the presence of an electric field. Reprinted with permission from Petera, J., Weatherley, L. R., Hume, A. P., and Gawrysiak, T. A. (2007). A finite element algorithm for particle/droplet trajectory tracking tested in a liquid–liquid system in the presence of an external electric field. *Computers in Chemical Engineering*, **31**, 1369–1388. Copyright Elsevier B.V.

hydrodynamics (Anderson and Pratt, 1978; Lewis and Pratt, 1953). Correspondingly, mathematical description of the motion of discrete drops is a significant step in the development of models for swarming drop systems and for electrostatic dispersions. The behavior of dispersions is significantly more complex than discrete drops for several reasons. Firstly, the swarms or clouds of drops comprise drops of a range of diameters and residence times. Secondly drop–drop interactions can result in coalescence and also change in drop trajectory. Ultimately, the design of equipment for the handling of liquid–liquid mixtures is determined by the drop size distribution and motion, and the physical properties of the two liquids. Frequent collision and coalescence generally increases intra-droplet mass transfer due to the gross hydrodynamic disturbances that occur when two droplets coalesce. Droplet reflection following collision of two drops is also important and affects the residence time distributions of the dispersion in a continuous system. Because of inter-phase momentum transfer, droplet motion will also influence the velocity field in the nondispersed phase. This can alter the hydrodynamic conditions at the interface and the residence time distribution in both phases. Residence time distributions of the drops and of the continuous phase are of some significance in determining overall performance of liquid–liquid extractors and reactors (Petera et al., 2005), as is well known.

As stated earlier, referring to Figure 6.4(c), electrostatic dispersion occurs when the forming drop at the nozzle acquires sufficient surface electrical charge that the effective

Figure 6.25 Electrostatic spraying of hexyl methyl immidazolium methyl sulfate (bmim-$MeSO_4$) into n-heptane.
Electrostatic spraying of hexyl-methyl immidazolium methyl sulfate (hmim MeSO4) into n-heptane *(Private communication: Kamiński, K, and Petera J (2014) with permission)*

interfacial tension at the surface of the drop is lowered to the extent that the interface becomes unstable and thus an electrostatic dispersion is created. An example of an electrostatic dispersion is shown in Figure 6.25.

Figure 6.25 shows electrostatic sprays in a liquid – liquid system at increasing applied voltage (from left to right) in a column developed and demonstrated by Kaminski et al. (2014) Each image shows electrostatic spraying of 1-butyl-3-methyl immidazolium methyl sulfate (bmim-$MeSO_4$) into an n-heptane continuous phase. Kaminski et al. (2014) also observed in electrically sprayed swarms of drops, drop size distributions reflect not only a reduction in drop size and a substantial narrowing of drop size distribution that occurs concurrently with size reduction but also acceleration of drops despite the reduction in size. This is an especially important observation since as a method of intensification, the application of electrical fields suggests scope not only for increase in mass transfer but for the improvement of contactor hydraulics. Control of drop motion will be possible in a way that may increase flooding limits, reduce column diameters (in the case of continuous column contactors), and reduce entrainment.

One of the significant theoretical challenges in this area is the modeling of the motion of dispersions of the electrostatically charged droplets in the continuous phase under various hydrodynamic conditions in the presence of an external electric field. Prior to considering dispersion modeling, it is worth summarizing early theoretical work on electrostatically induced instability of forming drops leading to dispersion.

The stability of a liquid–liquid interface in the presence of an external electrical field is strongly controlled by the balance of forces acting across the interface. The size of the electrostatic charge at the liquid–liquid surface can strongly contribute to the net forces acting. These may decrease the interfacial tension and lead to instability and breakup of liquid at the interface. The latter may occur when the forces acting exceed the intramolecular forces that are maintaining the integrity of the interface. A stability criterion was presented in a classic paper by Rayleigh (1882) relating the maximum Coulombic charge to the interfacial tension and dielectric constant at the point of interfacial breakup (see equation 6.67).

$$Q = \frac{\left(16\pi r^3 \tau \varepsilon\right)^{1/2}}{3 \times 10^9} \tag{6.67}$$

Hendricks and Schneider (1963) also developed a criterion (equation 6.68) for the onset of electrostatic spraying based on determining the minimum electrical potential, V, required to induce spraying on a single drop in the absence of space charge:

$$\frac{V^2}{20\pi(300)2r} > \tau \tag{6.68}$$

where r = drop radius and τ = interfacial tension.

Further study by Taylor (1964) proposed that an uncharged drop located in a uniform electrical field shatters when the field strength (voltage gradient) E_0 exceeds a defined value according to equation 6.69:

$$E_0\left(\frac{\varepsilon\varepsilon_0 r}{\tau}\right)^{1/2} = 0.458 \tag{6.69}$$

In the literature dealing with swarms of particles, such as in smoke plumes or in spray driers (Nijdam et al., 2003; Petera et al., 2007), two basic mathematical techniques may be used to approach the problem: Eulerian or Lagrangian. The Eulerian approach is known to be more appropriate for analysis of densely loaded systems in that the dispersed-phase cloud may be considered as a continuum in its own right. Conservation equations that are based on the analogous relationships that govern fluid motion generally may be written. In the case of the Lagrangian approach, discrete particle trajectories are determined. Therefore such an approach is better suited to relatively low concentration systems. One possible limit, quoted by Petera, is a dispersed-phase volume fraction of less than 10%. In such dilute systems, particle–particle interactions are significantly less likely and thus it may be inappropriate to considered the swarm of droplets as single continuum. The Lagrangian approach also has the advantage over other methods since other data can be forthcoming from the calculations and prediction of contactor performance. Additional data include residence time distribution, coalescence frequency, interaction with charged walls, and surface phenomena. Inter-phase coupling between drop phase and continuous phase may also be predicted by inclusion of the appropriate source terms in the conservation equations for the continuous phase. The source terms are those that result from drag forces acting on an individual particle/droplet.

A particle cloud model approach was developed by Litchford and Jeng (1991) and Baxter and Smith (1993), and this is useful for the description of the behavior of swarms or clouds of moving drops in an electrically intensified system, especially when turbulent fluctuations are present. In this approach, the concentration of particles within a cloud was represented by a Gaussian probability density function. Then the statistical evolution of a cloud of particles about a mean trajectory is tracked time-step by time-step. The major benefit of this approach stated in these papers is that far fewer trajectories need be determined for calculation of the source terms compared with use of real concentrations of particles or droplets.

The vital numerical and computational challenge to using the Lagrangian approach is the tracking of particle/droplet trajectories within the relevant computational grid or mesh while at the same time updating field variables, together with the particle position and velocity. All of the algorithms use a particle kinetic equation of the form:

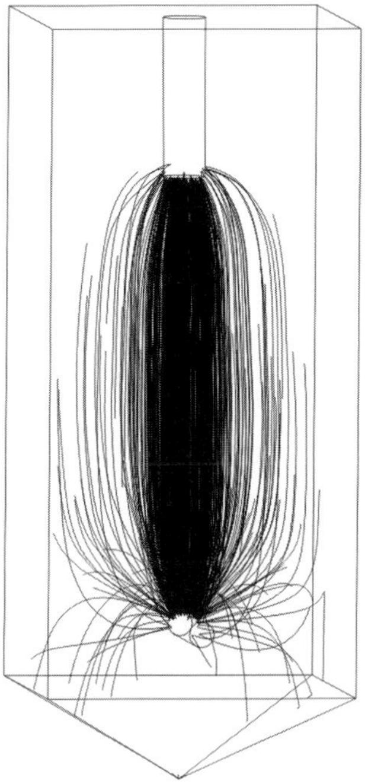

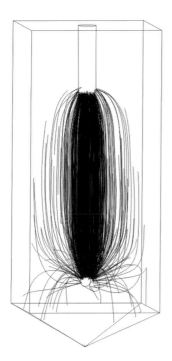

Figure 6.26 A snapshot of simulated trajectories in swarming particle/droplet motion in an experimental contactor. The distinct main core of trajectories are visible, accompanied by reflected particle trajectories giving rise to the back-mixing.
Reprinted with permission from Petera, J., Weatherley, L. R., Hume, A. P., and Gawrysiak, T. A. (2007). A finite element algorithm for particle/droplet trajectory tracking tested in a liquid-liquid system in the presence of an external electric field. *Computers in Chemical Engineering*, **31**, 1369–1388. Copyright Elsevier B.V.

$$\frac{d\mathbf{x}}{dt} = \mathbf{u}(\mathbf{x}_0, t) \tag{6.70}$$

Using this approach, Petera et al. (2007) showed that calculation of the simultaneous trajectories of multiple drops is possible, as shown in Figure 6.26**.** This shows the quantitative calculated trajectories for a known geometry and known system of water in silicone oil using the elegant finite element method described in their paper.

Figure 6.27 shows how the composite trajectories of a continuous forming cloud of sprayed electrically charged drops can provide information on the actual shape of the cloud, comparing the model predictions with an experimental cloud. In this case the cloud comprised of ethanol/water droplets spraying into preequilibrated n-decanol on the absence of mass transfer.

In order for modeling of clouds of drops to be a useful tool in design, further detail of the drop size distribution, residence time of drops, and mixing patterns in the surrounding continuous phase is required. The conceptual approach described by Baxter and Smith (1993) is informative in determining a strategy for calculation of key quantities, including interfacial area, holdup, and mixing parameters.

Figure 6.28 shows Baxter's simple concept for dynamic cloud dispersion in a one-dimensional flow field. Each circle represents a probability density function for the

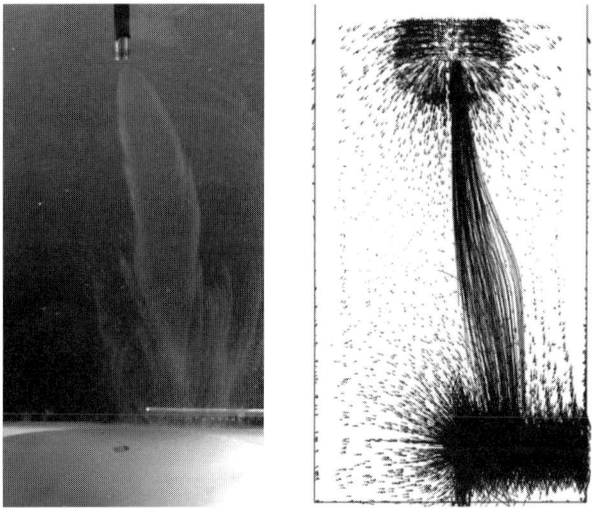

Figure 6.27 Comparison of a photograph of real droplet dispersion with simulated trajectories placed in the electrical field strength.
Reprinted with permission from Petera, J., Weatherley, L. R., Hume, A. P., and Gawrysiak, T. A. (2007). A finite element algorithm for particle/droplet trajectory tracking tested in a liquid-liquid system in the presence of an external electric field. *Computers in Chemical Engineering*, **31**, 1369–1388. Copyright Elsevier B.V.

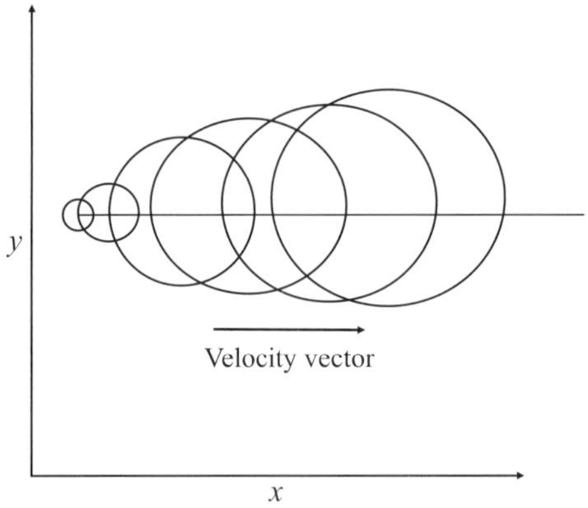

Figure 6.28 Lagrangian view of dispersion of a cloud in one-dimensional flow.
With permission from Baxter, L. L., and Smith, P. J. (1993). Dispersion of particles: the STP model. *Energy & Fuels*, **7**, 852–859. Copyright (1993) American Chemical Society

particles (or droplets) within a cloud as it expands during movement in the flow field from left to right. Based on the concept of multiple clouds continually forming, moving on the left to right trajectory, at the same time dispersing outwards, with overlap with earlier and later clouds, the drop population density and residence time distribution in any increment of the continuum will result from drops from numerous clouds.

The particle (or droplet) cloud model is based on the stochastic transport of particles approach, which uses statistical methods to trace the turbulent dispersion of particles about a mean trajectory. The mean trajectory is calculated from the ensemble average of the equations of motion for the particles represented by the cloud. The cloud enters the domain either as a point source or with an initial diameter and expands due to turbulent dispersion as it is transported through the domain until it exits. The distribution of particles in the cloud is defined by a probability density function (PDF) based on the position in the cloud relative to the cloud center. The value of the PDF represents the probability of finding particles represented by that cloud with a given residence time at a given location in the flow field. The average particle number density can be obtained by weighting the total flow rate of particles represented by that cloud. The cloud model was further developed by Petera et al. (2009) to account for the mass transport between the continuous phase and the dispersed phase in a cloud generated by electrostatic spraying and subjected to an external electrical field.

In respect of mass transfer, in outline, the driving force for mass transfer may be calculated on the basis of the concentration difference between that of the particle and that of the local continuous phase, interpolated along the cloud centerline (trajectory) for each time step. At each time step the amount of the mass transfer may be calculated and used to modify the local concentrations, taking into account the local density distribution of the droplets in the cloud and its population. This approach allowed the incorporation of coupling between the dispersed and continuous phases to determine the overall performance of a contactor involving clouds of electrostatically generated dispersed phase. The number of clouds (of the order of several hundred) and their populations (of order 10^4–10^7 in Petera's approach) may be assigned in advance to ensure consistency with the global flow rate of dispersed liquid fed to the system. The track of cloud diameters, droplet diameters and velocities, and the concentrations inside each droplet and continuous phase concentration field (apart from other fields) may be calculated and stored throughout the simulation. The model was verified for the application of electrostatic spraying techniques to the enzymatic hydrolysis of vegetable oils, which is a well-known mass transfer limited liquid–liquid reaction, as discussed later in the chapter.

Developing the theoretical analysis further, the crucial numerical issue in the Lagrangian approach is accurate particle/droplet trajectory tracking inside the computational grid/mesh used to analyze the continuum comprising the spray and surrounding liquid, while at the same time updating of field variables, particle position, and velocity.

The drop velocity in vector notation may be expressed as follows:

$$\frac{d\mathbf{x}}{dt} = \mathbf{u}(\mathbf{x}_0, t) \tag{6.71}$$

where $\mathbf{x}$ is the current particle position, $\mathbf{x}_0$ the initial position, and $\mathbf{u}$ the particle velocity vector (generally regarded as a function of time). Position is available through the first-order integration over a time interval $\Delta t_n = t_{n+1} - t_n$ according to:

$$\mathbf{x}_{n+1} = \mathbf{x}_n + \mathbf{u}(\mathbf{x}_0, t_n)\ \Delta t_n \tag{6.72}$$

Use of higher-order variants of equation 6.72, including Runge-Kutta methods of integration, is also possible.

The following section is reprinted (adapted) as an excerpt with permission from Petera, J., Weatherley, L.R., Rooney, D., and Kaminski, K. (2009). A finite element model of enzymatically catalyzed hydrolysis in an electrostatic spray reactor. *Computers and Chemical Engineering*, **33**, 144–161.

In the case of clouds of drops in the approach proposed by Petera et al. (2009), it is assumed that each individual statistical particle behaves in the same way within the same cloud although each one interacts with the continuous phase at a different position, giving a different contribution if belonging to a different cloud. As the model tracks the statistical evolution of a cloud of particles about a mean trajectory, the concentration of particles within the cloud is represented by a Gaussian probability density function (PDF) about each trajectory of the following form:

$$\mathrm{PDF}(\mathbf{x}) = \frac{1}{\left(\sqrt{2\pi}\ d_{\mathrm{cl}}\right)^3} \quad \exp\left[-\frac{1}{2}\left(\frac{\mathbf{x}-\mathbf{x}_{\mathrm{cl}}}{d_{\mathrm{cl}}}\right)^2\right] \tag{6.73}$$

where $\mathbf{x}$ is an arbitrary position within the cloud represented by the current center $\mathbf{x}_{\mathrm{cl}}$ (laying on the trajectory) and the current cloud diameter d_{cl}. The diameter itself grows along the center line (trajectory) due to dispersion in time and is assumed to be proportional to $\sqrt{D_{\mathrm{disp}}t}$, where D_{disp} is a dispersion coefficient (based on the formula resulting from a turbulence model) for droplets in the cloud and t is time.

$$d_{\mathrm{cl}} = \beta\sqrt{D_{\mathrm{disp}}t}, \beta = 0.1 - \text{emprical}\quad \text{constant} \tag{6.74}$$

On the basis of the particle density function, the droplet number density may be calculated using this equation:

$$n_{\mathrm{d}}(\mathbf{x}) = \sum_{\mathrm{ipa}=1}^{n_{\mathrm{clou}}} n_{\mathrm{ipa}}\mathrm{PDF}_{\mathrm{ipa}}(\mathbf{x}) \tag{6.75}$$

where n_{ipa} is the number of droplets in a cloud labeled by ipa (the droplet population in the cloud) and $n_{\mathrm{clou}}$is the number of clouds.

Similarly, the volumetric concentration of droplets is calculated from

$$f_{\mathrm{d}}(\mathbf{x}) = \sum_{\mathrm{ipa}=1}^{n_{\mathrm{clou}}} \frac{1}{6}\pi d_{\mathrm{ipa}}^3 n_{\mathrm{ipa}}\mathrm{PDF}_{\mathrm{ipa}}(\mathbf{x}) \tag{6.76}$$

and for the space charge field

$$\sigma(\mathbf{x}) = \sum_{\mathrm{ipa}=1}^{n_{\mathrm{clou}}} q_{\mathrm{ipa}} n_{\mathrm{ipa}}\mathrm{PDF}_{\mathrm{ipa}}(\mathbf{x}) \tag{6.77}$$

The numerical calculation of these variables can be performed only over a number of elements $n_{\mathrm{el}_{\mathrm{ipa}}}$ that are closer to the center line of each cloud than $3d_{\mathrm{cl}}$ (99 percent of the cloud population is situated in the circle of such radius).

Figure 6.29 shows the results of one set of computations based on the cloud model approach, taken from the paper of Petera et al. (2009). Both pictures show the droplet concentration distribution of a continuously refreshed cloud of fine drops produced by electrostatic spraying of an aqueous solution into vegetable oil.

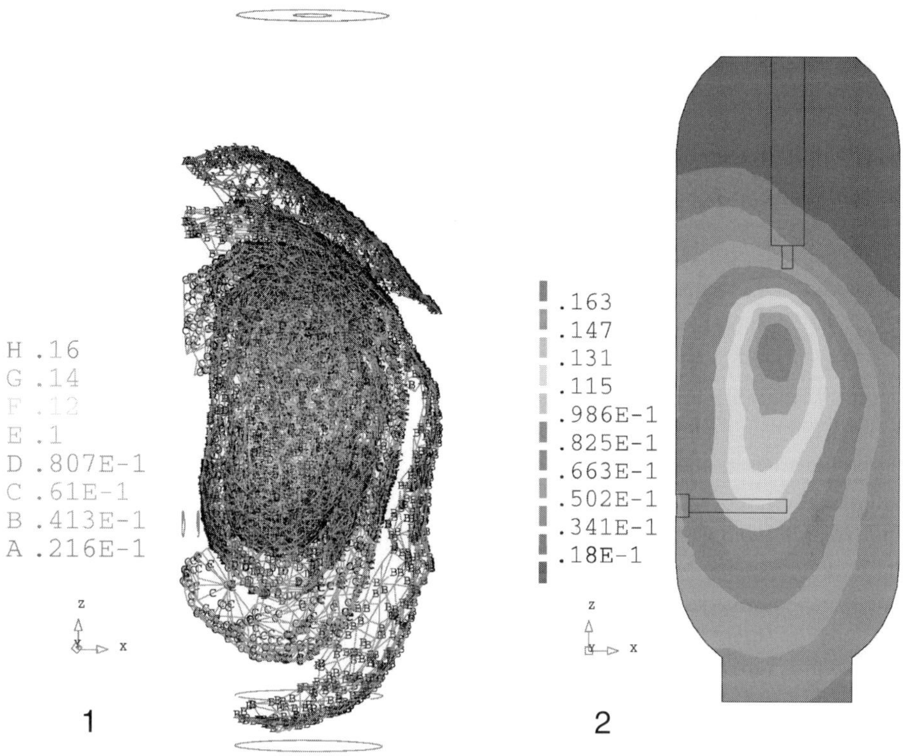

Figure 6.29 The droplet dispersion simulated by means of the cloud model. (1) 3D picture of clouds, (2) 2D section of the same clouds with the droplet concentration contours.
Reprinted with permission from Petera, J., Weatherley, L. R., Rooney, D., and Kaminski, K. (2009). A finite element model of enzymatically catalyzed hydrolysis in an electrostatic spray reactor. *Computers and Chemical Engineering*, **33**, 144–161. Copyright Elsevier B.V.

6.8 Mass Transfer

Mass transfer in electrically enhanced liquid–liquid systems has been explained in accordance with a number of mechanisms. The preceding section focuses on drop formation, size, motion, and spraying. All are integrally related to the enhancement of interfacial area for mass transfer. In most cases of practical interest, increase of specific interfacial area remains a primary mechanism for intensification.

There are a number of other mechanisms that are reviewed in the literature. These include improvement in interfacial mass transfer due to the enhanced relative velocity between moving drops and the continuous phase. This involves enhancement of shear forces at the liquid–liquid interface and hence higher film mass transfer in the continuous phase (Wham and Byers, 1987). Another mechanism analyzed experimentally by Wham and Byers is the role of droplet distortion and oscillations that are promoted by the electrical forces acting on an individual drop in an electrical field. This leads to unsteady stretching of the interface in an oscillatory cycle that results in improved conditions for mass transfer. The stretching of droplets under the influence of electrical field is illustrated earlier in the chapter in Figure 6.16.

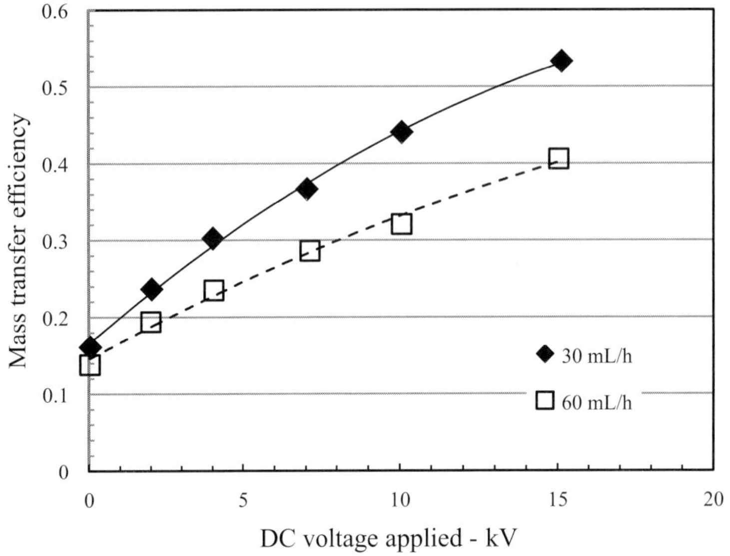

Figure 6.30 Total mass transfer efficiency as a function of applied voltage for the extraction of benzoic acid from water into mineral oil in a spray column; steady d.c. voltage.
Reprinted (adapted) with permission from He, W., Chang, J. S., and Baird, M. H. I. (1997). Enhancement of interphase mass transfer by a pulsed electric field. *Journal of Electrostatics*, **40–41**, 259–264. Copyright (1997) Elsevier B.V. All Rights Reserved

There are many experimental demonstrations of enhanced mass transfer in the presence of externally applied electrical fields. Figures 6.30 and 6.31 show data by He, Chang, and Baird (1997) for the extraction of benzoic acid from water into a mineral oil in a spray column. Large increases in mass transfer coefficient for both steady d.c. voltages and for pulsed d.c. voltages are observed. The use of pulsed fields for promotion of mass transfer is of some significance since the pulsation will promote droplet oscillation, which further promotes intra-droplet mass transfer through cyclic stretching of the drop interface. Another positive effect reported by He et al. (1997) is that overall energy input is less in the case of the pulsed systems.

Earlier data by Bailes and Thornton (1971) are plotted in Figure 6.32 and these show the importance of the field strength in determining the intensification of mass transfer. The relationship between overall mass transfer in an electrostatically enhanced spray column during the extraction of furfuraldehyde into n-heptane was compared for two different column heights. The increase in mass transfer coefficient for the shorter column is much greater compared with the longer column, providing strong evidence of high mass transfer rates at high nominal field strength (volts/cm).

6.9 Interfacial Disturbance

As referred to earlier, in Chapter 3 (Mass transfer), interfacial turbulence is an important phenomenon in mass transfer processes in liquid–liquid systems. Disturbances at a

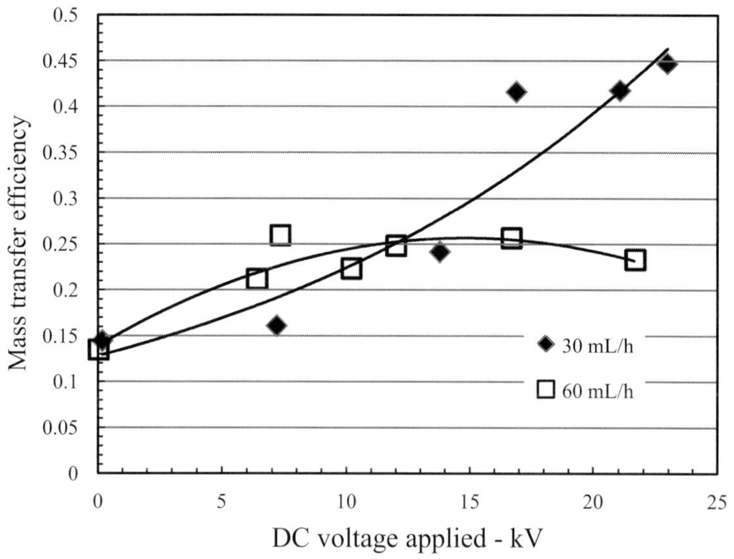

Figure 6.31 Total mass transfer efficiency as a function of applied voltage for the extraction of benzoic acid from water into mineral oil in a spray column; pulsed d.c. voltage.
Reprinted (adapted) with permission from He, W., Chang, J. S., and Baird, M. H. I. (1997). Enhancement of interphase mass transfer by a pulsed electric field. *Journal of Electrostatics*, **40–41**, 259–264. Copyright (1997) Elsevier B.V. All Rights Reserved

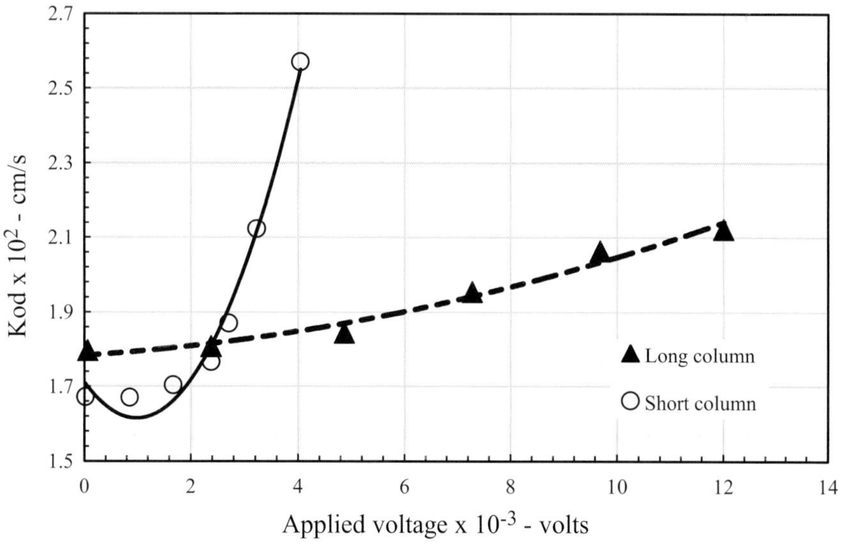

Figure 6.32 Variation of mass transfer coefficient with applied voltage for different column length. Data from Bailes and Thornton (1971)

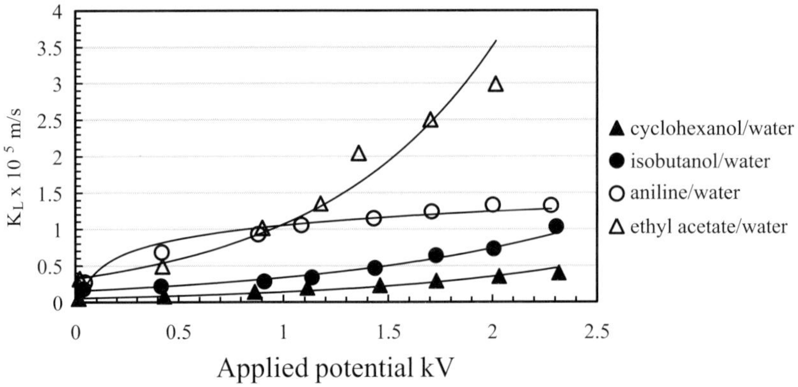

Figure 6.33 Effect of applied voltage on mass transfer coefficients for transfer across a plane interface: cyclohexanol/water, isobutanol/water, ethyl acetate/water, aniline/water.
Reprinted (adapted) with permission from Austin, L. J., Bancyk, L., and Sawistowski, H. (1971). Effect of electric field on mass transfer across a plane interface. *Chemical Engineering Science*, **26**, 2120–2121. Copyright (1971) Elsevier B.V. All Rights Reserved

liquid–liquid interface may be induced through the occurrence of local interfacial tension gradients and/or the presence of unstable density gradients often associated with mass transfer and high concentration gradients. Interfacial tension gradient driven convection is known as Marangoni convection (Quincke, 1888; Thomson, 1855). Marangoni-type interfacial flows modify the hydrodynamic behavior of the fluid layers near to the interface, thus affecting drop coalescence, jet break-up, and droplet drag coefficients during motion.

In the case of electrically charged drops, local variations in droplet surface charge density will lead to local variations in the effective interfacial tension and thus can result in Marangoni convection and increased mass transfer rate, as proposed in Stewart and Thornton's analysis (1967a, 1967b). Austin, Bancyk, and Sawistowski (1971) studied the application of an electric field across a plane interface of four partially miscible binary systems and established the effect on mass transfer, qualitatively linking the increase in mass transfer rate with visible interfacial disturbances promoted by an electrical field and thus Marangoni effects.

These authors determined similar visible disturbances when polarity of the applied electrical field was reversed but with significant differences in enhancement of mass transfer coefficient (see Figure 6.33). Also, the different enhancements observed in the various liquid–liquid systems evaluated suggested that the mechanism was not just hydrodynamic disturbance at the interface. Other transport phenomena induced by the presence of an electrical field such as ion transport and charge migration were suggested.

Studies by Gneist and Bart (2003) investigated the influence of d.c. and high–frequency a.c. fields on mass transfer from aqueous pendant and moving droplets into a continuous organic phase. Their results show that the application of a.c. field not surprisingly inhibits Marangoni effects whereas d.c. voltage enhances or initiates visual

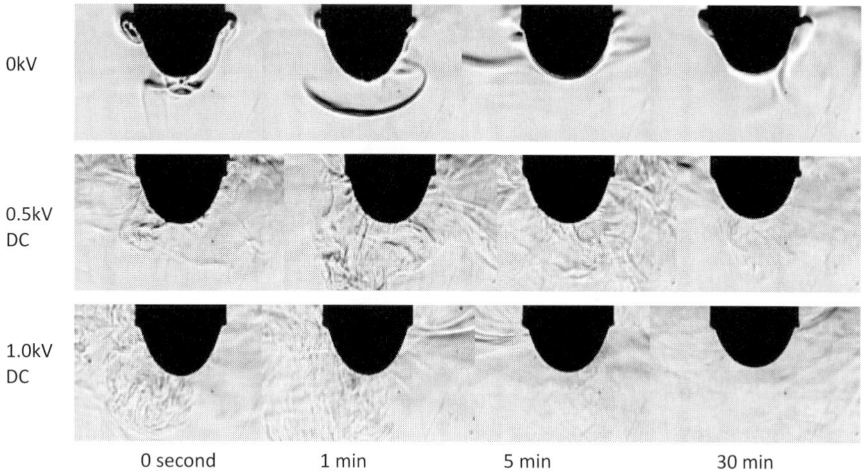

Figure 6.34 Interfacial turbulence in the transfer of ethanol from a pendant aqueous droplet (10 wt%) into a solute-free 1-decanol phase on 1 kV d.c. electric field. (Qiu, 2010)

Marangoni phenomena. Electrically induced Marangoni-type disturbances have been studied for pendant droplets forming into a second immiscible liquid in the presence of mass transfer, as shown in Figure 6.34 (Qiu, 2010). The drop phase comprises a 10% solution by weight of ethanol in water and the continuous phase comprises n-decanol as an extracting solvent.

The top row of images shows the effect of early mass transfer on interfacial disturbance, which reduces with respect to time from initial time to 30 minutes moving from left to right. The second and third rows show the effect of a d.c. voltage applied to the nozzle from which the pendant droplet is formed. The disturbance patterns are substantially increased and more finely grained as the voltage is increased from 0 kV to 1.0 kV. However, it is clear that the observed disturbance patterns are much greater initially, suggesting that the concentration difference at the interface at the time of drop formation exerts a major influence on the interfacial tension gradients rather than the presence of the electrical field.

Figure 6.35 shows a comparison of interfacial disturbance patterns at a planar interface for the same ethanol/water/n-decanol system as shown in the pendant droplet example. The difference between the two sets of images shows the presence of obvious disturbance patterns due to mass transfer effects only in Figure 6.35(a). In Figure 6.35(b), the disturbance patterns reflect the presence of an electrical field with much more finely grained disturbance patterns. Again, these are consistent with higher rates of mass transfer.

Figure 6.36 shows a further time sequence of Schleiren light patterns at the interface of a pendant drop of water and ethanol, suspended into n-decanol. The decay of the interfacial turbulence over the period of 30 minutes is very significant. These observations are consistent with the time-dependent nature of mass transfer analyzed by Anderson, Javed, and Thornton (1989) and Thornton and Batey (1989). They also underline the significance of the initial rate of mass transfer, as shown in Figure 6.37. It is believed this

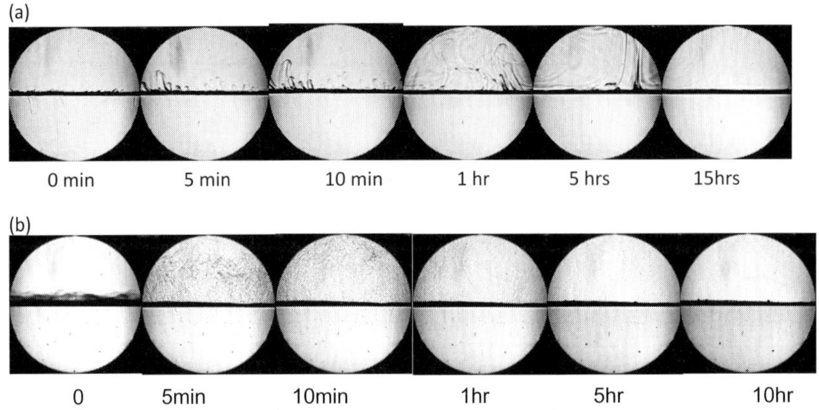

Figure 6.35 (a) Development of interfacial convection in the transfer of EtOH in the aqueous phase (5wt%) across the plane interface into a solute-free 1-decanol phase. (b) Development of interfacial convection in the transfer of EtOH in the aqueous phase (5wt%) across the plane interface into a solute-free 1-decanol phase in the presence of a 1 kV voltage applied across the interface. (Qiu, 2010)

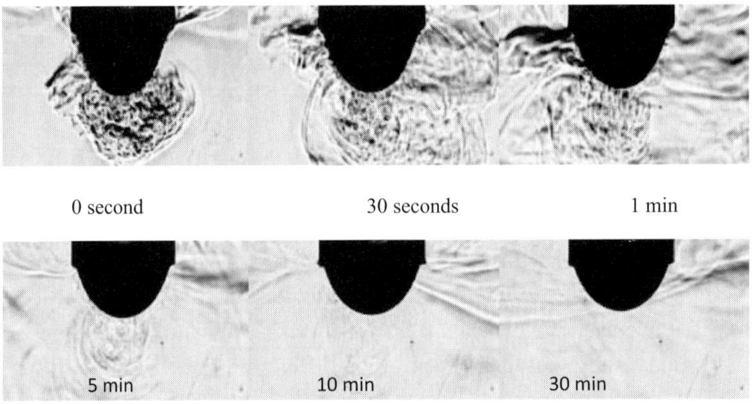

Figure 6.36 Interfacial turbulence in the transfer of ethanol from a pendant aqueous droplet (10wt% ethanol) into a solute-free 1-decanol phase on 1 kV d.c. electric field showing the strong time-dependent decay of the interfacial disturbances. (Qiu, 2010)

phenomenon is important in enhanced liquid–liquid extraction systems such as pulsed plate columns in which cyclic drop formation, detachment, and coalescence processes underpin the large increases in performance that may be observed.

More recent work by Kaminski, Weatherley, and Petera (2016) showed effective intensification of mass transfer in the presence of an electrical field for the extraction of ethanol from a solution in heptane into an ionic liquid solvent (1,3-methyl immidizolium methyl sulfate, see Figure 6.38). This appears to be one of the first demonstrations of the application of electrostatic fields to the intensification of mass transfer involving an ionic liquid.

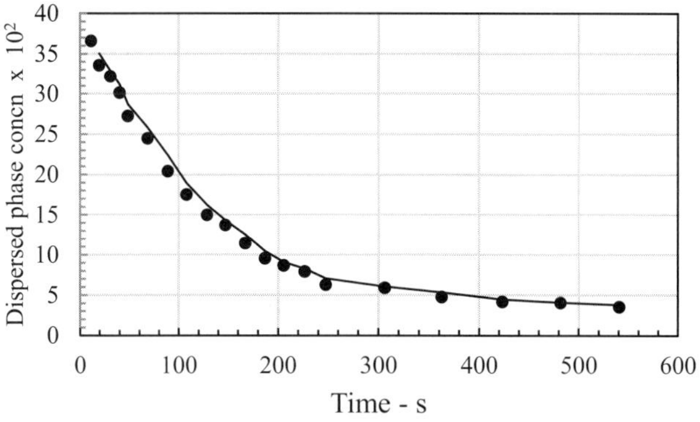

Figure 6.37 Time-dependent mass transfer from an uncharged pendant drop during the extraction of acetone from toluene into a water drop, showing dispersed-phase concentration versus time (5% initial solute concentration).

Reprinted (adapted) with permission from Anderson, T. J., Javed, K. H., and Thornton, J. D. (1989). Surface phenomena and mass transfer rates in liquid-liquid systems: Part 2. *American Institute of Chemical Engineers Journal*, **35**(7), 1125–1136. Copyright (1989) John Wiley and Sons – All Rights Reserved

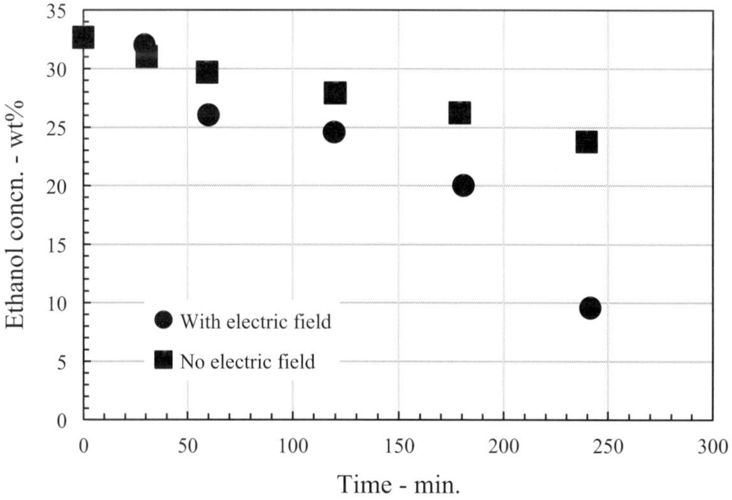

Figure 6.38 Electrically intensified mass transfer of ethanol from heptane into 1-butyl-methylimidazolium methyl sulfate.

Reprinted (adapted) with permission from Kamiński, K., Weatherley, L. R., and Petera, J. (2016) Application of numerical modelling to scaling-up of electrically induced extraction from an organic mixture using an ionic liquid. *Polish Academy of Sciences, Chemical and Process Engineering*, **37**(1), 133–148. https://creativecommons.org/licenses/by-nc-nd/4.0/legalcode

The intensification of mass transfer in the case of ionic liquids is significant since these liquid tend to be of high viscosity and thereby are associated with slow mass-transfer kinetics, which may limit their application using conventional contactor methods. Another constraint is the high inventory cost associated with ionic liquids and thus the development of intensified technologies that increase kinetics and improve recycle times may lead to a higher likelihood of commercial application.

6.10 Interfacial Mass Transfer: Further Theoretical Aspects

There has been little fundamental mathematical analysis of mass transfer in electrically enhanced liquid–liquid systems. Most published research focuses on experimental measurements of mass transfer in a range of systems as described in the previous section. The challenge in describing electrically enhanced liquid–liquid contact is complicated by the fact that much of the experience has been concerned with electrical dispersion and enhancement of mass transfer by increase in specific interfacial area value. As seen in Section 6.7, discussing electrical dispersion, progress has been made in describing clouds of electrically charged droplets in terms of motion, number, and size distribution (for example, see Figure 6.29). The slightly less complex problem is the description of enhanced transport rates due to interfacial disturbance in the presence of an electrical field. In both cases, description of mass transfer starts with consideration of the controlling hydrodynamic and transport equations. In this section we briefly summarize the fundamental controlling equations describing electrically enhanced mass transport and then provide two examples of experimental validation.

The fundamental approach to general description of mass transfer in liquid–liquid systems is discussed in Chapter 3. In the presence of electrical fields a similar, but modified, set of controlling equations is the starting point for quantitative analysis. The essential modification is the incorporation of electrical force terms into the equations. A full analysis is provided by Weatherley, Petera, and Qiu (2017) and a brief summary is provided here.

In the consideration of a two-phase liquid–liquid system comprising two bulk liquid phases and the interface, the incompressible Navier–Stokes equations govern the behavior of all three.

Firstly, there is the general equation of continuity that is generally applicable for a continuum:

$$\nabla \cdot \mathbf{u} = 0 \tag{6.78}$$

Secondly, there is the equation of momentum conservation expressed in terms of velocity vector and pressure field:

$$\rho\left(\frac{\partial \mathbf{u}}{\partial t} + \mathbf{u}\cdot\nabla\mathbf{u}\right) = \rho\mathbf{g} - \nabla p + \nabla\cdot\boldsymbol{\tau} + \mathbf{s}_{\mathrm{E}} \tag{6.79}$$

The stress tensor is written in the familiar form:

$$\boldsymbol{\tau} = 2\mu\mathbf{D} \tag{6.80}$$

The deformation tensor is written thus:

$$\mathbf{D} = 1/2\ [\nabla\mathbf{u} + (\nabla\mathbf{u})^{\mathrm{T}}] \tag{6.81}$$

The significant difference in the controlling equations is the incorporation of the following source term $\mathbf{s}_E$ written in terms of field strength $\mathbf{E}$, and σ, the local charge density as written in equation 6.82:

$$\mathbf{s}_E = \sigma \ \mathbf{E} \tag{6.82}$$

The change in the stress tensor at two different points on the interface can be expressed in terms of the pressure, the curvature of the interface, and the effective interfacial tension, γ_{eff}, according to equation 6.83:

$$-(p_2 - p_1)\mathbf{1} + (\boldsymbol{\tau}_2 - \boldsymbol{\tau}_1)\cdot\mathbf{n} = \gamma_{eff}\left(\frac{1}{R_1} + \frac{1}{R_2}\right) \ \mathbf{n} + \nabla\gamma_{eff} \tag{6.83}$$

The left-hand terms in equation 6.83 show the mechanical stresses resulting from the local flow and the right-hand terms include the curvature of the interface that is affected by the concentration gradient of the solute. Note the concentration gradient in the vicinity of the interface results in inhomogeneity of the interfacial tension.

Solution of equations 6.77–6.83 allows the complete description of the hydrodynamic conditions at a liquid–liquid interface in the presence of electrostatic charge and in the presence of an external electrical field. In order to achieve a solution, the relationship between local interfacial tension, composition, and electrical potential must be described and integrated into the solution algorithm.

We may now consider the description of transport across the liquid–liquid interface. The general approach is to establish a relationship between local solute concentration, C, and the interfacial tension, γ. The calculation process is necessarily iterative, establishing simultaneous solution of the sets of equations that relate concentration, interfacial tension, velocity vector, stress, and electrical field.

We can write the general equation for unsteady diffusional transport of the solute in terms of the diffusivity, concentration gradient, and local velocity as follows:

$$\frac{\partial C}{\partial t} + \mathbf{u}\cdot\nabla C = \nabla\cdot(D\nabla C) \tag{6.84}$$

The interfacial tension field, γ, is expressed in terms of the concentration, C, and the local electrical potential, Φ, as follows:

$$\gamma_{eff}(C, \Phi) = \gamma_0(C) - k_e(\Phi - \Phi_0)^2 \tag{6.85}$$

Equation 6.86 defines an effective interfacial tension, γ_{eff}, in terms of a baseline interfacial tension, γ_0, which in turn is related to concentration, C, as shown in equation 6.86:

$$\gamma_0(C) = \gamma_{00} - k_\gamma C \tag{6.86}$$

where k_γ is an adjustable parameter based on property data.

The above analysis was tested in a series of mass transfer experiments for the unsteady state extraction of ethanol from an aqueous solution into n-decanol across a plane interface in the presence a steady d.c. electrical field (Weatherley et al., 2017). The mass transfer was expressed as the increase in ethanol concentration with time in the decanol

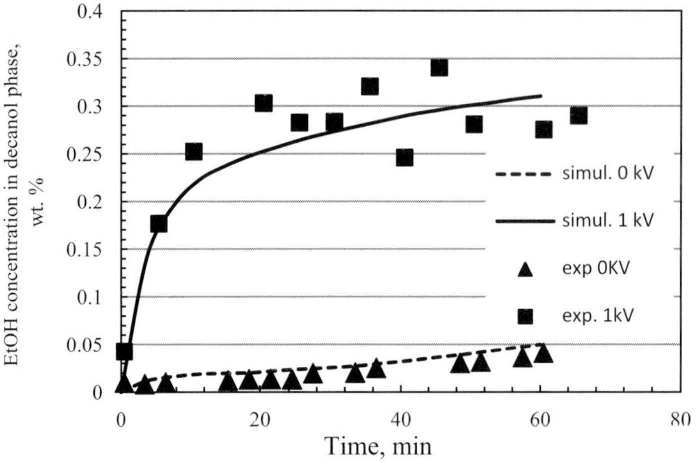

Figure 6.39 Time variation of ethanol concentration in a continuous decanol phase during mass transfer across a planar interface, showing the effect of electrical field at 1 kV upon extraction relative to the control at zero potential.

Reprinted (adapted) with permission Weatherley, L. R., Petera, J., and Qiu, Z. (2017). Intensification of mass transfer and reaction in electrically disturbed liquid-liquid systems. *The Chemical Engineering Journal*, **322**, 115–128. Copyright (2017) Elsevier B.V. All Rights Reserved

extract phase, as shown in Figure 6.39. The effect of the electrical field is very significant, as is shown by comparison with the progress of the mass transfer in the absence of an applied electrical field. It is also noteworthy that the finite element solution of the system equations (equations 6.78–6.86) provides an excellent fit with the experimental data.

In the case of electrostatic dispersions, as has already been stated, the drop hydrodynamics in these systems are significantly more complex than in the case of uncharged systems. The motion behavior of drops has a profound influence upon diffusion rates and rates of reaction, and, in the case of continuous equipment, upon the hydrodynamics. Accurate description of mass transfer by necessity is coupled with a quantitative description of the hydrodynamics.

Kaminski et al. (2014) present a concise summary of the controlling equations for convective mass transfer in an electrically intensified liquid–liquid system. The key to their approach is the use of velocity data from droplet motion simulations for incorporation into existing mass transfer relationships for two-phase systems.

Firstly, with reference to the continuous phase, the slip (or relative) velocity, v_{rel}, between a moving drop and the surrounding continuous fluid is formulated in a three-dimensional coordinate system in equation 6.87:

$$v_{rel} = \left((v_1 - u_{d1})^2 + (v_2 - u_{d2})^2 + (v_3 - u_{d3})^2\right)^{1/2} \tag{6.87}$$

The respective slip velocity components are $(v_1 - u_{d1})$, $(v_2 - u_{d2})$, and $(v_3 - u_{d3})$.

The Schmidt and Reynolds numbers based on the continuous phase properties are written in a familiar format:

$$\mathrm{Sc} = \frac{\mu_c}{\rho_c D_c}, \mathrm{Re} = \frac{v_{\mathrm{rel}} d \rho_c}{\mu_c} \tag{6.88}$$

Mass transfer coefficients are calculable using Sherwood number expressions for laminar and turbulent flows respectively, as follows:

$$\mathrm{Sh}_{\mathrm{lam}} = 0.664\mathrm{Re}^{1/2}\mathrm{Sc}^{1/3}, \ \mathrm{Sh}_{\mathrm{tur}} = \frac{0.037\mathrm{Re}^{0.8}\mathrm{Sc}}{1 + 2.44\left(\mathrm{Sc}^{2/3} - 1\right)\mathrm{Re}^{-0.1}} \tag{6.89}$$

These may be combined into a total Sherwood number:

$$\mathrm{Sh} = 2 + \left(\mathrm{Sh}_{\mathrm{lam}}^2 + \mathrm{Sh}_{\mathrm{tur}}^2\right)^{1/2} \tag{6.90}$$

The mass transfer flux between the droplet surface and the continuous phase may be expressed in the classical form:

$$N_A = \rho_c \frac{D_c}{l_\delta}(x_{\mathrm{Et}} - x_{\mathrm{Et,i}})\pi d^2 = \rho_c k_c (x_{\mathrm{Et}} - x_{\mathrm{Et,i}})\pi d^2 \tag{6.91}$$

where l_δ is the diffusion distance close to the interface:

$$l_\delta = \frac{1}{2} d\left(1 - \frac{d}{d+\delta}\right), \text{and } \delta = \frac{d}{\mathrm{Sh}} \tag{6.92}$$

We may now consider mass transfer between a droplet and the continuous phase.

The mass transfer equations for the interior of each droplet in the dispersed phase may be described. For very small drops in a liquid–liquid system, as in the present case, the mass transport coefficient can be calculated on the basis of rigid spherical droplets, from the classical Newman formula (equation 6.93). At very small drop sizes this formula has proved valid.

$$k_d = -\frac{d}{6t} \ln\left\{\frac{6}{\pi^2}\sum_{n=1}^{\infty}\left(\frac{1}{n^2}\right)\exp\left[-\frac{4D_d \pi^2 n^2 t}{d^2}\right]\right\} \mathrm{m/s} \tag{6.93}$$

The transport flux rates for the continuous phase side of the interface may be expressed in equation 6.94:

$$r_M = (k_c \rho_c a_d)(x_{\mathrm{Et}} - x_{\mathrm{Et,i}}) \tag{6.94}$$

where $x_{\mathrm{Et}} - x_{\mathrm{Et,\,i}}$ is the difference between transferring mass fraction in the bulk continuous phase and its value at the interface. The corresponding relationship for the dispersed phase may be written according to equation 6.95:

$$r_M = (k_d \rho_d a_d)(y_{\mathrm{Et,i}} - y_{\mathrm{Et}}) = (k_d \rho_d a_d)(k_e x_{\mathrm{Et,i}} - k_e x_{\mathrm{Et,\,e}}) \tag{6.95}$$

where $a_d = n_d \pi d^2$ is calculated locally on the basis of the droplet number density distribution, n_d, calculated independently from the cloud's characteristics.

The distribution coefficient used in equation 6.95 is defined as $k_e = y^*/x$ and $x_{\mathrm{Et,\,e}} = y_{\mathrm{Et}}/k_e$

From the equations above we obtain the total rate of mass transfer r_M:

$$r_M = K_{\mathrm{tot}}\ a_d(x_{\mathrm{Et}} - x_{\mathrm{Et,e}}), r_{M\,a} = \frac{r_M}{a_d} = K_{\mathrm{tot}}\ (x_{\mathrm{Et}} - x_{\mathrm{Et,e}})\ \left[\mathrm{kg}/\left(\mathrm{m}^2\ \mathrm{s}\right)\right] \tag{6.96}$$

where the overall mass transfer coefficient, K_{tot}, may be written:

$$K_{tot} = \frac{1}{\frac{1}{k_c\rho_c} + \frac{1}{k_e k_d \rho_d}} \tag{6.97}$$

The mass transfer rate in the dispersed phase may be written:

$$\frac{d(y_{Et} m_d)}{dt} = \pi d^2 r_{Ma} \tag{6.98}$$

For a single droplet, the expression for the mass of a droplet, m_d, represents a cloud of droplets of the same diameter. This equation is in principle a first-order differential equation and can be solved numerically within the overall algorithm, giving the concentration evolution inside a droplet as the result of mass transfer. The resultant mass transported to/from the continuous phase is calculated as the sum of local contributions from different clouds. This gives the evolution of the concentration of the components involved in the reaction.

Experimental validation of the above type of analysis for electrically dispersed systems is sparsely evident in current literature. The example summarized here concerns simultaneous mass transfer and reaction in an electrically enhanced liquid–liquid reactor in which electrostatic sprays are used to intensify overall rates (Petera et al., 2009). The liquid system studied was a natural oil/water system in which a triglyceride ester was hydrolyzed catalytically in the presence of an enzyme. The principle involved electrostatically spraying lipase held in aqueous solution as a continuous spray into the immiscible triglyceride acyl ester.

The theoretical approach is based upon on fundamental relationships, as above, governing the electrical field, the potential distribution, the equations of motion for both continuous and dispersed phases (taking into account mutual coupling in the hydrodynamics), and, most importantly, the mass transfer between the phases. The additional challenge addressed was the concurrent incorporation of reaction kinetics into the system of equations. In this case study, classical Michaelis–Menten kinetics were used to integrate the reaction rate (at the molecular level) with the mass transfer rate equations operating at the liquid–liquid interface. The particle cloud modeling technique described earlier in the chapter is based on the stochastic transport of particles, using statistical methods to trace the turbulent dispersion of particles about a mean trajectory. The cloud model was further developed for describing the mass transfer between continuous and the dispersed phases. The solution of the system of equations uses finite element methods based on an iterative suite of algorithms. The concentration driving force for mass transfer (e.g. ΔC) was calculated on the basis of the concentration difference between that of the particle and that of the local continuous phase, interpolated along the centerline (trajectory) of the cloud at each time step (e.g. Δt) in the calculation algorithm. At each time step the amount of mass exchange and reaction was thus calculated and used to modify the local concentrations, taking into account the local density distribution of the droplets in the cloud and its population, thus coupling the compositional changes of both the dispersed and continuous phases.

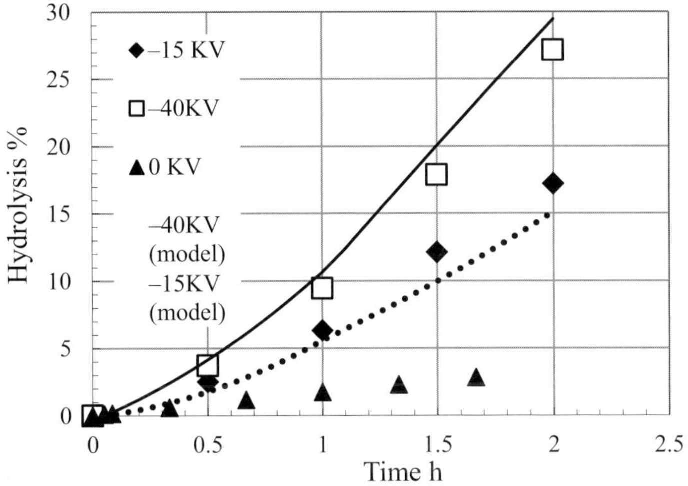

Figure 6.40 Quantitative model comparison of mass transfer and reaction in an electrically dispersed liquid–liquid reactor for the enzymatic hydrolysis of a triglyceride ester.
Reprinted with permission from Petera, J., Weatherley, L. R., Rooney, D., and Kaminski, K. (2009). A finite element model of enzymatically catalyzed hydrolysis in an electrostatic spray reactor. *Computers and Chemical Engineering*, **33**, 144–161.

Experimental validation of the system model equations is shown in Figure 6.40, which compares experimental rate data for enzymatic hydrolysis in an electrically intensified liquid–liquid reactor with model predictions based on the above theoertical approach, using finite element software to effect solution of the complect array of governing equations (Petera et al., 2009).

6.11 Applications and Scale-Up

Laboratory-scale electrically enhanced liquid–liquid extraction is documented for a range of products and solvents. An updated summary of earlier examples (Weatherley, 1993) is shown in Table 6.2. Product solutes include a range of metal ions, organic acids, and alcohols.

In principle, commercial application of electrically enhanced techniques could be significant. The technique, if scaleable, would provide scope for reductions in equipment size, reductions in process residence time, simplification of process flowsheets, energy savings, and improvement in product recovery and purity.

Large-scale application of electrostatically intensified liquid–liquid extraction has not been reported to any extent and most research has focused on understanding fundamentals of the technique. One major impediment to application to bulk product applications is concern at the potential for electrostatic spark production and resulting fire or explosion hazard. Many solvent extraction separation processes use volatile,

Table 6.2 Liquid–liquid systems: electrical enhancement.

Continuous phase	Dispersed phase	Transferring component	Reference
Amine extractants	Acidic aqueous solutions	Uranium, cobalt	Warren, Prestridge, and Sinclair (1978), Warren and Prestridge (1979)
Hydroxyoximes in kerosene	Acid sulfate solutions, ammoniacal solutions	Copper	Warren et al. (1978), Warren and Prestridge (1979), Bailes and Guymer (1976)
D2EHPA kerosene	Acidic aqueous solutions	Zinc	Warren et al. (1978), Warren and Prestridge (1979)
Aliphatic/aromatic mixtures	n-Methylpyrrolidone	Aromatics	Mehner, Mueller, and Hoehfeld (1971)
Toluene/n-butyl acetate	Aqueous solution	Benzoic acid	Bailes (1971)
Carbon tetrachloride	Aqueous solution	Acetic acid	Kowalski and Ziolowski (1981)
Tri-n-butyl phosphate in heavy distillate	Aqueous solution	Ethanol	Laughland, Millar, and Weatherley (1987)
LIX65 in Escaid 100	Aqueous solution	Copper	Martin and Vignet (1983)
Kerosene	Aqueous solution	Acetic acid	Glitzenstein, Tamir, and Oren (1995)
Iso-butanol	Water	Water	Iyer and Sawistowski (1974)
2-Ethylhexanol	Water	Water	Wham and Byers (1987)
Cyanex	Aqueous	Nickel–cobalt recovery	Briggs, Cheng, and Ibana (2000)
Heptane	1-Butyl-3-methylimidazolium methylsulfate	Ethanol	Kamiński et al. (2016)
Toluene	Water	Acetone	Anderson et al. (1989)
Cyclohexanol	Water	Water	Iyer and Sawistowski (1974), Austin, Bancyk, and Sawistowki (1971)
Iso-butanol	Water	Water	Iyer and Sawistowski (1974), Austin, Bancyk, and Sawistowki (1971)
Aniline	Water	Water	Iyer and Sawistowski (1974), Austin, Bancyk, and Sawistowki (1971)
Ethyl acetate	Water	Water	Iyer and Sawistowski (1974), Austin, Bancyk, and Sawistowki (1971)

inflammable solvents thus virtually ruling out the application of electrostatic fields in process equipment. Counter to these concerns is the advantage of a low-energy intensification process with short residence time operation with low inventory.

Blanketing the equipment with inert gas (e.g. N_2, Ar) may be possible but would add to both capital and operating cost that may offset the advantages of intensification. Additional instrumentation would also be required to ensure maintenance of an inert atmosphere. Therefore, future applications of electrical intensification may focus on high-value/low-volume product processing requiring small-scale equipment, which is easily contained and with feasible process safety management. Another focus for future applications would be processes that use solvents of relatively high flash point, low volatility, and low vapor pressure. Examples include liquid carbon dioxide and ionic liquids. As discussed earlier in the chapter, certain ionic liquids exhibit appropriate electrical properties that make them easy to electrically disperse, at the same time having negligible vapor pressure and low flammability risk.

Another area of some uncertainty is the amount of energy required for electrically intensified processing compared to other technologies. There are very few data available. Thornton and Brown (1966) studied the extraction of benzoic acid from toluene into water in the presence of electrical fields. Enhancements in extraction rate of between 200 and 300% were reported. It was also claimed, based on water flow rate, that the specific energy consumption was as low as 0.133 watts per liter per hour of water flow rate. For a flow rate of 200 L h^{-1} this translated into a power consumption of approximately 30 W, which is approximately one-third of the power consumption estimated for an equivalent mechanically agitated system.

References

Anderson, T. J., Javed, K. H., and Thornton, J. D. (1989). Surface phenomena and mass transfer rates in liquid–liquid systems: Part 2. *American Institute of Chemical Engineers Journal*, **35**(7), 1125–1136.

Anderson, W. J. and Pratt, H. R. C. (1978). Wake shedding and circulatory flow in bubble and droplet-type contactors. *Chemical Engineering Science*, **33**, 995–1002.

Austin, L. J., Bancyk, L., and Sawistowski, H. (1971). Effect of electric field on mass transfer across a plane interface. *Chemical Engineering Science*, **26**, 2120–2121.

Bailes, P. J. (1981). Solvent extraction in an electrostatic field. *Industrial and Engineering Chemistry Research*, **20**, 564–570.

Bailes, P. J. and Guymer, P. (1976). UK Patent application No. 1737/76.

Bailes, P. J. and Thornton, J. D. (1971). Proceedings of ISEC 71, April 19-23, The Hague, 1431-1439, Society of Chemical Industry, London.

Barletta, M. and Gisario, A. (2009). Electrostatic spray painting of carbon fibre-reinforced epoxy composites. *Progress in Organic Coatings*, **64**, 339–349.

Baxter, L. L. and Smith, P. J. (1993). Turbulent dispersion of particles: The STP model. *Energy & Fuels*, **7**, 852–859.

Briggs, M. K., Cheng, C. Y., and Ibana, D. C. (2000). An electrostatic solvent extraction contactor for nickel-cobalt recovery. *Minerals Engineering*, **13**(12), 1281–1288.

Coffee, R. A. (1980). Electrodynamic spraying. In J. O. Walker, ed., *Spraying Systems for the 1980s*. London: BCPC Monographs, No. 24, pp. 95–107.

Cross, J. (1987). *Electrostatics: Principles, Problems and Applications*. Bristol: Adam Hilger.

Felici, N. J. (1984). Conduction and electrification in dielectric liquids: two related phenomena of the same electrochemical nature. *Journal of Electrostatics*, **15**, 291–297.

Gangu, S. A. (2013).Towards in situ extraction of fine chemicals and biorenewable fuels from fermentation broths using Ionic liquids and the intensification of contacting by the application of electric fields. Ph.D. thesis, The University of Kansas.

Glitzenstein, A., Tamir, A., and Oren, Y. (1995). Mass transfer enhancement of acetic acid acoss a plane kerosene / water interface by an electric field. *Canadian Journal of Chemical Engineering*, **73**, 95–102.

Gneist, G. and Bart, H. J (2002). Electrostatic drop formation in liquid–liquid systems. *Chemical Engineering Technology*, **25**(9), 899–904.

Gneist, G. and Bart, H.-J. (2003). Influence of high-frequency AC fields on mass transfer in solvent extraction. *Journal of Electrostatics*, **59**, 73–86.

He, W., Chang, J. S., and Baird, M. H. I. (1997). Enhancement of interphase mass transfer by a pulsed electric field. *Journal of Electrostatics*, **40–41**, 259–264.

Hendricks, C. D. and Schneider, J. M. (1963). Stability of a conducting droplet under the influence of surface tension and electrostatic forces. *American Journal of Physics*, **31**, 450–453.

Hume, A. P., Petera, J., and Weatherley, L. R. (2003).Trajectories of charge drops in a liquid–liquid system. *The Chemical Engineering Journal*, **95**, 171–177.

Hume, A. P., Petera, J., and Weatherley, L. R. (2004). Trajectories of charge drops in a liquid–liquid system – the effect of geometrical scale-up. *Industrial and Engineering Chemistry Research*, **43**, 2264–2270.

Iyer, P. V. R. and Sawistowski, H. (1974). Effect of Electric Field Across a Plane Interface. Proceedings of the International Solvent Extraction Conference, Lyon, France. SCI, London, pp. 1029–1046.

Kamiński, K., Krawczyk, M., Augustyniak, J., Weatherley, L. R., and Petera, J. (2014). Electrically induced liquid–liquid extraction from organic mixtures with the use of ionic liquids. *The Chemical Engineering Journal*, **235**, 109–123.

Kamiński, K., Weatherley, L.R., and Petera, J. (2016) Application of numerical modelling to scaling-up of electrically induced extraction from an organic mixture using an ionic liquid. *Polish Academy of Sciences, Chemical and Process Engineering*, **37**(1), 133–148.

Kowalski, W. and Ziolowski, Z. (1981). Increase in mass transfer in extraction columns by means of an electric field. *International Chemical Engineering*, **21**(2), 323–327.

Laughland, G. J., Millar, M. K., and Weatherley, L. R. (1987). Electrostatically enhanced recovery of ethanol from fermentation liquor by solvent extraction. *IChemE Symposium Series*, **103**, 263–279.

Law, S. E. (2001). Agricultural electrostatic spray application a review of significant research and development during the 20th century. *Journal of Electrostatics*, **51–52**, 25–42.

Levich, V. G. (1962). *Physicochemical Hydrodynamics*. Englewood Cliffs, NJ: Prentice-Hall.

Lewis, J. B. and Pratt, H. R. C. (1953). Oscillating droplets. *Nature*, **171**, 1155.

Litchford, R. J. and Jeng, S. M. (1991). Efficient statistical transport model for turbulent particle dispersion in sprays. *American Institute of Aeronautics and Astronautics Journal*, **29**, 1443.

Martin, L. and Vignet, P. (1983). Electrical field contactor for solvent extraction. *Separation Science and Technology*, **18**(14/15), 1455–1471.

McGranaghan, G. J. and Robinson, A. J. (2014). The mechanisms of heat transfer during convective boiling under the influence of AC electric fields. *International Journal of Heat and Mass Transfer*, **73**, 376–388.

Mehner, W., Mueller, E., and Hoehfeld, G. (1971). The Lurgi multi-stage liquid–liquid extractor. In *Proceedings of the International Solvent Extraction Conference, The Hague*, London: SCI, Vol. 2, 1265–1276.

Nijdam, J. J., Baoyu, G., Fletcher, D. F., Langrish, T. A. G. (2003). Lagrangian and Eulerian models for simulating turbulent dispersion and agglomeration of droplets within a spray. Proceedings of the Third International Conference on CFD in the Minerals and Process Industries, CSIRO, Melbourne, Australia, 10–12 December 2003, pp. 377–382.

Panofsky, W. K. and Phillips, M. (1969). *Classical Electricity and Magnetism*, 2nd ed., Reading, MA: Addison-Wesley.

Petera, J., Rooney D., and Weatherley, L. R. (1998). Particle and droplet trajectories in a non-linear electrical field. *Chemical Engineering Science*, **53**(22), 3781–3792.

Petera, J., Strzelecki,W., Agrawal,D., and Weatherley, L. R. (2005) Charged droplet and particle-mixing studies in liquid–liquid systems in the presence of non-linear electrical fields. *Chemical Engineering Science*, **60**(1), 135–149.

Petera, J., Weatherley, L. R., Hume, A. P., and Gawrysiak, T. A. (2007). A finite element algorithm for particle/droplet trajectory tracking tested in a liquid–liquid system in the presence of an external electric field. *Computers in Chemical Engineering*, **31**, 1369–1388.

Petera, J., Weatherley, L. R., Rooney, D., and Kaminski, K. (2009). A finite element model of enzymatically catalyzed hydrolysis in an electrostatic spray reactor. *Computers and Chemical Engineering*, **33**, 144–161.

Qiu, J. (2010). Intensification of liquid–liquid contacting processes. Ph.D. thesis, The University of Kansas.

Quan, X., Gao, M., Cheng, P., and Li, J. (2015). An experimental investigation of pool boiling heat transfer on smooth/rib surfaces under an electric field. *International Journal of Heat and Mass Transfer*, **85**, 595–608.

Quincke, G. (1888). Ueber periodische Ausbreitung von Flüssigkeitsoberflächen und dadurch hervorgerufene Bewegungserscheinungen. *Annalen der Physik und Chemie*, **35**, 593.

Rayleigh, J. W. S. (1882). The equilibrium of liquid conducting masses charged with electricity. *Philosophical Magazine (Series 5)*, **14**, 184–186.

Scheele, G. F. and Meister, B. J. (1968). Drop formation at low velocities in liquid–liquid systems: prediction of drop volume. *AIChE Journal*, **14**(1), 9–15.

Serway, R. A. (1996). *Physics for Scientists and Engineers with Modern Physics*, 4th ed., p. 687.

Singh, M., Ghanshyam, C., Mishra, P.K., and Chak, R. (2013). Current status of electrostatic spraying technology for efficient crop protection. *AMA* – Agricultural Mechanization in Asia, Africa and Latin America, **44**(2), 46–53.

Stewart, G. and Thornton, J. D. (1967a) Charge and velocity characteristics of electrically charged droplets. Part I. Theoretical considerations. *Institution of Chemical Engineers Symposium Series*, **26**, 29–36.

Stewart, G. and Thornton, J. D. (1967b) Charge and velocity characteristics of electrically charged droplets. Part II. Preliminary measurements of droplet charge and velocity. *Institution of Chemical Engineers Symposium Series*, **26**, 37–41.

Takamatsu, T., Hashimoto, Y., Yamaguchi, M., and Katayma, T. (1981). Theoretical and experimental studies of charged drop formation in a uniform electric-field. *Journal of Chemical Engineering of Japan*, **14**(3), 178–182.

Takamatsu, T., Yamaguchi, M., and Katayma, T. (1982). Formation of single charged drops in liquid media under a uniform electric field. *Journal of Chemical Engineering of Japan*, **15**(5), 349–355.

Takamatsu, T., Yamaguchi, M., and Katayma, T. (1983). Formation of single charged drops in liquid media in a non-uniform electric field. *Journal of Chemical Engineering of Japan*, **16**(4), 267–272.

Taylor, G. I. (1964). Disintegration of water drops in an electric field. *Proceedings of the Royal Society of London*, **A280**, 383–397.

Taylor, G. I. (1966). Studies in electrohydrodynamics. I. The circulation produced in a drop by electrical field. *Proceedings of the Royal Society of London Ser. A, Mathematical and Physical Sciences*, **291**, 159–166.

Thomson, J. (1855). On certain curious motions observable at the surfaces of wine and other alcoholic liquors. *Philosophical Magazine*, **10**, 330–333.

Thornton, J. D. (1976). Electrically enhanced liquid–liquid extraction. *Birmingham University Chemical Engineering Journal*, **27**, 6–13.

Thornton, J. D. and Batey, W. (1989). Partial mass-transfer coefficients and packing performance in liquid–liquid extraction. *Industrial and Engineering Chemistry Research*, **28**, 1096–1101.

Thornton, J. D. and Brown, B. A. (1966). Liquid/fluid extraction process. UK patent 1,205,562.

Warren, K. W. and Prestridge, F. L. (1979). US Patent 4,161,439, July 1979.

Warren, K. W., Prestridge, F.L., and Sinclair, B. L. (1978). Electrostatic separators may supplant mixer settlers. *Mining Engineering*, **30**(4), 355–357.

Weatherley, L.R. (1993). Electrically enhanced mass transfer. *Heat Recovery Systems and CHP*, **13**, 515–537.

Weatherley, L. R., Petera, J., and Qiu, Z. (2017). Intensification of mass transfer and reaction in electrically disturbed liquid–liquid systems. *The Chemical Engineering Journal*, **322**, 115–128.

Wham, R. M. and Byers, C. H. (1987). Mass transfer from single droplets in imposed electric fields. *Separation Science and Technology*, **22**, 447–466.

Yang, Q., Ben, Q., Li, B. Q., and Ding, Y. (2013). 3D phase field modeling of electrohydrodynamic multiphase flows. *International Journal of Multiphase Flow*, **57**, 1–9.

7 Intensification of Liquid–Liquid Coalescence

7.1 Introduction

There are many areas in the chemical industry, the food industry, the hydrometallurgical industries, and the oil industry that depend heavily upon liquid–liquid mixtures for successful reaction, product separation, and recovery. In diffusion-based liquid–liquid separation processes, such as partitioning and solvent extraction, the two liquids are efficiently dispersed in each other in order to promote turbulent mass transfer and to generate high interfacial areas per unit volume. In some cases similar requirements are necessary for efficient heat transfer between the phases. A highly relevant feature of many real industrial processes is the presence or addition of surfactant agents that help promote reduction in drop size, and may also increase the stability of the liquid–liquid mixture. In many process operations, separation of the two liquid phases is essential following completion of the mass transfer or reaction. In some cases the two liquid phases may reach equilibrium with the partitioning of the target molecule completed. Product recovery and recycle of the solvent phase requires highly effective separation of the liquid phases. Any entrainment or residual contamination of the separated liquids by the desired solute may constitute an economic loss or present an environmental challenge if either of the liquids is to be recycled or disposed of.

The most widely used practice is to conduct phase separation or disengagement in gravity settlers, whereby the lighter phase rises preferentially toward the free surface and forms a coalesced phase. Alternatively, the heavy phase, if dispersed, will descend and coalesce into the lower layer of that phase. The provision of a large interfacial area into which the coalescing drops of dispersion can merge is a major requirement in many cases. Rates of coalescence in many systems can be relatively slow, especially if there is only a small density difference between the two liquids and therefore long residence times may be necessary. Another factor affecting residence time requirement is the fact that even when a drop reaches an interface there is usually a finite dwell time during which the drop sits at the interface prior to actual coalescence. The physical factors and mechanisms involved in the kinetics of coalescence are discussed later in this chapter. Both these factors lead to large equipment volumes that are associated with high capital cost, large inventories of solvent hold-up, and increase in safety risk, especially in the case of highly inflammable or toxic solvents.

The fundamental goal of coalescence is the separation of liquid–liquid dispersions into two uniform liquid phases that contain minimal amounts of the other phase. In many liquid–liquid systems one of the liquids is dispersed as droplets, with the other phase being present as a continuous liquid. The separation usually requires the merging of the dispersed droplets to form a second uniform continuous liquid phase. As in the case of solid–liquid separation, density difference between the two phases is a critical parameter upon which the separation is highly dependent. In contrast to many solid–liquid separations, the merging of the droplets in liquid–liquid systems involves different mechanisms with the rider that rate may also be determined by factors such as collision frequency, presence of surface charge, film drainage, and the presence of surfactants.

The industrial application of liquid–liquid systems covers separation processes such as solvent extraction, and chemical reaction processes such as in phase-transfer catalysis. The equipment used in industrial processes is discussed elsewhere, as the focus of the main theme of this book is intensification. Whether in intensified processes or in conventional liquid–liquid processes, there are two fundamental requirements for viable applications. These are: (1) efficient liquid–liquid contacting under conditions that promote fast kinetics (i.e. mass transfer), (2) separation of the two liquid phases. These requirements apply regardless of the type of hardware used for the contacting, be it columns, mixer settlers, or other devices such as centrifugal contactors. In basic terms there is a conflict between the two requirements, since high rates of mass transfer may require creation of very fine dispersions of one liquid phase in the second in order to create high interfacial areas. On the other hand, fine dispersions may be difficult to separate. For example, in the case of separation based on gravitational settling and density difference these may require very long settling times so any advantage with respect to fast mass transfer kinetics and equipment size reduction may be offset by the need for large settling equipment.

The consequences of poor phase separation may be very significant for three main reasons. Firstly, in an extraction process, the efficiency of extraction is highly dependent on ensuring that as high a percentage as possible of the desired molecule ends up in only one of the liquid phases. Thermodynamic properties as manifest in values of distribution coefficient and selectivity are the major factors normally regarded as being critical in determining theoretical maxima of degrees of separation. Poor phase separation will very likely result in some of the desired product being entrained in the reject or raffinate stream, thus representing a loss to the process regardless of favorable distribution properties.

Secondly, many of the solvents used in industrial separation processes are costly, and frequently contribute to a significant portion of the capital investment in a chemical manufacturing process. Even small losses may add up to substantial financial costs over a period of time. This is also true for the costly homogeneous catalysts deployed in reactions based on phase-transfer catalysis.

Thirdly, the majority of solvents used in extraction and reactive processes present some degree of environmental hazard if released. Therefore, the ability to avoid any significant entrainment loss in, for example, an aqueous raffinate stream is important.

Another area of industrial interest in coalescence phenomena is in oil and gas production. Much of the crude oil produced both offshore and onshore emerges from

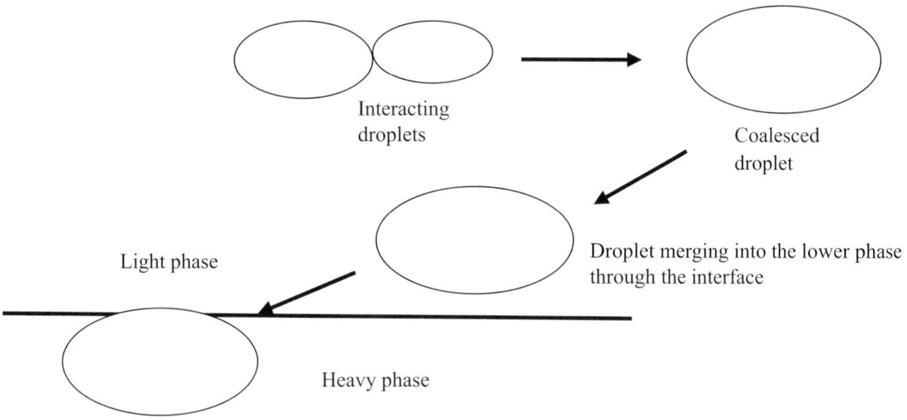

Figure 7.1 Droplet growth resulting from merging of drops and droplet disappearance due to coalescence into a continuous liquid phase.

the well head associated with large amounts of water, some of which may be of high salinity. The separation of produced water from crude oil is therefore of central importance in ensuring that oil pumping systems are only pumping oil and not oil and water mixtures. In addition, the desalting of crude oil through the separation of saline water fractions and through water washing is critical to reduce corrosion risks in pipelines and other oil production plant hardware. Intensification of liquid–liquid and gas–liquid operations is a particular priority in the case of offshore oil and gas production platforms, where space for process plant is at a premium compared with onshore facilities. There are large financial incentives to reduce weight and deck area requirements of separation plant offshore (Urdahl et al., 1998).

Phase separation of two immiscible liquids is an area where intensified technologies have made significant advances. The goal of this chapter is to review the some of the fundamental mechanisms that underpin intensified coalescence. The translation of these mechanisms into equipment design and application is discussed elsewhere. Cusak (2009) provides a concise overview of the principles and some of the available technologies that are based on gravity settling as a means of coalescence.

There are two basic phenomena that can result in coalescence. In a dispersion of droplets, drop–drop interactions can result in coalescence into a larger drop. This may be followed by further drop–drop encounters, resulting in even larger drops. The creation of larger drops changes the buoyancy so if the drops are of the denser phase then the drops will settle at a higher rate than the parent drops. The second phenomenon is the encounter between falling or rising drops and a plane interface existing between the two coalesced liquid phases. The encounter (Figure 7.1) may result in merging of the drop into the continuous phase or it may not.

Figure 7.2(a) depicts the movement of more dense drops descending toward a plane interface formed between the two immiscible liquid phases. Figure 7.2(b) shows a photograph of coalescing drops close to an interface, showing in this case the increase in drop size as the interface is approached. The small drops of the light phase that are

(a) (b)

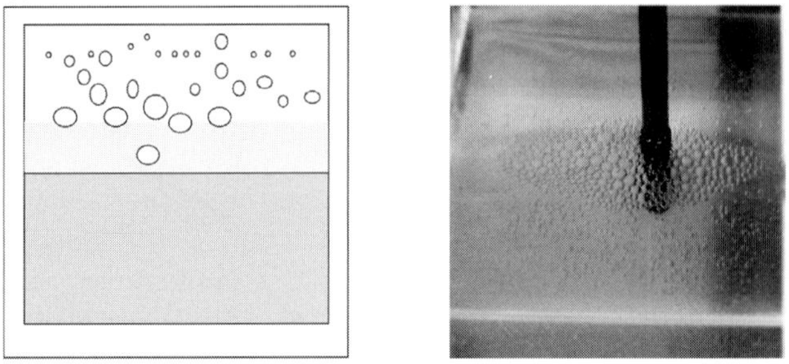

Figure 7.2 Depiction of gravity coalescence and the expected increase in drop size as the drops approach and merge into the coalescence zone, for sodium methoxide dispersed in canola oil. (Qiu, 2010)

dispersed below the interface are coalescing into larger drops as they rise and eventually coalesce.

In the simplest analysis, separation of two immiscible liquid phases is described by virtue of density difference according to Stoke's law, which, assuming spherical droplet size, may be written as:

$$v = \frac{(\rho_h - \rho_l) g d^2}{18\mu} \tag{7.1}$$

where v = settling velocity, ρ_h and ρ_l = density of heavy and light phases respectively, μ = viscosity of the continuous phase, g = acceleration due to gravity, d = drop diameter of the dispersed phase.

This may be further refined to account for the nonsphericity of moving droplets using the Rybczynski–Hadamard correction for the force acting on a moving droplet as follows:

$$v = \frac{(\rho_h - \rho_l) g d^2}{18\mu} \frac{\mu_c + \mu_d}{2\mu_c + 3\mu_d} \tag{7.2}$$

In the simplest conceptualization of gravity coalescence in process equipment, Cusak depicts the mechanism as shown in Figure 7.3, which shows a side view of a settler or coalescing vessel. The mixed phase enters as a continuous stream at the top left-hand side. The presence of eddy currents is controlled to promote drop–drop interaction and interaction between settling drops and the interface that forms as settling and coalescence proceed. As the liquids flow through the vessel in the general direction of the horizontal axis, the volumes of the clarified upper and lower phases are seen to grow as coalescence proceeds. The two clarified liquid layers may be separated by a "dispersion" wedge that diminishes in volume as the liquid flow proceeds from left to right. The kinetics of the process are generally slow, requiring significant residence times and hence large vessel volumes.

The corresponding situation in a vertical coalescer is shown in Figure 7.4.

While some limited mixing is desirable to promote drop–drop encounters, the occurrence of significant eddy currents in either of the liquid phases can work to the

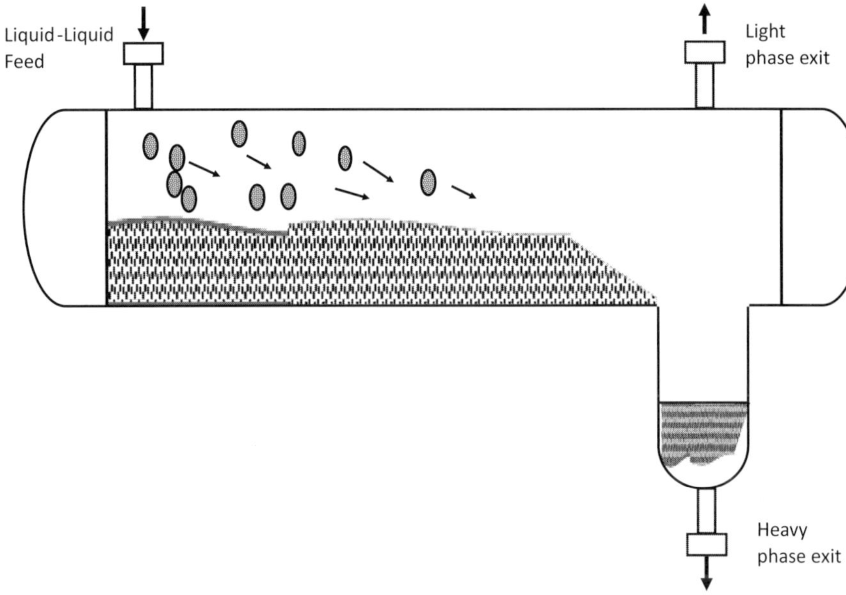

Figure 7.3 Principle of gravity settling shown in a horizontal settler.

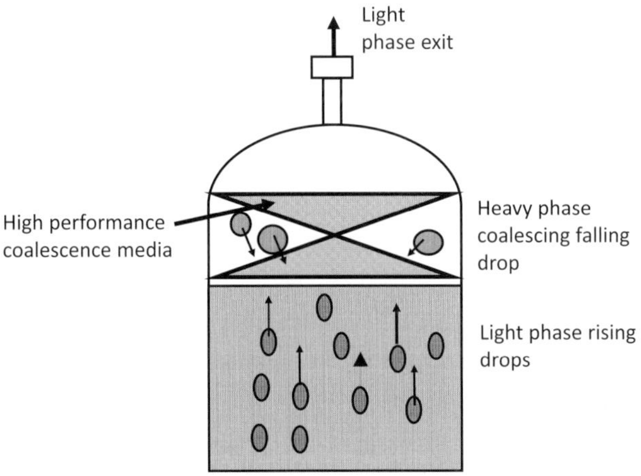

Figure 7.4 Arrangement of a vertical coalescer.

opposite effect, resulting in re-dispersion and consequential entrainment in the liquid outlets. As a result, the control of eddy current flows is a significant factor in the development of improved designs for gravity settling. One means by which this may be achieved is by the insertion of plates into the gravity settling chamber to create a series of channels through which the mixed liquid phases flow. The channels promote laminar flow of the phases that enhances the ability of the dispersed phase droplets to reach an interface into which they can merge. Subject to suitable surface properties, the existence of a solid surface also can enhance coalescence of droplets that settle or rise on to the

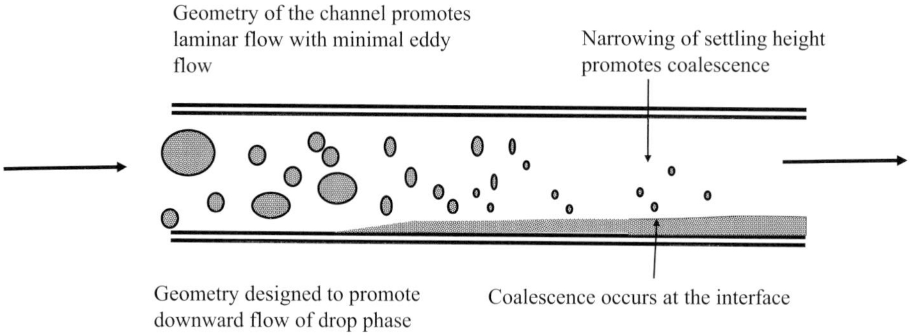

Figure 7.5 Mechanisms involved in enhanced gravity settling.

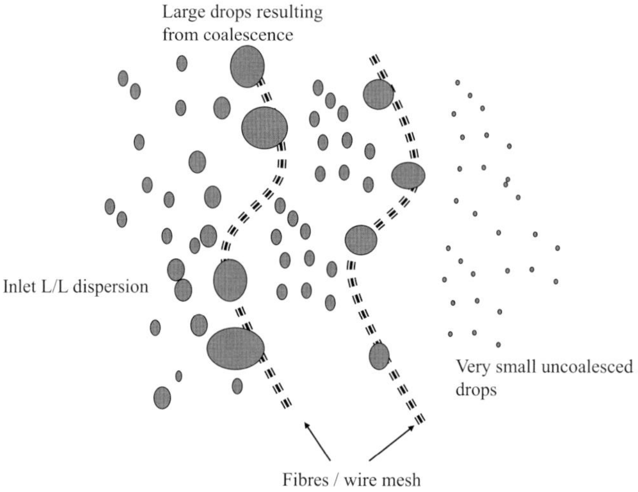

Figure 7.6 Enhanced gravity coalescence using fibers and other surfaces.

surface due to a self-spreading. The surface of the plates must be carefully selected to ensure interfacial tensions between liquid and surface to promote coalescence. The concept is shown in Figure 7.5 (Cusak, 2009) and the effect of reduced settling height combined with a laminar flow regime is illustrated.

The exploitation of drop–solid surface interactions may be taken a stage further by the use of, for example, fibers, wire mesh, and other materials such as packings that actively enhance coalescence through surface energy forces or interfacial tension forces, as illustrated in Figure 7.6 (Cusak, 2009). The mechanism involves the collision of numerous dispersed drops with the fibers, resulting in larger drops that will eventually detach from the surface, descend (or rise) rapidly, and coalescence into the main body of the separated liquid phase.

Figure 7.7 also shows the situation when coalescence occurs in a mesh of fibers or filaments. Here the role of the cross-over points between the fibers is illustrated. The region where two of the filaments crossover each other provides a "crevice" in which small

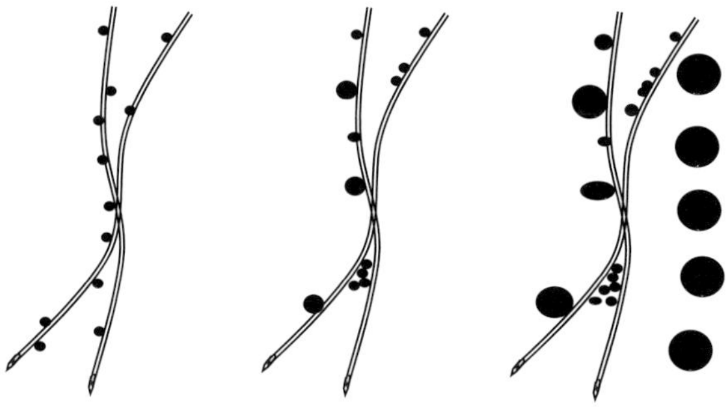

Figure 7.7 Situation when coalescence occurs in a mesh of fibers or filaments.

droplets collect and merge with other small drops. Once the drop size exceeds a certain value, the drop resulting from the coalescence will detach and move to undergo further coalescence either with another droplet or into a plane interface of coalesced phase.

Interactions between the droplet phase and engineered solid surfaces to exploit the mechanism of enhanced coalescence, as illustrated in Figures 7.6 and 7.7, are amply demonstrated in process equipment available on the market, as discussed in further detail in Chapter 1. Dusec™ Coalescers, referred to in Chapter 1, break up secondary dispersions where droplets are so small they do not readily wet a surface or settle under gravity. They are constructed in cartridge form with the liquid flowing from the center radially outwards and are made from a selection of fiber materials.

7.2 Interfacial Drainage, Drop Size, and Drop–Drop Interactions

Eow and Ghadiri (2003) provide a review of the basic mechanisms of interfacial drainage and interfacial rupture that comprise the two main hydrodynamic mechanisms involved in gravity coalescence. The mechanism of coalescence, either drop to drop or drop to liquid interface, is generally considered to occur in two stages. Firstly, as the drops approach the interface, a film of the opposite phase is created between the two coalescing drops (see Figure 7.8). Drop deformation is also considered to be an important factor, as depicted in the right-hand diagram in Figure 7.8, where the drop deformation is exaggerated, showing the formation of flat lower surface of the drop as it descends toward the plane interface. This deformation occurs partly due to the oscillatory distortion that occurs with a moving droplet, but also due to the forces that increase as the gap between the descending drop and interface decreases, leading to the flattening of the lower surface of the droplet. The flattening is considered to be a major factor in increasing the contact time at the interface.

As the film thickness between the two liquid entities decreases, viscous forces become dominant and lead to reduction in flow of the liquid out of the film. Reduction in viscosity at increased temperatures may lead to faster rates of interfacial drainage and

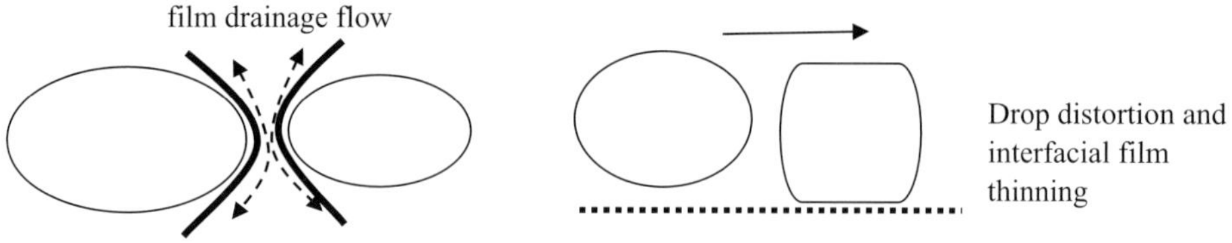

Figure 7.8 Film drainage.

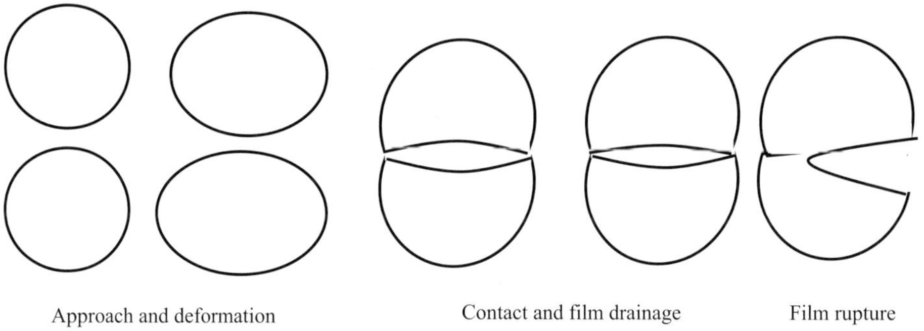

Figure 7.9 Binary droplet interaction in the approach to coalescence.
Reprinted (adapted) with permission from Gebauer, F., Villwock, J., Kraume, M. and Bart, H.-J. (2016). Detailed analysis of single drop coalescence: Influence of ions on film drainage and coalescence time. *Chemical Engineering Research and Design*, **115**, 282–291.

thus enhance the probability of coalescence. Specifically, Dreher et al. (1999) show that increases in coalescence time for the case of Newtonian liquids are generally associated with increases in viscosity of the continuous phase.

Gebauer et al. (2016) show a simple representation of the sequence of occurrences as two droplets approach and coalesce (see Figure 7.9). The steps identified may be listed as (1) approach, (2) contact, (3) film drainage, (4) film rupture, and (5) coalescence.

At some point, the drainage ceases and the thinning of the film between the two adjacent interfaces stops. After a time as a static film, rupturing occurs and coalescence proceeds. Film rupture is described by Charles and Mason (1960) in terms of creation of a "hole" in the film leading to rupture of the film and hence to coalescence. This may occur at multiple points in the film, depending on the system properties and the size of the drop. Another complicating factor is that the occurrence of film rupture may not be unique to a single film thickness value for the same system – this is indicative of a degree of randomness.

Another analysis of flow phenomena following film rupture is presented by Weheliye, Dong, and Angeli (2017), who identified two main regimes for the expansion of the neck of liquid connecting two coalescing drops. In the case of a thin neck, for example immediately following film rupture, viscous or inertial forces may dominate and slow the coalescence rate. Aarts et al. (2005, 2008) show the growth of the neck between a coalescing droplet and the interface with the bulk continuous liquid phase (Figure 7.10).

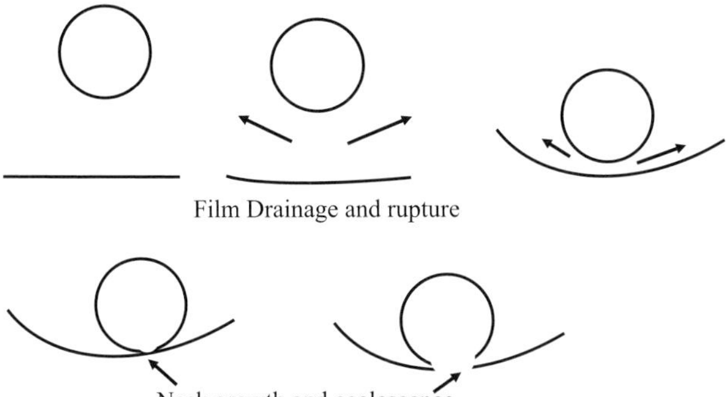

Figure 7.10 Stages in the coalescence of a drop approaching and merging into an interface. Reprinted (adapted) with permission from Aarts, G., and Lekkerkerker, H. N. (2008). Droplet coalescence: drainage, film rupture and neck growth in ultralow interfacial tension systems. *Journal of Fluid Mechanics*, **606**, 275–294. Copyright (2008) Cambridge University Press

Aarts states criteria for the two regimes of flow associated with film rupture and drainage. When viscous forces dominate the flow regime the capillary number Ca approaches unity, where Ca is defined in equation 7.3:

$$C_a = \frac{\mu V_{\text{visc}}}{\gamma} = 1 \tag{7.3}$$

Therefore neck velocity may be estimated from:

$$V_{\text{visc}} = \frac{\gamma}{\mu} \tag{7.4}$$

The crossover time of transition from viscous flow to inertial flow in the film is defined as

$$t_{\text{crossover}} = \frac{\mu^3}{\rho \gamma^2} \tag{7.5}$$

$$V_{\text{inertial flow}} = \frac{1.2}{\sqrt{t}} \left(\frac{R\gamma}{\rho} \right)^{1/4} \tag{7.6}$$

A further point underlined by Aarts and Lekkerkerker (2008) is the difference between solid spheres approaching an interface compared with a liquid droplet. In the latter case fluid circulates in the drop since it has a finite viscosity, which tends to speed up drainage compared to a solid sphere. Internal circulation is an important factor influencing coalescence rate and is discussed later when the effect of surfactant addition is considered.

The time in the precoalescence state after thinning stops is variable and dependent on the balance of kinetic and interfacial forces. There are other factors that contribute to the so-called rest time. These include temperature, drop size, the shape of the interface, fluid properties, vibration, and contamination. Another factor affecting coalescence, described by Chatterjee, Nikolov, and Wasan (1996), is oscillation. Oscillations may be generated as a drop approaches the interface. Secondly, oscillations are produced as result of actual coalescence and these are thought to impact subsequent coalescence events. In particular, Chatterjee concludes that the rest time following halting of interfacial drainage may be strongly influenced by oscillations.

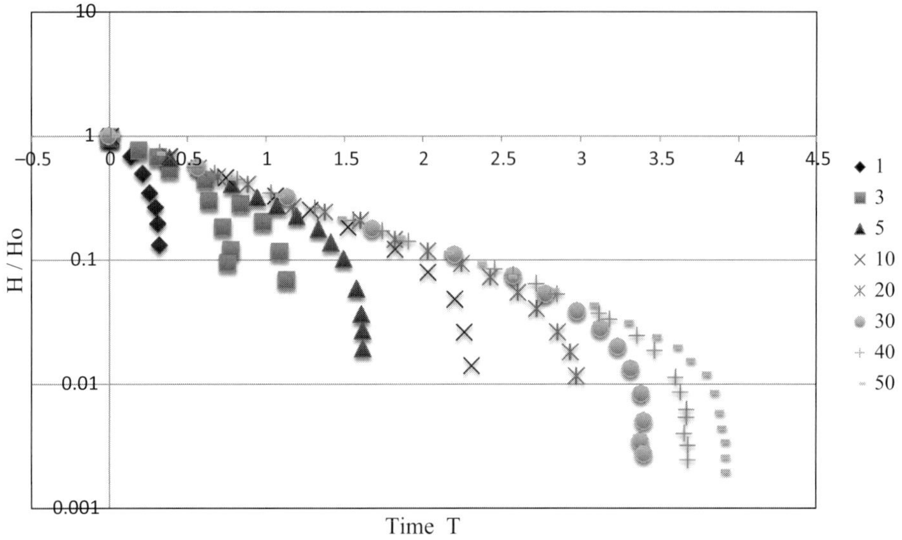

Figure 7.11 Effect of van der Waals forces on fractional film thickness for viscous drainage during the approach of two equal-sized spherical drops, according to equation 7.8.
Reprinted (adapted) with permission from Jeelani, S. A. K. and Hartland, S. (1993). Effect of velocity fields on binary and interfacial coalescence. *Journal of Colloid and Interface Science*, **156**, 467–477. Copyright (1993) Elsevier B.V. All Rights Reserved

Jeelani and Hartland (1993) summarize the mechanisms involved in coalescence after collision has occurred. Their description states that collision of two drops leads to flattening of the drop surface that results in viscous drainage in a thin planar film whose area expands and contracts. An alternative mechanism involves inertial flow in a planar film of expanding area, which eventually results in film rupture. The coalescence time is calculable from a knowledge of the applied force and film radius. Jeelani and Windhab (2009) present diagrammatic representation of different film drainage mechanics.

Another factor is an increasing role for repulsive electrostatic forces present due to double layer effects (Hartland, Yang, and Jeelani, 1994). These forces can reduce the rate of interfacial drainage. On the other hand, the influence of van der Waals forces may increase as the film thins and these can promote disruption of the film, thus increasing the probability of coalescence (Brown and Hanson, 1965; Charles and Mason, 1960). Jeelani and Hartland (1993) describe in detail the influence of van der Waals forces on interfacial drainage and coalescence (Figure 7.11). Figure 7.11 shows the enhancement in film thickness reduction that is predicted when van der Waals forces are taken into account. The parameter in Figure 7.11 reflects the magnitude of the expanding interfacial film in the shear field and is defined by Jeelani and Hartland as follows:

$$P_{\mathrm{d}} = \frac{\mu k H_0 r^2}{2\sigma h_0} \tag{7.7}$$

Jeelani and Hartland (1993) then analyzed the kinetics of interfacial drainage and the role of van der Waals forces as follows. The variation in film thickness with time may be expressed in terms of the total forces, which amount to applied forces due to fluid motion and gravitational forces, plus the van der Waals force.

The van der Waals force f_W is described in terms of the Hamaker A_m constant thus:

$$f_W = \frac{A_m r}{12h^2} \quad (7.8)$$

The rate of interfacial drainage is represented by:

$$-\frac{dh}{dt} = \frac{8f_t h}{3\pi\mu n r^2} \quad (7.9)$$

where f_t is the total force acting on the film.

Incorporation of both force terms provide an expression for the rate of drainage as follows:

$$-\frac{dh}{dt} = \frac{8f}{3\pi\mu n^2 r^2} h + \frac{4A_m}{18\pi\mu r n^2} \frac{1}{h} \quad (7.10)$$

Jeelani and Hartland presented the integrated result for film thickness with time as follows:

$$T = -\frac{1}{2} \ln \left[\frac{1+H^2}{1+H_0^2}\right] \quad (7.11)$$

Where H is the dimensionless film thickness and T is dimensionless time, defined as follows:

$$H = h\sqrt{12f/rA_m} \quad 7.12$$

$$T = 8ft/(3\pi\mu n^2 r^2) \quad (7.13)$$

Drop size can also influence rate of coalescence between individual drops and between a discrete drop and a continuous phase. Gebauer et al. (2016) show a strong relationship between drop size and coalescence time for droplets of toluene in water, as shown in Figure 7.12, with increase in drop size clearly correlating to increase in coalescence

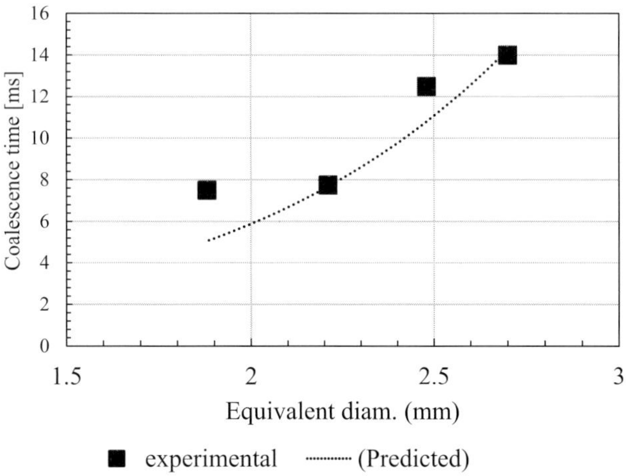

Figure 7.12 Coalescence time as a function of drop diameter for toluene drops in water. Reprinted (adapted) with permission from Gebauer, F., Villwock, J., Kraume, M. and Bart, H.-J. (2016). Detailed analysis of single drop coalescence: Influence of ions on film drainage and coalescence time. *Chemical Engineering Research and Design*, **115**, 282–291. Copyright (2016) Elsevier B.V. All Rights Reserved

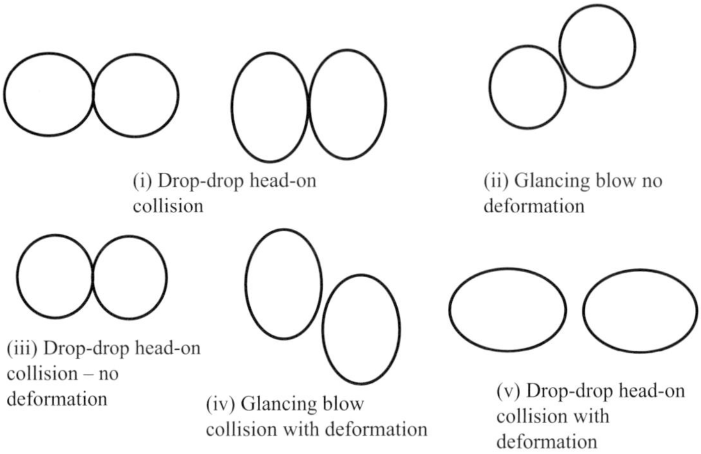

Figure 7.13 Different collision regimes of drop–drop interactions: (i) horizontal drop–drop head-on collision with vertical deformation; (ii) glancing-blow collision with no deformation; (iii) drop–drop head-on collision with no deformation; (iv) glancing-blow collision with deformation; and (v) horizontal drop–drop head-on collision with horizontal deformation.
Reprinted (adapted) with permission from Borrell, M., Yoon, Y., and Leal, L. G. (2004). Experimental analysis of the coalescence process via head-on collisions in a time-dependent flow. *Physics of Fluids*, **16**(11), 3945–3954. Copyright (2004) AIP Publishing, All Rights Reserved

time. A roughly linear relationship between drop size and coalescence rate was also shown by Dreher et al. (1999). With interfacial tensions of approximately 29 mN/m they observed that coalescence time increased monotonically with continuous-phase viscosity. The addition of a high molecular weight viscosity enhancer, poly-iso-butylene, to the continuous phase had little effect, within experimental error, on the coalescence time of the drops with a diameter of approximately 2.5 mm. Interfacial tension itself had a significant effect on coalescence time, showing a reduction with respect to increase in the interfacial tension. Decrease in coalescence times was also explained by changes in purity of the system. Larger drops tend to display larger contact area with the interface, leading to slower rates in drainage and thus slower rates of coalescence. Another factor is the ease with which drops are deformed (Chen et al., 1998; Mackay and Mason, 1963). Droplets that are less deformable, or have greater rigidity, tend to coalesce more rapidly than those that are less stable in terms of shape. This is explained by higher rates of film drainage available on account of increased film width. This may also be expressed in terms of droplet geometric stability: less stable drops exhibit greater rest time and thus increased coalescence time.

The nature of the drop–drop collision is a factor influencing the outcome of an encounter between two colliding drops. Borrell, Yoon, and Leal (2004) compared head-on collisions with glancing-blow collisions. Surprisingly, it was found that coalescence occurred in the same fashion. In the case of head-on collisions the velocity gradient between the approaching drops is time dependent whereas in the case of glancing-blow collisions it was predicted that on account of change in drop shape, the velocity gradient would be constant. Figure 7.13, reproduced from Borell et al. (2004), shows five situations in which it would be expected that different coalescence behaviors would be exhibited: (i) horizontal drop–drop head-on collision with vertical

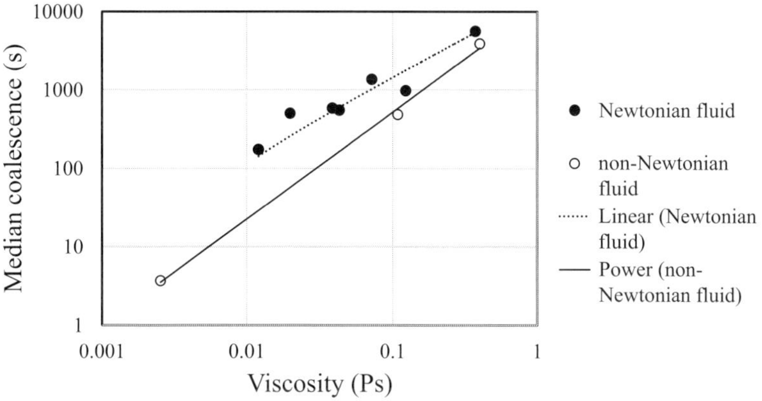

Figure 7.14 Median coalescence time versus continuous phase viscosity for a range of fluids with interfacial tension of approximately 29 mN m^{-1} and drops of a diameter of approximately 2.5 mm.

Reprinted (adapted) with permission from Dreher, T. M., Glass, J., O'Connor, A. J., and Stevens, G. W. (1999). Effect of rheology on coalescence rates and emulsion stability. *AIChE Journal*, **45**(6), 1182–1190. Copyright (1999) John Wiley and Sons – All Rights Reserved

deformation; (ii) glancing-blow collision with no deformation; (iii) drop–drop head-on collision without deformation; (iv) glancing-blow collision with deformation; and (v) horizontal drop–drop head-on collision with horizontal deformation. It was concluded that axisymmetric film drainage occurring in a head-on collision is approximately the same as in a non-axisymmetric glancing collision.

The influence of the viscosity and rheological properties of the continuous phase on rate of coalescence was studied by Dreher et al. (1999). The continuous phase fluids used to determine the data shown in Figure 7.14 included both Newtonian fluids and non-Newtonian fluids. The Newtonian fluids were mixtures of n-decane and Hyvis 3 polybutene fluids mixed in varying proportions to effect a range of viscosities. Non-Newtonian behavior was achieved by the addition of poly-iso-butylene. These data appear to show a monotonic increase in coalescence time with respect to viscosity that is consistent with slower rates of interfacial drainage, which may be expected as viscosity increases. The trend appears to be very similar for both Newtonian and non-Newtonian continuous phase fluids at the drop size of 2.5 mm and interfacial tension of 29 mN m^{-1}.

Analysis of the fluid flow conditions in the film showed that in the case of the drops and interfaces present in the Dreher study, elasticity of the continuous phase did not have a significant effect on the drainage velocity and hence coalescence rate. This was attributed to a high degree of interfacial mobility. This may explain the very similar trends illustrated in Figure 7.14 where the Newtonian and non-Newtonian cases are compared.

7.3 Probability Theory Applied to Coalescence Modeling

Many analyses have been conducted to determine the rate of coalescence, expressed as "coalescence efficiency," and its relationship to drop size, hold-up, and velocity.

The studies and analysis by Coulaloglou and Tavlarides (1977) are among the most cited and in some cases still provide the best fit to observed data. Many published studies are concerned with the effect of unsteady-state coalescence and break-up phenomena in stirred tanks upon drop size distribution. Therefore the treatment here is limited to a summary analysis of the controlling equations and a display of some results.

Most quantitative theoretical studies utilize "population balance equations" that relate convective flow terms over volume boundaries and describe number and density distributions of drops of a particular diameter. The latter are further quantified using source terms for drop birth rate, and for drop death rate where "death" is defined either as drop breakage or coalescence.

Kamp and Kraume (2016) summarize the population balance approach as follows.

Mathematical definitions of the various terms are listed here:

$f(d_p,t)$ is the time-dependent number density distribution;
$n_d(d_p)$ is the number of daughter drops;
$g(d_p)$ is the breakage rate;
$\xi\left(d_p,d'_p\right)$ is the collision frequency;
$\lambda\left(d_p,d'_p\right)$ is the coalescence efficiency;
$c_{1,\,b}$, $c_{2,\,b}$, $c_{1,\,c}$ are numerical parameters;
β is the drop size distribution function;

$$\frac{df(d_p,t)}{dt} = \int_{d_p}^{d_{p,\max}} n_d(d'_p)\cdot\beta(d_p,d'_p)\cdot g(d'_p)\cdot f(d'_p,t)dd'_p + \frac{1}{2}\int_0^{d_p}\xi(d''_p,d'_p)\cdot\lambda(d''_p,d'_p)\cdot f(d'_p,t)\cdot f(d''_p,t)dd'_p - f(d'_p,t)\int_0^{d_{p,\max}-d_p}\xi(d_p,d'_p)\cdot\lambda(d_p,d'_p)\cdot f(d'_p,t)\cdot dd'_p$$

where $d''_p = \left(d_p^3 - d'^3_p\right)^{1/3}$

and d_p, d'_p, and d''_p represent droplet diameters.

The simplest model based on a probability analysis was presented by Coulaloglou and Tavlarides (1977) as follows.

Coalescence rate may be defined for two drops as the product of collision frequency and a coalescence probability term, as shown in equation 7.14.

$$F\left(d_p,d'_p\right) = \xi\left(d_p,d'_p\right)\cdot\lambda\left(d_p,d'_p\right) \tag{7.14}$$

Then the coalescence probability is expressed by Coulaloglou and Tavlarides. The collision frequency is given by:

$$\xi_{C\&T} = c_{1c}\frac{\varepsilon^{1/3}}{1+\varphi}\left(d_p+d'_p\right)^2\left(d_p^{2/3}+d'^{2/3}_p\right)^{1/2} \tag{7.15}$$

Collision probability (efficiency) is given by:

$$\lambda_{\mathrm{C\&T}} = \exp\left(-c_{2,\mathrm{c}}\frac{\mu_{\mathrm{c}}\rho_{\mathrm{c}}v_{\mathrm{rel}}^{3}}{\gamma^{2}\sqrt{8\left(d_{\mathrm{p}}+d'_{\mathrm{p}}\right)}}\left(\frac{d_{\mathrm{p}}d'_{\mathrm{p}}}{d_{\mathrm{p}}+d'_{\mathrm{p}}}\right)^{4}\right) \tag{7.16}$$

Breakage rate is given by:

$$g_{\mathrm{C\&T}} = C_{1,\mathrm{b}}\frac{\varepsilon^{1/3}}{(1+\varphi)d_{\mathrm{p}}^{2/3}}\exp\left(-c_{2,\mathrm{b}}\frac{\gamma(1+\varphi)^{2}}{\rho_{\mathrm{d}}\varepsilon^{2/3}d_{\mathrm{p}}^{5/3}}\right) \tag{7.17}$$

Daughter drop size distribution is given by:

$$\beta_{\mathrm{C\&T}}\left(V_{\mathrm{p}},V'_{\mathrm{p}}\right) = \frac{1}{V_{\sigma}\sqrt{2\pi}}\exp\left(-\frac{\left(V_{\mathrm{p}}-V_{\mu}\right)^{2}}{2V_{\sigma}^{2}}\right) \tag{7.18}$$

where

$$V_{\sigma} = \frac{V'_{\mathrm{p}}}{c_{\beta}n_{\mathrm{d}}},\quad V_{\mu} = \frac{V'_{\mathrm{p}}}{n_{\mathrm{d}}},\quad V_{\mathrm{p}} = \frac{\pi}{6}d_{\mathrm{p}}, n_{\mathrm{d}} = 2$$

Kamp and Kraume (2016) compared the predictions of Coulaloglou and Tavlarides analysis and prediction of coalescence with other models based on film drainage (Henschke, Schlieper, and Pfennig, 2002; Prince and Blanch, 1990; Sovova, 1981), with models based on energy dissipation rate (Sovova, 1981), and one model based on critical approach velocity (CAV) (Liao and Lucas, 2010). Results of Kamp and Kraume (2016) show that collision of droplets does not necessarily end in coalescence, since repulsion or agglomeration may also result. Typically, as illustrated in equations 7.14–7.18, the probability of coalescence after droplet collision is quantified by a coalescence probability (efficiency) term in the population balance equations. The study of discrete drops showed that the collision velocity v_{coll} has a greater influence on the coalescence efficiency.

Kamp and Kraume show experimental measurements of coalescence efficiency for drops in the size range 1.5–2.9 mm as a function of collision velocity. The model equations of Coulaloglou and Tavlarides (1977) show a reasonably good representation of the efficiency/relative collision velocity data based on drops of 1 mm diameter.

Drop diameter is also important and Figure 7.15 illustrates the strong dependence of coalescence efficiency on drop diameter. The predictions based on the equations developed by Coulaloglou and Tavlarides (1977), other models based on film drainage, energy dissipation rate, and those based on critical approach velocity are compared. The early model of Coulaloglou and Tavlarides shows good prediction relative to the others, though with limited data at drop sizes less than 1.5 mm.

7.4 Electrically Enhanced Coalescence

As described earlier, the kinetics of liquid–liquid coalescence exert a major influence on the size and economics of the process and so there is a major incentive to intensify the

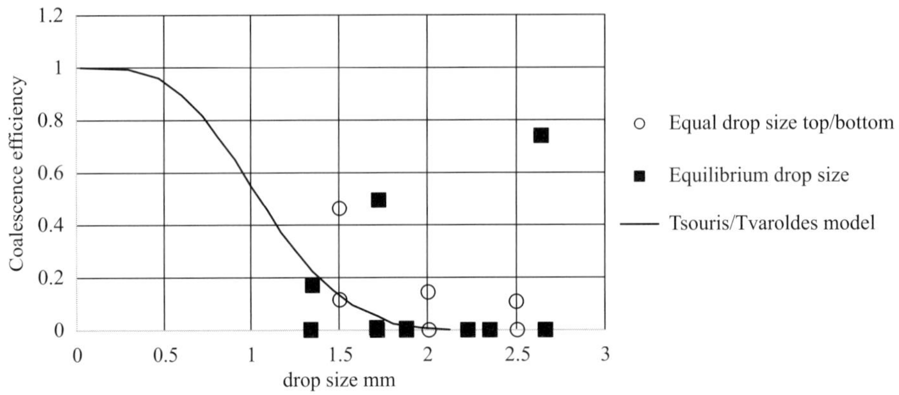

Figure 7.15 Coalescence probability versus drop diameter for coalescence efficiency model of Coulaloglou and Tavlarides (1977).
Experimental data from Kamp and Kraume (2016)

various rate processes that contribute to the overall time for complete and efficient phase separation. Increase in the settling velocity or the rise velocity is one major route to intensification. Another part of the coalescence process that can be used to intensify the rate is by reducing the dwell time during which the drops remain at the interface prior to coalescence. The role of interfacial phenomena in coalescence has led to a significant focus on the exploitation of electrical effects for the enhancement of the kinetics. One primary effect available through the application of electrostatic fields is by placing electrical charge on the dispersed drops. Then by creating an electrical field across both dispersion and coalesced phase, the charged drops may be accelerated due to the electrical forces acting. This increases the velocity of the drops and increases the momentum of individual drops. The increase in velocity of the drops rising or falling toward the interface means that the time needed to reach the phase boundary will be reduced. Secondly, since the momentum of the drops is increased, the energy dissipated as the drop collides with the interface will be greater than in the case of the presence of gravitational forces only, and therefore the dwell time at the interface will be significantly reduced. Similar effects can be exploited for uncharged drops, providing an external electrical field can be applied to the system. Enhancement of coalescence using high-voltage electrical fields has been used at an industrial scale for the enhanced separation of water in oil dispersions that occur in the oil industry during crude oil production.

Typically electro-coalescence is achieved by applying an electric field between a high-voltage electrode and a ground electrode. The liquid–liquid emulsion or droplet dispersion is located in the region between the two electrodes. For favorable applications of this technique, the continuous phase should have a low dielectric constant and also exhibit strong insulating characteristics. These criteria are met for many hydrocarbon liquids. Water droplets or those of similarly conducting liquids in high-voltage electric fields may be attracted to each other and eventually coalesce. The mechanism is generally explained as being due to electric field-induced electrostatic forces between noncharged, polarized drops, sometimes referred to as dipole coalescence. The second mechanism is that of electrophoretic acceleration of charged water drops. Drops of opposite charge may not necessarily attract and coalesce. Depending on the external field strength approaching drops may in fact repel each other, resulting in acceleration of the drops toward the electrodes. If contact with an electrode is made, charge reversal may occur, leading to a cycle of repulsion and attraction with drops apparently bouncing between the two electrodes.

As stated above, one of the significant limitations to the use of electrical fields is that the nondispersed phase must have a low conductivity compared with the droplet phase. This is essential for an effective electrical field to be established across the continuum of coalescing phases. Eow et al. (2001) describe two of the significant mechanisms thought to be involved in electrically intensified coalescence: (1) dipole-induced coalescence and (2) migratory coalescence. Dielectric coalescence occurs when uncharged droplets approach each other, the force of attraction, F, between the drops being described in terms of the electrical field strength, E, the drop radius, r, and the dielectric properties of the two liquids, ε_1, and the distance between the drops, d (equation 7.19):

$$F = \frac{24\pi\varepsilon_0\varepsilon_1 r^6 E^2}{(d + 2r)^4} \tag{7.19}$$

The approach of the drops to within a distance at which the force F becomes significant is promoted by effects such as Brownian motion, convection, and flocculation. Under many circumstances, dipole-induced coalescence is thought to be the major mechanism involved. Equation 7.19 is only valid for drops of the same size and as the drops approach closely, higher order terms become significant and this expression will cease to be accurate.

Migratory coalescence is also considered to be a contributing mechanism (Williams and Bailey, 1986) in electrically enhanced coalescence in which the drops are electrically charged and therefore under certain conditions will be attracted to each other. The charge on the approaching droplets may either be due to electrical double layer effects, or charge acquired due to contact with a charged electrode. Eow et al. (2001) describe the electrical double layer charge in such cases by:

$$q_{\mathrm{d}} = \frac{4\pi r^2 \varepsilon_1 \varepsilon_0 \zeta}{\delta_{\mathrm{DL}}} \tag{7.20}$$

where the double-layer thickness is given by:

$$\delta_{\mathrm{DL}} = \left(\frac{K_{\mathrm{dif}}\varepsilon_1\varepsilon_0}{C_{\mathrm{m}}}\right)^{1/2} \tag{7.21}$$

where ζ is the zeta potential on the drop, K_{dif} is the diffusion coefficient, and C_{m} is the conductivity of the continuous phase.

If the drops come in contact with a live electrode then additional charge acquisition may occur, according to equation 7.22.

$$q_{\mathrm{i}} = \left(\frac{\pi^2}{6}\right) 4\pi r^2 \varepsilon_1 \varepsilon_0 E_0 \tag{7.22}$$

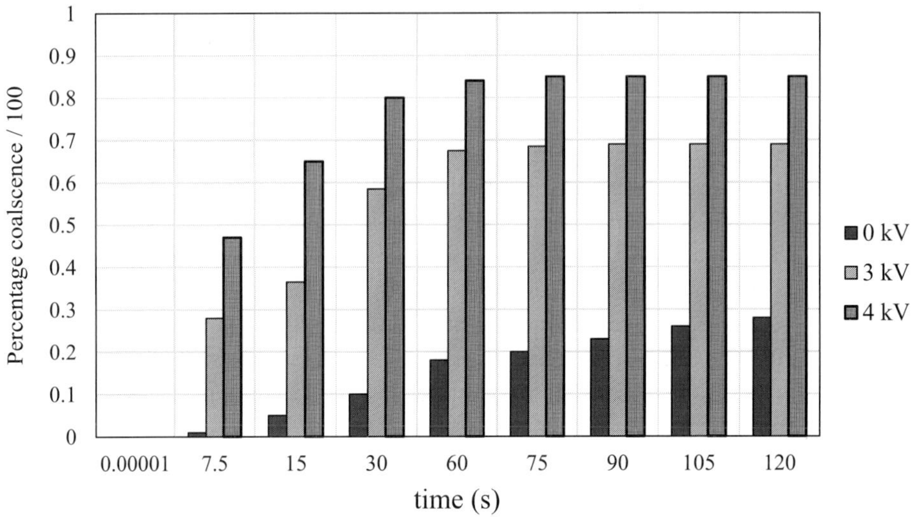

Figure 7.16 Electrically enhanced coalescence of water drops in silicone oil between two parallel plate electrodes: percentage coalescence versus time.
Data calculated and adapted from Williams and Bailey (1986)

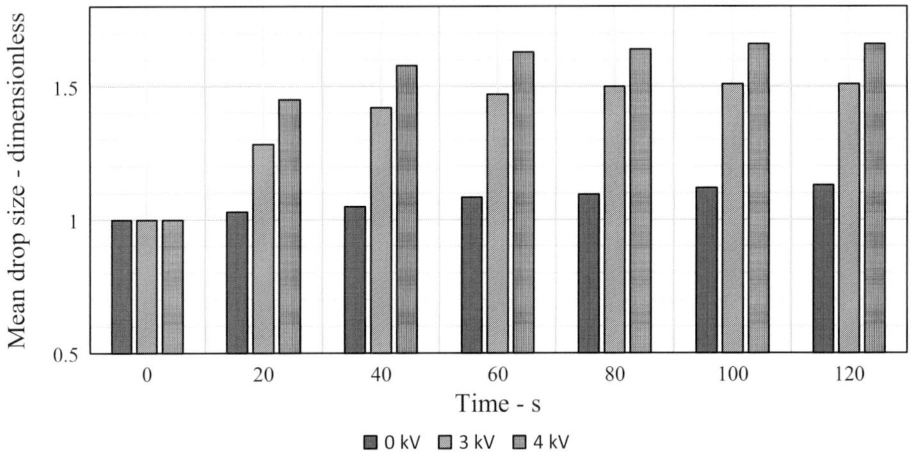

Figure 7.17 Electrically enhanced coalescence of water drops in silicone oil between two parallel plate electrodes: normalized drop diameter versus time.
Data calculated and adapted from Williams and Bailey (1986)

Migratory coalescence also requires a significant relaxation time of the charge on the drops, where relaxation time, τ, is defined as follows:

$$\tau = \frac{\varepsilon_1 \varepsilon_0}{C_m}$$

Figures 7.16 and 7.17 show the influence of an applied electrical d.c. field on the coalescence of water drops dispersed in silicone oil between two parallel plate electrodes (Williams and Bailey, 1986). The degree of coalescence was measured by tracking the obscuration of a helium–neon laser beam shone through a coalescing emulsion. The reduction in obscuration is a direct measure of the degree of coalescence, as shown in Figure 7.16 where the data of Williams and Bailey are recalculated and presented as a percentage of coalescence. Figure 7.17 shows the corresponding increase in drop size with time after the introduction of the electric field. Other mechanisms may also impact the effectiveness of intensified coalescence in the presence of an electrical field. These include electro-hydrodynamic instability and also the dielectric breakdown of the inter-drop film (Eow et al., 2001; Zhang, Basaran, and Wham, 1995).

Accurate and quantitative mathematical modeling techniques for electrically intensified coalescence are only at early stages of development (Wallau, Schlawitschek, and Arellano-Garcia, 2016). There are many variables that control the behavior. These include the type of applied electrical field (d.c. or variable waveform such as a.c.), pulsation frequency, electrical field strength, electrode polarity and geometry, phase ratio, electrical permittivity, conductivity, and other fluid properties such as density, viscosity, and surface tension.

7.5 Surfactants

The influence of the presence of electrolyte species and surfactants on liquid–liquid coalescence is well known but is an important method for intensification of liquid–liquid separation. There are several possible mechanisms by which surfactants can enhance coalescence of droplets that are in close proximity. The most basic mechanism involved is that the surfactant entity adsorbs onto the surface of the drops, enhancing the electrostatic charge on the surface. This creates conditions under which drops in close proximity may be attracted, collide, and coalescence (Figure 7.18).

Weheliye et al. (2017) suggested that when surfactants are present in a dispersed liquid–liquid system, they distribute along the interfaces. In the case of drop–interface coalescence, it is suggested that the uneven distribution of surfactants along the interface can deform the interface and change the tangential stresses. This is believed to influence the internal velocity fields inside the droplet and thus influence the coalescence rate.

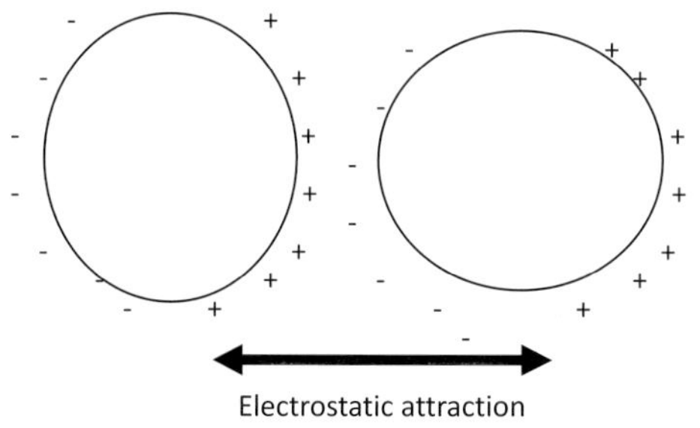

Figure 7.18 Idealized depiction of two adjacent drops with surface charges induced by surfactant adsorption.

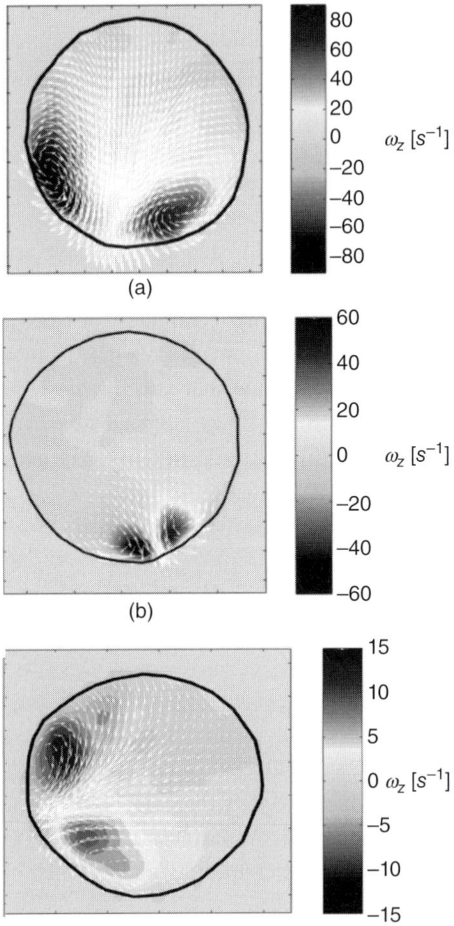

Figure 7.19 the effect of increasing surfactant concentration f, on vorticity of coalescing drops in the liquid-liquid system water-glycerol mixtures and Exxsol D80, in the presence of Span 80. (a) f = 1 × 10−4 (b) f = 2 × 10−4 (c) f = 5 × 10−4
((From Weheliye (2017) Permission approved through open access Creative Commons Attribution License (CC BY)) https://doi.org/10.1016/j.ces.2016.12.009))

Weheliye et al. (2017) show the effects of internal circulation within drops on the coalescence rate, and in particular the influence of surfactants on the internal droplet hydrodynamics. It was shown that propagation of vortices inside the droplet was a strong function of surfactant concentration, with faster vortex propagation at low surfactant concentrations. Vortex intensity was significantly decreased as surfactant concentration was increased, as illustrated in the velocity and vorticity maps shown in Figure 7.19.

The Weheliye study concluded that, prior to rupture, the interface under the droplet deforms as it sits close to the interface. Deformation was observed to increase with increasing surfactant concentration. Also, as the surfactant concentration increased, the time for film drainage and for rupture to occur also increased.

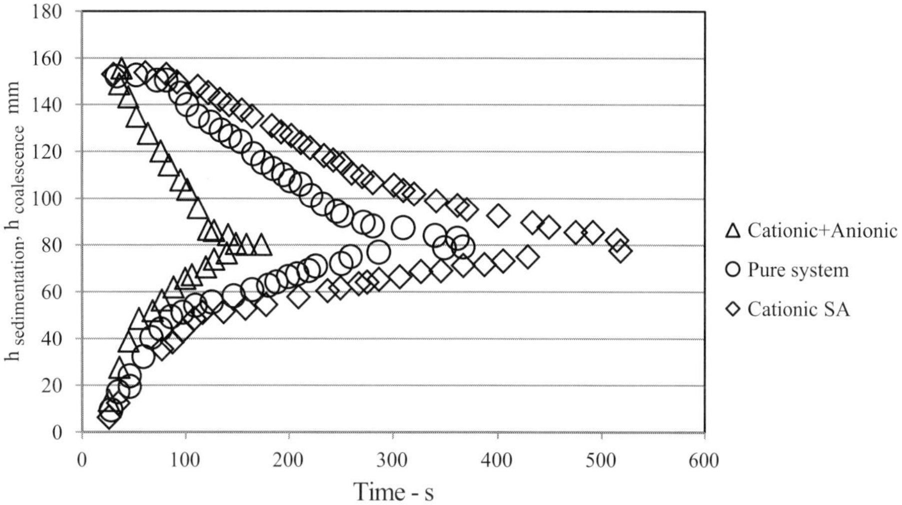

Figure 7.20 Variation of interface position (height) for coalescing drops on decane/paraffin drops in water, showing the effect of surfactant addition. Pure system as circles; cationic surfactant only as diamonds; anionic surfactant only as squares; both surfactants as triangles.
Reprinted (adapted) with permission from May, K., Jeelani, S. A. K., and Hartland, S. (1998). Influence of ionic surfactants on separation of liquid–liquid dispersions. *Colloids and Surfaces A: Physicochemical and Engineering Aspects*, **139**, 41–47. Copyright (1998) Elsevier B.V. All Rights Reserved

The nature of the surfactant is an important factor in determining the forces acting between two adjacent drops. May, Jeelani, and Hartland (1998) demonstrated this in a study of coalescence of drops of decane/paraffin oil dispersed in a continuous phase of water. The addition of cationic surfactant tetra-decyl tri-methyl ammonium bromide showed large increases in coalescence times, as shown by the triangular points in Figure 7.20. Similarly, the addition of anionic surfactant also resulted in slower coalescence times compared with the pure system. This was readily explained by the repulsive forces resulting from adjacent drops acquiring charge of the same polarity as a result of surfactant adsorption.

The addition of both cationic and anionic surfactants together resulted in a large reduction of coalescence time, as shown by the triangular points in Figure 7.20. This was indicative of the creation of a dipole on each drop, resulting in attraction of oppositely charged surfaces and thus increased probability of collision and coalescence, as depicted in Figure 7.18.

May et al. (1998) summarize the controlling equations, starting with the equation relating force, *f*, exerted between two charged droplet surfaces, where q_1 and q_2 are the electrostatic charges on each adjacent drop, *E* is the dielectric constant of the fluid, and E_0 the permittivity constant.

$$f = -\frac{2\pi q_1 q_2}{E E_0 a} \tag{7.23}$$

The force *f* is repulsive when the charges are either both negative or both positive. When one of the charges is positive and the other negative, the force is attractive.

$$a = \pi r_{\mathrm{f}}^2 \tag{7.24}$$

The surface charge may be expressed in terms of the concentration of surfactant at the surface, Γ, in the case of each type by equations 7.25 and 7.26 for cationic and anionic surfactants, respectively.

$$q_1 = \pm\, aIND_1\Gamma_1 \tag{7.25}$$

$$q_2 = \pm\, aIND_2\Gamma_2 \tag{7.26}$$

Substitution of equations 7.24–7.26 into equation 7.23 results in equation 7.27, expressing the force between the two surfaces in terms of area, surfactant concentrations, and electrical properties.

$$f = \frac{2\pi aN^2I^2D_1D_2\Gamma_1\Gamma_2}{EE_0} \tag{7.27}$$

where r_f = radius of curvature; a – area; I = Avogadro number; D_1, D_2 = fraction of surfactant dissociated; Γ_1, Γ_2 = surfactant concentration (cationic and anionic, respectively).

The surfactant concentrations, Γ_1, Γ_2, can be obtained by using the integrated form of the Gibbs adsorption equation to express these in terms of the measureable bulk phase concentrations c_1, c_2, interfacial tensions σ_{*1}, σ_{*2}, characteristic constants for each surfactant k_1 and k_2, and interfacial tension σ_0 for the pure liquid–liquid system.

$$\Gamma_1 = \frac{k_1(\sigma_0 - \sigma_{*1})}{RT} c_1 e^{-k_1 c_{11}} \tag{7.28}$$

$$\Gamma_2 = \frac{k_2(\sigma_0 - \sigma_{*2})}{RT} c_2 e^{-k_2 c_{22}} \tag{7.29}$$

Equation 7.24 for the area of the film, a, may be expressed in terms of properties (Derjaguin and Kussakov, 1939) thus:

$$a = \frac{\pi \Delta\rho g \phi_0^4}{12\sigma} \tag{7.30}$$

where ϕ_0 = drop diameter; $\Delta\rho$ = density difference.

Another possible mechanism through which surfactant addition may intensify coalescence of drops, is by modification of fluid properties of the interfacial film. In particular, local interfacial tension modification is highly likely, giving rise to significant changes in the flow behavior during interfacial drainage. Allan, Charles, and Mason (1960) observed that a small quantity of Span-80 surfactant in the heptane phase could significantly increase the rest time of the water drop at the interface. Due to the reduction in the interfacial tension, the water drop sits in a marked deformation or depression at the interface before coalescence.

Bak and Podgorska (2012) show data for drop size of toluene drops dispersed in aqueous solutions of Tween 20. Figure 7.21 shows the dramatic effect of small concentrations of the surfactant on drop size and drop-size distribution, with reduction in mean drop size and narrowing of distribution as the Tween 20 concentration is increased. This demonstrates clearly that presence of surfactants can act against efficient coalescence by reducing interfacial tension and hence promoting dispersion.

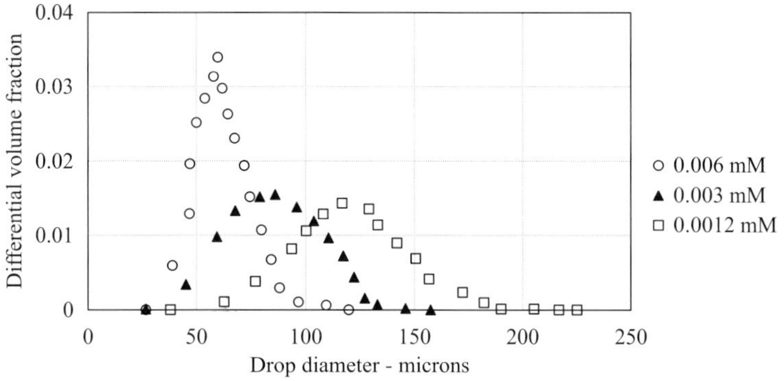

Figure 7.21 Influence of Tween 20 concentration on drop-size distribution.
Reprinted (adapted) with permission from Bak, A. and Podgorska, W. (2012). Investigation of drop breakage and coalescence in the liquid–liquid system with nonionic surfactants Tween20 and Tween80. *Chemical Engineering Science*, **74**, 181–191. Copyright Elsevier B.V.

7.6 Electrolytes

In spite of extensive studies, there are few definitive general rules that dictate the influence of electrolytes on coalescence and there is significant system-to-system variation. The nature of the dispersion is one important factor. Chen et al. (1998) describe opposite effects of electrolyte addition for aqueous drops in an organic continuous phase versus organic drops in an aqueous continuous phase. In the former case, increasing the concentration of electrolyte salts increased the rate of coalescence, while in the latter case the reverse was observed, with a decrease in rate. Valence of the ions is also an important factor, with opposite effects reported for coalescence of aqueous drops compared with coalescence of organic drops. In the case of organic drops, rate of coalescence was seen to be proportional to the valence. In the case of aqueous drops the rate was reported to be inversely proportional to valence.

Another important factor is the polarity of the organic phase. One general observation is that in the case of nonpolar organic liquids, the addition of electrolytes has marginal or insignificant effect on rate of coalescence. This may be explained by the fact that the electrolyte will be almost entirely present only in the aqueous phase, with minimal association in the organic phase. In contrast, in the case of polar solvents, in most cases there is a degree of electrolyte partitioning between the two phases. This will lead to modification of intermolecular forces in each liquid, which will impact the forces close to the liquid–liquid interface and thus influence film drainage and film rupture. Changes in electrostatic forces in the region of the film will also occur due to the presence of ions in each phase and thus will also impact these phenomena. In the case of polar organic phases, the addition of electrolytes may also significantly impact the distribution of other components, for example solutes that are the target of recovery by extraction. Small changes in distribution behavior, and thus relative compositions of each phase, in particular can lead to significant changes in interfacial tension, with consequences for coalescence rates.

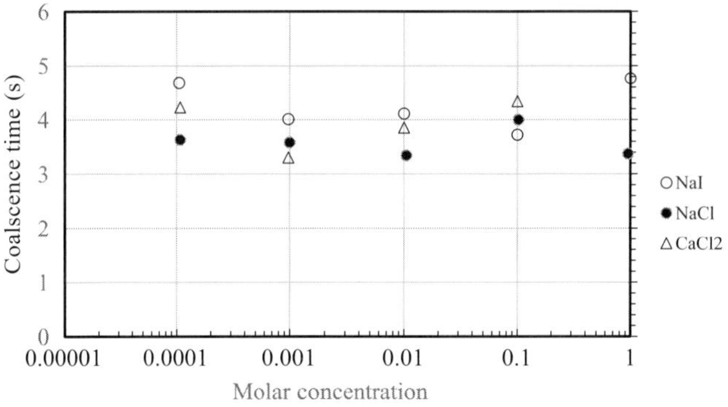

Figure 7.22 Coalescence rates of drops of heptane in water in the presence of sodium iodide, sodium chloride, and calcium chloride.

Reprinted (adapted) with permission from Stevens, G. W., Pratt, H. R. C., and Tai, D. R. (1990). Droplet coalescence in aqueous electrolyte solutions. *Journal of Colloid and Interface Science*, **136**(2), 470–479. Copyright (1990) Elsevier B.V. All Rights Reserved

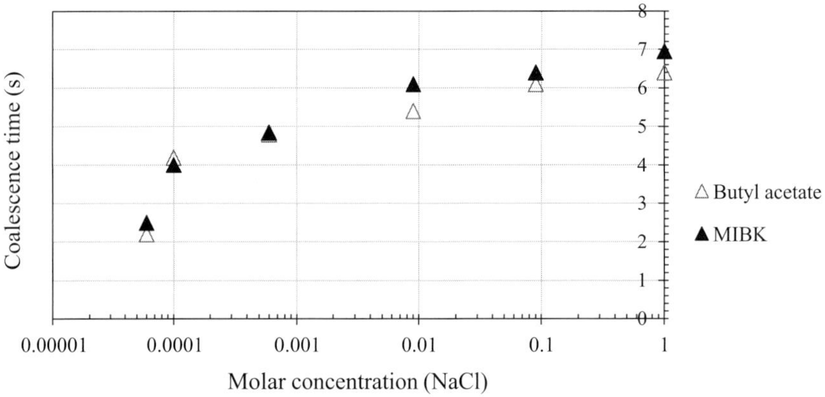

Figure 7.23 Comparison of MIBK and butyl acetate coalescence rates with varying sodium chloride concentration.

Reprinted (adapted) with permission from Stevens, G. W., Pratt, H. R. C., and Tai, D. R. (1990). Droplet coalescence in aqueous electrolyte solutions. *Journal of Colloid and Interface Science*, **136**(2), 470–479. Copyright (1990) Elsevier B.V. All Rights Reserved

Measurements by Stevens, Pratt, and Tai (1990) confirm the large difference in behavior of polar liquids in water compared with nonpolar liquids. Examples of their data are shown in Figures 7.22 and 7.23.

Figure 7.22 shows that the addition of electrolytes had negligible effect on coalescence time for drops of the highly nonpolar liquid heptane. In contrast, Figure 7.23 shows the strong negative relationship between electrolyte concentration and coalescence time for drops of two highly polar liquids, methyl isobutyl ketone and butyl acetate.

Stevens et al. (1990) examined in some detail the possible role of electro-viscous forces when electrolytes are added with the expectation that enhancement of

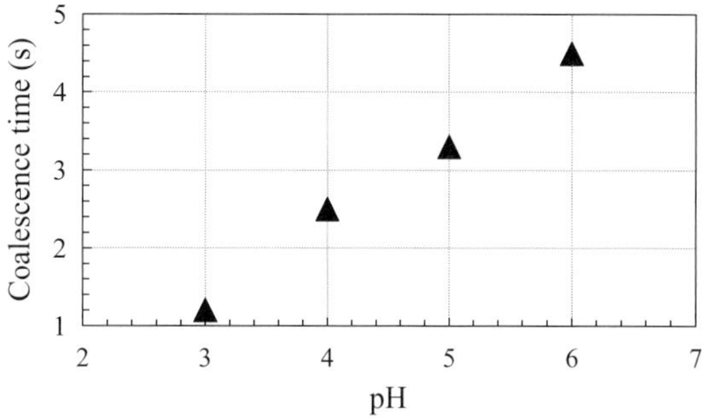

Figure 7.24 Coalescence time of butyl acetate drops at varying pH in the presence of sodium chloride.

Reprinted (adapted) with permission from Stevens, G. W., Pratt, H. R. C., and Tai, D. R. (1990). Droplet coalescence in aqueous electrolyte solutions. *Journal of Colloid and Interface Science*, **136**(2), 470–479.

coalescence may be expected if such forces are dominant. Zeta potential measurements made on drop dispersions in water of both polar and nonpolar liquids were similar, therefore dispelling a significant role of electrical forces. Comparison of drop count/time data for both types of dispersed liquids suggested strongly that viscous and inertial forces are dominant in both cases. However, it was also suggested that formation of interfacial tension gradients in the case of the polar drops and possible dipole interactions could result in increase in the effective viscosity of the inter-drop film, thus slowing the coalescence rate.

Another significant factor in the case of polar drops is the value of pH, as shown in data from Stevens et al. in Figure 7.24. This data show a dramatic reduction of coalescence time as the pH of the system is lowered by addition of hydrochloric acid. Several possible explanations are offered for this. One, the increase in interfacial tension results from acid addition. Another possibility is the creation of dipole–dipole interactions due to molecular restructuring around the interface. In this case, the polar organic molecules align to the polar aqueous phase, with this effect being intensified as the pH is lowered. Such an alignment would likely result in increase in surface viscosity, and hence an increase in stability.

Another and more recent approach reviewed in some detail by Gebauer et al. (2016) is the effect of the nature of the ionic species that is added to the liquid–liquid mixture. Here the Hofmeister classification (1888) of precipitation effects of various anions and cations is relevant. The classification was originally based on the ability of ion addition to either stabilize or disrupt the interaction of proteins. Ions that exhibited a kosmotropic effect were those that resulted in stabilization of protein–water mixtures. Ions that exhibited a chaotropic effect were those that disrupted the stability of protein–water mixtures. Hydrophobic effects that are associated with good conditions for coalescence of water–organic mixtures are promoted by the presence of kosmotropes. Ionic kosmotropes are usually smaller ions with higher charge density, thereby exhibiting greater polar behavior and enhanced hydration. In contrast, some of the larger ions behave as

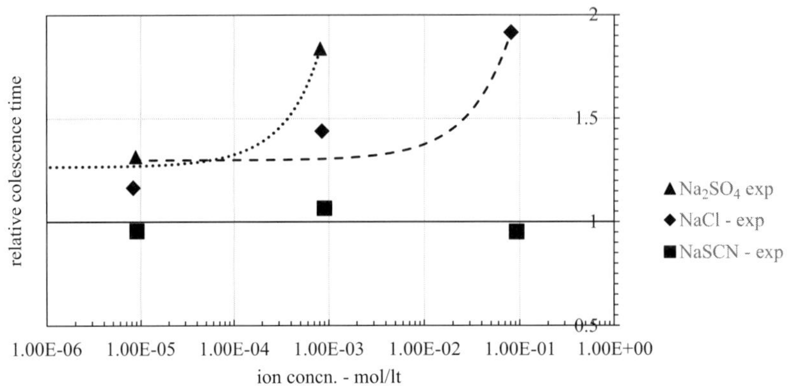

Figure 7.25 Effect of increasing salt concentration on the coalescence time with fitted correlations for Na_2SO_4, NaCl, and NaSCN (points – experimental; lines – prediction)
Reprinted (adapted) with permission from Gebauer F., Villwock J., Kraume M. and Bart Hans-Jörg. (2016). Detailed analysis of single drop coalescence–Influence of ions on film drainage and coalescence time. *Chemical Engineering Research and Design*, 115, 282–291. Copyright Elsevier B.V

chaotropes that tend to stabilize dispersed liquid–liquid mixtures. The effects of both kosmotropes and chaotropes on coalescence kinetics is far from straightforward. There is evidence that suggests that kosmotropes can promote self-organization close to the interface that hinders film drainage. It was also confirmed that faster film drainage can occur in the presence of certain chaotropic anions.

Gebauer notes a summary version of the original version of the kosmotrope–chaotrope spectrum by ion species. For anions:

$$F^- < SO_4^{2-} < HPO_4^{2-}CH_3COO^- < Cl^- < NO_3^-,$$
$$Br^- < ClO_4^- < SCN^- < Cl_3CCOO^-$$

7.7 Phase Inversion for Enhanced Coalescence

Phase inversion as a means of intensifying coalescence of liquid–liquid dispersions is described in some detail by Hadjiev and co-workers (1989, 1995, and 2005). The principle of the technique is illustrated in Figure 7.26. Phase inversion in this context may be achieved in certain liquid–liquid systems by dispersing the initial mixture through a nozzle or perforated plate into a continuum of the original continuous phase. Thus large drops are formed and comprise an inner assembly of drops (the initial dispersed liquid) surrounded by an outer phase that is still part of the large new large drop. There are some similarities with a liquid membrane systems described in Chapter 4. Therefore the "large" carrier drops contain the primary dispersion. Hadjiev describes these large drops as a micro-decanter exhibiting greatly reduced distances to the interface via which coalescence occurs.

Drop–drop coalescence takes place inside the carrier drop, which results in an increase in the relative velocity of the two phases and hence a reduction in the coalescence time. A conceptual diagram is shown in Figure 7.27. The leftmost image depicts the large formed

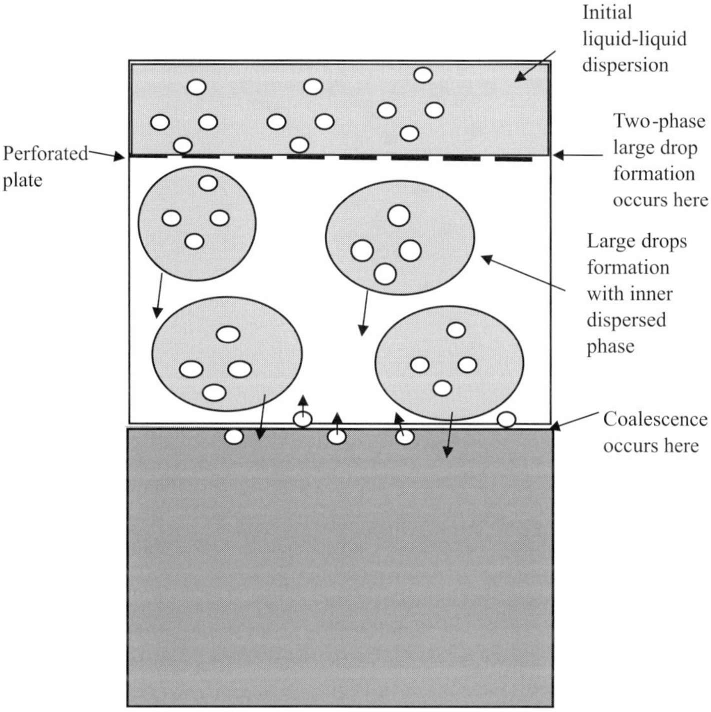

Figure 7.26 Principle of phase inversion as a means of intensifying coalescence.

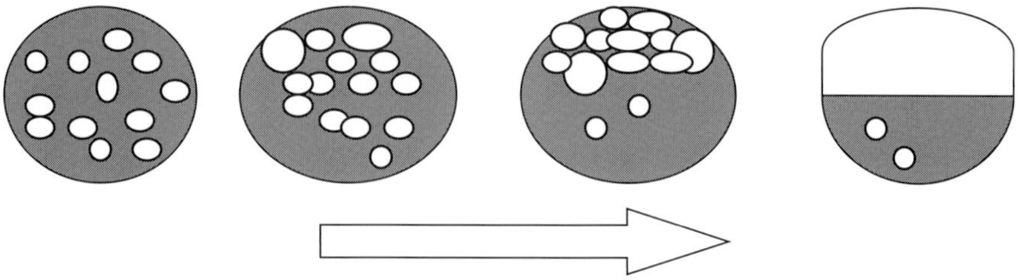

Figure 7.27 Steps in settling and coalescence of the internal drop phase in phase inversion coalescence.

"carrier" droplet as would emerge from the nozzle with the dispersed small "carried" drops shown within the carrier drop. Moving from left to right we see the less dense drops rising within the carrier drop and eventually coalescing, as depicted in the rightmost image. At some point the light phase now collected in the upper part of the carrier drop merges with the main continuous phase, thus effecting the completion of coalescence. Typically the large two-phase drops that are formed have diameters in the range 4–6 mm, with the inner drops having diameters in the 100 micron range.

Hadjiev and Aurelle (1995) also developed a simple modeling tool that predicted the relationship between coalescence efficiency and the size of the large multiphase drops

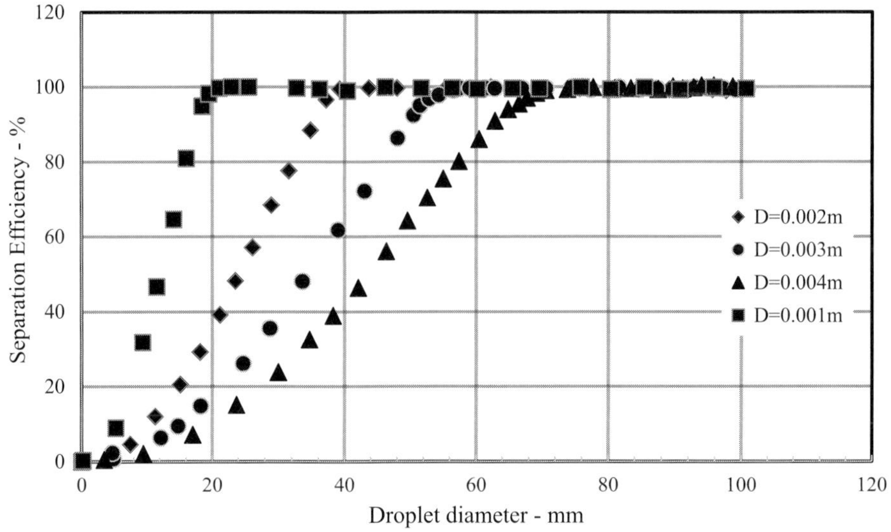

Figure 7.28 Theoretical prediction of the separation efficiency of dispersed drops as a function of the carrier (large) drop diameter, *D*.
Reprinted (adapted) with permission from Hadjiev, D. and Aurelle, Y. (1995). Phase inversion: a method for separation of fine liquid–liquid dispersions. *The Chemical Engineering Journal*, **58**, 45–51. Copyright (1995) Elsevier B.V. All Rights Reserved

formed at the perforated plate. Their predictions are shown in Figure 7.28 and indicate a clear inverse relation between the size of the carrier droplets and the coalescence efficiency. Another important parameter is the height of the liquid layer prior to the interface at that final coalescence occurs. It was shown that the coalescence efficiency increases significantly as the residence time in the pre-coalescence layer increases.

7.8 Ultrasonics

The application of ultrasonic fields to the enhanced separation of particles and liquid drops in suspension has attracted significant attention in recent years. There have been numerous experimental studies quantifying the relative interactions of the various forces and properties that influence separation of dispersions in the presence of ultrasound. Mandralis and Feke (1993) and more recently Luo et al. (2016) provide useful analysis of the fundamental phenomena pertaining to these separations. Drops or particles in liquid suspension respond to resonant acoustic fields if there is a finite difference in the acoustic contrast factor of the drops and the surrounding fluid. The acoustic contrast factor is used to describe the relationship between the densities and the sound velocities (or, equivalently because of the form of the expression, the densities and compressibilities) of the two phases. The magnitude of the acoustic contrast factor is central to the feasibility of deploying ultrasonic standing waves for phase separation. In the latter context, the acoustic contrast factor is the number that, depending on its sign, tells

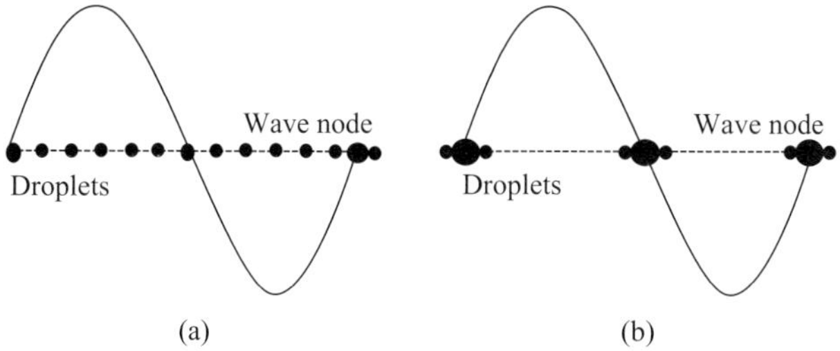

Figure 7.29 Schematic of displacement effect for water droplets in oil. (a) The initial distribution of water droplets; (b) the motion of droplet under ultrasonic irradiation.
Reprinted with permission from Luo, X., He, L.,Wang, H., Yan, H., and Qin, Y. (2016). An experimental study on the motion of water droplets in oil under ultrasonic irradiation. *Ultrasonics Sonochemistry*, **28**, 110–117. Copyright Elsevier B.V.

whether a given type of particle in a given medium will be attracted to the pressure nodes or antinodes.

The acoustic contrast factor is defined in equation 7.31 in terms of the density and compressibility of the two media:

$$\varphi = \frac{5\rho_d - 2\rho_c}{2\rho_d - \rho_c} - \frac{\beta_d}{\beta_c} \tag{7.31}$$

where ρ_d and ρ_c are the densities of the dispersed and continuous phases, respectively, and β_d and β_c are the compressibilities of the dispersed and continuous phases, respectively.

For a positive value of ϕ the particles are attracted to the pressure nodes in the fields, and vice versa in the case of a negative value. The impact of acoustic forces that may be available is highly dependent on the size of the drops or particles in addition to density and compressibility. Under the right conditions, acoustically induced forces may be several orders of magnitude greater than those of gravity, thus enabling significant intensification of drop motion and hence coalescence rate.

Luo et al. (2016) summarize how ultrasonic fields impact the forces acting on suspended droplets. Two main forces may be present, subject to meeting the acoustic contrast criterion. Firstly, there is a primary acoustic force promotes the agglomeration of the suspended drops at the pressure nodes or antinodes that results from the ultrasonic field. There may also be present a secondary acoustic force that is an attractive force when two drops in the suspension have the same compressibility. Figure 7.29 depicts the relationship between drop sizes and position as influenced in the presence of a standing wave ultrasonic field.

The primary acoustic force acting on the small suspended drops is generally accepted as acting parallel to the direction of the propagated acoustic field and in one dimension is expressed by equations 7.32 and 7.33:

$$F_1 = 4\pi a^3 \kappa E_{ac} F \sin(2\kappa x) \tag{7.32}$$

where:

$$F = \frac{\rho_r + (2/3)(\rho_r - 1)}{(1 + 2\rho_r)} - \frac{1}{(3\sigma^2 \rho_r)} \tag{7.33}$$

where a = drop radius; K = wave number of the acoustic field; x = distance between the drop and the pressure antinode; F = acoustic contrast factor as defined in equation 7.33; ρ_r = density ratio; σ = the ratio of the speed of sound through the drop phase to that through the continuous phase.

The two images in Figure 7.29 show in simple terms the overall effect of the ultrasonic field on the drop sizes and the positions of the drops in the continuum. The drops move toward the wave antinode as shown in Figure 7.29(a). Figure 7.29(b) shows the increase of drop sitting at the antinode due to coalescence.

Figure 7.30, also from Luo et al. (2016), shows the motion of a single water drop suspended in oil with time under the influence of a standing wave ultrasonic field. The main variable is the energy intensity of the applied field. The amplitude of oscillation of the drops in the y direction (vertical) decreases very significantly as the energy intensity is increased. This shows that the drop assumes a much more stable and lower position in the continuum. This effect is one that is exploited to enhance coalescence and achieve fractionation of multiple drops in a dispersion.

Figure 7.31 provides additional detail on the movement of multiple suspended drops in a standing wave ultrasonic field, showing the movement of drops toward the nodes and antinodes created by the ultrasonic field. The direction of motion is dictated by the acoustic contrast factor. The lighter drops are congregating toward the antinode, which would be the case for a negative acoustic contrast factor. The darker drops show the opposite case, where the acoustic contrast factor is positive and thus the drops congregate toward the nodes in the field.

The net effect of application of an ultrasonic field parallel to the direction of drop movement is shown in Figure 7.32, from Luo, Cao, and He (2017). These data show the relationship between coalescence time and ultrasound energy intensity for two different drop sizes in a water drop in oil system. The results are very significant and show the potential to reduce coalescence times by up to 50% by the application of an ultrasonic field.

Earlier data by Nii, Kikumoto, and Tokuyama (2009) show results for the ultrasonic separation of mixtures of water and canola oil with initial oil drop sizes in the range of 0.4 μm. Their results in Figure 7.33 show de-emulsification behavior under two regimes of ultrasound application. Figure 7.33(a) shows the rate of decrease of absorbance following a 3 minute application of the ultrasonic field. The respective sets of points show the positive influence of the ultrasound power ranging from zero to 45 W. Figure 7.33(b) shows similar data but at a constant power input of 12 W and that is applied for varying periods of time. These data are interesting since they appear to show that the application of the ultrasonic field appears to promote enhanced coalescence that continues to be exhibited even after the power is switched off. The results also show no evidence of emulsification at these power levels.

The above section summarizes the separation that is achievable in the case of standing wave ultrasonic fields that are applied parallel to the direction of flow. The movement of the drops or particles is dictated largely by the relative magnitude of ultrasonically induced forces and drag forces (Tolt and Feke, 1988). In addition to application of standing ultrasonic energy for de-emulsification in a direction, parallel to

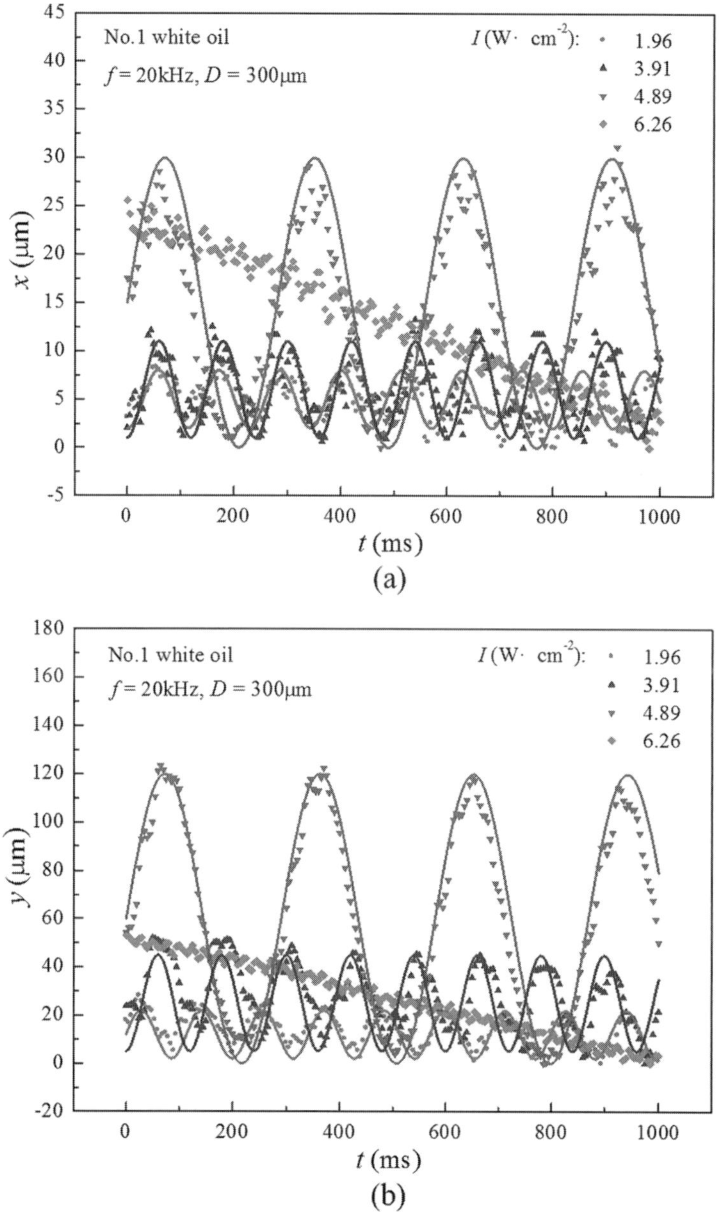

Figure 7.30 Displacement pattern of a single water drop in oil, showing the effect of ultrasound energy intensity on drop motion with time as a standing wave.

Reprinted with permission from Luo, X., He, L.,Wang, H., Yan, H., and Qin, Y. (2016). An experimental study on the motion of water droplets in oil under ultrasonic irradiation. *Ultrasonics Sonochemistry*, **28**, 110–117. Copyright Elsevier B.V.

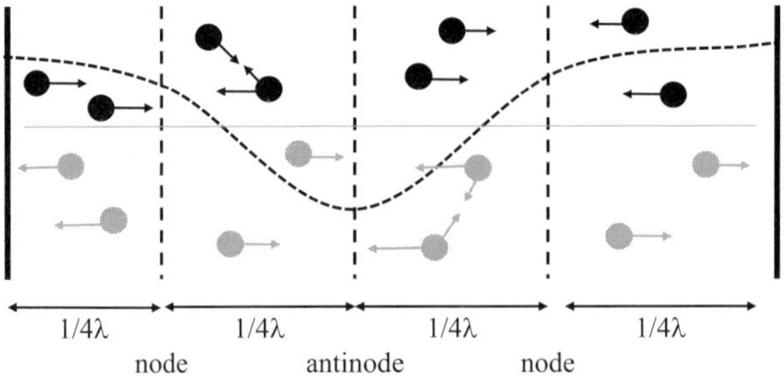

Figure 7.31 Motion of suspended drops to the nodes and antinodes in a standing wave ultrasonic field.
Reprinted (adapted) with permission from Luo, X., Cao, J., He, L., et al. (2017). An experimental study on the coalescence process of binary droplets in oil under ultrasonic standing waves. *Ultrasonics Sonochemistry*, **34**, 839–846. Copyright Elsevier B.V.

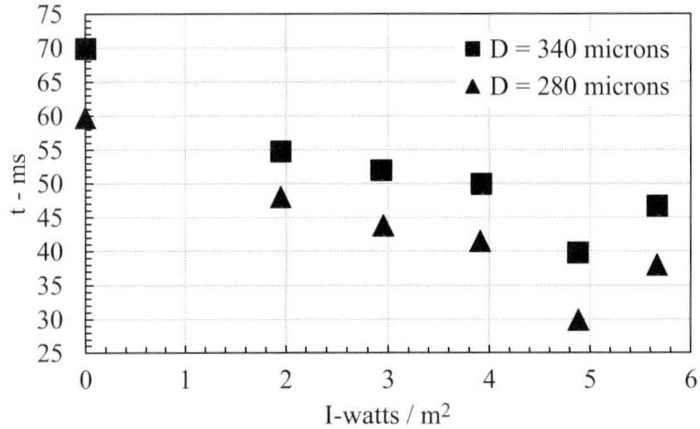

Figure 7.32 Coalescence time of binary drops under ultrasonic irradiation of different intensities: frequency = 20 kHz.
Reprinted (adapted) with permission from Luo, X., Cao, J., He, L., et al. (2017). An experimental study on the coalescence process of binary droplets in oil under ultrasonic standing waves. *Ultrasonics Sonochemistry*, **34**, 839–846. Copyright Elsevier B.V.

flow of the continuum, other hydrodynamic configurations have been explored, especially for applications to segregate drops and particles into specific size ranges. The deployment of a drifting standing wave to enhance the transport of drops or particles through the continuum has also been reported (Whitworth, Grundy, and Coakley, 1991), but the efficiency of separation is still dependent on the ultrasonic forces exceeding the drag forces on the drops. An improved development is described by Mandralis and Feke (1992, 1993), which used a planar ultrasonic field applied transversely across the flowing mixture of drop/particles and the continuous phase. Figure 7.34 shows the principle of the technique. The mixture, comprising solid particles or dispersed drops in suspension, are fed in laminar flow at the left-hand side of a rectangular channel. The ultrasonic field is applied perpendicular to the direction of fluid flow by means of a transducer shown on the lower edge of the channel and equipped with a reflector surface on the upper edge of the channel. Application of the ultrasonic field in this manner results in the dispersed entities taking up position in the

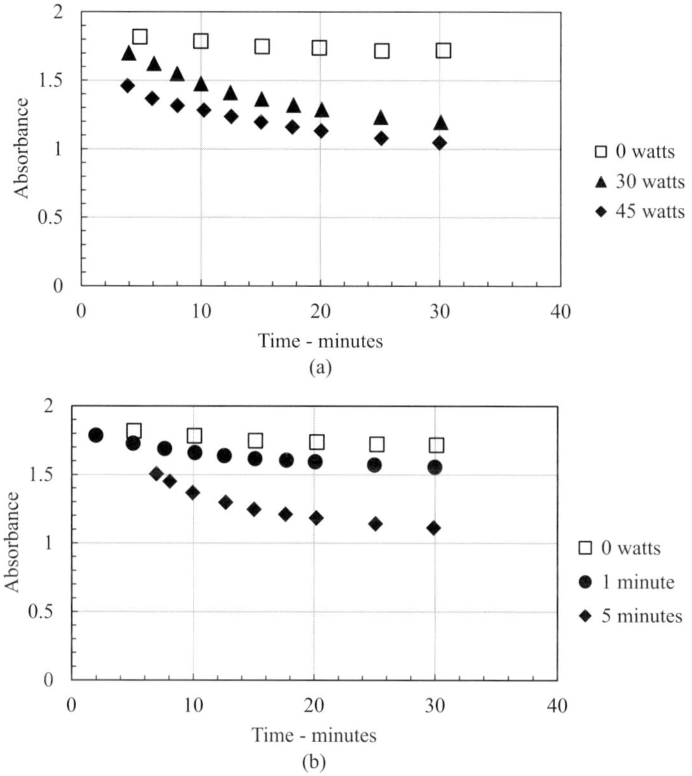

Figure 7.33 (a) Absorbance change with time after the start of irradiation for canola oil in water mixtures. (b) Absorbance change with time after the starting irradiation for canola oil in water mixtures, with ultrasound switched off after the listed irradiation time.

Reprinted (adapted) with permission from Nii, S., Kikumoto, S., and Tokuyama, H. (2009). Quantitative approach to ultrasonic emulsion separation. *Ultrasonics Sonochemistry*, **16**, 145–149. Copyright Elsevier B.V.

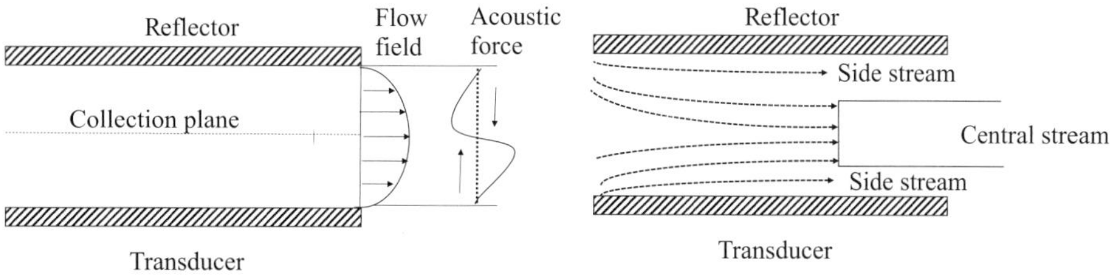

Figure 7.34 Ultrasonically intensified particle separation using a divided flow field and a transverse ultrasonic field.

Reprinted (adapted) with permission from Mandralis, Z. I. and Feke, D. L. (1993). Continuous suspension fractionation using acoustic and divided-flow fields. *Chemical Engineering Science*, **48**(23), 3897–3905. Copyright (1993) Elsevier B.V. All Rights Reserved vertical plane depending on the magnitude and sign of their respective acoustic contrast factors (see equation 7.31). For example, positive contrast particles will collect close to the center of the channel, while negative contrast particles or drops will tend to congregate close to the wall, as shown in the right-hand image in Figure 7.34. Mandralis and Feke (1992, 1993) report that this approach can be used not only for complete phase separation but also for fractionation of particles or droplets into different size ranges. The acoustic forces arising from the application of ultrasound are a function of the size, density, and compressibility of the dispersed entities and so can be controlled by altering the frequency and intensity of the applied field. In addition, the flow velocity can be adjusting to further tune the degree of separation. The exit section of the channel in such a case would be designed to split the exit flows into a number of split streams, each containing the dispersed entities within a specific size range. High-resolution fractionations and very low power requirements are claimed for small-scale applications.

The mechanism for enhanced liquid–liquid coalescence described above has focused on the motion response of particles or suspended drops to the applied ultrasonic field. Another significant phenomenon strongly associated with ultrasound is cavitation. The phenomenon of cavitation occurs especially when high intensity ultrasound fields are applied to a liquid. This results in an oscillating cycle of localized compression alternating with very low pressure. In the low pressure part of the cycle the high intensity ultrasound generates very small bubbles in the liquid. As the bubbles expand to absorb the ultrasound energy that is being applied, they reach a critical volume at which they become unstable and collapse as the high pressure part of the cycle takes effect. The implosion of the bubbles or voids can result in large increases in localized temperature and pressure, and even in the creation of local liquid jets (hielscher.com, 2017).

The application of ultrasonic fields to liquids can result in cavitation. Gardner and Apfel (1993) discuss the potential impact of ultrasound-induced cavitation on liquid–liquid mixtures. Cavitation can promote coalescence by greatly enhancing the breaking of the inter-droplet film between two drops in close proximity (see Section 7.2), and thus accelerate the film drainage rate. On the other hand, as is emphasized by Antes et al. (2017), use of high powered ultrasound can result in surface rupture and the creation of microjets producing fine emulsions, quite the opposite of the desired effect for coalescence. For cavitation to be effective for enhanced coalescence, the energy level of the applied ultrasound must be sufficient to breakdown the inter-droplet films without causing the drops to break up.

7.9 Membranes and Filaments

The application of membrane technology is highly developed for a wide range of separations, from those at the molecular level, such as reverse osmosis and ultrafiltration, through to those at the macroscopic level, such as microfiltration and conventional solid–liquid filtration. Separation of liquid–liquid mixtures using membranes is also well established. Membrane-based methods for intensifying liquid–liquid separations are attractive because they can operate at higher energy efficiencies than other techniques. Secondly, a range of membrane materials are available that can be designed for

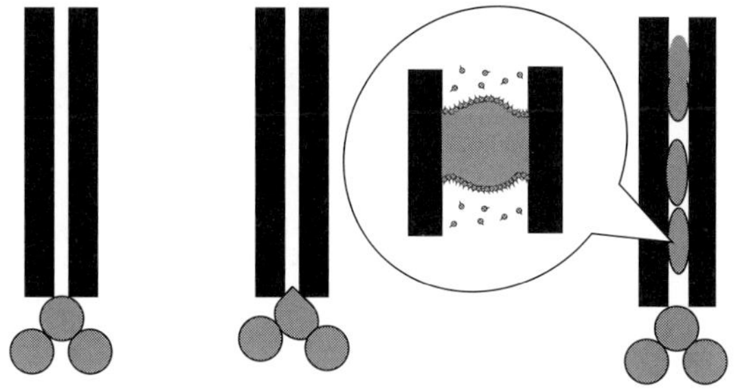

Figure 7.35 (a) Emulsion drops approaching a membrane pore entrance below the critical pressure. (b) Emulsion drops entering the membrane pore entrance at the critical pressure and showing deformation. (c) Emulsion drops entering and inside the pore structure above the critical pressure, resulting in separation into water and oil phases.
Reprinted with permission from Kukizaki, M. and Goto, M. (2008). Demulsification of water-in-oil emulsions by permeation through Shirasu-porous-glass (SPG) membranes. *Journal of Membrane Science*, **322**, 196–203. Copyright Elsevier B.V.

specific applications, for example for either oil separation from aqueous media or vice versa. One significant difference compared with applications such as microfiltration, is the importance of the hydrophilic or hydrophobic properties of the membrane. For coalescence, differences in the surface properties of membranes are used to exploit differences in wettability of the membrane surface exhibited by the two liquid phases present in an emulsion or liquid–liquid dispersion. As is the case for the majority of membrane processes, transmembrane pressure also plays an important role when it comes to the separation of two liquid phases. Kukizaki and Goto (2008) present a summary of one set of mechanisms in play during the separation of a liquid–liquid dispersion, as shown in Figure 7.35. The scheme shows dispersed water drops (water in oil) that also contain surfactant at the entrance of a membrane pore. A certain critical membrane pressure is required for the drops to enter the membrane pore and this is as the critical pressure, P_c. Figure 7.35(a) shows the case where the transmembrane pressure, P, is less than the critical pressure and thus the drops experience difficulty in entering the pore. At or above the critical pressure the drops can enter the pore, as shown in Figure 7.35(b) and (c). The surface properties of the pore and outer surface of the membrane are critical in determining the behavior inside the pore. The drops may be deformed as they enter the pore and it is postulated that the deformation can result in the stripping off of the surfactant layer that may be present. As the surfactant layer is removed and the drop enters the pore, the water drops interact strongly with the hydrophilic surface of the pore and this interaction promotes the coalescence of adjacent drops in the pore.

Figure 7.36 shows an alternative mechanism described by Kocherginsky, Tan, and Lu (2003). Here a large drop is shown approaching the entrance to a pore in the membrane in Figure 7.36(a). Again there is surfactant shown as present. Once the drop is in contact with the hydrophilic surface close to the pore, the drop may be too large to

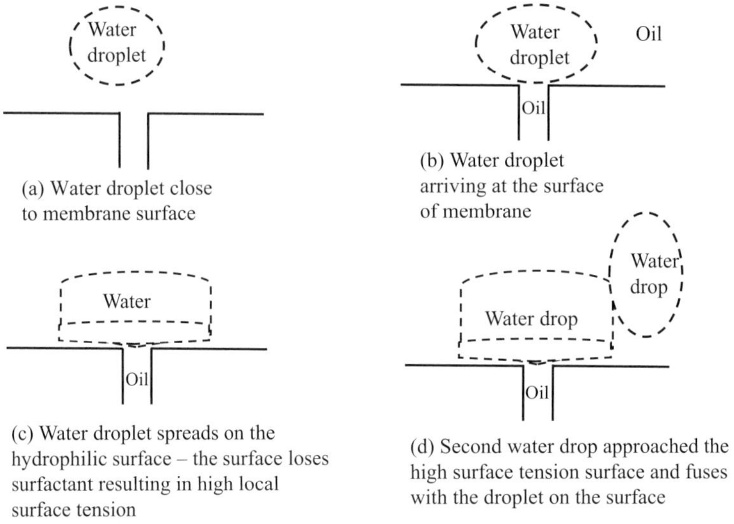

Figure 7.36 Schematic of the steps in demulsification at the surface of hydrophilic nitrocellulose membrane. (a) Water drop near to the membrane surface. (b) Water drop arriving at the surface of the membrane. (c) Water drop spreading on the hydrophilic surface of the membrane. (d) Second water drop approaching the high surface tension surface, leading to coalescence with the initial drop. Reprinted (adapted) with permission from Kocherginsky, N. M., Tan, C. L., and Lu, W. F. (2003). Demulsification of water-in-oil emulsions via filtration through a hydrophilic polymer membrane. *Journal of Membrane Science*, **220**, 117–128. Copyright Elsevier B.V.

enter the pore and as shown in Figure 7.36(b) and (c) the drop spreads out, at the same time losing its surfactant surface layer. The loss of the surfactant gives rise to a high interfacial tension that creates an environmental for coalescence with other approaching drops close to the membrane surface, thus resulting in larger drops being formed. This takes place outside the pore structure of the membrane.

Kocherginsky et al. (2003) provide experimental evidence that supports the mechanism shown in Figure 7.36 using hydrophilic nitrocellulose membranes and for cuprophane hydrophilic hollow fiber membranes (Table 7.1). Their work showed effective separation of emulsions comprising water and kerosene in the presence of di-2-ethylhexyl phosphoric acid and SPAN 80 in the presence of sodium hydroxide solution. The study concluded that de-emulsification with hydrophilic membranes is feasible only if the membrane pore diameters are smaller than the drop size in the emulsion. It was also reported that de-emulsification efficiency increases inversely with the pore diameter. Transmembrane pressure was also shown to have a significant effect on de-emulsification efficiency, also with an inverse relationship showing that lower transmembrane pressure led to higher efficiency. Comparison was made with separation of the same emulsions using hydrophobic PVDF (polyvinylidene fluoride) hollow fiber membranes. In this case very poor separation was observed.

There are a range of other membrane materials that have been demonstrated as effective tools for liquid–liquid phase separation. Rohrbach et al. (2014) list materials

Table 7.1 Novel membrane materials developed for liquid–liquid separations.

Membrane material	Reference
Shirasu porous glass (SPG)	Kukizaki and Goto (2008), Nakashima and Kuroki (1981), Nakashima, Shimizu, and Kukizaki (2000)
Nanofibrillated cellulose (NFC) hydrogel	Rohrbach et al. (2014)
Carbon nanotube (CNT) integrated polymer composite membrane with a polyvinyl alcohol barrier	Maphutha et al. (2013)
Hydrophobic–oleophilic kapok filter	Huang and Lim (2006)
Nitrocellulose flat membrane HAWP04700	Kocherginsky et al. (2003)
Cuprophane hollow fiber	Kocherginsky et al. (2003)
PVDF (polyvinylidene fluoride) hollow fiber XUMP-053	Kocherginsky et al. (2003)
Super-oleophobic hydrogel-coated mesh	Xue et al. (2011)
Carbon nanotube-based membranes	Lee and Baik (2010)
Super-hydrophobic silica aerogels	Venkateswara Rao, Hegde, and Hirashima (2007)
Super-hydrophobic and super-oleophilic coating mesh film	Feng et al. (2004)

such as polydimethylsiloxane (PDMS)-coated nanowire membranes, polytetrafluoroethylene (PTFE)-coated metal mesh, carbon-based porous materials, and a range of cross-linked polymer gels. It is pointed out that many of the current materials are prone to clogging and fouling, especially those materials that are oleophilic. In such cases oil that causes clogging or fouling can be difficult to remove and may ultimately present an environmental problem. Increased transmembrane pressures resulting from oil build-up in the pores of the membrane or on the surface lead to higher pumping energy requirements and cost. In light of such issues, there has been a move to develop hydrophilic materials that exclude the oil or solvent phase.

Table 7.1 lists a selection of novel membrane materials that show significant promise for improved and intensified separation of liquid–liquid mixtures. Advances have been made in a number of areas, which include the development of materials with highly controlled degrees of hydrophobicity and hydrophilicity. Application of carbon nanotube technology is another notable area, together with the application of hydrogels polymers as a means of creating "super-hydrophobic" materials. Innovations in the physical structures of the membrane materials also feature significantly in the literature, with a focus on surface morphology in addition to pore structure. The success of these developments has been the ability to create super-hydrophobic surfaces having a very high contact angle with water. It is clear from the various literatures that surface roughness at the micro-level is a major factor. Feng et al. (2004) provided one of the earlier demonstrations of this approach but also highlighted the development of surfaces that incorporate both super-hydrophobic and super-oleophilic properties. This is achieved by using a simple spray drying technique to coat a stainless steel micromesh with a PTFE coating to create the membrane.

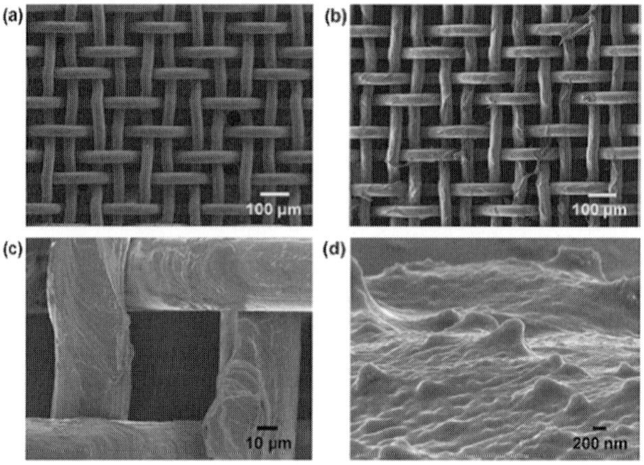

Figure 7.37 SEM images of the PAM hydrogel-coated mesh prepared from a stainless steel mesh with an average pore diameter of about 50 μm. (a) Large-area view of the uncoated stainless steel mesh. (b) Large-area view of the PAM hydrogel-coated stainless steel mesh. (c) Enlarged view of a single pore of the PAM hydrogel-coated stainless steel mesh. (d) The higher magnification image of one single wire on hydrogel-coated stainless steel mesh, in which the nanostructured papillae can be clearly observed.

Reprinted with permission Xue, Z., Wang, S., Lin, L., et al. (2011). A novel superhydrophilic and underwater superoleophobic hydrogel-coated mesh for oil/water separation. *Advanced Materials*, **23**, 4270–4273.

Xue et al. (2011) discussed the development of new super-hydrophilic and super-oleophobic membranes by fabricating a hydrogel-coated mesh comprising of a "rough" hydrogel coating combined with microporous metal-based substrates. The highly effective separation of various oil/water mixtures was demonstrated using these materials. Systems for which enhanced separation was reported include vegetable oil/water; gasoline/water; diesel/water; and crude oil/water. The novelty of the materials lies in the presence of a super-hydrophobic interface that prevents oil-based fouling of the membrane surface. Figure 7.37 includes micrographs from Xue et al. that show the polyacryamide gel coated stainless steel mesh structure of the membrane.

Another approach is described by Lee and Baik (2010), also based on a stainless steel mesh support. A technique was developed to synthesize vertically aligned multi walled nanotubes on the surface of the mesh. This also created a dual structure with micro-scale pores. The low surface energy of the carbon was exploited to amplify both hydrophobicity and oleophilicity.

Use of carbon nanotube technology is also described by Maphutha et al. (2013). The fabrication and evaluation of a carbon nanotube–infused polymer composite containing a PVA barrier is described. Effective rejection of oil from a wastewater stream is reported using this membrane system. It was claimed that the structure of this membrane provided increased mechanical strength but retained highly effective oil/water separation.

Kukizaki and Goto (2008) also quantified the mechanisms of coalescence observed when using Shirasu porous glass membranes. These novel membranes were synthesized by the phase separation of primary Na_2O–CaO–MgO–Al_2O_3–B_2O_3–SiO_2 type glasses and subsequent treatment by acid leaching. They showed that the membrane pore diameter is a critical property influencing phase-separation performance and is primarily determined by the phase-separation conditions. The pore diameters ranged from 40 nm to 20 μm, with uniform size determined by the conditions of phase separation. High mechanical strength is also reported. Although these membranes are hydrophilic, Kukizaki et al. determined the importance of transmembrane pressure and its influence on the mechanism of coalescence of water drops. Their study examined the separation of an aqueous phase comprising dilute sodium chloride solution, and an organic phase of kerosene containing a lipophilic surfactant, together with tetraglycerin-condensed ricinoleic ester. The critical pressure was defined as the minimum transmembrane pressure required to permeate the droplets through a membrane. Results of the study showed that at transmembrane pressures above the critical value, aqueous droplets larger than the membrane pore size were demulsified. This may be explained by the hydrophilicity of the membrane surface, with coalescence explained by a similar mechanism proposed by Kocherginsky et al. (2003) shown in Figure 7.36. On the other hand, at pressures below the critical pressure, the larger aqueous drops were retained at the surface of the membrane due to the physical filtration barrier as determined by the pore size.

Figure 7.38 (Kukizaki and Goto, 2008) shows the strong experimental relationship between de-emulsification efficiency, α, and transmembrane pressure for the system described above. The occurrence of the maximum tends to confirm the mechanism described above. The de-emulsification efficiency is defined thus:

$$\alpha = \frac{10^4 V_s}{\phi_d V_{perm}} \tag{7.34}$$

where V_s = total volume of the combined liquid phases; φ_d = aqueous phase fraction in the feed; and V_{perm} = volume of the permeate liquid.

Figure 7.39, reproduced from Kukizaki and Goto (2008), shows the impressive coalescence that is possible for a water in oil emulsion using a SPG porous glass membrane. The initial emulsion had a mean drop diameter of 2.3 μm. The membrane mean pore diameter was 0.86 μm and the transmembrane pressure was 392 kPa.

Huang and Lim (2006) evaluated kapok fiber as a filter material for coalescence of oil/water mixtures. Kapok is a natural vegetable fiber exhibiting hydrophobic properties and was demonstrated as highly effective for the separation of diesel/water and hydraulic oil/water mixtures. Oil/water mixtures in the range of 5–15% oil were evaluated and successfully separated by pumping the mixtures through a packed bed of fibers. The proposed mechanism for the separation is based on the hydrophobicity of the kapok fibers, which allows preferential flow of aqueous phase through a bed of fibers relative to the oil phase. Separation efficiencies in excess of 99% are reported though experiments showed that hydraulic transport resistances increased with the advance of the oil through the bed of fibers, also with increase in fiber packing density, and with increased oil/water ratios.

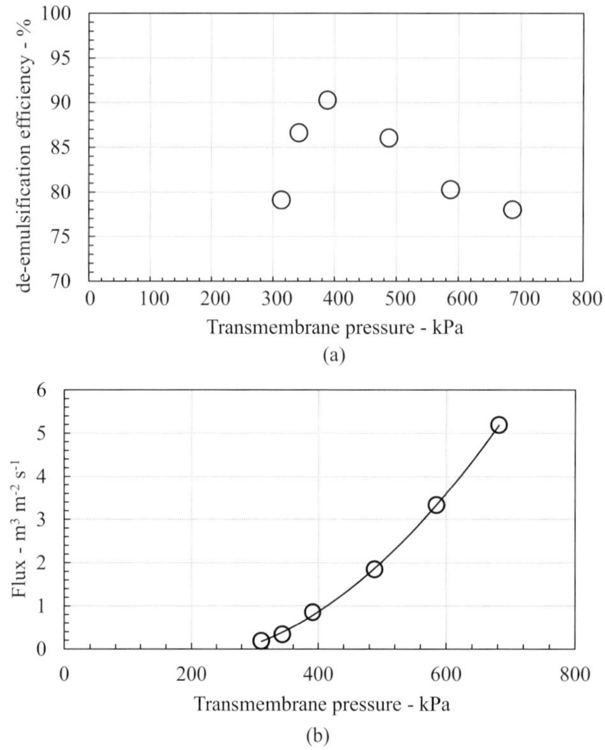

Figure 7.38 Experimental relationship between de-emulsification efficiency, α, and transmembrane pressure for the system dilute sodium chloride solution/kerosene/lipophilic surfactant + tetraglycerin-condensed ricinoleic ester.

Reprinted (adapted) with permission from Kukizaki, M. and Goto, M. (2008). Demulsification of water-in-oil emulsions by permeation through Shirasu-porous-glass (SPG) membranes. *Journal of Membrane Science*, **322**, 196–203. Copyright Elsevier B.V. All Rights Reserved

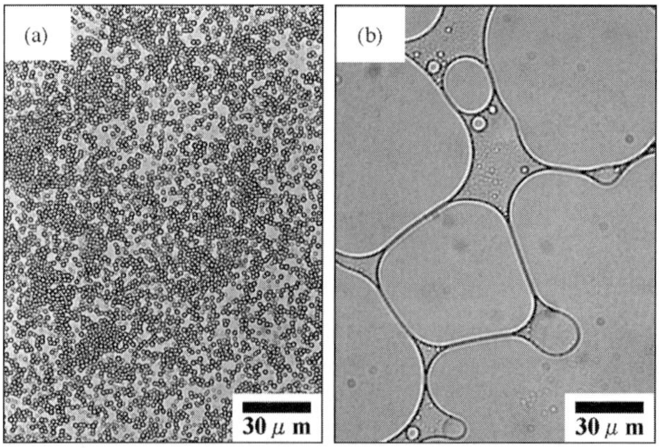

Figure 7.39 Micrographs showing comparison of a water in oil emulsion on (a) the retentate side of an SPG porous glass membrane with (b) that on the permeate side.

Reprinted (adapted) with permission from Kukizaki, M. and Goto, M. (2008). Demulsification of water-in-oil emulsions by permeation through Shirasu-porous-glass (SPG) membranes. *Journal of Membrane Science*, **322**, 196–203. Copyright Elsevier B.V. All Rights Reserved

References

Aarts, G., and Lekkerkerker, H. N. (2008). Droplet coalescence: drainage, film rupture and neck growth in ultralow interfacial tension systems. *Journal of Fluid Mechanics*, **606**, 275–294.

Aarts, G., Lekkerkerker, N., Guo, H., Wegdam, G. H., and Bonn, D. (2005). Hydrodynamics of droplet coalescence. *Physical Review Letters*, **95**, 164503.

Allen, R. S., Charles, G. E., and Mason, S. G. (1960). The approach of gas bubbles to a gas/liquid interface. *Journal of Colloid Science*, **16**(2), 150–165.

Antes, F. G., Diehl, L. O., Pereira, J. S. F., et al. (2017). Effect of ultrasonic frequency on separation of water from heavy crude oil emulsion using ultrasonic baths. *Ultrasonics Sonochemistry*, **35**, 541–546.

Bak, A. and Podgorska, W. (2012). Investigation of drop breakage and coalescence in the liquid–liquid system with nonionic surfactants Tween20 and Tween80. *Chemical Engineering Science*, **74**, 181–191.

Borrell, M., Yoon, Y., and Leal, L. G. (2004). Experimental analysis of the coalescence process via head-on collisions in a time-dependent flow. *Physics of Fluids*, **16**(11), 3945–3954.

Brown, A. H. and Hanson, C. (1965). Effect of oscillating electric fields on coalescence in liquid + liquid systems. *Transactions of the. Faraday Society,* **61**, 1754.

Charles, G. E. and Mason, S. G. (1960). The coalescence of liquid drops with flat liquid/liquid interfaces. *Journal of Colloid Science*, **15**(3), 236–267.

Chatterjee, J., Nikolov, A. D., and Wasan, D. T. (1996). Study of drop-interface coalescence using piezoimaging. *Industrial and Engineering Chemistry Research*, **35**, 2933–2938.

Chen, C. T., Maa, J. R., Yang, Y. M., and Chang, C. H. (1998). Effects of electrolytes and polarity of organic liquids on the coalescence of droplets at aqueous-organic interfaces. *Surface Science*, **406**(1–3), 167–177.

Coulaloglou, C. A. and Tavlarides, L. L. (1977). Description of interaction processes in agitated liquid–liquid dispersions. *Chemical Engineering Science*, **32**, 1289–1297.

Cusack, R. (2009). Rethink your liquid–liquid separations. *Hydrocarbon Processing*, **June**, 53–60.

Derjaguin, B. and Kussakov, M. (1939). Anomalous properties of thin films. *Acta Physicochim. U.R.S.S.*, **10**, 25.

Dreher, T. M., Glass, J., O'Connor, A. J., and Stevens, G. W. (1999). Effect of rheology on coalescence rates and emulsion stability. *AIChE Journal*, **45**(6), 1182–1190.

Eow, J. S. and Ghadiri, M. (2003). The behaviour of a liquid/liquid interface and drop-interface coalescence under the influence of an electric field. *Colloids and Surfaces A: Physicochemical and Engineering Aspects*, **215**, 101–123.

Eow, J. S., Ghadiri, M., Sharif, A., and Williams, T. J. (2001). Electrostatic enhancement of the coalescence of water droplets in oil: a review of the current understanding. *Chemical Engineering Journal*, **84**(3), 173–192.

Feng, L., Zhang, Z., Mai, Z., et al. (2004). A super-hydrophobic and super-oleophilic coating mesh film for the separation of oil and water. *Angewandte Chemie International Edition*, **43**, 2012–2014.

Gardner, E. A. and Apfel, R. E. (1993). Using acoustics to study and stimulate the coalescence of oil drops surrounded by water. *Journal of Colloid Interface Science*, **159**, 226–237.

Gebauer, F., Villwock, J., Kraume, M. and Bart, H.-J. (2016). Detailed analysis of single drop coalescence: Influence of ions on film drainage and coalescence time. *Chemical Engineering Research and Design*, **115**, 282–291.

Hadjiev, D. and Aurelle, Y. (1995). Phase inversion: a method for separation of fine liquid–liquid dispersions. *The Chemical Engineering Journal*, **58**, 45–51.

Hadjiev, D. and Kuychoukov, G. A. (1989). Separator for liquid–liquid dispersions. *The Chemical Engineering Journal*, **41**, 113–116.

Hadjiev, D. and Paulo, J. B. A. (2005). Extraction separation in mixer–settlers based on phase inversion. *Separation and Purification Technology*, **43**, 257–262.

Hartland, S., Yang, B., and Jeelani, S. A. K. (1994). Dimple formation in the thin film beneath a drop or bubble approaching a plane surface. *Chemical Engineering Science*, **49**(9), 1313–1322.

Henschke, M., Schlieper, L. H., and Pfennig, A. (2002). Determination of a coalescence parameter from batch-settling experiments. *Chemical Engineering Journal*, **85**, 369–378.

www.hielscher.com/ultrasonic-cavitation-in-liquids-2.htm, 2017, Ultrasonic Cavitation in Liquids

Hofmeister, F. (1888). *Archives of Experimental Pathology and Pharmacology*, **24**, 247–260.

Huang, X. and Lim, T.-T. (2006). Performance and mechanism of a hydrophobic–oleophilic kapok filter for oil/water separation. *Desalination*, **190**, 295–307.

Jeelani, S. A. K. and Hartland, S. (1993). Effect of velocity fields on binary and interfacial coalescence. *Journal of Colloid and Interface Science*, **156**, 467–477.

Jeelani, S. A. K. and Windhab, E. J. (2009). Drop coalescence in planar extensional flow and gravity. *Chemical Engineering Science*, **64**, 2718–2722.

Kamp, J. and Kraume, M. (2016). From single drop coalescence to droplet swarms: Scale-up considering the influence of collision velocity and drop size on coalescence probability. *Chemical Engineering Science*, **156**, 162–177.

Kocherginsky, N. M., Tan, C. L., and Lu, W. F. (2003). Demulsification of water-in-oil emulsions via filtration through a hydrophilic polymer membrane. *Journal of Membrane Science*, **220**, 117–128.

Kukizaki, M. and Goto, M. (2008). Demulsification of water-in-oil emulsions by permeation through Shirasu-porous-glass (SPG) membranes. *Journal of Membrane Science*, **322**, 196–203.

Lee, C. and Baik, S. (2010). Vertically-aligned carbon nano-tube membrane filters with superhydrophobicity and superoleophilicity. *Carbon*, **48**, 2192–2197.

Liao, Y. and Lucas, D. (2010). A literature review on mechanisms and models for the coalescence process of fluid particles. *Chemical Engineering Science*, **65**, 2851–2864.

Luo, X., He, L., Wang, H., Yan, H., and Qin, Y. (2016). An experimental study on the motion of water droplets in oil under ultrasonic irradiation. *Ultrasonics Sonochemistry*, **28**, 110–117.

Luo, X., Cao, J., He, L., et al. (2017). An experimental study on the coalescence process of binary droplets in oil under ultrasonic standing waves. *Ultrasonics Sonochemistry*, **34**, 839–846.

MacKay, G. D. M. and Mason, S. G. (1963). Some effects of interfacial diffusion on the gravity coalescence of liquid drops. *Journal of Colloid Science*, **18**(7), 674–683.

Mandralis, Z. I. and Feke, D. L. (1992). Fractionation of suspensions using synchronized ultrasonic and flow fields. *Amercial Institute of Chemical Engineers Journal*, **39**, 197–206.

Mandralis, Z. I. and Feke, D. L. (1993). Continuous suspension fractionation using acoustic and divided-flow fields. *Chemical Engineering Science*, **48**(23), 3897–3905.

Maphutha, S., Moothi, K., Meyyappan, M., and Iyuke, S. E. (2013). A carbon nanotube-infused polysulfone membrane with polyvinyl alcohol layer for treating oil-containing waste water. *Scientific Reports*, **3**(1509), 1–6.

May, K., Jeelani, S. A. K., and Hartland, S. (1998). Influence of ionic surfactants on separation of liquid–liquid dispersions. *Colloids and Surfaces A: Physicochemical and Engineering Aspects*, **139**, 41–47.

Nakashima, T. and Kuroki, Y. (1981). Effect of composition and heat treatment on the phase separation of NaO–B_2O_3–SiO_2–Al_2O_3–CaO glass prepared from volcanic ashes. *Nippon Kagaku Kaishi*, **8**, 1231.

Nakashima, T., Shimizu, M., and Kukizaki, M. (2000). Particle control of emulsion by membrane emulsification and its applications. *Advanced Drug Delivery Reviews,* **45**, 47–56.

Nii, S., Kikumoto, S., and Tokuyama, H. (2009). Quantitative approach to ultrasonic emulsion separation. *Ultrasonics Sonochemistry*, **16**, 145–149.

Prince, M. J. and Blanch, H. W. (1990) Bubble coalescence and break-up in air-sparged bubble columns. *Amercial Institute of Chemical Engineers Journal*, **36**, 1485–1499.

Qiu, J. (2010). Intensification of liquid–liquid contacting processes. Ph.D. thesis, The University of Kansas.

Rohrbach, K., Li, Y., Zhu, H., et al. (2014). A cellulose based hydrophilic, oleophobic hydrated filter for water/oil separation. *Chemical Communications*, **50**, 13296–13299.

Sovova, H. (1981). Breakage and coalescence of drops in a batch stirred vessel – II comparison of model and experiments. *Chemical Engineering Science,* **36**, 1567–1573.

Stevens, G. W., Pratt, H. R. C., and Tai, D. R. (1990). Droplet coalescence in aqueous electrolyte solutions. *Journal of Colloid and Interface Science*, **136**(2), 470–479.

Tolt, T. L. and Feke, D. L. (1988). Analysis and application of acoustics to suspension processing, in ASME 23rd Intersociety Energy Conversion Engineering Conference, 4, 327.

Urdahl, O., Berry, P., Wayth, N., et al. (1998). The development of a new compact electrostatic coalescer concept. SPE 48990, SPE Annual Technical Conference and Exhibition held in New Orleans, Louisiana, 27-30 September 1995.

Venkateswara Rao, A. V., Hegde, N. D., and Hirashima, H. (2007). Absorption and desorption of organic liquids in elastic superhydrophobic silica aerogels. *Journal of Colloid and Interface Science.* **305**, 124–132.

Wallau, W., Schlawitschek, C., and Arellano-Garcia, H. (2016). Electric field driven separation of oil–water mixtures: model development and experimental verification. *Industrial and Engineering Chemistry Research*, **55**, 4585–4598.

Weheliye, W. H., Dong, T., and Angeli, P. (2017). On the effect of surfactants on drop coalescence at liquid/liquid interfaces. *Chemical Engineering Science,* **161**, 215–227.

Whitworth, G., Grundy, M., and Coakley, W. (1991). Transport and harvesting of suspended particles using modulated ultrasound. *Ultrasonics*, **29**(6), 439–444.

Williams, T. J. and Bailey, A. G. (1986). Changes in the size distribution of a water-in-oil emulsion due to electric field induced coalescence. *IEEE Transactions On Industry Applications*, **Ia-22**(3), May/June, 536–541.

Xue, Z., Wang, S., Lin, L., et al. (2011). A novel superhydrophilic and underwater superoleophobic hydrogel-coated mesh for oil/water separation. *Advanced Materials*, **23**, 4270–4273.

Zhang, X., Basaran, O. A., and Wham, R. M. (1995). Theoretical prediction of electric field-enhanced coalescence of spherical drops. *AIChE Journal*, **41**(7), 1629–1639.

8 Ionic Liquid Solvents and Intensification

8.1 General Introduction to Ionic Liquids

Ionic liquids are organic salts. A number of general features mark them out as being potentially highly valuable for the intensification and improvement of liquid–liquid processes. Ionic liquids are liquid at or near room temperature ($T < 100$ °C). Many that have been identified and characterized to date exhibit very low melting points and very low vapor pressures at normal temperatures. Other positive features in many cases include low flammability and a wide liquid range. The liquid range can be as high as 300 °C (e.g. −90 °C to +200 °C). Ionic liquids offer almost unlimited versatility in terms of chemical structure, which may be tuned to meet chemistry-based or process-based requirements such as solubility, solvating power, thermodynamic activity, and thermophysical properties. Ionic liquids have an important place in the development of intensified technology and chemical processes. One significant drawback is that many ionic liquids, although having useful chemical properties also exhibit moderate to high viscosity, which may prove problematical when large-scale processes using these liquids are being considered. Difficulties with pumping, mixing, together with the challenge of slow diffusion rates may impede commercial application. The properties of ionic liquids when compared with molecular solvents show significant differences. The main difference is that solvent properties can be controlled by selection of the anion and cation that comprise the ionic liquid entity.

Figure 8.1 shows some examples of the structures of several cations used in ionic liquid synthesis. The important feature to note is the presence of the R group or in some cases both an R group and a different functional group R'. These represent functional groups that may be taken from a large range of possibilities. Common examples include alkyl groups such as methyl, ethyl, and butyl, but may also include carbonyl groups, ester groups, and ether groups.

Figure 8.1 shows the basic molecular structures of four organic cations used in ionic liquids: (a) R,R'-imidazolium; (b) R-pyridinium; (c) R,R'-pyrrolidinium; (d) R-tributyl ammonium.

Figure 8.2 show a generic structure for the imidazolium cation, with one of the two attached functional groups shown as C_nH_{2n+1}, where n is shown as having possible values as high as 18. Variations of the second position (off the left-hand nitrogen) and variation of the value n provides some insight into the number of possible structures based on the cation structure alone.

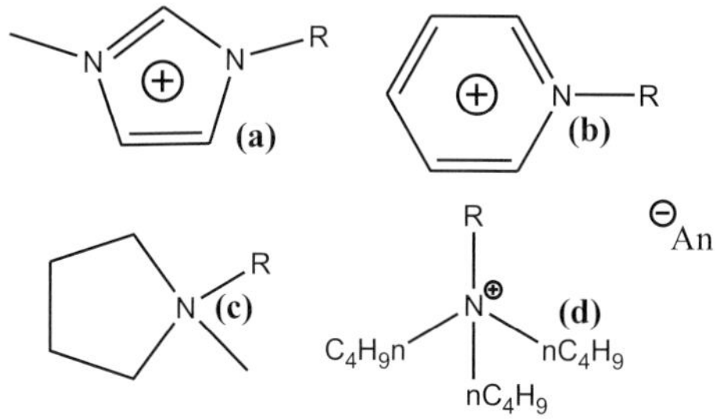

R= nC_2H_5, nC_4H_9, nC_6H_{13}, & $nC_{10}H_{21}$
An^-= [Tf_2N], [BF_4], [Br]

Figure 8.1 Some cation structures of ionic liquids.

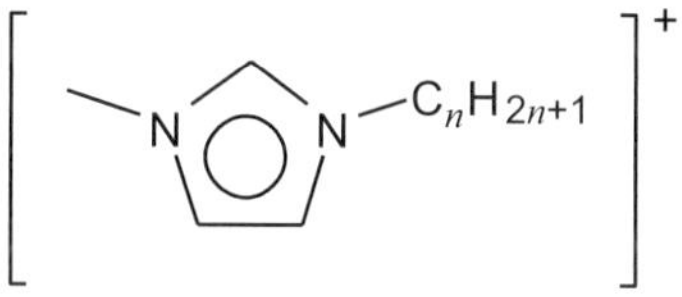

$[C_n\text{-mim}]^+$, $n = 0\text{-}18$

Figure 8.2 Generic structure of the cation RC_nH_{2n+1}–imidazolium.

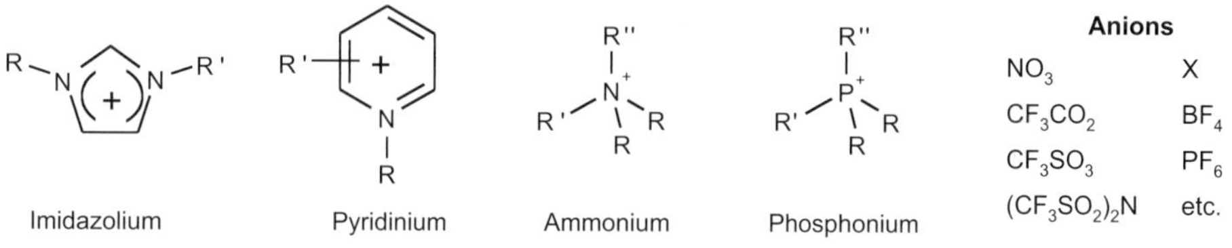

Figure 8.3 Examples of ionic liquid cations and some possible anions.

Figure 8.3 shows the cations together with some possible anions that may comprise an ionic liquid. Also to be noted is the presence of 4 "R" groups shown on the ammonium and phosphonium cations. The number of possible combinations of choices of base structure (e.g. imidazolium), R groups, and anions is therefore extremely large, with some authors quoting number of possible structures to be as high as 10^{18} and higher (Holbrey and Seddon, 1999).

Examples of ionic liquids that have been synthesized and characterized are shown in Figure 8.4, and include both cations and anions that have been successfully matched to produce a liquid.

With so many potential combinations there is significant potential to engineer the molecular structure of ionic liquids to achieve desired physical and chemical properties. These properties are listed by Krawczyk, Kamiński, and Petera (2012) among many others, and include viscosity, density, solubility (of gases, liquids, and solids), H-bond acceptor/donor properties, and coordination properties.

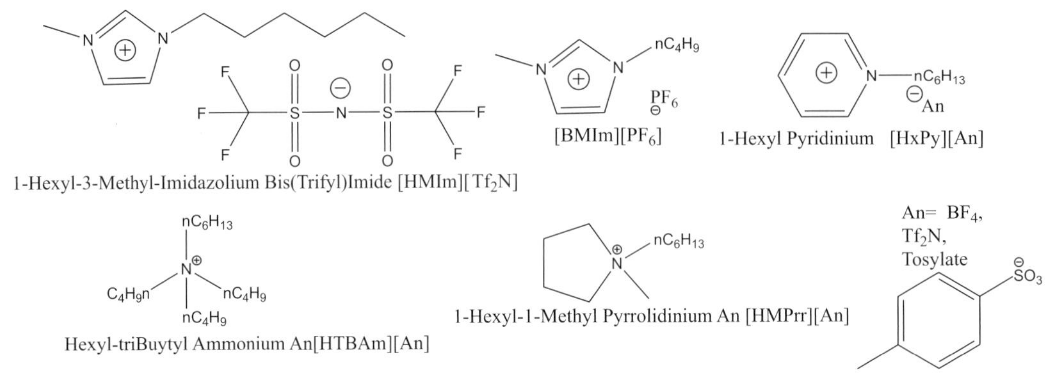

Figure 8.4 Further examples of possible ionic liquid structures.

Welton (1999) summarized the main physical properties of ionic liquids that have stimulated interest from both the chemistry community and the chemical engineering community. Ionic liquids that have been successfully synthesized and studied show some general advantageous properties. These include good solvating power for a wide range of both organic and inorganic compounds. Welton also highlighted that unusual combinations of reagents can be dissolved in the same phase. Their coordination properties are also highlighted in Welton's review, where it is stated that ionic liquids have the potential to be highly polar but at the same time behave as non-coordinating solvents. Perhaps the most important property in the context of liquid–liquid systems is that a number of ionic liquids are shown to be immiscible with a variety of organic solvents and thus provide a polar alternative to aqueous solutions for extraction and reactions in two-phase systems. Immiscibility with water and aqueous solutions is also possible using hydrophobic ionic liquids. On account of the lack of vapor pressure over a wide range of temperatures, ionic liquids are well suited for application in high-vacuum environments. This latter property also impacts flammability hazard; many ionic liquids show highly elevated flash points compared with equivalent organic solvents.

Generally, ionic liquids may be considered as potential replacements for many volatile organic compounds used in numerous organic reactions. These include reactions that involve the use of both homogeneous transition metal catalysts and stoichiometric reactions that use no catalyst. A review by Dupont, de Souza, and Suarez (2002) suggested that ionic liquid solvents are promising candidates for coupling reactions, halogenations, photochemical reactions, and numerous organic syntheses. These latter include classical organic reactions, such as Diels–Alder cycloaddition reactions, alkylations, hydrogenations, fluorinations, hydroformylations, and dimerizations. One of the main mechanisms responsible for the attractiveness of some ionic liquids as reaction media is based on the fact that reaction paths that involve charge-separated intermediates are accelerated on account of the lowering of the activation barrier in the presence of ionic liquids compared with conventional solvents.

Marsh (2006) briefly reviewed some of the potential drawbacks of ionic liquids, with chemical stability and cost of ionic liquids highlighted as potentially negative. For example, some ionic liquids based on aluminum fluorides are readily hydrolyzed by water and therefore may be applied only in a dry environment. Ionic liquids based on tetrafluoroborates and hexafluorophosphates are air and water stable but decompose slowly to HF, flagging significant safety and environmental concerns. On the other hand, anions based on $(CF_3SO_2)_2N^-$, $(C_2F_5SO_2)_2N^-$ (amides), and $(CF_3SO_2)_3C^-$ (methides) are stable. Other questions relate to toxicity, biodegradeability, and bioaccumulation.

8.2 Ionic Liquids and Intensification

In keeping with the main theme of the book, the focus in this section continues with the application of ionic liquids to intensification in liquid–liquid systems. The area of focus is on the intensification of classical organic chemical reactions and the application of the principles of green chemistry (Anastas and Warner, 1998). Application of ionic liquid technology to phase-transfer catalysis, and to separations involving liquid–liquid extraction is also considered. The potential role of ionic liquids for the exploitation of biocatalytic processes requires understanding of their potentially toxic effects on living biomass and on the activity of enzymes. A related issue is that of general toxicity and the potential environmental impacts that may be associated with industrial deployment of ionic liquid chemistry. The degradability of ionic liquids is an important part of environmental assessment that is considered later in the chapter. There are many other areas, in addition to liquid–liquid systems, where large and successful research efforts have resulted in significant breakthroughs in the application of ionic liquids, such as to gas–liquid systems, and in solid–liquid systems. Discussion of these will be included as they relate to liquid–liquid systems.

Firstly, we consider the potential role of ionic liquids in the important separation challenge of azeotrope breakage. There are two main approaches that have been adopted for the exploitation of ionic liquids for addressing the challenge of azeotrope breakage. The first is replacement of distillation by liquid–liquid extraction. The second is use of an ionic liquid as an azeotrope-breaking agent by addition during distillation to modify the thermodynamic properties (i.e. vapor pressures and relative volatilities) of the mixture in order to perform an effective extractive distillation. Separation of azeotropic mixtures by distillation is an important challenge in many industrial chemical processes. Ethanol/water separation is perhaps the best-known example and serves to illustrate the need for more environmentally friendly techniques. Breakage of the ethanol/water azeotrope by classical means has used highly undesirable reagents such as benzene or chloroform as extracting agents that are added to the mixture to modify the activity coefficients in favor of a clean separation of ethanol and water. This method of extractive distillation is still used to achieve azeotrope breakage on an industrial scale but requires significant energy input in order to recycle the extracting agent and to achieve a high degree of separation from the final products. The toxic nature of the extracting agents further adds to the technological requirements for containment, control, and environmental monitoring.

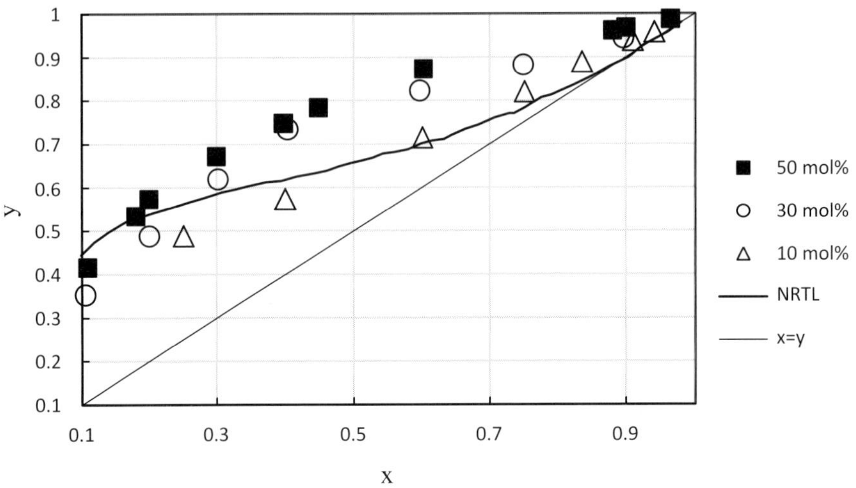

Figure 8.5 Influence of [emim] [BF_4] on ethanol/water VLE at 363.15 K. Data from Arlt (2002) and Marsh (2006)

The concept of using an ionic liquid as an effective extracting agent has been considered and demonstrated in a number of published research papers over the last 20 years. For example, the azeotrope ethanol/heptane has been analyzed thermodynamically (Domańska and Marciniak, 2008; Foco et al., 2006). Studies of activity coefficient data at infinite dilution for methyl imidazolium-based ionic liquids suggested low activity coefficient behavior with respect to ethanol and high activity coefficient values for heptane. One of the preferred anions is methyl sulfate and this was based not only on favorable activity coefficient data, but also on account of the ease with which these alkyl sulfate imidazolium derivatives may be synthesized economically and also avoiding the use of chlorine (Holbrey et al., 2002).

Figure 8.5 shows calculated and experimental volatility data for mixtures of ethanol and water in the presence of various concentrations of ethyl methyl imidazolium tetrafluoroborate (Jork et al., 2004; Marsh, 2006), plotted as fraction of ethanol in the vapor phase versus fraction of ethanol in the liquid phase. These data show comparison of experimental data with the nonrandom two-liquid model (NRTL) prediction for the pure ethanol/water system and clearly shows that the addition of the [emim] [BF_4] has a major effect on the equilibria, taking the mixture out of the azeotrope region.

These data suggest that azeotrope breakage by modification of the vapor liquid equilibrium (VLE) properties through the use of ionic liquids as an extracting agent in place of more hazardous organic solvents is feasible. Other separation techniques may also be considered for the separation of azeotropic mixtures, such as pervaporation and reverse osmosis.

Krawczyk et al. (2012) studied the separation of an azeotropic mixture of ethanol and heptane by liquid–liquid extraction and showed that good separation was achievable using butyl methyl imidazolium methyl sulfate [bmim] [$MeSO_4$]. This provided further confirmation of the work of Pereiro and Rodrıguez (2008) who demonstrated the suitability of [bmim] [$MeSO_4$] for the clean separation of heptane from ethanol.

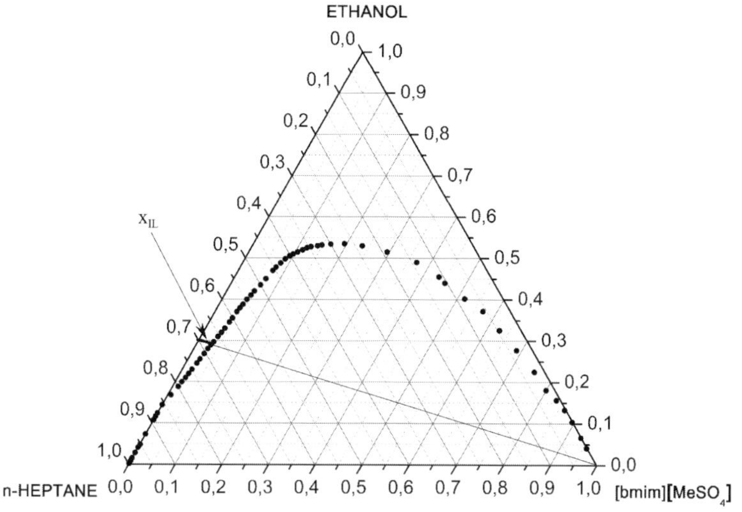

Figure 8.6 Equilibrium data [mass fraction] for ternary system of ethanol/heptane/[bmim] [$MeSO_4$]. Reprinted with permission from Kaminski, K., Krawczyk, M., Augustyniak, J., Weatherley, L. R., and Petera, J. (2014). Electrically induced liquid-liquid extraction from organic mixtures with the use of ionic liquids. *Chemical Engineering Journal*, **235**, 109–123. Copyright Elsevier B.V. All Rights Reserved

Measurements of the solute distribution ratio, selectivity, and extraction factor suggested this particular ionic liquid as a good candidate for this separation. The composition data shown on the phase diagram (Figure 8.6) confirm that [bmim] [$MeSO_4$] and heptane are practically immiscible and that ethanol prefers the ionic liquid phase (Kaminski et al., 2014). The calculated selectivity values were shown to be high enough to suggest the practical feasibility of [bmim] [$MeSO_4$] being used to break the heptane/ethanol azeotrope. Another important conclusion is that as the heptane and the ionic liquid are virtually totally immiscible, the heptane product stream is free of ionic liquid, thereby avoiding the need for purification for recovery of any ionic liquid.

Another significant separation challenge for the petrochemical industry is the separation of aromatics from alkanes. Aromatic hydrocarbons are a valuable commodity with an increasing incentive for their production. The separation challenge is on account of the similar molecular weights and the closeness of vapor pressure behavior, which means that conventional techniques for separation such as distillation are difficult. These mixtures also exhibit azeotropic behavior, further complicating the separation. Liquid–liquid extraction using specialist solvents such as sulfolane or using liquefied gases such as propane have remained as the principal techniques for the separation of alkanes and aromatics. Current liquid–liquid extraction techniques are limited to the feasible separation of mixtures containing less than 20% aromatics (Weissermel and Arpe, 2003). Typical ethylene crackers currently use feeds containing up to 25% aromatic hydrocarbons but would perform more efficiently if the aromatics were removed (Meindersma, Hansmeier, and de Haan, 2010).

Research into the use of ionic liquids for alkane/aromatic separation shows promising results and in thermodynamic terms may prove advantageous. Garcia et al. (2012)

summarize the challenges referred to above, but also demonstrate the effective use of ionic liquids for alkane/aromatic hydrocarbon separation.

A significant justification for deploying ionic liquids is based on the nonvolatility of ionic liquids and the ease of product recovery through simple flash distillation. Additional justification for exploring ionic liquids for these applications is based on the flexibility of their solvation properties. These are highly tunable in accordance with the molecular structure of the ionic liquid. Garcia et al. (2011) conducted a comprehensive experimental and theoretical evaluation of liquid–liquid equilibrium data for the system heptane/toluene in contact with four individual ionic liquids: [bpy] [Tf_2N], [2bmpy] [Tf_2N], [3bmpy] [Tf_2N], and [4bmpy] [Tf_2N]. Distribution coefficients and separation factors for behavior at 313 K and atmospheric pressure were compared with literature data for the equivalent "conventional" extraction system: toluene/heptane/sulfolane. In each of the four cases, the ionic liquids exhibited distribution coefficients that were significantly greater than those for the case of sulfolane. On the other hand, the separation factors were seen to be lower, especially at the lower toluene concentrations. The distribution coefficients and the separation factors for each of the ionic liquids were also compared. Minimal differences were observed when the [bpy] [Tf_2N] was compared with the [2bmpy] [Tf_2N] and when the [3bmpy] [Tf_2N] was compared with the [4bmpy] [Tf_2N]. In all cases the NRTL predictions were in line with the observed equilibrium data.

In a later paper, Garcia et al. (2012) observed that only a few ionic liquids studied so far exhibit both favorable extractive capacity and selectivity that is better than the conventional solvent sulfolane. In their reported research, Garcia et al. explored the possibility of improving the performance of ionic liquids relative to sulfolane, by mixing pairs of ionic liquids to investigate positive synergistic effects, such as are well known for a number of conventional solvents used in liquid–liquid extraction. The combination of [bpy] [BF_4] and [bpy] [Tf_2N] for toluene/heptane separation was explored on the basis that [bpy] [BF_4] alone exhibited high selectivity for toluene relative to sulfolane, and that [bpy] [Tf_2N] alone exhibited high extractive capacity relative to sulfolane. Combination of the two ionic liquids was expected to achieve higher values of both selectivity and extractive capacity relative to sulfolane. Another potential advantage of a mixed ionic liquid solvent in this case was the offsetting of the higher viscosity of [bpy] [BF_4] by combination with the lower viscosity [bpy] [Tf_2N]. Higher stability of the [bpy] [Tf_2N] was also seen as an advantage. The overall outcome of the study showed that mixed ionic liquids exhibited properties that were intermediate between those of the pure liquids. The distribution ratio with respect to toluene and the separation factor were higher than those of sulfolane for the mixture of [bpy] [BF_4] and [bpy] [Tf_2N] at a mole fraction of [bpy] [BF_4]of 0.7.

Other ionic liquid systems have been evaluated for aromatic/alkane separations. In a wider study, Mokhtarania, Musavib, and Parvini (2014) compared the performance of [bmim] [NO_3] (1-butyl-3-methylimidazolium nitrate) with that of [omim] [NO_3] (1-methyl-3-octylimidazolium nitrate) for the liquid–liquid extraction of hexane, heptane, and octane from mixtures with toluene. As a general finding, the selectivity values for each of these ternary systems in all cases were greater than unity, and was highest in the case of the [bmim] [NO_3]. The selectivity values increased with the chain length of the alkane. In each case, the observed data were consistent with that calculated using the

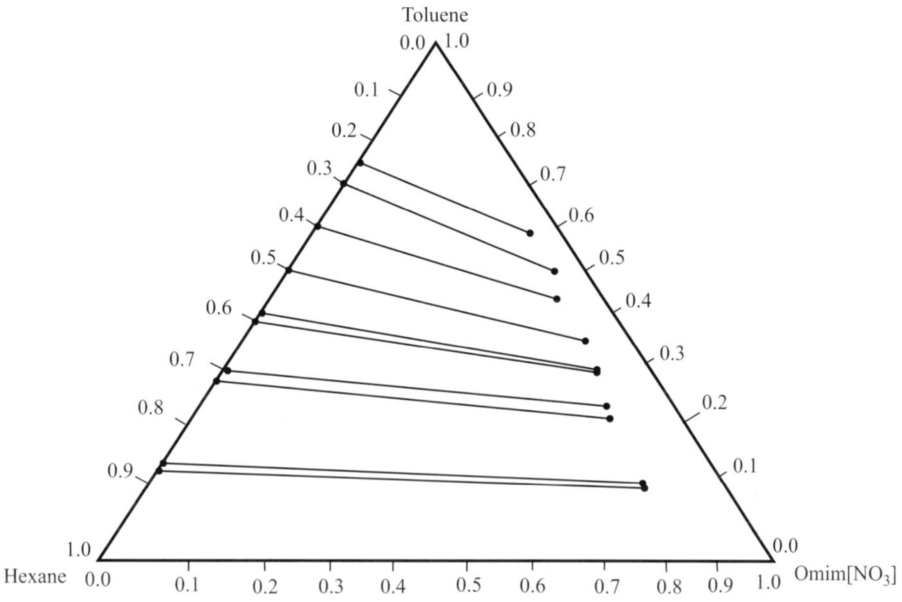

Figure 8.7 Experimental and calculated liquid–liquid equilibria of the ternary system [omim] [NO_3] and hexane/toluene at 298.15 K.
Reprinted (adapted) with permission from Mokhtarania, B., Musavib, J., and Parvini, M. (2014). Extraction of toluene from alkane using [Bmim] [NO3] or [Omim] [NO3]ionic liquid at 298.15 K and atmospheric pressure. *Fluid Phase Equilibria*, **363**, 41–47. Copyright (2014) Elsevier B.V. All Rights Reserved

NRTL model. It was concluded that both ionic liquid solvents would be viable candidates for separation of these alkanes from toluene. A sample of their data is shown in Figure 8.7 for the hexane/toluene [omim] [NO_3] system and illustrates the complete insolubility of the ionic liquid in one of the two liquid phases across a significant range of compositions.

In a later study, Larriba et al. (2016) presented data on the use of 1-allyl-3-methylimidazolium dicyanamide ([amim] [DCA]) and 1-benzyl-3-methylimidazolium dicyanamide ([bzmim] [DCA]) for the separation of aromatics from alkanes. These data show that unlike many other ionic liquids considered for this category of separation, [amim] [DCA] also has other properties, including viscosity and density, that are similar to those of sulfolane. This significantly raises the prospect of commercial applications. A sample of selectivity data by Larriba et al. (2016) is shown in Figure 8.8.

Each of the ionic liquid solvents show selectivity values for toluene relative to the n-heptane selectivities that are considerably higher than the values for sulfolane. In practical applications, this raises the prospect of significantly better product purity of the toluene extracted if these ionic liquids were to be deployed as extractants in place of sulfolane. The ionic liquid [amim] [DCA] shows the highest value of selectivity and this is attributed to the presence of the allyl group on the ionic liquid that further reduces the solubility of the heptane. The ionic liquid [bzmim] [DCA] showed toluene/n-heptane selectivities slightly lower than those of [amim] [DCA]. The values were shown to be similar to those observed for [emim] [DCA] (ethyl methyl immidazolium dicyanamide). In the case of [bmim] [DCA],

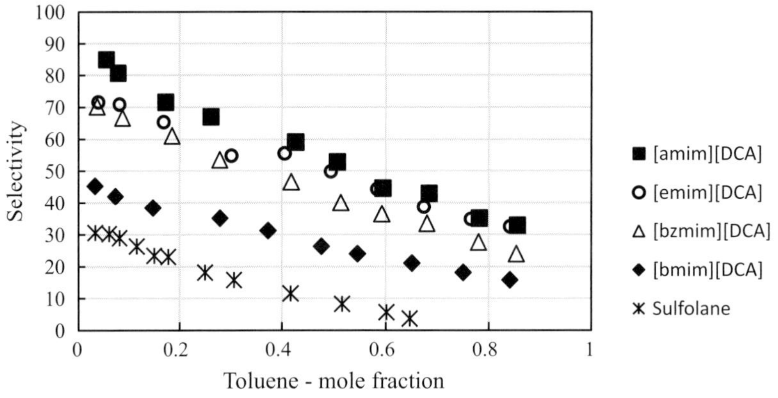

Figure 8.8 Toluene/n-heptane selectivities for the system n-heptane/toluene and ionic liquid solvent.

Reprinted (adapted) with permission from Larriba, M., Navarro, P., Gonzalez-Miquel, M., et al. (2016). Dicyanamide-based ionic liquids in the liquid-liquid extraction of aromatics from alkanes: Experimental evaluation and computational predictions. *Chemical Engineering Research and Design*, **109**, 561–572. Copyright (2016) Elsevier B.V. All Rights Reserved

comparison showed significantly lower values of selectivity, which may be explained by the higher solubility of n-heptane in [bmim] [DCA].

Other examples showing the potential role of ionic liquids for azeotrope breakage are listed by Marsh (2006), with the entrainer shown in parentheses: ethanol/water ([emim] [BF_4]); acetone/methanol ([emim] [BF_4]); acetic acid/water ([emim] [BF_4]); THF/water ([emim] [BF_4]) and ([bmim] [BF_4]); cyclohexane/benzene ([emim] [$(CF_3SO_2)_2N$]).

8.3 Ionic Liquids as Reaction Media

Ionic liquids have excellent potential as reaction media on account of their inherent catalytic properties. Effective homogeneous catalysis requires the ability to achieve high concentration of the catalyst and at the same time achieve high solubility of both gas and liquid substrates. The flexible properties of ionic liquids allow enhancement and optimization of the solubility of both homogeneous catalysts and reactants in the liquid. In this section, discussion of ionic liquids as reaction media is confined to those examples involving liquid–liquid systems. Of particular interest is the role of ionic liquids for the improvement of phase-transfer catalysis. General discussion of phase-transfer catalysis and its importance as part of process intensification is given in Chapter 9. However, a well-known example of phase-transfer catalysis in an industrial process made possible by the application of ionic liquid technology will be briefly described here. The example is of the now well-known Difasol process described in a review by Plechkova and Seddon (2008). The chemical reaction involved is the dimerization of butene to iso-octenes, as shown in Scheme 8.1 (Behr et al., 2015).

Octenes are an important group of organic intermediates used in the manufacture of PVC phthalate plasticizers. The original process developed was the Dimersol process.

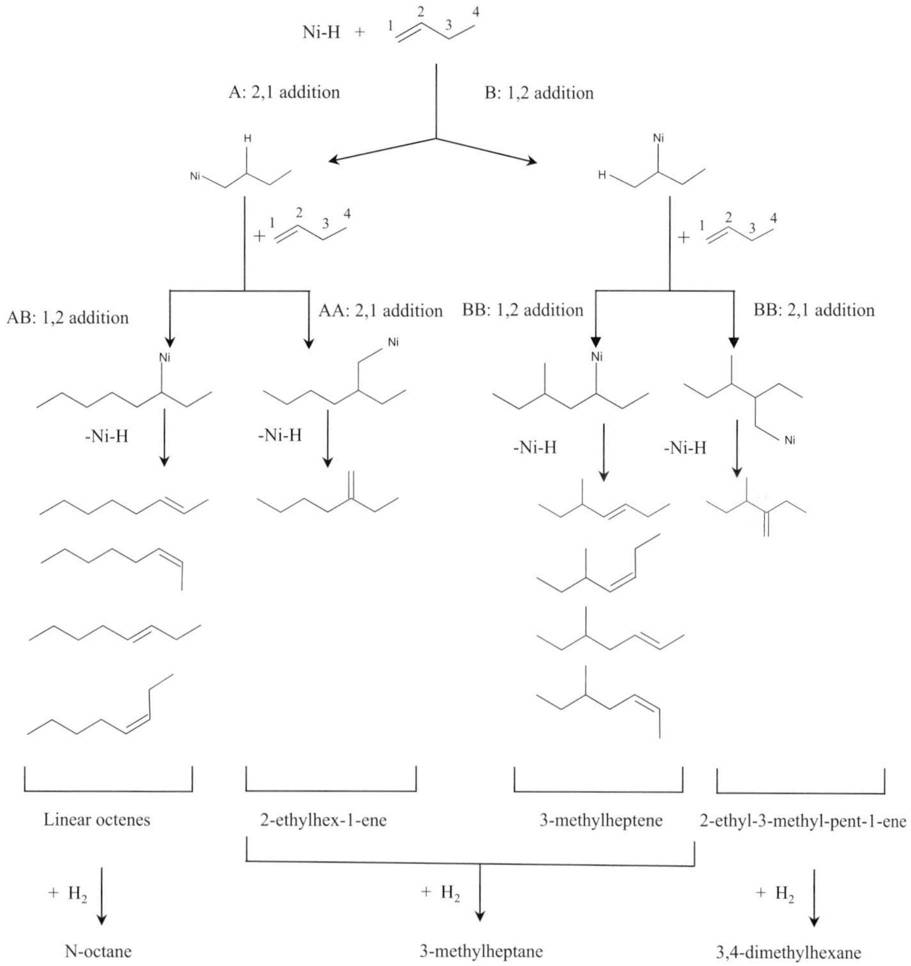

Scheme 8.1 Dimerization of butene to iso-octenes.
Reprinted (adapted) with permission from Behr, A., Bayrak, Z., Peitz, S., Stochniol G., and Maschmeyer, D. (2015). Oligomerization of 1-butene with a homogeneous catalyst system based on allylic nickel complexes. *RSC Advances*, **5**, 41372–41376. Copyright (2015) Royal Society of Chemistry – All Rights Reserved

The original Dimersol process was designed to perform dimerization of alkenes, for example propene (Dimersol-G) and butenes (Dimersol-X), and upgrade these into the more valuable branched hexenes and octenes. Plechkova and Seddon (2008) reported this is as an important industrial process, with 35 plants in operation worldwide at that time, with each plant producing between 20 000 and 90 000 tonnes per year of dimer, with a total annual production of 3 500 000 tonnes. The longer-chain alkenes produced in the dimerization process can be hydroformylated to alcohols such as isononanols that may be converted into dialkyl phthalates for use in the production of poly-vinyl chloride plasticizers.

The reaction chemistry implemented in the original Dimersol process used a cationic nickel complex homogeneous catalyst ($[PR_3NiCH_2R']$ $[AlCl_4]$), dissolved in either

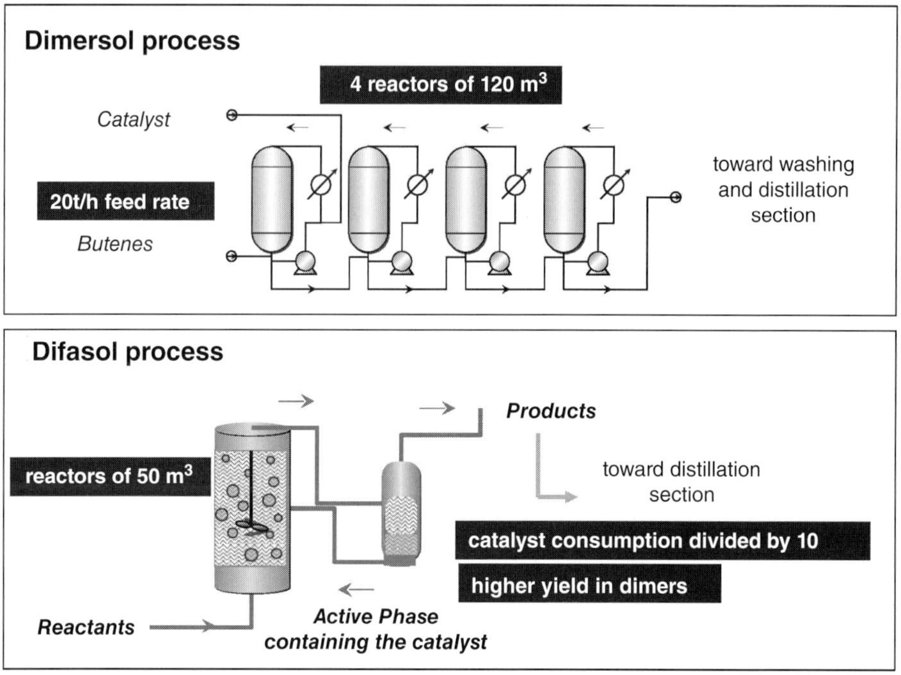

Figure 8.9 Schemes of the Dimersol and Difasol processes for the isomerization of butenes to iso-octenes.
Reprinted (adapted) with permission from Chauvin, Y. (2006). Olefin metathesis: the early days (Nobel Lecture). *Angewandte Chemie International Edition*, **45**, 3740–3765. Copyright (2006) John Wiley and Sons – All Rights Reserved

aromatic solvents or halogenated hydrocarbons (Chauvin et al., 1974). Both of the latter are classed as significant potential environmental hazards and have known carcinogenic effects. An alternative process based on ionic liquid chemistry is the IFP Difasol process that uses a chloroaluminate(III) ionic liquid as the catalyst in a two-phase liquid–liquid system. This chemistry was developed and pioneered at IFP (France). The reaction can be performed as a biphasic system in a temperature range of between −15 °C and +5 °C, since the products form a second layer that is easily recovered with the catalysts remaining in the ionic liquid phase. The activity of the catalyst proved to be significantly greater in both solvent-free and conventional solvent systems. Improved selectivity for desirable dimers was also noted. This process has been patented as the Difasol process and can be retro-fitted into existing Dimersol plants.

The two processes are compared as separate processes in Figure 8.9 (Chauvin, 2006). In the original Dimersol process based on 20 tonnes per hour of butene feed, four 120 m^3 stirred tank reactors operating in series were required to achieve the desired production rate of octene isomers. Not shown is the downstream separation facilities including distillation columns and washing equipment. In contrast, the Difasol process involves much smaller volumes and a single 50 m^3 reactor is required for the same production rate. This represents a reduction in equipment size and material inventory of nearly 90% based on comparison of the reactors alone.

Table 8.1 Comparison of Dimersol, Difersol, and combined Dimersol/Difasol processes for octane production from butenes. Data from Olivier-Bourbigou and Hugues (2002).

Operating conditions	Dimersol alone	Difasol alone	Combined
Butene conversion, wt%	70	65–70	80–85
Dimer selectivity, wt%	85	92–95	90–92
Octene yield, %	60	65–70	>75

In comparison with the homogeneous Dimersol process, the advantages of the biphasic Difasol system are listed by Seddon (2008):

- Improved use of the catalyst leading to reduced cost and need for catalyst disposal.
- Improved selectivity for the required dimer.
- Higher yield of dimers is achieved in a single step even with a low concentration alkene feed.
- Could be applied to the possible extension of the Dimersol process for higher less-reactive alkenes.
- Reactor size is much smaller than in the homogeneous system.

Olivier-Bourbigou and Hugues (2002) describe a combined process of Dimersol and Difersol that further adds to the efficiency through the use of ionic liquids. When the Difasol reactor, which involves the use of an ionic liquid, was added to an existing Dimersol reactor, the process became even more efficient. With reference to Figure 8.9, in the combined process, the effluent from the final Dimersol reactor is partially vaporized to separate unconverted C4 components from the octenes product. The vapor phase (i.e. the unconverted C4s) is then sent to the Difasol process for further production of octane products. The combined process achieved up to 25% increase in octane yield with reduced catalyst consumption and a reduction in the contamination of the ionic liquid (Table 8.1).

8.4 Toxicity

The toxicity of ionic liquids is a major issue if they are to be considered as green solvents and effective replacements for conventional volatile organic solvents. For a number of years ionic liquids have been touted as environmentally friendly but this assertion has come under challenge in recent times. The commonly accepted notion that ionic liquids have low toxicity has been shown in some cases to be incorrect (Swatloski, Holbrey, and Rogers, 2003).

Toxicity may be considered from two angles: (1) toxicity to the ecosystem (plants, animals, insects) and (2) toxicity to microorganisms relevant to biochemical engineering processes and biological products. Toxicity to the ecosystem may have broad ranging impacts on the acceptability of the deployment of ionic liquid chemistry at a commercial scale and has to be judged on equal terms with any other group of chemical products or reagents. Toxicity to microorganisms may also have significant wider

Table 8.2 LC_{50} data (mg L^{-1}) for a range of ionic liquids on *Daphnia magna* (Bernot et al., 2005).

Cation	R^1	R^2	Anion	LC_{50}
R^1,R^2-imidazolium	C_4H_9	CH_3	PF_6^-	19.9, 24
	C_4H_9	CH_3	Cl^-	14.8, 6.5
	C_4H_9	CH_3	Br^-	8.03
	C_4H_9	CH_3	BF_4^-	10.68
	$C_{12}H_{25}$	CH_3	Cl^-	0.0043
	$C_{16}H_{33}$	CH_3	Cl^-	0.0034
	$C_{18}H_{37}$	CH_3	Cl^-	0.0017
R^1,R^2-pyrodinium	C_4H_9	H	Cl^-	20
R^1,R^1,R^1,R^2-phosphonium	C_4H_9	C_2H_5	$(EtO)_2PO_2^-$	11
	C_6H_{13}	$C_{14}H_{29}$	Cl^-	0.072
R^1,R^1,R^1,R^2-ammonium	C_8H_{17}	CH_3	$[Tf_2N]^-$	0.2
	$\{C_2H_4O(C_2H_4O)_4Me\}_2C_{14}H_{29}$	CH_3	$MeSO_4^-$	1

environmental impacts, but also may impact the degree that biochemical processes such as fermentation can be improved by integration with processes such as separations that use ionic liquid solvents. In both cases, real experimental data are required if the impact of toxicological effects are to be properly characterized. This is especially challenging when attempting to quantify chronic (long-term) toxicity effects.

Under the first category relating to toxicity (impact on ecosystems), some of the early work on the toxicity of ionic liquids centered on release of ionic liquids into the aquatic environment. The acute and chronic toxicity of [bmim] [PF_6] and [bmim] [Cl] on *Daphnia magna* was studied by Bernot et al. (2005). The observations were used as a preliminary indicator for assessing acute toxicity of ionic liquids. LC_{50} data are listed in Table 8.2 for a range of ionic liquids, showing the concentration of ionic liquid resulting in 50% mortality.

The data in Table 8.2 show a strong trend relating shorter alkyl-substituted chain lengths with lower toxicity values and longer alkyl chains relating to higher toxicity.

Other research with algae confirm a similar toxicity trend for many ionic liquids. For example, Cho et al. (2007, 2008) used *Pseudokirchneriella subcapitata* to study the effect of different head groups, side chains, and anions of a range of ionic liquids on algal growth rate and photosynthetic activity. It was shown that the toxic influence on growth rate was more significant than the toxic influence on photosynthesis. Another factor brought out in these studies was the potential hydrolytic degradation of perfluorate anions (used in some ionic liquids) into fluoride ion, F^-, thereby increasing the toxicity of the ionic liquid. It was also shown for this microalgae species that the toxicity was sensitive to the anion species, giving a ranking as follows: $[SbF_6]^- > [PF_6]^- > [BF_4]^- > [CF_3SO_3]^- > [C_8H_{17}OSO_3]^- > [Br]^- \sim [Cl]^-$.

Understanding the toxicity of ionic liquids to microorganisms is important since the potential of using ionic liquids as host media for biotransformation and biocatalysis is significant. Application of ionic liquids to the wider area of extractive fermentations also raises important questions regarding toxicity if recycling of live and active biomass or enzymes is to be achieved. Solvent biocompatibility is an essential requirement for successful biphasic biocatalysis. There are numerous studies on the toxicology of ionic liquids with regard to human cells and animal cells, in addition to studies on aquatic organisms and algae as already discussed (Landry et al., 2005; Sipes, Knudsen, and Kuester, 2008; Stepnowski et al., 2004; Wang et al., 2007).

A review of recent literature suggests that little toxicity data exist for hydrophobic ionic liquids that are candidates for in situ extraction from fermentation media involving microorganisms of commercial interest. Gangu (2013) reviewed toxicity data for microbial systems involving ionic liquid reactions and separations involving target molecules of commercial potential, such as isomers of naphthalene dihydrodiol.

Another example is the production of lactic acid by bacterial fermentation (Matsumoto, Mochiduki, and Kondo, 2004). The compatibility of nine different lactic acid-producing bacteria with imidazolium-based ionic liquids for online extraction was assessed. Typically, the production of lactic acid and cell count with time are the markers used to determine possible toxic effects of ([bmim] [PF_6], [hmim] [PF_6], and [omim] [PF_6]). Comparison was made with a number of conventional organic solvents. The measure of toxicity in this study was in terms of the ratio of lactic acid produced in the presence of the ionic liquid relative to a control in aqueous medium only. The smallest effect was found with the bacterium *Lactobacillus delbruekii* in the presence of [hmim] [PF_6], where a relative activity value of 0.94 was recorded. Other data for n-butyl-, n-hexyl-, and n-octyl hmim liquids showed overall values in the range 0.04–0.95. It is notable that in the case of *Lactobacillus delbruekii* the recorded toxic effects were not seen to be a strong function of chain length. This is in contrast to other data, for example those shown in Table 8.2 for *Daphnia magna*, where a very strong effect of chain length on toxic effect was observed.

Other studies by Baumann, Daugulis, and Jessop (2005) investigated ionic liquid solvent applications for xenobiotic degradation in a two-phase partitioning bioreactor. This type of bioreactor may be useful for the breakdown of biologically recalcitrant organic molecules. Generally it requires large volume bioreactors involving substrates that are both water-insoluble and cytotoxic. Phenol degradation by three separate bacteria, *Achromobacter xylosoxidans* Y234, *Pseudomonas putida* ATCC 1172, and *Sphingomonas aromaticivorans* B0695, was determined in the presence of several ionic liquids with favorable partitioning properties. Measurements of cell viability and metabolic rate showed that $P_{6,6,6,14}$] [Tf_2N] was found to be biocompatible with all three bacteria, while ionic liquids [$P_{6,6,6,14}$] [$C_{10}H_{19}O_2$] and [$P_{6,6,6,14}$] [C_2N_3] were biocompatible with only *A. xylosoxidans* and *P. putida*. In the cases of [$P_{6,6,6,14}$] [Cl], [$P_{2,4,4,4}$] [$C_4H_{11}O_4P$], and [$P_{1,4,4,4}$] [TOS], and with all the bacteria, zero cell growth was observed.

There is also a significant body of literature on the toxicity effect of a range of ionic liquids on the bacterium *Escherichia coli*. Ganske and Bornscheuer (2006) investigated the growth of *E. coli* in buffer solution in the presence of several ionic liquids, including [bmim] [BF_4] and [bmim] [PF_6]. The toxic responses were compared with those obtained with several conventional organic solvents, including methanol, ethanol, and dimethyl sulfoxide. Experiments were conducted at concentrations between 0.1 and 2.5 wt% and showed that growth inhibition of the *E. coli* occurred at concentrations in the range 0.7–1 wt%. At concentrations greater than 4 wt% zero growth was observed. Overall it was concluded that each of these ionic liquids showed greater toxic effect on *E.coli* compared with any of the conventional solvents evaluated.

An extensive study of toxic effects of ionic liquids on *Escherichia coli* is also presented by Lee et al. (2005). This study evaluated some 13 ionic liquids, comparing their toxicity to *E.coli.* with that of eight conventional organic solvents. Overall, there was a significant difference noted between hydrophilic ionic liquids such as [emim] [BF_4] and hydrophobic liquids such as [bmim] [PF_6], with hydrophilic species generally exhibiting greater toxicity in tests on solid cultures. In the case of suspended cultures, the opposite effect was observed and attributed to the availability of higher transport rates in the case of suspended media. Alkyl chain length was also seen as a significant factor in determining toxicity. This was illustrated in the case of [emim] [BF_4] with an EC_{50} value of 35 000 mg L^{-1} compared with 9000 mg L^{-1} in the case of[bmim] [BF_4]. This latter value was reduced to 4000 mg L^{-1} in the case of the more hydrophobic PF_6 anion. In the case of the [Tf_2N] anion, the EC_{50} value reduced further to 150 mg L^{-1}.

Comparison with the toxicity of conventional organic solvents revealed significant variations. Data for acetonitrile, ethanol, acetone, and DMSO showed significantly lower toxicities compared with most or all of the ionic liquids. On the other hand, solvents, including chloroform, chlorohexane, and 1-chlorophenol, were shown to be more toxic to *E.coli.* than most of the ionic liquids studied.

Detailed evaluation of the mechanisms involved in the toxic effect of ionic liquids on *E.coli* focuses on membrane disruption. Cornmell et al. (2008) applied Fourier transform infrared spectroscopy (FT-IR) to the determination of intracellular uptake of ionic liquids and found this to be a useful technique that exploits the distinctive differences in IR spectra shown by ionic liquids compared with cell components. Application of FT-IR provided an effective means to identify for the presence of ionic liquids, primarily in the lipid membranes, attributable to membrane disruption. Ionic liquids showing greater toxicity were found to accumulate in the lipid membranes faster compared to liquids exhibiting lower degrees of toxicity.

There are numerous other similar studies showing the variable toxic effects of ionic liquids on a range of different bacteria and microorganisms. These include *Bacillus cereus* and the yeast *Pichia pastoris* (Ganske and Bornscheuer, 2006); *Pseudomonas. fluorescens*, *Bacillus subtilis*, *Staphylococcus. aureus*, and *Saccharomyces cerevisiae* (Docherty and Kulpa, 2005); *Clostridium butyricum* DSM 10703 (Rebros et al., 2009).

Notably, *S. cerevisiae* was least affected by the presence of ionic liquids (Ganske and Bornscheuer, 2006). A relatively unexplored area is identify techniques for enhancing biocompatibility and immobilization is an area that perhaps has future potential as a means of addressing solvent toxicity issues related to ionic liquids. One study by Yang et al. (2009) investigated the tolerance of yeast cells immobilized in calcium alginate beads to ionic liquids of the category[R-mim] [PF_6]. As observed in other work referred to previously in regards to the effect of chain length, activity retention in the immobilized yeast decreased significantly at long chain lengths for the cation. The activity retention was even lower in the case of Br^- anions and this may be attributed to the dehydrating influence of the water-miscible bromide ions. Comparison of activity retention of immobilized biomass relative to suspended free cells showed that retention was significantly higher in the case of immobilized cells. This was attributed to the protection of immobilized cells by hydrophilic calcium alginate beads from the action of hydrophobic ionic liquids. Water saturation of the ionic liquid phase prior to contact with the biomass was also observed to have a positive influence on activity retention. Another strategy was reported by Coleman et al. (2012) with the synthesis of chiral imidazolium ionic liquids incorporating amino acid ester and dipeptidyl functional groups. These liquids were reported as having significantly higher biocompatibility than many other liquids studied, with minimum inhibitory concentrations of greater than 2 mM.

In common with other research on ionic liquids, it is clear that further investigations are required to provide a clearer understanding of the relationship between chemical structure of the ionic liquid and its toxicity. Key factors include the chain length of attached functional groups, the stability of the ionic liquid, and the target biomass, be this microorganisms or higher orders of living cells. There is also a degree of uncertainty regarding the toxic response of different strains of the same bacterium that may show opposite responses to exposure to the same liquid.

Toxicity effects of ionic liquids on higher organisms are also the subject of significant research, though this is still at a relatively undeveloped stage. An early example was by Rogers and Seddon (2003) who presented an investigation of the toxicity of $[C_4mim]Cl$ and $[C_8mim]Cl$ to *Caenorhabditis elegans*, which is a well-understood free-living soil roundworm. It was found that $[C_4mim]Cl$ and $[C_8mim]Cl$ showed little evidence of acute toxicity to *Caenorhabditis elegans*. On the other hand $[C_{14}mim]Cl$, with a higher alkyl chain length, exhibited some degree of toxicity. Other studies have been conducted on zebrafish (Prett et al., 2006) with imidazolium-, pyridinium-, and pyrodinium-based ionic liquids (using various anions that are not highly lethal), with LC_{50} values higher than 100 mg L^{-1}. Another study, this time using rats, was published by Landry et al. (2005). They used 1-butyl-3-methylimidazolium chloride, which showed little negative impact on body mass, activity, and general health when used at concentrations of 175 mg kg^{-1}. LC_{50} values for mortality were recorded at exposure levels above 550 mg kg^{-1}.

Among the more recent studies is one from Oliveira et al. (2016), which presents an evaluation of the toxicity of four protic ionic liquids against various microorganisms and also studies their biodegradability. The liquids included *N*-methyl-2-hydroxyethylammonium acetate (m-2-HEAA), *N*-methyl-2-hydroxyethylammonium propionate (m-2-HEAPr), *N*-methyl-2-hydroxyethylammonium butyrate (m-2-HEAB), and *N*-methyl-2-hydroxyethylammonium pentanoate (m-2-HEAP). Their antimicrobial activity was determined against two bacteria, *Staphlylococcus aureus ATCC-6533* and *Escherichia coli CCT-0355*, the yeast *Candida albicans* ATCC-76645, and the fungi *Fusarium sp. LM03*. The results are generally consistent with other earlier data, showing a longer alkyl chain tending to increase the negative impact of the liquids in terms of toxicity. The findings showed m-2-HEAA to be the least toxic and m-2-HEAP proved to be the most toxic. All of the liquids that were evaluated demonstrated low biodegradability, which is studied next.

8.5 Degradability

One advantage of many ionic liquids is chemical stability. This is an important feature, since the cost of ionic liquids dictates that recycling is necessary if economic application is to be attained. Therefore, effective ionic liquids should be chemically robust. However, this may be inconsistent with a desire for easy chemical or biological breakdown as part of final disposal. Studies of the degradation of ionic liquids has focused on chemical breakdown and on biological breakdown. Pham, Chul-Woong Cho, and Yun (2010) reviewed in some depth many of the factors and potential mechanisms involved in both the chemical and biological breakdown of several ionic liquids. Pham et al. list several groups who have focused on the oxidative and thermal degradation of ionic liquids in aqueous media (Awad et al., 2004; Baranyai et al., 2004; Berthon et al., 2006; Itakura et al., 2009; Li et al., 2007; Morawski et al., 2005; Siedlecka et al., 2008a, 2008b, 2009; Stepnowski and Zaleska, 2005).

Within this group, Stepnowski and Zaleska (2005) and Morawski et al. (2005), in particular, showed that a combination of UV light and a photocatalyst such as titanium dioxide showed significant effectiveness in the oxidation of imidazolium-based ionic liquids. Slightly later work by Li et al. (2007) demonstrated effective oxidation of 1,3-dialkylimidazolium ionic liquids in the presence of mixtures of hydrogen peroxide and acetic acid. Results from others in the above-listed papers showed a role for Fenton-type peroxidations for the breakdown of some ionic liquids in aqueous media (Siedlecka and Stepnowski, 2009; Siedlecka et al., 2008a, 2009b). In many cases, structure dependence is promoted as a major feature of many chemical oxidations of ionic liquids, with alkyl chain length showing significant influence. On the other hand, the type of anion seemed to have less influence. Pham et al. note some apparent contradictions in the effect of alkyl chain length, suggesting the necessity for further research.

Pham et al. (2010) also makes an important distinction between chemical degradation of ionic liquids and biodegradation. The former requires oxidizing agents and possibly catalysts, together with external energy input, whereas biodegradation involves the microbial breakdown of the chemical compound, often through a large number of intermediate steps. Some progress has been made in designing ionic liquids that have biodegradable side chains (Gathergood and Scammells, 2002; Gatherwood, Garcia, and Scammells, 2004;, Gatherwood, Scammells, and Garcia, 2006). The early approach looked at side chains containing sites that could be candidates for enzymatic hydrolysis, side chains containing oxygen, and side chains containing phenyl rings that would be vulnerable to attack by oxygenases.

In the context of biodegradability, Perica et al. (2013) made an important distinction between protic ionic liquids and aprotic ionic liquids. The structure of aprotic ionic liquids is based on the more traditional ionic liquid cations, which would include the ubiquitous imidazolium and pyridinium cations and those having longer alkyl chain substitutions, in combination with anions such as chloride, bromide, tetrafluoroborate, and hexafluorophosphate. The protic group of ionic liquids, on the other hand, is based on a different structure, such as cations using substituted amines (monoethanolamine,

Table 8.3 Percentage biodegradation after 5 days, 14 days, and 28 days (Perica et al., 2013).

	% Biodegradation		
Ionic liquid	5 days	14 days	28 days
2-HEAF	11	61	86
2-HEAB	64	69	95
2-HDEAF	4	4	13
2-HDEAA	21	46	69
2-HDEAPr	45	62	68
2-HDEAB	47	66	78
2-HDEAiB	46	71	79
2-HDEAPe	38	57	69
2-HTEAB	28	32	59
2-HTEAPe	23	38	57
[bmim] [Cl]	0.23	0.66	1.17
[omim] [Cl]	0.29	0.35	1.33
[BPy] [Cl]	0.31	0.46	0.61

Table 8.4 Chemical oxygen demand (COD) and biological oxygen demand (BOD) for the ionic liquids methyl-2-hydroxyethylammonium acetate (m-2-HEAA), *N*-methyl-2-hydroxyethylammonium propionate (m-2-HEAPr), *N*-methyl-2-hydroxyethylammonium butyrate (m-2-HEAB), and *N*-methyl-2-hydroxyethylammonium pentanoate (m-2-HEAP). Data from Oliveira et al. (2016).

Ionic liquid	COD, mg L^{-1}	BOD, mg L^{-1}	COD/BOD
m-2-HEAA	28760	271.10	106.10
m-2-HEAPr	23280	333.00	69.90
m-2-HEAB	20540	322.20	63.70
m-2-HEAP	15060	255.00	59.00

MEA, diethanolamine, DEA, or triethanolamine, TEA) and anions based on organic acids such as formate, propionate, butyrate, etc. Structures based on these smaller, lower molecular weight groups implies perhaps easier degradation and possibly lower long-term environmental impact (Cota et al., 2007).

Table 8.3 (Perica et al., 2013) shows comparison of degradation of both protic and aprotic ionic liquids measured at 5 days, 14 days, and 28 days. The biodegradation data were determined as detailed by Perica et al. using a standard method based on oxygen consumption at 22 °C for the ionic liquid in an aqueous mineral nutrient solution at an initial concentration of 100 mg L^{-1}. The biodegradation was initiated by inoculation with a mixed culture of microorganisms from an urban biological treatment plant.

Data in Table 8.3 show a sharp distinction between the degradation values observed in the cases of protic ionic liquids (2-HEAF to 2-HTEAPe) in the table, and the values for the aprotic liquids ([bmim] [Cl], [omim] [Cl], and [Bpy] [Cl]) at the end of the table.

Biodegradability was partially confirmed in more recent work on protic ionic liquids by Oliveira et al. (2016) (Table 8.4). Of particular note here is that a comparison of chemical oxygen demand (COD) and biological oxygen demand (BOD) values are presented for the four protic ionic liquids m-2-HEAA, m-2-HEAPr, m-2-HEAB, and m-2-HEAP. In spite of significant evidence of biodegradation, the dominant mechanism for breakdown is assessed in terms of chemical oxidation.

Another aspect of the ecotoxicity of ionic liquids is their influence on microbes that are relied upon for the biodegradation of other xenobiotic compounds present in effluent streams. Wells and Coombe (2006) include consideration of this in their toxicity and biodegradation study of 10 important ionic liquids (Table 8.5). Part of the study focused on the inhibition of microbial activity for biological oxidation of glucose and inhibition of glutamate. The results shown in Table 8.5 show that at low levels of exposure significant reduction on oxidative degradation is in evidence. The authors suggest that the ecotoxicity may not be caused by action on particular species-specific receptors but by a more general mechanism. The ionic liquids exhibiting the highest levels of toxicity in this situation were as follows: imidazolium $C_{12}H_{25}CH_3Cl^-$, imidazolium $C_{16}H_{33}CH_3Cl^-$, imidazolium $C_{18}H_{37}CH_3Cl^-$, phosphonium $C_6H_{13}C_{14}H_{29}Cl^-$, and ammonium $\{C_2H_4O(C_2H_4O)_4Me\}_2C_{14}H_{29}$ CH_3MeSO_4.

The toxicity correlation was observed in microbial tests as biochemical oxygen demand evaluations could not be successfully conducted on these liquids due to the significant bacterial inhibition at concentrations of less than 10 mg L^{-1}. These observations by Wells and Coombe are consistent with those of others such as Stock et al. (2004), based on acetylcholinesterase enzyme assays.

8.6 Role of Ionic Liquids in Biocatalysis

Whole cell biocatalysis and enzymatic biotransformations are increasingly important categories of chemical production methods and are highly significant for the pharmaceutical and fine chemical industries in particular. An additional dimension to their importance is the on-going interest in the use of renewable feedstocks from biomass and from natural oils and fats for the production of highly functionalized molecules. A report by the US Department of Energy (Werpy and Peterson, 2004) suggested that the majority of chemical products in the future that are derived from renewable biomass will be manufactured using some form of biocatalytic reaction engineering. Another key dimension is the capability provided by biocatalysis for high-yielding chiral syntheses. In addition, environmental advantages may accrue from the use of biocatalytic reaction engineering that operates at close to ambient temperature and pressure. Another significant rationale behind the interest in biotransformations is the wide choice of microorganisms, including genetically modified versions, which is available. Biocatalysis can be divided into two categories: (1) whole-cell biocatalysis in which the enzyme catalyzes the reaction inside the whole cell and (2) enzymatic catalysis involving pure enzymes that have been extracted and purified, typically from microorganisms, from plant or mammalian cells, or from marine organisms.

There are several aspects that require analysis when considering the potential role of ionic liquids in either whole-cell biocatalysis or enzymatic biocatalysis. Several

Table 8.5 Effect of ionic liquid exposure on glucose/glutamate inhibition. Data from Wells and Coombe (2006).

Ionic liquid	R^1	R^2	Anion	Mol. wt. cation	100 mg L^{-1}	10 mg L^{-1}	1 mg L^{-1}
imidazolium	C_4H_9	CH_3	PF_6^-	139	9	8	28
imidazolium	C_4H_9	CH_3	Cl^-	139	0	18	15
imidazolium	$C_{12}H_{25}$	CH_3	Cl^-	251	97	59	3
imidazolium	$C_{16}H_{33}$	CH_3	Cl^-	307	100	100	16
imidazolium	$C_{18}H_{37}$	CH_3	Cl^-	335	100	100	100
pyridinium	C_4H_9	–	Cl^-	136	4	13	21
phosphonium	C_4H_9	C_2H_5	$(EtO)_2PO_2^-$	231	19	15	16
phosphonium	C_6H_{13}	$C_{14}H_{29}$	Cl^-	483	100	100	78
ammonium	C_8H_{17}	CH_3	$N(SO_2CF_3)_2$	368	23	37	26
ammonium	$\{C_2H_4O(C_2H_4O)_4Me\}_2C_{14}H_{29}$	CH_3	$MeSO_4$	696	100	100	47

phenomena can have major impacts on the feasibility of using ionic liquids in biological processes. These include, substrate inhibition, product inhibition, toxicity to biomass, enzyme compatibility, mass transfer, and partition behavior. The first two of these phenomena, substrate inhibition and product inhibition, can provide significant impediments to efficient reaction kinetics and to high product yields. In simple terms, live biomass and enzymes can significantly reduce their activity in the presence of high concentrations of either the substrate or the final products. The means of overcoming the problem is as follows. In the case of product inhibition, the continuous separation of the product during the reaction is required and this must be performed in a manner that does not deactivate or kill the biomass. This technique is often referred to as extractive fermentation or extractive biotransformation. In the case of substrate inhibition, the challenge is to provide a means of releasing substrate to the biomass or the enzyme in a controlled manner that optimizes the concentration according to the required reaction kinetics.

Both substrate inhibition and product inhibition are suited to the use of an immiscible solvent that is in contact with the aqueous reaction mixture. Figure 8.10 shows generic flow schemes for the two cases of (a) in situ product removal and (b) in situ substrate dosing, both based on a liquid–liquid system. In both cases the reactor, shown here in batch configuration, contains biomass or formulated enzyme in aqueous solution containing nutrients.

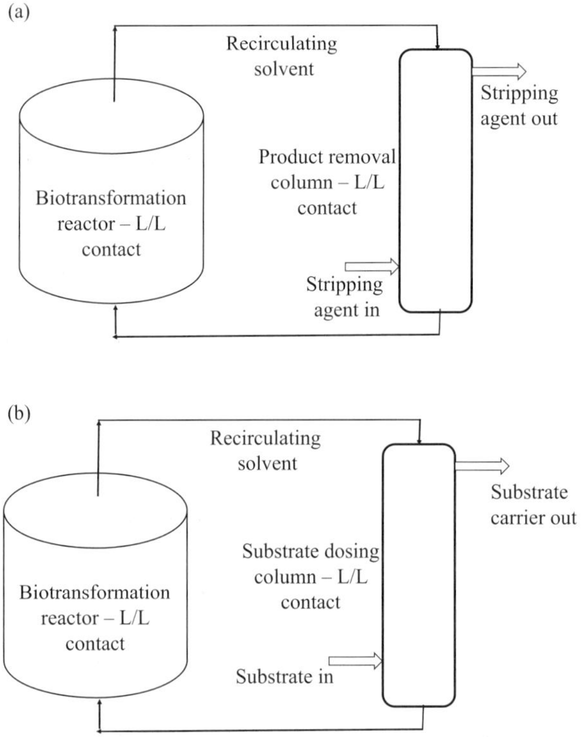

Figure 8.10 Principles of (a) in situ product removal and (b) in situ substrate dosing.

In the case of Figure 8.10(a), the reactor is also loaded initially with substrate. As this is converted to product, the solvent phase is fed into the reactor and continuously extracts the product, and thus controlling the product concentration. The loaded solvent is separated (not shown) and fed to a stripping unit where the product is removed from the extracting solvent, and the extracting solvent is then recycled back to the reactor for further product pick-up.

In the case of Figure 8.10(b), the substrate is introduced into the organic solvent, either as a direct feed or, as shown in the figure, by extraction from the dosing column. The solvent, now loaded with substrate, is fed to the reactor where the substrate is released by partitioning into the aqueous reaction mixture. The release of substrate into the reactor is controllable by means of the feed concentration in the solvent stream, the circulation rate of the solvent, liquid–liquid mass transfer conditions in the reactor, and the partitioning properties of the system with respect to the substrate.

In both scenarios a critical requirement is biocompatibility between the recirculating solvent and the biomass/enzyme present in the reactor. Toxicity data are therefore essential in order to determine if a solvent that shows favorable partition properties is not going to kill or deactivate the biocatalysts. Ionic liquids, on account of the huge range of possible structures that they offer, are an attractive prospect to meet toxicity criteria and other criteria.

There are numerous recent examples of organic syntheses where the reaction is improved when the biocatalyst is hosted in a biphasic system involving an ionic liquid. Gangu (2013) provided a detailed review and examples, some of which are now summarized.

The asymmetric reduction of organic ketones forms a significant group of reactions where significant progress in the successful application of ionic liquid chemistry has been made. Brautigam and Bringer-Meyerb (2007) presented a detailed study of biphasic whole-cell biocatalysis involving the asymmetric reduction of (a) ethyl 4-chloroacetoacetate, (b) α-phenacyl chloride, and (c) 1-(4-chlorophenyl)ethanone (4-chloroacetophenone) to the corresponding enantiomeric alcohols (Scheme 8.2). The following water-immiscible ionic liquids [bmim] [PF_6], [hmim] [PF_6], [bmim] [Tf_2N], [hmim] [Tf_2N], [bmpi] [Tf_2N], [hmpi] [Tf_2N], [bmpi] [E3FAP], [hmim] [E3FAP], and [EWTMG] [E3FAP] were evaluated.

In this example the hexafluorophosphate (PF_6) liquids showed reasonably good biocompatibility with the *E. coli*, with 70% cell viability reported. On the other hand, the liquids with the [E_3FAP] anion were very poor, showing less than 5% cell viability. A further remarkable result of the Brautigam and Bringer-Meyerb (2007) study was the large increase in yields in the case of reaction shown inScheme 8.2(a) when the reaction was conducted in an ionic liquid medium rather than an aqueous medium. Yields greater than 95% are reported for the reaction conducted in [BMPl] [Tf_2N] and [HMPl] [Tf_2N] ionic liquids compared with less than 8% in aqueous solution. The higher yields were attributed to higher partition coefficients for the products, thereby reducing the impact of product inhibition.

It should be emphasized that there are wide differences in the effects of ionic liquids on biocatalytic reactions and there are many reports of negative effects on

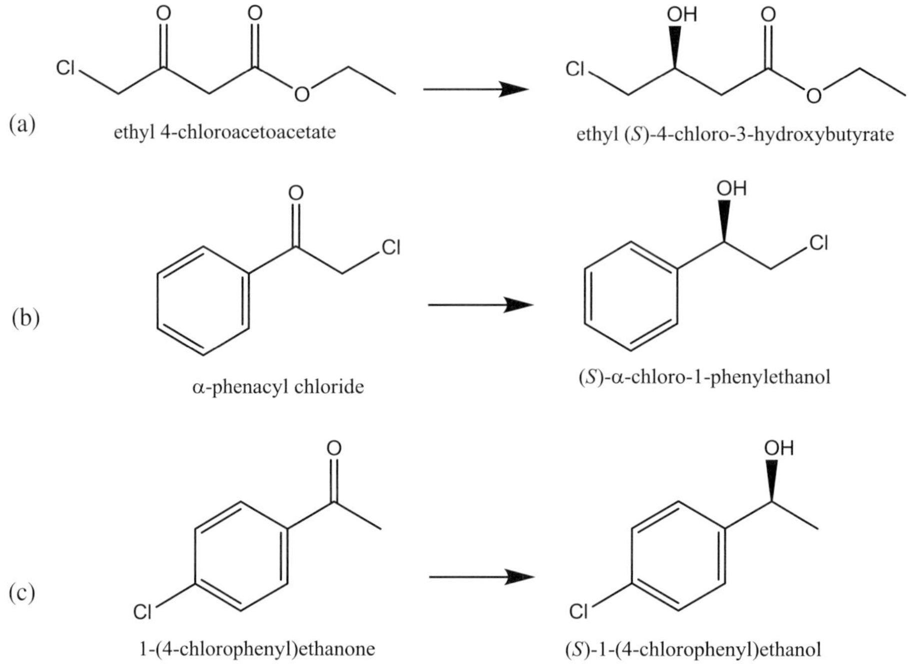

Scheme 8.2 Asymmetric reduction of (a) ethyl 4-chloroacetate, (b) α-phenacyl chloride, and (c) 1-(4-chlorophenyl)ethanone by whole cell biocatalysis in the presence of *Escherichia coli* (Brautigam and Bringer-Meyer (2007)).

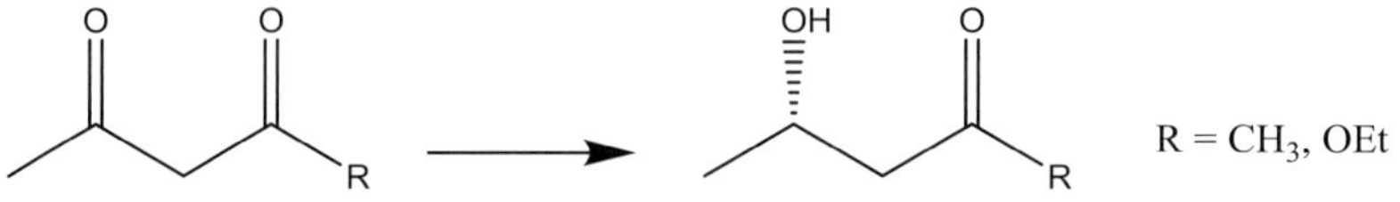

Scheme 8.3 Asymmetric reduction of aliphatic ketones to corresponding alcohols by immobilized baker's yeast (Howarth et al., 2001).

account of toxicity or poor partitioning properties with respect to both product molecules and substrates.

Another important reaction performance criteria influenced by the presence of ionic liquid solvents in biocatalytic reaction media is enantiomeric excess. This is the numerical parameter used to define chiral purity, reflecting the degree to which one enantiomer product is present relative to the other. Enantiomeric excess (ee) is usually given on a scale from 0% for a racemic mixture to 100% for a pure enantiomer.

An example of the positive effect of the presence of an ionic liquid was shown for the asymmetric reduction of the two aliphatic ketones, ethyl acetoacetate and pentane-2,4-dione (Scheme 8.3) to the corresponding alcohols in a biotransformation using immobilized baker's yeast as the enzyme source (Howarth, James, and Dai, 2001). For the reduction of pentane-2,4-dione (Scheme 8.3: R = CH_3), the enantiomeric excess increased from 74% in aqueous medium to 95% in the biphasic ionic liquid

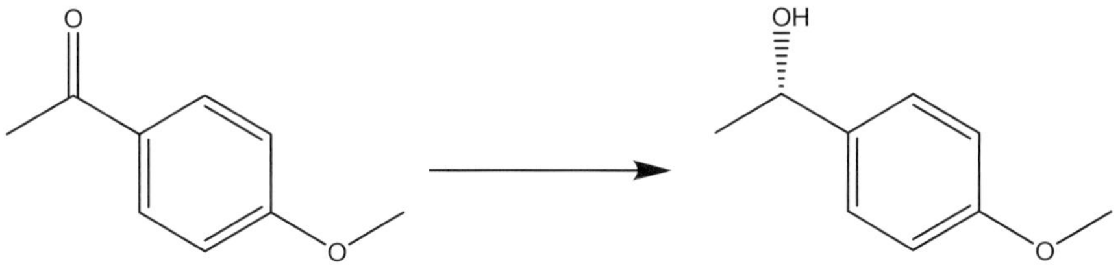

Scheme 8.4 Asymmetric bioreduction of 1-(4-methoxyphenyl)ethanone by immobilized yeast *Rhodotorula sp.* AS2.2241 (Wang, Zong, and Lou, 2009).

[bmim] [PF_6] medium, but the product yield was observed to decrease from 90% to 22%. For the reduction of ethyl acetoacetate (Scheme 8.3: R = OCH_3) in the presence of the ionic liquid [bmim] [PF_6] the product yield rose from to 60% compared with 43% using hexane. On the other hand, the enantiomeric excess decreased from 91% to 76% in the presence of the ionic liquid.

A third example is shown in Scheme 8.4 for the asymmetric reduction of 1-(4-methoxyphenyl)ethanone to (*S*)-1-(4-methoxyphenyl)ethanol by immobilized yeast *Rhodotorula sp.* AS2.2241.

The biotransformation was successfully demonstrated by Wang et al. (2009) in the presence several ionic liquids using a novel strain of *Rhodotorula sp.* AS2.2241 in a biphasic system comprising aqueous solution and an immiscible ionic liquid. The study compared the performance of various ionic liquids for this biotransformation, including [R-mim] [PF_6], where R was varied from n-C_4H_{11} to n-C_7H_{15}, and [R-mim] [Tf_2N], where R was C_2H_5. Good results were obtained in the case of [bmim] [PF_6] that showed low cell inhibition and thus high degrees of cell viability during the reaction. The cell viability values in the presence of this ionic liquid were reported at approximately 90%, which is significantly greater than the 30% viability observed in the case of just aqueous medium. In this case the partition coefficient for the ketone substrate into the [bmim] [PF_6] ionic liquid is high, with a value of 45.9 reported, thus providing successful control of the concentration of the substrate in the aqueous phase.

Data from the paper by Wang are shown in Table 8.6, listing the conversions obtained under the same conditions but with a different ionic liquid, comparing the effect of carbon number and comparing the PF_6^- anion ionic liquids with the Tf_2N^- anion ionic liquids. The partition coefficients under reaction conditions are also listed and show a relatively minor effect in the case of the PF_6^- anion ionic liquids. The high values in all cases demonstrate the ability to control conditions in the bioreactor and reduce substrate inhibition. The conversions obtained are variable, though with a general trend of higher conversion with increase in the partition coefficient value. The maximum conversion of 69.5% was obtained with the [bmim] [PF_6]/water biphasic medium. Another very significant finding of this work was that high values of enantiomeric excess were recoded for each of the six ionic liquids studied, with ee

Table 8.6 Effect of various water-immiscible ionic liquids on the asymmetric reduction of 1-(4-methoxyphenyl)ethanone catalyzed by immobilized *Rhodotorula sp.* AS2.2241 cells (from Wang et al. (2009)). Conditions: 10 mM 1-(4-methoxyphenyl)ethanone/ionic liquid, 2.0 mL Tris-HCl buffer (50 mM, pH 7.5), 20% (w/v) glucose, 0.32 g mL^{-1} beads, at 25 °C).

Medium	Partition coefficient for 1-(4-methoxyphenyl)ethanone	% conversion
$[C_4mim] [PF_6]$	45.9	69.5
$[C_5mim] [PF_6]$	44.5	55.1
$[C_6mim] [PF_6]$	43.7	51.6
$[C_7mim] [PF_6]$	42.1	38.2
$[C_2mim] [Tf_2N]$	36.8	66.4
$[C_4mim] [Tf_2N]$	31.3	58.3

values greater that 99% in the case of each of the ionic liquids listed in Table 8.6. Also in the case of the [bmim] $[PF_6]$/water biphasic medium, a high optimum substrate loading was observed, with values approximately eight-times higher compared with just aqueous medium. It was also noted that the systems exhibited significantly greater good operational stability of the biocatalyst with reduced loss of activity after multiple cycles.

References

Anastas, P. T. and Warner, J. C. (1998). *Green Chemistry: Theory and Practice*. Oxford: Oxford University Press.

Awad, W. H., Gilman, J. W., Nyden, M., et al. (2004). Thermal degradation studies of alkyl-imidazoliumsalts and their application in nanocomposites. *Thermochimica. Acta*, **409**, 3–11.

Baranyai, K. J., Deacon, G. B., MacFarlane, D. R., Pringle, J. M., and Scott, J. L. (2004). Thermal degradation of ionic liquids at elevated temperatures. *Australian Journal of Chemistry*, **57**, 145–147.

Baumann, M. D., Daugulis, A. J., and Jessop, P. G. (2005). Phosphonium ionic liquids for degradation of phenol in a two-phase partitioning bioreactor. *Applied Microbiology and Biotechnology*, **67**, 131–137.

Behr, A., Bayrak, Z., Peitz, S., Stochniol G., and Maschmeyer, D. (2015). Oligomerization of 1-butene with a homogeneous catalyst system based on allylic nickel complexes. *RSC Advances*, **5**, 41372–41376.

Bernot, R. J., Brueske, M. A., Evans-White, M. A., and Lambert, G. A. (2005). Acute and chronic toxicity of imidazolium-based ionic liquids on *Daphnia magna*. *Environmental Toxicology and Chemistry*, **24**, 87–92.

Berthon, L., Nikitenko, S. I., Bisel, I., et al. (2006). Influence of gamma irradiation on hydrophobic room-temperature ionic liquids $[BuMeIm]PF_6$ and $[BuMeIm](CF_3SO_2)_2N$. *DaltonTransactions*, 2526–2534.

Brautigam, S., and Bringer-Meyerb, S. W.-B. D. (2007). Asymmetric whole cell biotransformations in biphasic ionic liquid/water-systems by use of recombinant *Escherichia coli* with intracellular cofactor regeneration. *Tetrahedron: Asymmetry*, **18**, 1883–1887.

Chauvin, Y. (2006). Olefin metathesis: the early days (Nobel Lecture). *Angewandte Chemie International Edition*, **45**, 3740–3765.

Chauvin, Y., Gaillard, J. F., Quang, D. V., and Andrews, J. W. (1974) The IFP dimersol process for the dimerization of C3 and C4 olefinic cuts. *Chemistry & Industry*, 375–378.

Cho, C.-W., Pham, T. P. T., Jeon, Y.-C., et al. (2007). Toxicity of imidazolium salt with anion bromide to a phytoplankton *Selenastrum capricornutum*: effect of alkyl chain length. *Chemosphere*, **69**, 1003–1007.

Cho, C.-W., Pham, T. P. T., Jeon, Y.-C., et al. (2008). Microalgal photosynthetic activity measurement system for rapid toxicity assessment. *Ecotoxicology*, **17**, 455–463.

Coleman, D., Spulak, M., Garcia, M. T., and Gathergood, N. (2012). Antimicrobial toxicity studies of ionic liquids leading to a 'hit' MRSA selective antibacterial imidazolium salt. *Green Chemistry*, **14**, 1350–1356.

Cornmell, R. J., Winder, C. L., Gordon, J. T. G., Goodacre, R., and Stephens, G. (2008). Accumulation of ionic liquids in *Escherichia coli* cells. *Green Chemistry*, **10**, 836–841.

Cota, I., Gonzalez-Olmos, R., Iglesias, M., and Medina, F. (2007). New short aliphatic chain ionic liquids: synthesis, physical properties, and catalytic activity in aldol condensations. *Journal of Physical Chemistry B*, **111**, 12468–12477.

Docherty, K. M. and Kulpa, C. F. (2005). Toxicity and antimicrobial activity of imidazolium and pyridinium ionic liquids. *Green Chemistry*, **7**, 185–189.

Domańska, U. and Marciniak, A. (2008). Activity coefficients at infinite dilution measurements for organic solutes and water in the ionic liquid 1-butyl-3-methylimidazolium trifluoromethanesulfonate. *Journal of Physical Chemistry B*, **112**, 11100–11105.

Dupont, J., de Souza, R. F., and Suarez, P. A. Z. (2002). Ionic liquid (molten salt) phase organometallic catalysis. *Chemical Reviews*, **102**, 3667–3692.

Foco, G. M., Bottini, S. B., Quezada, N., de la Fuente, J. C., and Peters, C. J. (2006). Activity coefficients at infinite dilution in 1-alkyl-3-methylimidazolium tetrafluoroborate ionic liquids. *Journal of Chemical Engineering Data*, **51**, 1088–1091.

Gangu, S. A. (2013). Towards in situ extraction of fine chemicals and biorenewable fuels from fermentation broths using ionic liquids and the intensification of contacting by the application of electric fields. Ph.D. thesis, The University of Kansas.

Ganske, F. and Bornscheuer, U. T. (2006). Growth of *Escherichia coli, Pichia pastoris* and *Bacillus cereus* in the presence of the ionic liquids [BMIM] [BF_4] and [BMIM] [PF_6] and organic solvents. *Biotechnology Letters*, **28**(7), 465–469.

García, G., García, S., Torrecilla, J. S., and Rodríguez, F. (2011). *N*-Butylpyridinium bis-(trifluoromethylsulfonyl)imide ionic liquids as solvents for the liquid–liquid extraction of aromatics from their mixtures with alkanes: Isomeric effect of the cation. *Fluid Phase Equilibria*, **301**, 62–66.

García, S., Larriba, M., García, J., Torrecilla, J. S., and Rodríguez, F. (2012). Liquid–liquid extraction of toluene from n-heptane using binary mixtures of *N*-butylpyridinium tetrafluoroborate and *N*-butylpyridinium bis(trifluoromethylsulfonyl) imide ionic liquids. *Chemical Engineering Journal*, **180**, 210–215.

Gathergood, N. and Scammells, P. J. (2002). Design and preparation of room-temperature ionic liquids containing biodegradable side chains. *Australian Journal of Chemistry*, **55**, 557–560.

Gathergood, N., Garcia, M. T., and Scammells, P. J. (2004). Biodegradable ionic liquids: part I. Concept, preliminary targets and evaluation. *Green Chemistry*, **6**, 166–175.

Gathergood, N., Scammells, P. J., and Garcia, M. T. (2006). Biodegradable ionic liquids. Part II: the first readily biodegradable ionic liquids. *Green Chemistry*, **8**, 156–160.

Holbrey, J. and Seddon, K. R. (1999). Ionic liquids. *Clean Products and Processes*, **1**, 223–236.

Holbrey, J. D., Reichert, W. M., Swatloski, R. P., et al. (2002). Efficient, halide free synthesis of new, low cost ionic liquids: 1,3-dialkylimidazolium salts containing methyl and ethyl-sulfate anions. *Green Chemistry*, **4**, 407–413.

Howarth, J., James, P., and Dai, J. (2001). Immobilized baker's yeast reduction of ketones in an ionic liquid, [bmim]PF_6 and water mix. *Tetrahedron Letters*, **42**, 7517–7519.

Itakura, T., Hirata, K., Aoki, M., et al. (2009). Decomposition and removal of ionic liquid in aqueous solution by hydrothermal and photocatalytic treatment. *Environmental Chemistry Letters*, 7(4), 343–345.

Jork, C., Seiler, M., Beste, Y., and Arlt, W. (2004). Influence of ionic liquids on the phase behavior of aqueous azeotropic systems. *Journal of Chemical Engineering Data*, 49, 852–857.

Kaminski, K., Krawczyk, M., Augustyniak, J., Weatherley, L. R., and Petera, J. (2014). Electrically induced liquid–liquid extraction from organic mixtures with the use of ionic liquids. *Chemical Engineering Journal*, **235**, 109–123.

Krawczyk, M., Kamiński, K., and Petera, J. (2012). Experimental and numerical investigation of electrostatic spray liquid–liquid extraction with ionic liquids. *Chemical Process Engineering*, **33**(1), 167–183.

Landry, T. D., Brooks, K., Poche, D., and Woolhiser, M. (2005). Acute toxicity profile of 1-butyl-3-methylimidazolium chloride. *Bulletin of Environmental Contamination and Toxicology*, **74**, 559–565.

Larriba, M., Navarro, P., Gonzalez-Miquel, M., et al. (2016). Dicyanamide-based ionic liquids in the liquid–liquid extraction of aromatics from alkanes: Experimental evaluation and computational predictions. *Chemical Engineering Research and Design*, **109**, 561–572.

Lee, S.-M., Chang, W.-J., Choi, A.-R., and Koo, Y.-M. (2005). Influence of ionic liquids on the growth of *Escherichia coli*. *Korean Journal of Chemical Engineering*, **22**(5), 687–690.

Li, X., Zhao, J., Li, Q., Wang, L., and Tsang, S.C. (2007). Ultrasonic chemical oxidative degradations of 1,3-dialkylimidazolium ionic liquids and their mechanistic elucidations. *Dalton Transactions*, 1875–1880.

Marsh, K. N. (2006). Room Temperature Ionic Liquids – A Review of Properties and Applications. Presented at Process Intensification and Innovation Process Conference, Christchurch, New Zealand, September 24–29, 2006.

Matsumoto, M., Mochiduki, K., and Kondo, K. (2004). Toxicity of ionic liquids and organic solvents to lactic acid-producing bacteria. *Journal of Bioscience and Bioengineering*, **98**, 343–347.

Meindersma, G. W., Hansmeier, A. R., and de Haan, A. B. (2010). Ionic liquids for aromatics extraction. Present status and future outlook. *Industrial and Engineering. Chemistry Research*, **49**, 7530–7540.

Mokhtarania, B., Musavib, J., and Parvini, M. (2014). Extraction of toluene from alkane using [Bmim] [NO_3] or [Omim] [NO_3]ionic liquid at 298.15 K and atmospheric pressure. *Fluid Phase Equilibria*, **363**, 41–47.

Morawski, A. W., Janus, M., Goc-Maciejewska, I., Syguda, A., and Pernak, J. (2005). Decomposition of ionic liquids by photocatalysis. *Polish Journal of Chemistry*, **79**, 1929–1935.

Oliveira, M. V. S., Vidal, B. T., Melo, C. M., et al. (2016). (Eco)toxicity and biodegradability of protic ionic liquids. *Chemosphere*, **147**, 460–466.

Olivier-Bourbigou, H. and Hugues, F. (2002). In R. D. Rogers, K. R. Seddon, and S. Volkov, eds., *Green Industrial Applications of Ionic Liquids* (NATO Science Series II: Mathematics, Physics and Chemistry). Dordrecht: Kluwer, Vol. 92, pp. 67–84.

Pereiro, A. B. and Rodrıguez, A. (2008). Azeotrope-breaking using [BMIM] [$MeSO_4$] ionicliquid in an extraction column. *Separation and Purification Technology*, **62**, 733–738.

Perica, B., Sierra, J., Martí, E., et al. (2013). (Eco)toxicity and biodegradability of selected protic and aprotic ionic liquids. *Journal of Hazardous Materials*, **261**, 99–105.

Pham, T. P. T., Chul-Woong Cho, C. W., and Yun, Y. S. (2010). Environmental fate and toxicity of ionic liquids: A review. *Water Research*, **44**, 352–372.

Plechkova, N. V. and Seddon, K. R. (2008). Applications of ionic liquids in the chemical industry. *Chemical Society Reviews*, **37**, 123–150.

Prett, C., Chiappe, C., Pieraccini, D., et al. (2006). Acute toxicity of ionic liquids to the zebrafish (*Danio rerio*). *Green Chemistry*, **8**, 238–240.

Rebros, M., Gunaratne, H. Q. N., Ferguson, J., Seddon, K. R., and Stephens, G. A. (2009). High throughput screen to test the biocompatibility of water-miscible ionic liquids. *Green Chemistry*, **11**, 402–408.

Rogers, R. D. and Seddon, K. R. (2003). Ionic liquids –solvents of the future? *Science*, **302**, 792–793.

Siedlecka, E. M., Mrozik, W., Kaczyński, Z., and Stepnowski, P. (2008a). Degradation of 1-butyl-3-methylimidazolium chloride ionic liquid in a Fenton-like system. *Journal of Hazardous Materials*, **154**, 893–900.

Siedlecka, E. M., Golębiowski, M., Kumirska, J., and Stepnowski, P. (2008b). Identification of 1-butyl-3-methylimidazolium chloride degradation products formed in Fe(III)/H_2O_2 oxidation system. *Chemia Analityczna (Warsaw)*, **53**, 943–951.

Siedlecka, E. M., Gołebiowski, M., Czupryniak, K. J., Ossowski, T., and Stepnowski, P. (2009). Degradation of ionic liquids by Fenton reaction; the effect of anions as counter and background ions. *Applied Catalysis B: Environmental*, 91, 573–579.

Sipes, I. G., Knudsen, G. A., and Kuester, R. K. (2008). The effects of dose and route on the toxicokinetics and disposition of 1-butyl-3-methylimidazolium chloride in male F-344 rats and female B6C3F1 mice. *Drug Metabolism and Disposition*, **36**(2), 284–293.

Stepnowski, P. and Zaleska, A. (2004). Comparison of different advanced oxidation processes for the degradation of room temperature ionic liquids. *Journal of Photochemistry and Photobiology A: Chemistry*, 170, 45–50.

Stepnowski, P., Skladanowski, A., Ludwiczak, A., and Laczynska, E. (2004). Evaluating the cytotoxicity of ionic liquids using human cell line HeLa. *Human and Experimental Toxicology*, **23**(11), 513–517.

Stock, F., Hoffmann, J., Ranke, J., et al. (2004). Effects of ionic liquids on the acetycholinesterase: A structure–activity relationship consideration. *Green Chemistry*, **6**, 286–290.

Swatloski, R. P., Holbrey, J. D., and Rogers, R. D. (2003). Ionic liquids are not always green: hydrolysis of 1-butyl-3methylimidazolium hexafluorophosphate. *Chemistry Communication*, **5**, 361–363.

Wang, W., Zong, M.-H., and Lou, W.-Y. (2009). Use of an ionic liquid to improve asymmetric reduction of 4-methoxyacetophenone catalyzed by immobilized *Rhodotorula sp.* AS2.2241 cells. *Journal of Molecular Catalysis B: Enzymatic*, **56**, 70–76.

Wang, X., Ohlin, C. A., Lu, Q., et al. (2007). Cytotoxicity of ionic liquids and precursor compounds towards human cell line HeLa. *Green Chemistry*, **9**, 1191–1197.

Weissermel, K. and Arpe, H.J. (2003). *Industrial Organic Chemistry*, 4th ed. Weinheim: Wiley-VCH.

Wells, A. S. and Coombe, V. T. (2006). On the freshwater ecotoxicity and biodegradation properties of some common ionic liquids. *Organic Process Research & Development*, **10**, 794–798.

Welton, T. (1999). Room temperature ionic liquids: Solvents for synthesis and catalysis. *Chemical Reviews*, **99**, 2071–2083.

Werpy, T. and Petersen, G. (2004). *Top Value Added Chemicals From Biomass*, Vol. 1. US Department of Energy.

Yang, Z.-H., Zeng, R., and Wang, Y., et al. (2009).Tolerance of immobilized yeast cells in imidazolium-based ionic liquids. *Food Technology and Biotechnology*, **47**, 62–66.

9 Liquid–Liquid Phase-Transfer Catalysis

9.1 Introduction

Phase-transfer catalysis involves chemical reactions that occur in a two-phase liquid–liquid system and it has been shown to provide an effective method for organic synthesis. Phase-transfer catalytic reactions can facilitate high conversions and high reaction selectivity. They also provide opportunities to develop economic and environmentally benign reaction chemistry. However, phase-transfer catalytic reactions may not be as efficient as homogeneous reactions on account of mass transfer limitations. Intensification of liquid–liquid operations therefore has a potentially important role in the application of phase-transfer catalysis to industrial processes.

In this chapter the basic mechanisms involved in phase-transfer catalysis, and the related suite of reactions that involve catalytic transfer hydrogenations, are briefly described and reviewed. Phase-transfer catalysis (PTC) and transfer-hydrogenation catalysis are important and growing areas in the overall field of catalysis and chemical reaction engineering. Reactions based on phase-transfer catalysis are potentially used in organic synthesis, in the pharmaceutical and agrochemical industries, in polymer synthesis, and in an increasing number of "green" chemistry applications. A phase-transfer catalysis system requires two liquid phases, each of which contains one of the substrate species and one phase contains a soluble catalyst species. Makosza (2000) summarizes the nature of phase-transfer catalytic systems utilizing a heterogeneous two-phase system, one phase being a reservoir of reacting anions or base for generation of organic anions, with organic reactants and catalysts (source of lipophilic cations) located in the second phase. The reacting anions are continuously introduced into the organic phase in the form of lipophilic ion pairs with lipophilic cations supplied by the catalyst.

Phase-transfer catalysis is useful in the case of reactions in which the reactants are mutually insoluble. The addition of a complexing agent, which may also be an active catalyst itself, essentially binds with one of the substrates and facilitates transport to the other phase. The reaction takes place either at the interface or in one of the bulk phases. The complexing agent is released and returns to the first phase to "collect" more substrate for transport. Typical agents for achieving phase-transfer catalysis include organic quaternary ammonium compounds, containing tetraalkyl ammonium cations. Compounds specifically include hexadecyl tributyl phosphonium bromide, tetrabutyl ammonium bromide, and methyl trioctyl ammonium chloride. *N*-Methyl-*N*,

N-dioctyloctan-1-ammonium chloride, otherwise known as Aliquat 336, is also a highly effective phase-transfer catalyst.

Benjamin (2013) summarizes the basic requirements of an effective phase-transfer catalyst as follows:

(1) The ability of the catalyst to form a relatively stable ion pair with the reactant.
(2) A favorable partitioning in the organic phase.
(3) A return to the interface so the process can be repeated.

A number of key benefits are also noted by Makosza (2000), emphasizing the role of phase-transfer catalysis in the development and application of green chemistry, listed as follows:

- elimination of organic solvents
- elimination of dangerous, inconvenient, and expensive reactants and their substitution with relatively economic and less hazardous compounds such as NaOH, KOH, K_2CO_3
- high reactivity and selectivity of the active species
- high yields and purity of products
- simplicity of the procedure
- low investment cost
- low energy consumption
- possibility to mimic countercurrent processes
- minimization of industrial waste.

Makosza (2000) divides reactions to which PTC is applicable into two major categories:

1 reactions of anions that are available as salts, for example, sodium cyanide, sodium azide, sodium acetate, etc.
2 reactions of anions that should be generated in situ, such as alkoxides, phenolates, *N*-anions of amides or heterocycles, etc. and particularly carbanions.

In the former case, the salts are used as aqueous solutions or in the form of powdered solids, whereas the organic phase contains organic reactants neat (when liquid) or in appropriate solvents. Since the phases are mutually immiscible the reaction does not proceed unless the catalyst, usually a tetraalkyl ammonium salt, Q^+X^-, is present. The catalyst continuously transfers reacting anions into the organic phase in the form of lipophilic ion pairs produced according to ion-exchange equilibria, where they react further, for example with alkyl halides affording nucleophilic substitution. A variety of other reactions with participation of inorganic anions, such as addition, reduction, and oxidation, may be efficiently executed using this methodology.

The general scheme shown in Figure 9.1 illustrates the overall mechanism for phase-transfer catalysis.

In Figure 9.1 the active anion required for the reaction is depicted as Y^- and is present as the sodium salt. This is highly polar and very hydrophilic, and therefore does not dissolve to any significant extent in the organic phase that contains the substrate shown here as R–X.

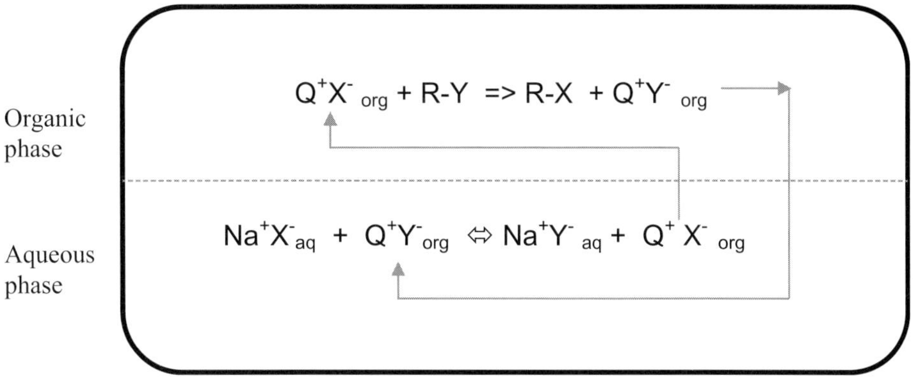

Figure 9.1 Generalized scheme showing the transport of species across the liquid–liquid interface in a phase-transfer catalysis.

$$Na^+X^-(aq) + Q^+Y^-(org) \Leftrightarrow Na^+Y^-(aq) + Q^+X^-(org)$$
$$Q^+X^-(org) + \text{R-Y} \Rightarrow \text{R-X} + Q^+Y^-(org)$$

The phase-transfer catalyst in this case is shown as Q^+Y^-, which distributes between both phases. The addition of this to the aqueous phase facilitates an ion exchange between the X^- aqueous ion and the Y^- organic anion to produce the organo-ion complex Q^+X^- that thus reacts with the organic substrate R–Y, resulting in the product R–X. The other reaction product is Q^+Y^-, which can now transfer back into the aqueous phase for the next cycle of transport of X^-.

A significant issue that arises in the application of phase-transfer catalysis is mass-transfer limitation, particularly when dealing with polymerization reactions (Tsai and Lee, 1987a, 1987b, 1987c; Tsai, Jeng, and Tsai, 1990; Tsai, Lee, and Jeng, 1988). In the context of process intensification, application of phase-transfer catalysis certainly provides scope for enhancing reaction chemistry but also provides opportunity for the application of intensification techniques for mass-transfer enhancement.

9.2 Examples in Organic Synthesis

There are many examples of classical organic syntheses, including those of industrial importance, for which phase-transfer catalysis has been demonstrated, including alkylations, nitrations, sulfonations, etherifications, oxidations (including epoxidations), and condensation polymerizations. In many of these syntheses the principle of phase-transfer catalysis has been successfully demonstrated but some are mass-transfer limited, especially in cases where one or both of the phases exhibit significant viscosity. This is particularly the case for polymerization reactions. A key requirement for phase-transfer catalysis to be commercially successful is a combination of fast reaction kinetics and efficient mass transfer.

9.2.1 Synthesis of Phenyl Alkyl Acetonitriles and Aryl Acetonitriles

Examples of industrial important that are highlighted by Makorsza include the synthesis of phenyl alkyl acetonitriles from phenyl acetonitrile in the presence of sodium hydroxide and the tetraalkyl ammonium halide phase-transfer catalysts. The scheme proposed is as follows (note that "int" means "in the interfacial region"):

$$\mathrm{PhCH_2CN(org) + NaOH \Leftrightarrow PhC^{-}CN\,Na^{+}(int) + H_2O}$$

$$\mathrm{PhC^{-}HCN\,Na^{+}(int) + Q^{+}X^{-}(org) \Leftrightarrow PhC^{-}HCN\,Q^{+}(org) + Na^{+}X^{-}(aq)}$$

$$\mathrm{PhC^{-}HCN\,Q^{+}(org) + R\text{-}X(org) \rightarrow \underset{\text{phenyl alkyl acetonitrile}}{PhRCHCN(org)} + Q^{+}X^{-}(org)}$$

A related examplc is thc alkylation of aryl acetonitriles to α-aryl alkane nitriles:

$$\mathrm{ArCH_2CN + R\text{-}X + B^{-} \rightarrow ArRCHCN + X^{-} + BH}$$

The scheme is analogous to that above for phenyl alkyl nitriles, again conducted in the presence of sodium hydroxide and a phase-transfer catalyst. This synthesis is of importance in the pharmaceutical industry where organic nitriles appear in many pharmaceutical products, ranging from antidiabetic drugs to anticancer agents to antipsychotics (Fleming et al., 2010). The phase-transfer catalysis route is claimed to utilize much less energy compared with conventional multi-stage synthesis, with easy product recovery via phase separation.

9.2.2 Synthesis of *p*-Chlorophenyl Acetonitrile

Yadav and Jadhav (2003) provide a specific example of phase-transfer catalysis involving inorganic salts and organic salts in an aqueous/organic system, in this case for the synthesis of *p*-chlorophenyl acetonitrile. The reaction mixture in this example comprises the organic substrate *p*-chlorobenzyl chloride, the salts potassium iodide and potassium cyanide, and the phase-transfer catalyst, represented here generically as $Q^{+}X^{-}$.

The reactions occurring in each phase may be broken down as follows:

For the upper organic phase:

$$\mathrm{R^{+}X^{-}(org) + Q^{+}I^{-}(org) \rightarrow R^{+}I^{-}(org) + Q^{+}X^{-}(org)}$$
$$\mathrm{R^{+}I^{-}(org) + Q^{+}CN^{-}(org) \rightarrow R^{+}CN^{-}(org) + Q^{+}I^{-}(org)}$$

For the lower aqueous phase:

$$\mathrm{Q^{+}X^{-}(aq) + K^{+}I^{-}(aq) \rightarrow Q^{+}I^{-}(aq) + K^{+}X^{-}(aq)}$$
$$\mathrm{Q^{+}X^{-}(aq) + Na^{+}CN^{-}(aq) \rightarrow Q^{+}CN^{-}(aq) + Na^{+}X^{-}(aq)}$$

where R^{+} is the *p*-chlorobenzyl cation, Q^{+} and X^{-} are the catalyst cation and anion, respectively, I^{-} is the iodide anion, and CN^{-} is the cyanide anion.

The steps in the reaction and movement of species across the interface are shown in Figure 9.2. The quaternary cation Q^{+} is in the aqueous phase as two ion pairs $Q^{+}I^{-}$ and

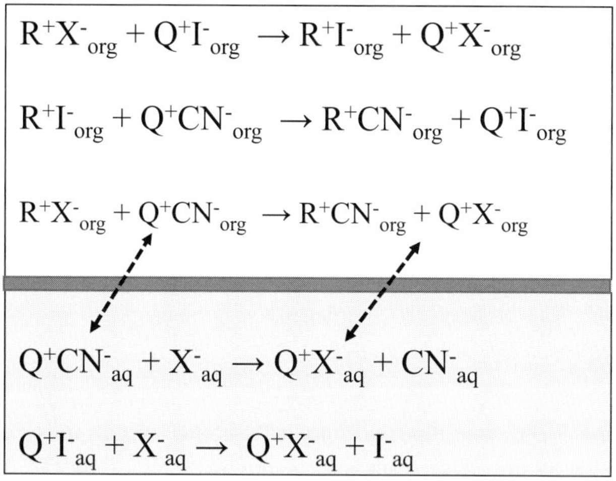

Figure 9.2 Mechanism of liquid–liquid phase-transfer catalyzed synthesis of *p*-chlorophenyl acetonitrile by potassium iodide. (Yadav and Jadhav, 2003)

Q^+CN^-. These are rapidly partitioned into the organic phase due to the high affinity of the cation for the organic solvent. The other reactant, in this case *p*-chlorobenzyl chloride is reacted to the iodide form and in the process regenerates the catalyst Q^+X^-. The next stage in the sequence is the reaction of the *p*-chlorobenzyl iodide with the ion pair Q^+CN^- to yield the desired product, *p*-chlorophenyl acetonitrile. Overall reaction rates are enhanced; that is explained by the in situ formation of the iodide ion pair/ *p*-chlorobenzyl iodide. The presence of potassium iodide provides co-catalytic action that is central to the enhancement of the reaction.

9.2.3 Transfer Hydrogenation

One significant example of transfer hydrogenation catalysis as an intensification technique is the liquid-phase hydrogenation of acetophenone (Figure 9.3) (Casagrande et al., 2002; Gonzalez-Galvez et al., 2013; Zhu, 2014).

Asymmetric transfer hydrogenation of ketones is an important step in the production of chiral alcohols and other intermediates used in the production of a number of important drugs. Deployment of phase-transfer hydrogenation in a liquid–liquid system avoids using hydrogen gas at high pressure associated with conventional gas/liquid hydrogenation of ketones. The two-phase catalytic hydrogenation typically uses 2-propanol as the hydrogen source (proton and hydride) though there is more recent work showing the possible use of sodium formate as the hydrogen source (Demmans, Ko, and Morris, 2016; Zhu, 2014). The latter authors highlight the use of 2-propanol as the hydrogen source that results in acetone as the co-product. Under certain conditions the acetone may be reduced back to 2-propanol, thus lowering the reaction yield. On the other hand, the use of sodium formate as a hydrogen source results in CO_2 as the

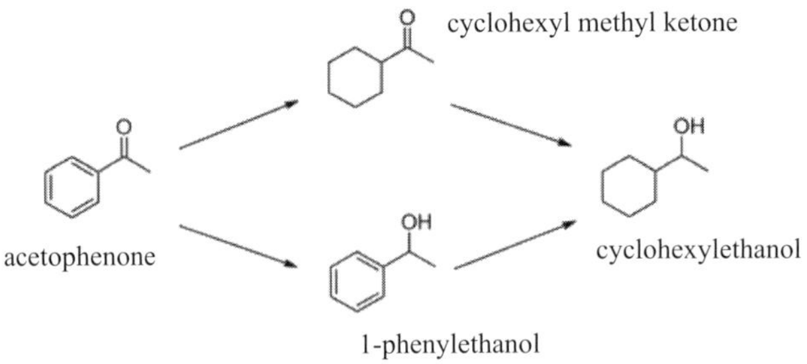

Figure 9.3 Liquid-phase hydrogenation of acetophenone. After Gonzalez-Galvez et al. (2013)

co-product, which may be continuously removed from the system thus enhancing the yield of the forward reaction. Demmans also mentions the possibility of conducting a separate hydrogenation of the CO_2 to regenerate formate.

The transfer hydrogenation of acetophenone is particularly important because of the industrial application of the two possible reaction products: 1-phenylethanol and 1-cyclohexylethanol. In fact, 1-cyclohexylethanol is used in the manufacture of some polymers and 1-phenylethanol is used in the pharmaceutical and perfume industries. Conventionally, hydrogenation of acetophenone is performed in the liquid phase, at low hydrogen pressure, using a transition metal catalyst supported on zeolites or as oxides. There are many papers showing the role of heterogeneous catalysts for the hydrogenation of acetophenone.

Zhu (2014) highlights a novel approach to the hydrogenation of acetophenone that avoids the use of gaseous hydrogen by exploiting liquid phase hydrogenation by phase-transfer catalysis using sodium formate as hydrogen source and a ruthenium-based Noroyu catalyst. The findings show the potential advantages of phase-transfer catalysis for this important industrial organic chemical reaction. The avoidance of using gaseous hydrogen is a significant advantage for both safety and for kinetic reasons. The alleviation of substrate inhibition was also demonstrated. The use of sodium formate as hydrogen source bypasses the requirement of high temperature and high pressure in traditional hydrogenation and could be considered a good example of green chemistry. It is also claimed that this particular synthesis method provides a potential route for the large-scale production based on aqueous transfer hydrogenation.

Another important example of the use of phase-transfer phenomena to intensify the hydrogenation of acetophenone is provided by Wang et al. (2014). The authors successfully demonstrated transfer hydrogenation of acetophenone in the presence of a water-soluble ruthenium complex $RuCl_2(TPPTS)_2$ as catalyst, where TPPTS is $P(m\text{-}C_6H_4SO_3Na)_3$. The addition of surfactants to enhance the effective concentration of the catalyst at the aqueous/organic interface is a significant innovation and significantly improved the reaction conditions. The results showed the reaction to be accelerated by the presence of double long-chain cationic surfactants that promoted the formation of vesicles. The structural principle of a surfactant at a liquid–liquid interface is well known. The structure comprises an entity created by the surfactant in a

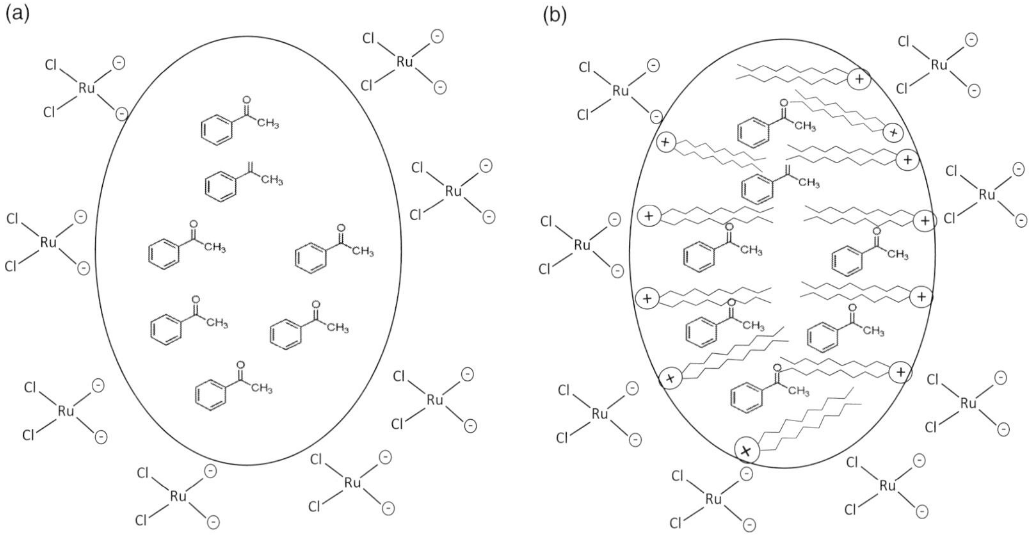

Figure 9.4 Schematic diagram of the location of aqueous phase ruthenium-based homogeneous catalyst close to the interface with an organic droplet containing acetophenone.

two-phase medium, usually aqueous/organic. The hydrophilic head of the surfactant is situated on one side of the interface, for example in the outer aqueous phase, and the hydrophobic tails of the surfactant form a bilayer with the organic phase as the interior phase (for example), the inner interface of which is in contact with the hydrophobic tails of a second layer of surfactant. The hydrophobic heads of this inner surfactant layer form a shell around an inner droplet of aqueous phase. Wang et al. exploited this phenomenon, as shown in Figure 9.4. The figure shows the active ruthenium catalyst located in high enriched concentration at the biphasic interface. The proposed mechanism is electrostatic attraction of the ruthenium-based catalytic complex to the interface that promotes faster reaction kinetics. The electrostatic attraction is attributed to the electrostatic field at the hydrophilic interface of the vesicle. A second crucial aspect to the enhancement of the overall reaction conditions is the solubilization of the water-insoluble acetophenone substrate in the hydrophobic interior core of the vesicles.

The catalyst used was the ruthenium complex $RuCl_2(TPPTS)_2$, along with a cationic surfactant CTAB (cetyl tetra-ammonium bromide) and with 2-propanol as the hydrogen source. The nature of the surfactant was found to be critical and it was shown that no kinetic enhancement was observed when either an anionic surfactant (sodium decyl sulfate) or a non-ionic surfactant (Tween 80) were used. The conversion was greatly enhanced only in the case of the addition of the cationic surfactant CTAB. Another significant factor is the importance of both the concentration and the chain length of the alkyl groups in the surfactant. In general, surfactants having long chain lengths at each end of the molecule promoted faster kinetics. The proposed mechanism is based on the hypothesis that the resulting structure of the vesicle provides an environment to host higher concentrations of the acetophenone substrate. The surfactant $DDAB_{16}$ (dihexadecyl dimethyl ammonium bromide) was shown to be effective in enhancing conversion

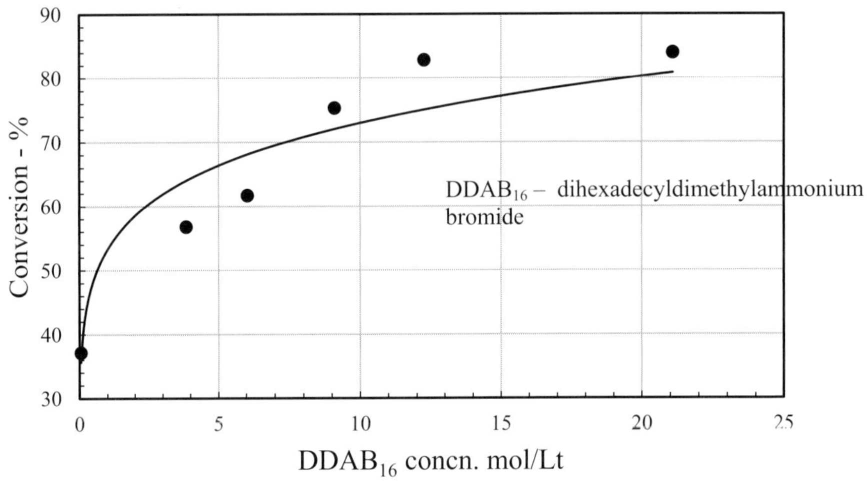

Figure 9.5 Influence of dihexadecyl dimethyl ammonium bromide concentration on the two-phase transfer hydrogenation of acetophenone. Reaction conditions: [Ru] = 9.7×10^{-4} mol L^{-1}; TPPTS:Ru = 3:1; substrate:catalyst:KOH = 250:1:14; 2-propanol:water = 1:1; temperature = 60 °C; reaction time = 6 h.

in the transfer hydrogenation. It is significant that faster rates of reaction were observed at higher concentrations of surfactant, up to a limiting value (see Figure 9.5).

Another recent approach to asymmetric transfer hydrogenation was also studied by Demmans et al. (2016), demonstrating the effectiveness of an iron(II) catalyst in the biphasic asymmetric transfer hydrogenation (ATH) of ketones to enantio-enriched alcohols that employs water and potassium formate as the proton and hydride source, respectively. Of particular note is the addition of a precatalyst, [FeCl(CO)(P–NH–N–P)]BF_4 (where P–NH–N–P = (*S*,*S*)$PPh_2CH_2CH_2NHCHPhCHPhNCHCH_2PPh_2$), to the organic phase together with the substrate. The precatalyst is activated by base to produce a system that is claimed to rival the activity of the best ruthenium biphasic catalysts. Mass transfer limitation of the biphasic system involving 2-propanol was addressed in the Demmans study by the addition of the phase-transfer catalyst tetrabutyl ammonium bromide. This requirement may be avoided if sodium formate is used as the hydrogen source. Furthermore, the study showed a strong positive influence of sodium formate concentration on the overall conversion. In-situ removal of CO_2 proved to have a significant effect on conversion.

Figure 9.6 shows the structure of the iron(II) catalyst precursor. The iron-based catalyst is preferable to the more commonly used catalysts for this family of reactions, ruthenium, rhodium, and iridium. The lower costs and lesser toxicity are significant environmental and economic advantages, providing the reaction performance is equal or better. The mechanism proposed by Demmans et al. is shown in Figure 9.7.

There are a number of other factors that affect reaction performance, whether expressed in terms of conversion, turn-over frequency, or enantiomeric excess.

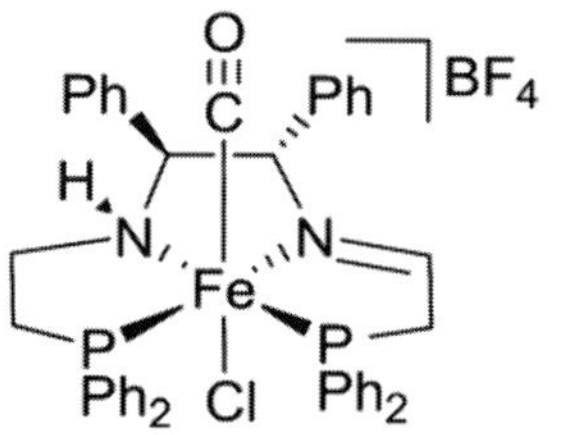

Figure 9.6 Structure of the iron(II) catalyst precursor *trans*-(*S*,*S*)-[FeCl(CO)(P–NH–N–P] BF_4. From Demmans, K. Z., Ko, O. W. K., and Morris, R. H. (2016). Aqueous biphasic ironcatalyzed asymmetric transfer hydrogenation of aromatic ketones. *RSC Advances*, **6**(91), 88580–88587. Published by The Royal Society of Chemistry (Creative Commons Attribution 3.0 Unported Licence)

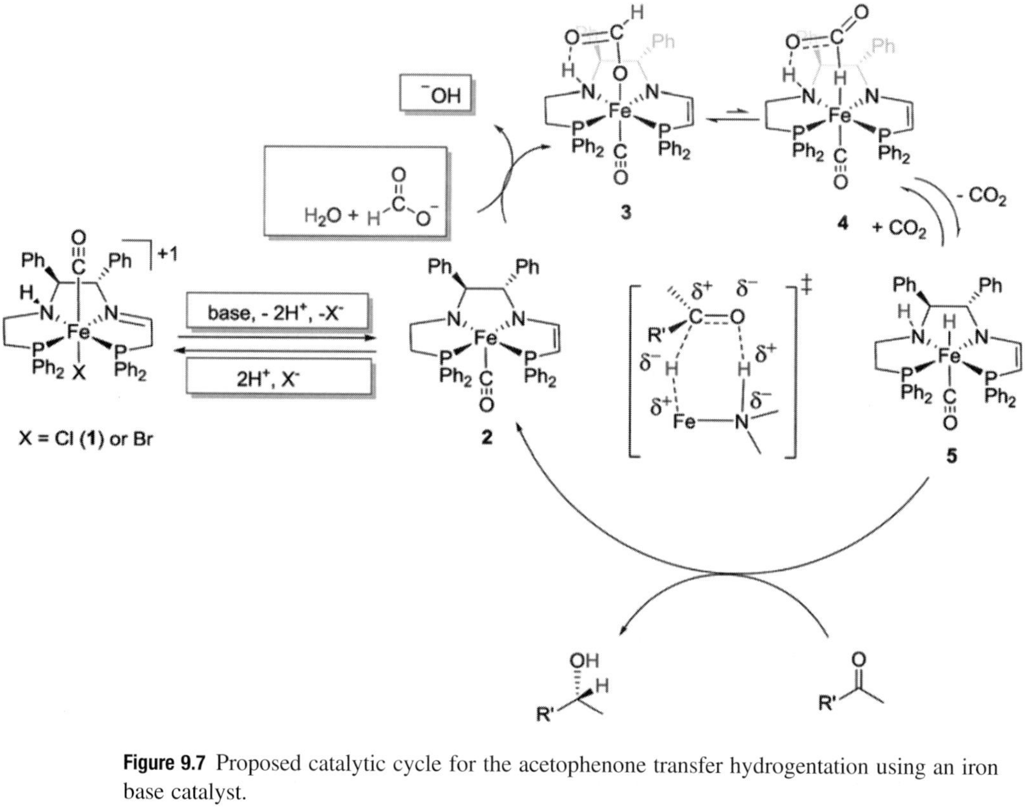

Figure 9.7 Proposed catalytic cycle for the acetophenone transfer hydrogentation using an iron base catalyst.
From Demmans, K. Z., Ko, O. W. K., and Morris, R. H. (2016). Aqueous biphasic ironcatalyzed asymmetric transfer hydrogenation of aromatic ketones. *RSC Advances*, **6**(91), 88580–88587. Published by The Royal Society of Chemistry (Creative Commons Attribution 3.0 Unported Licence)

Demmans et al. determined the influence of: (1) the presence of a phase-transfer catalyst (comparing two different co-ions (Br^- and BF_4^-); (2) the presence of potassium hydroxide (to remove HCl, which is a by-product of catalyst activation); (3) sodium formate concentration; and (4) acetophenone : catalyst concentration.

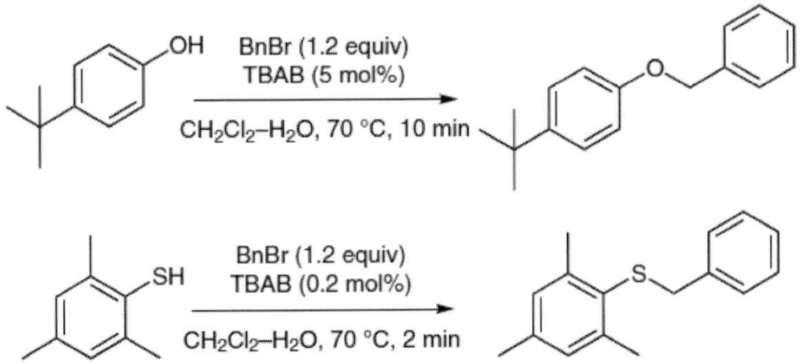

Figure 9.8 Phase-transfer catalysis for alkyl oxidation and sulfonation reactions in a liquid–liquid system using a tetrabutyl ammonium bromide phase-transfer catalyst.
From Reichart, B., Kappe, T., and Glasnov, T. N. (2013). Phase-transfer catalysis: mixing effects in continuous-flow liquid/liquid O- and S-alkylation processes. *Synlett*, **24**, 2393–2396 with permission Copyright Georg Thieme Verlag KG

Conversions in the range 5–25% were observed, with the highest conversion obtained at the highest concentration of sodium formate. Overall, the conversion increased with the concentration of formate in the aqueous phase. The addition of the phase-transfer catalyst, tetrabutyl ammonium bromide (TBAB), resulted in a decrease in conversion, with a small increase in catalyst activity reported in the absence of the TBAB. By contrast, the addition of the phase-transfer catalyst with a different anion (tetrabutyl ammonium tetrafluoroborate, TBA-BF_4) resulted in a significant increase in conversion from 9% to 18% as reported in the paper. The possible explanation proposed for the difference is deactivation of the catalyst by bromide ions on account of coordination with open sites in the catalyst structure. An optimal concentration of catalyst was observed in the case of TBA-BF_4 phase-transfer catalyst, with decrease in activity at higher concentrations. The presence of potassium hydroxide appeared to exert no effect on catalyst activity, though at higher concentrations of potassium hydroxide significant reductions in enantiomeric excess were observed. Enantiomeric excess values (ee%) in the range 69–80% were observed.

9.2.4 Alkylations

The successful application of phase-transfer catalysis for alkyl oxidation and sulfonation reactions in a liquid–liquid system using a tetrabutyl ammonium bromide phase-transfer catalyst, as shown in Figure 9.8, has been described (Reichart, Kappe, and Glasnov, 2013).

The addition of the TBAB catalyst gave impressive increases in overall product yield, with the yield increasing from 5% to over 70% for the phase-transfer *O*-alkylation of 4-tert-butylphenol. This result was obtained in a small microwave-activated batch reactor. Further work that is particularly noteworthy was conducted using a range of flow microreactors (see Figure 9.9): a packed bed reactor, a stainless steel

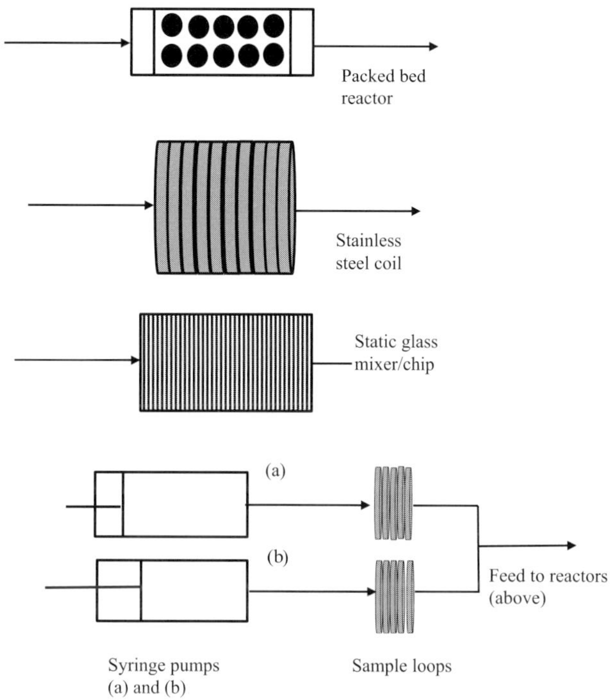

Figure 9.9 Flow microreactors (packed bed reactor, stainless steel coil, and static glass mixer reactor) demonstrated for *O*-alkylation reactions in each of the flow reactors and for phase-transfer *S*-alkylation of 2,4,6-trimethylbenzenethiol.
From Reichart, B., Kappe, T., and Glasnov, T. N. (2013). Phase-transfer catalysis: mixing effects in continuous-flow liquid/liquid O- and S-alkylation processes. *Synlett*, **24**, 2393–2396 with permission Copyright Georg Thieme Verlag KG

coil, and a static glass mixer reactor. The results are even more impressive and show product yields of over 90% for the same *o*-alkylation reaction in each of the flow reactors. Similar results are reported for the phase-transfer *S*-alkylation of 2,4,6-trimethylbenzenethiol.

The successful alkylation of 1,1-bis(methoxy-NNO-azoxy)alkanes by phase-transfer catalysis using DMSO (dimethyl sulfoxide) as solvent and tetrethyl ammonium bromide as the phase-transfer catalyst is described by Zyuzin (2013). The aqueous liquid was concentrated aqueous alkali. Two reaction schemes are shown in Figure 9.10.

Figure 9.10(a) shows the alkylation reaction using alkyl iodide as the second substrate. The reaction is shown in two parts. The first part is a mono alkylation (R^1) and the second part results in the di-alkylated product. Successful syntheses were achieved for methyl, ethyl, propyl, and butyl alkylations.

A novel three-phase liquid–liquid–liquid phase-transfer catalysis system for the benzylation of cresols is reported by Yadav and Badure (2008). This is shown in Figures 9.11, 9.12, and 9.13.

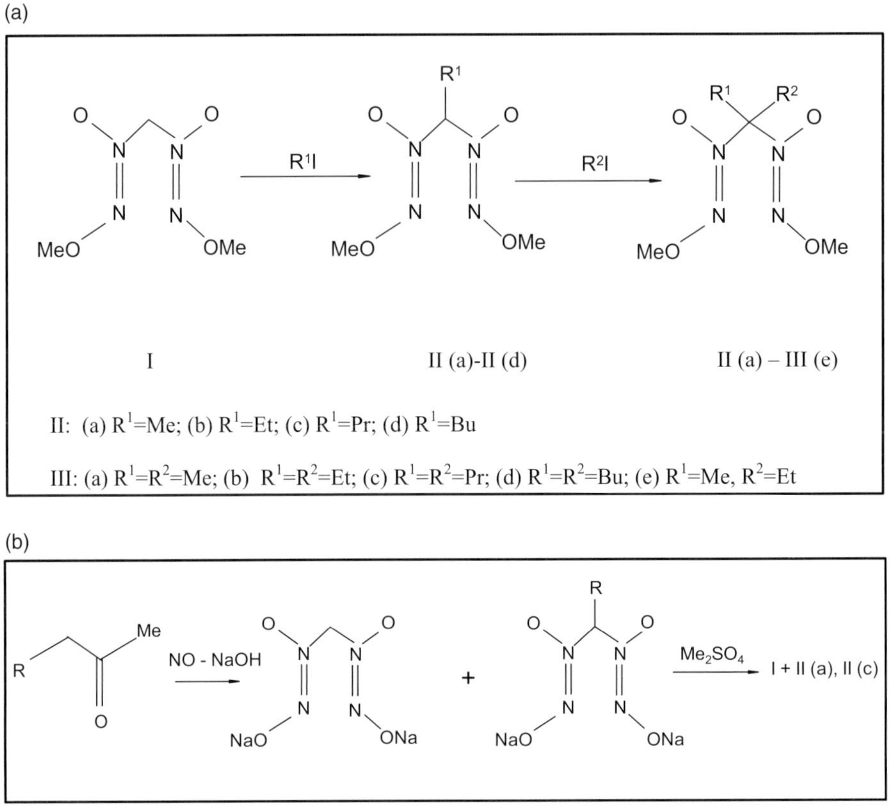

Figure 9.10 Alkylation of 1,1-bis(methoxy-NNO-azoxy)alkanes by phase-transfer catalysis using DMSO (dimethyl sulfoxide) as solvent and tetrethyl ammonium bromide as the phase-transfer catalyst. (Zyuzin, 2013)

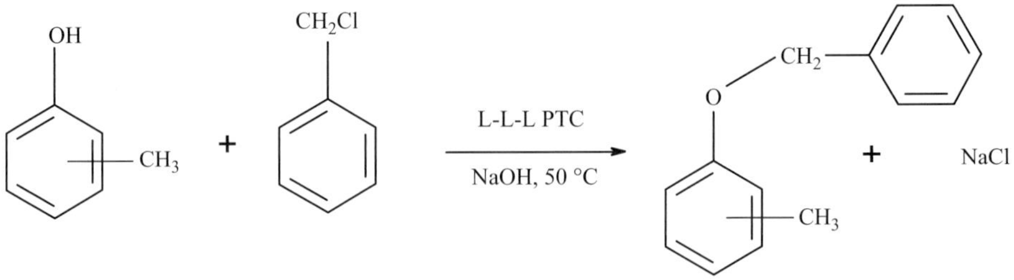

Figure 9.11 Reaction scheme for the *O*-benzylation of cresols with benzyl chloride.
Reprinted with permission from Yadav, G. D. and Badure, O. V. (2008). Selective engineering in O-alkylation of m-cresol with benzyl chloride using liquid-liquid-liquid phase transfer catalysis. *Journal of Molecular Catalysis A: Chemical*, **288**, 33–41. Copyright Elsevier B.V.

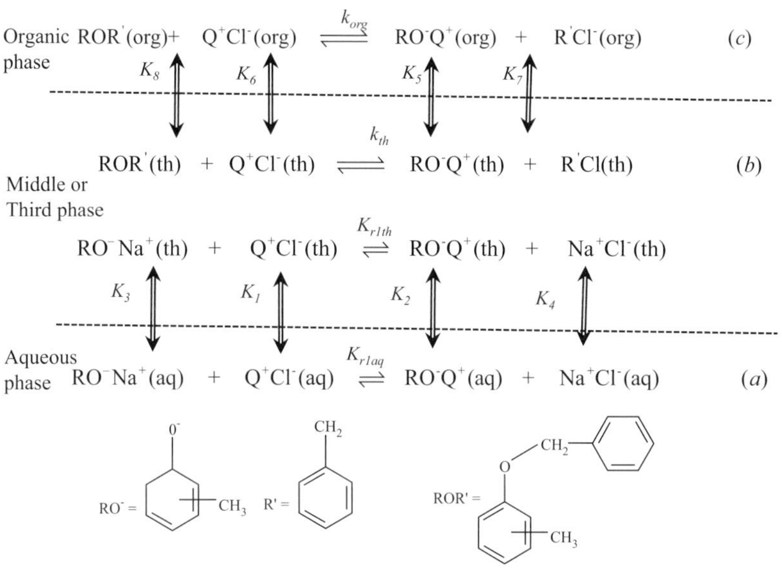

Figure 9.12 L-L-L phase-transfer catalysis mechanism for the etherification of cresols.
Reprinted with permission from Yadav, G. D. and Badure, O. V. (2008). Selective engineering in O-alkylation of m-cresol with benzyl chloride using liquid–liquid–liquid phase transfer catalysis. *Journal of Molecular Catalysis A: Chemical*, **288**, 33–41.

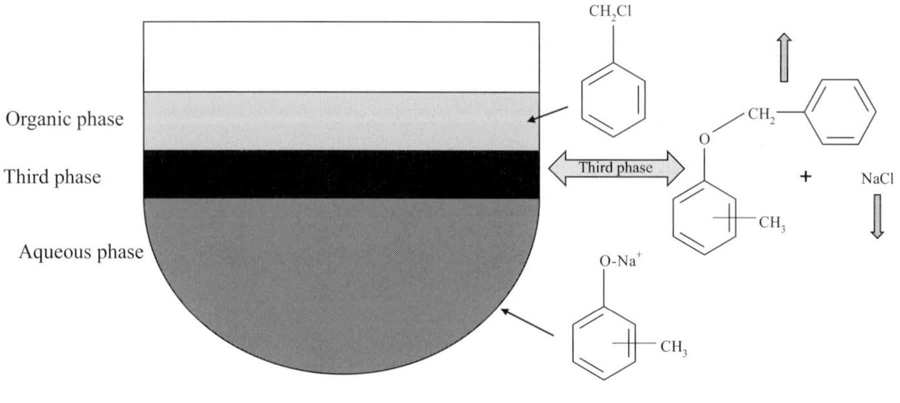

Figure 9.13 L-L-L phase-transfer catalysis mechanism for the etherification of cresols, showing three liquid phases.
From Yadav, G. D. and Badure, O. V. (2008). Selective engineering in O-alkylation of m-cresol with benzyl chloride using liquid–liquid–liquid phase transfer catalysis. *Journal of Molecular Catalysis A: Chemical*, **288**, 33–41.

The configuration of the three phases is as follows:

1 the top phase is the least dense organic phase containing the initial benzyl chloride substrate;
2 the middle phase, shown as the "third" phase, acts as an extracting phase for the final product;
3 the bottom phase is the most dense aqueous phase and contains the initial cresol substrate as the sodium salt.

9.2.5 Oxidations

The use of hydrogen peroxide for organic oxidations is attracting significant interest as an alternative to gas–liquid phase systems for organic oxidations, with hydrogen peroxide replacing air, gaseous oxygen, organic hydroperoxides, or peroxy acids (Meshechkina et al., 2012). Hydrogen peroxide is relatively cheap, is easily transported, and is relatively benign.

Here we look at two examples from recent literature that highlight the promise of hydrogen peroxide for organic oxidations. These examples also highlight the important potential role of phase-transfer catalysts in the exploitation of hydrogen peroxide.

The first example relates to the oxidation of cyclopentene, which can be summarized as in Figure 9.14, by Meshechkina et al. (2012).

Reaction 1 in Figure 9.14 shows the oxidation of cyclopentene to 1,2-epoxycyclopentane. Reaction 2 in Figure 9.14 shows the subsequent hydrolysis reaction of the 1,2-epoxycyclopentane to1,2-cyclopentanediol. Five phase-transfer catalysts, listed below, were compared and reaction performance determined as the percentage yield of the desired product 1,2-epoxycyclopentane (Meshechkina et al., 2012). Reactions were conducted in the presence of an organic solvent (1,2-dichloroethane) and with a sodium tungstate catalyst (Na_2WO_4). The five phase-transfer catalysts were: tetraethyl ammonium bromide (TEAB), tetraethyl ammonium iodide (TEAI), tetrabutyl ammonium bromide (TBAB), Katamin AB ($[C_{14}H_{29}N(CH_3)_2C_6H_5CH_2]Cl$), and Adogen 464 ($[CH_3(C_9H_{19})_3N]Cl$).

In the cases of TBAB, Katamin, and Adogen 464, very high yields of 1,2-ECP, in the range 85–95%, were obtained. In the cases of TEAB and TEAI, lower yields were observed, in the range 27–42%.

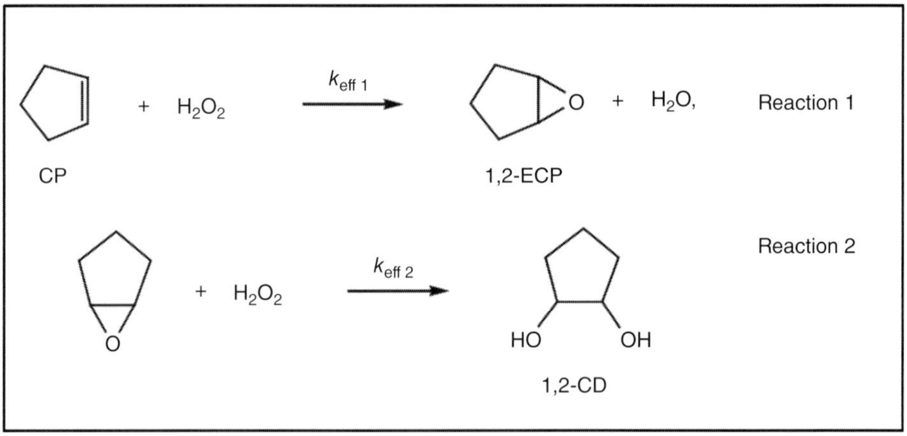

Figure 9.14 The oxidation of cyclopentene to 1,2-epoxycyclopentane (1,2-ECP) and the subsequent hydrolysis reaction to 1,2-cyclopentanediol (1,2-CD).

A second part of this study then focused on the relative yield of 1,2-ECP and 1,2-CD and the effect of the choice of organic solvent. The solvents examined, in addition to 1,2-dichloroethane, were *N,N*-dimethyl formamide, n-butanol, trichloroethylene, tetrachloroethylene, and xylene. The yields of 1,2-ECP and 1,2-CD in the presence of the Adogen 464 PTC were observed to be strongly influenced by the organic solvent, with the highest yield of 1,2-CD observed in the case of xylene (93.4%) and the lowest yield in the case of *N,N*-dimethyl formamide (76.6%). A strong correlation between the dielectric constant of the organic solvents and the relative yields of the 1,2-ECP and 1,2-CD was also ascertained, which showed that the highest 1,2-CD yield corresponded to the lowest dielectric constant.

The second, and more recent, example is from Chatel et al. (2014) who demonstrated the successful oxidation of cyclic alcohols to ketones using phase-transfer catalysis (Figures 9.15 and 9.16). The specific reaction studied was the oxidation of cyclohexanol to cyclohexanone using hydrogen peroxide and tungstic acid as catalyst. The phase-transfer catalysts comprised a range of ionic liquids that were used as co-catalysts with tungstic acid, and with Aliquat 336.

The tungstic acid catalyst is believed to form a complex with hydrogen peroxide, according to the following equation:

$$H_2WO_4 + H_2O_2 \overset{H_2O}{\rightarrow} H_2\left[WO(O_2)_2(OH)_2\right]$$

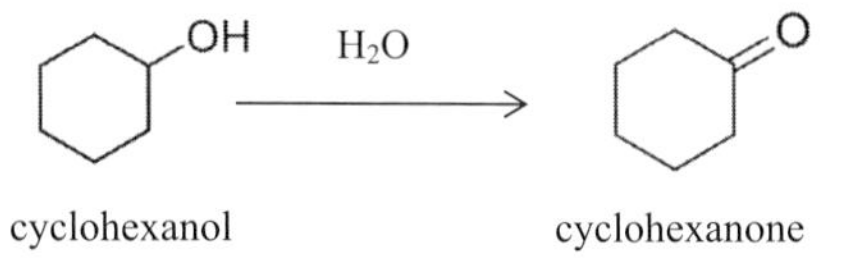

Figure 9.15 Oxidation of cyclic alcohols to ketones using phase-transfer catalysis. (Chatel et al., 2014)

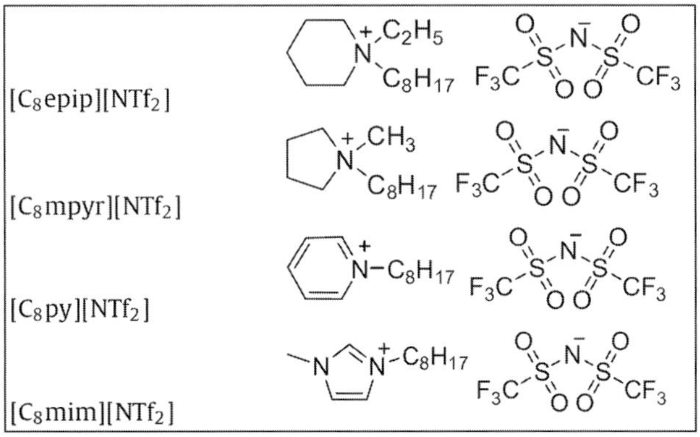

Figure 9.16 Ionic liquids used to enhance performance of cyclohexanol oxidations by phase-transfer catalysis.

Reprinted with permission from Chatel, G., Monniera, C., Kardosa N., et al. (2014) Green, selective and swift oxidation of cyclic alcohols to corresponding ketones. *Applied Catalysis A: General*, **478**, 157–164. Copyright Elsevier B.V.

The phase-transfer catalyst acts to transport the complex to the organic phase in order for the reaction to cyclohexanone to proceed. Four ionic liquids were evaluated as phase-transfer catalysts (Figure 9.16).

In summary, the use of tungstic acid alone as catalyst in the hydrogen peroxide/cyclohexanol system yielded very limited cyclohexanone product. A maximum figure of 14% yield was reported. Use of ionic liquids as phase-transfer catalysts gave significant improvements, the largest improvement reported being in the case of [C_8mim] [NTf_2], with a yield of 58% reported. With further optimization of the tungstic acid/ionic liquid ratio this was increased to a yield of 82% using [C_8mim] [NTf_2] as the phase-transfer co-catalyst. This was further increased to 94% for the same mixture using a microwave-heated reactor. A further and final part of the study concerned the use of Aliquat 336 as the phase-transfer catalyst with tungstic acid. In this case, product yields up to 97% were reported using conventional heating. Using microwave heating, the maximum yield in the Aliquat 336 system was 94%, but with a reaction time of 2.5 min compared with 15 min in the case of the conventionally heated reactor. A further set of experiments using low-frequency ultrasound to intensify reaction conditions was reported; a product yield of 98% was reported with a reaction time of 15 min.

This latter work is of special significance in that it provides an excellent illustration of the intensification of a reaction in a liquid–liquid system not only by virtue of applying a phase-transfer catalyst very effectively, but it also highlights the importance of unconventional methods of energy addition and agitation. The role of microwave energy in reducing the reaction time for high product yield is not fully explained, but is confirmed in a number of cases involving hydrogen peroxide (Bogdal and Lukasiewicz, 2000). In later work, Bogdal and Loupy (2008) showed as a partial explanation that microwave energy may greatly improve the conditions for the complexation reaction involving tungstic acid and hydrogen peroxide.

9.2.6 Nitrations

Wang et al. (2011) demonstrated the intensification of nitration of aromatics by phase-transfer catalysis with the addition of surfactants.

These workers showed the important effect of adding a surfactant to the liquid–liquid phase-transfer catalysis system for the regioselective nitration of *m*-xylene. This work served to illustrate the advantages of phase-transfer catalysis compared with conventional systems involving heterogeneous catalysis, such as zeolites. The system studied is worth looking at in further detail.

The reaction system is depicted in the reactions below and Figure 9.17, showing the two principle reactions involved that differ for cationic surfactant catalysts and anionic surfactant catalysts. The first is the exchange of nitrate ion (from nitric acid) for the organic anion X^- from the cationic surfactant catalyst to produce the organic compatible species $Q^+NO_3^-$ that is transported into the organic phase.

$$Q^+X^-(\text{org}) + HNO_3(\text{aq}) \rightarrow Q^+NO_3^-(\text{org}) + HX(\text{aq})$$
$$Q^+X^-(\text{org}) + HNO_3(\text{aq}) \rightarrow Q^+X^-(HONO_2)(\text{org})$$

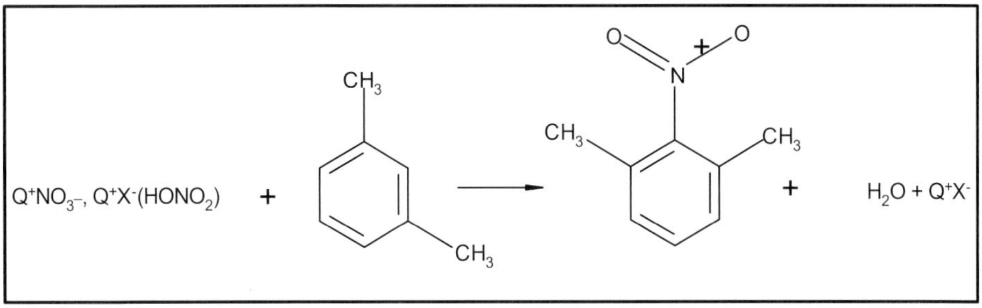

Figure 9.17 Regioselective nitration of aromatics using phase-transfer catalysis.

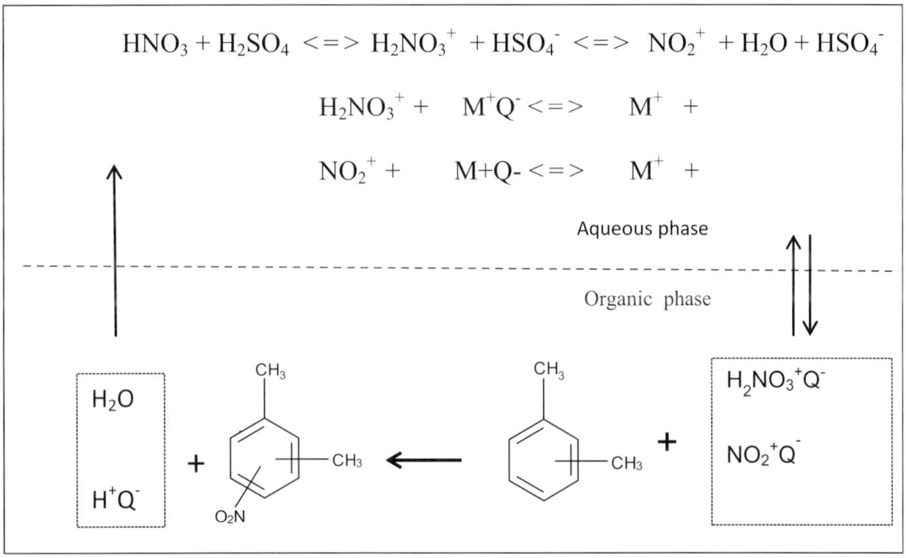

Figure 9.18 Regioselective nitration of aromatics using phase-transfer catalysis: mechanism. Reprinted with permission from Wang, P., Lu, M., Zhu, J., Song, Y., and Xiong, X. (2011). Regioselective nitration of aromatics under phase-transfer catalysis conditions. *Catalysis Communications*, **14**, 42–47. Copyright Elsevier B.V.

The second reaction is in parallel to the first, but producing a different complex, $Q^+X^-(HONO_2)$, that is also compatible with the organic phase and thus transports into the organic phase where the nitration actually occurs.

A possible scheme for the nitration in this case is as shown in Figure 9.17.

Quoting from the paper:

> A significant improvement in the selectivity and conversion of the nitration of xylene was observed: the ratio of 4-nitro-*m*-xylene to 2-nitro-*m*-xylene was unprecedented increased up to 91.3%:7.7%, the ratio of 4-nitro-*o*-xylene to 3-nitro-*o*-xylene was also increased to 71.1%:27.2%; both the conversions were over 96%.

A further scheme is for the case of the anionic surfactant M^+Q^- proposed by Wang et al. and is shown in Figures 9.18 and 9.19.

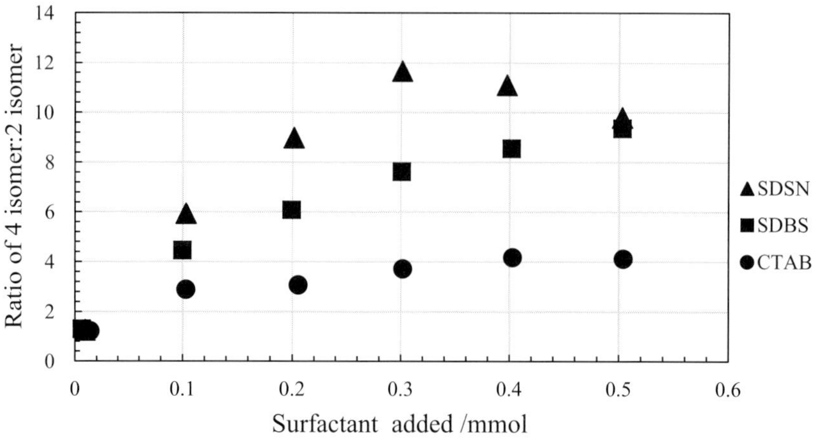

Figure 9.19 Regioselective nitration of aromatics using phase-transfer catalysis: mechanism, showing the effect of surfactant addition on product distribution of *m*-xylene in nitration with different surfactants. Reaction conditions: 0.08 mol *m*-xylene, 0.08 mol of 65% HNO_3, 0.05 mol of 98% H_2SO_4, 3 h at 50 °C.

Reprinted with permission from Wang, P., Lu, M., Zhu, J., Song, Y., and Xiong, X. (2011). Regioselective nitration of aromatics under phase-transfer catalysis conditions. *Catalysis Communications*, **14**, 42–47. Copyright Elsevier B.V.

In the aqueous phase we see the formation of the organic-compatible complexes $H_2NO_3{}^+Q^-$ and $NO_2{}^+Q^-$. Then in the organic phase again we see the nitration reaction occurring, with the reformation of the surfactant catalyst M^+Q^- that recycles back into the aqueous phase.

Overall, significant improvements in the selectivity and conversion for the nitration of xylene are reported, with conversion yields of more than 96%. Of even greater significance, as shown in Figure 9.19, is the large increase in the regioselectivity, as reflected in the greatly enhanced yield ratio of 4-nitro-*m*-xylene to 2-nitro-*m*-xylene as the amount of surfactant is increased.

9.2.7 Organic Polymerizations

One of the major benefits of process intensification is the provision of opportunities to exploit recent developments in organic catalytic chemistry through a combination with innovation in process engineering. Several examples are noteworthy. The use of microwave energy in combination with phase-transfer catalysis was the subject of a review by Bogdal et al. (2010), in which enhancement of many organic reactions was discussed. These include *o*-alkylation, alkyl ester synthesis, aromatic ester synthesis, aliphatic ether synthesis, aromatic ether synthesis, *N*- and *C*-alkylations, condensation reactions, Heck and Suzuki cross-coupling reactions, organic oxidations and epoxidations, and polymerizations. The authors conclude that the coupling of microwave technology and phase-transfer catalysis creates a clean, selective, and efficient methodology for performing certain organic reactions, with substantial

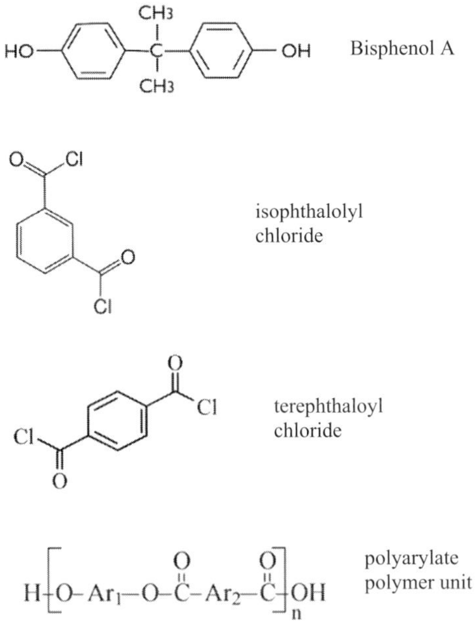

Figure 9.20 Polymerizations using phase-transfer catalysis: the components involved in the phase-transfer polycondensation for polyarylate synthesis. (Wolinksi and Wronski, 2009)

improvements in terms of mild conditions and simplicity of operating procedures, leading to faster and cleaner reactions compared with conventional heating.

Wolinksi and Wronski (2009) provided another notable example in the field of phase-transfer catalysis for a polymerization reaction. In common with many multi-phase polymerizations, these reactions can experience significant mass transfer limitations. The reaction of interest was the reaction of bisphenol A with isophthalolyl chloride and terephthaloyl chloride to give polyarylate polymers. Figure 9.20 shows the components of the reaction.

This is an interfacial polycondensation reaction of polyarylate and was demonstrated in a Taylor-Couette reactor (see Chapter 5). The generic reaction scheme for the polycondensation is outlined in Figure 9.21

Results of the polymerization conducted in the Taylor-Couette reactor are shown in Figure 9.22. The effect of the rotational Reynolds number on the reaction of bisphenol A is very substantial and confirms the intensification of the reaction when conducted in the Taylor-Couette reactor. These data also show the importance of the initial monomer ratio on the change in bisphenol A in the aqueous phase.

9.2.8 Pseudo-Phase-Transfer Catalysis

The term pseudo-phase-transfer catalysis was first described by Yadav and Kadam (2012) and the example of the benzoin condensation is the first of its kind. Specifically, the reaction of interest was the benzoin condensation reaction from benzaldehyde (Figure 9.23).

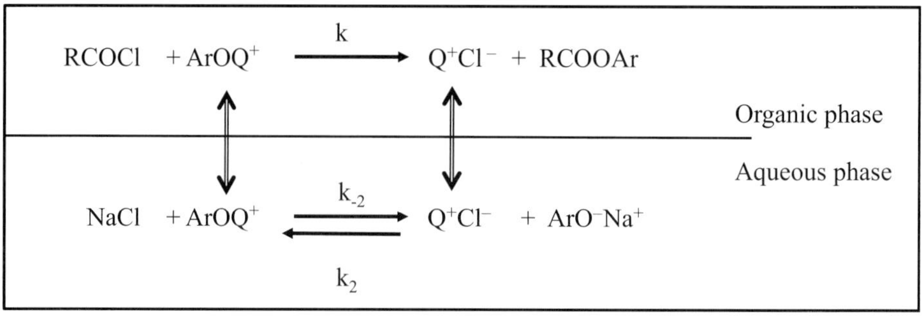

Figure 9.21 General scheme of a polycondensation using phase-transfer catalysis. (Wolinksi and Wronski, 2009)

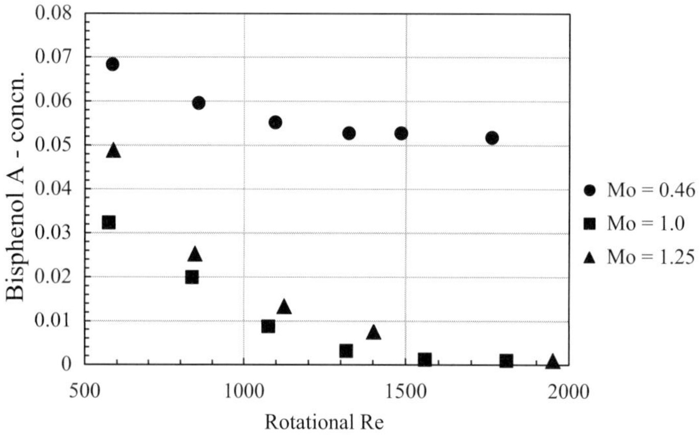

Figure 9.22 Influence of the rotational Reynolds number (Re) and the molar ratio of monomer (Mo) on the change in bisphenol A in the aqueous phase.

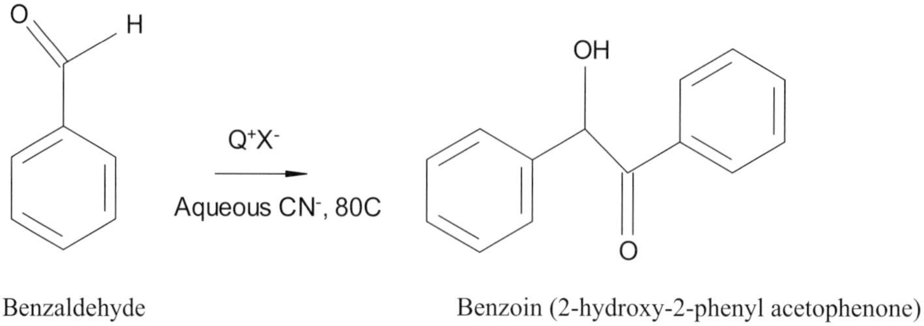

Figure 9.23 The benzoin condensation reaction. (Yadav and Kadam, 2012)

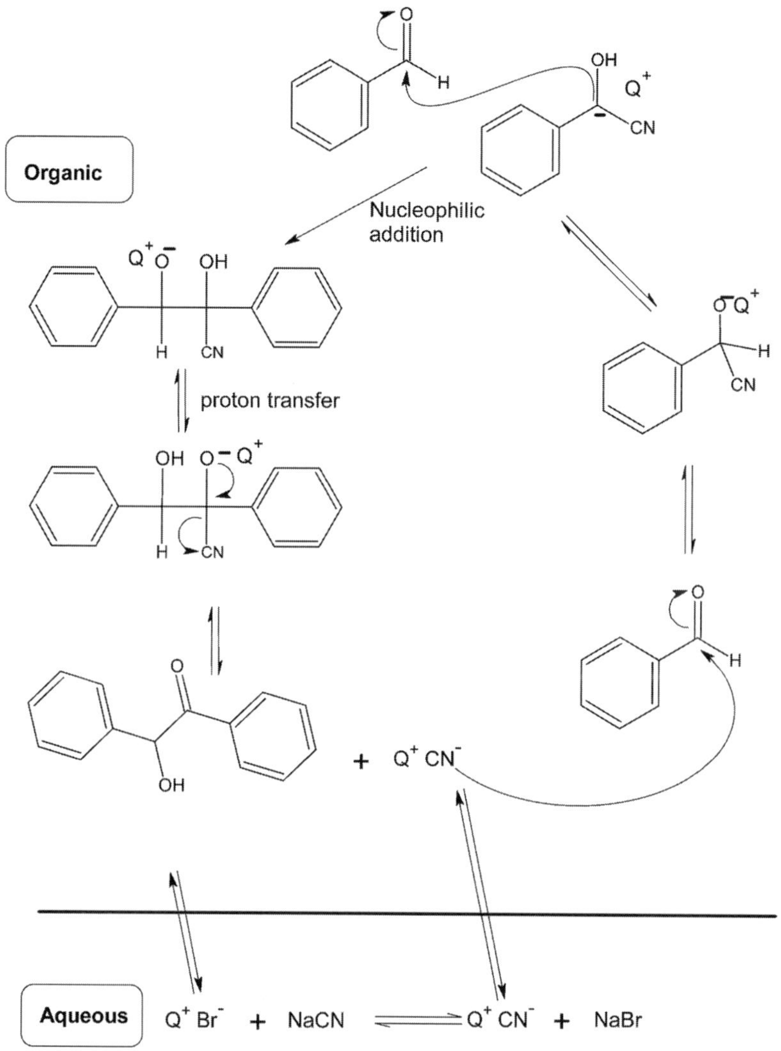

Figure 9.24 The benzoin condensation reaction: proposed mechanism.
Reprinted with permission from Yadav, G. D. and Kadam, A. A. (2012). Atom-efficient benzoin condensation in liquid–liquid system using quaternary ammonium salts: pseudo-phase transfer catalysis. *Organic Process Research and Development*, **16**, 755–763. Copyright (2012) American Chemical Society

The phase-transfer catalyst was sodium cyanide and a quaternary ammonium bromide salt that, in the aqueous phase, forms the ion pair Q^+CN^-. This transfers to the organic phase which initially comprised benzaldehyde in dodecane. The conversion to benzoin occurs, but the ion-pair catalyst is regenerated in the organic phase and, after the reaction has started, regeneration of the catalyst becomes self-sustaining in the organic phase (Figure 9.24).

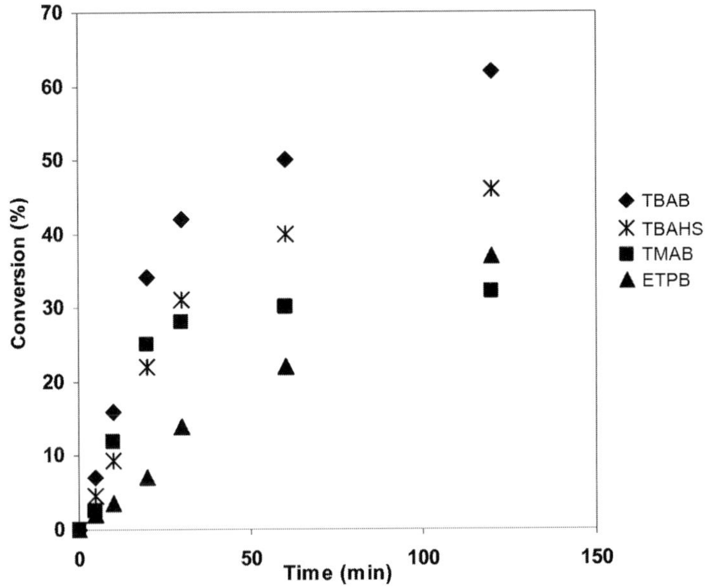

Figure 9.25 Benzoin condensation reaction: influence of phase-transfer catalyst on kinetics. Effect of various catalysts: benzaldeyhde, 0.012 mol; dodecane, 0.2 cm^3; nacn, 0.0005 mol; toluene, 10 cm^3; water, 10 cm^3; temperature, 80 °C; agitation, 900 rpm. TBAB is tetra-n-butyl ammonium bromide; TBAHS is tetrabutyl ammonium hydogensulfate; TMAB is tetramethyl ammonium bromide; ETPB is ethyl triphenyl phosphonium bromide.

Reprinted with permission from Yadav, G. D. and Kadam, A. A. (2012). Atom-efficient benzoin condensation in liquid–liquid system using quaternary ammonium salts: pseudo-phase transfer catalysis. *Organic Process Research and Development*, **16**, 755–763. Copyright (2012) American Chemical Society

The study showed that TBAB was the most effective PTC compared with TBAHS, TMAB, and ETPB (Figure 9.25).

The work also showed that because of the self-regenerating nature of the catalyst exclusively in the organic phase, the reaction is kinetically controlled and is also dependent upon on a high distribution coefficient for the cyanide ion pair between aqueous and organic phases.

References

Benjamin, I. (2013). Dissociation and transport of ion pairs across the liquid/liquid interface. Implications for phase transfer catalysis. *Journal of Physical Chemistry B*, **117**, 4325–4331.

Bogdal, D. and Loupy, A. (2008). Application of microwave irradiation to phase-transfer catalyzed reactions. *Organic Process Research & Development*, **12**, 710–722.

Bogdal, D, and Lukasiewicz, M. (2000). Microwave-assisted oxidation of alcohols using aqueous hydrogen peroxide. *Synlett*, **1**, 143–145.

Bogdal, D., Galica, M., Bartus, G., Wolinski, J., and Wronski, S. (2010). Preparation of polymers under phase-transfer catalytic conditions. *Organic Process Research & Development*, **14**, 669–683.

Casagrande, M., Storaro, L., Talon, A., et al. (2002). Liquid phase acetophenone hydrogenation on Ru/Cr/B catalysts supported on silica. *Journal of Molecular Catalysis A: Chemical*, **188**, 133–139.

Chatel, G., Monniera, C., Kardosa N., et al. (2014) Green, selective and swift oxidation of cyclic alcohols to corresponding ketones. *Applied Catalysis A: General*, **478**, 157–164.

Demmans, K. Z., Ko, O. W. K., and Morris, R. H. (2016). Aqueous biphasic iron-catalyzed asymmetric transfer hydrogenation of aromatic ketones. *RSC Advances*, **6**(91), 88580–88587.

Fleming, F. F., Yao, L., Ravikumar, P. C., Funk, L., and Shook, B. C. (2010). Nitrile-containing pharmaceuticals: efficacious roles of the nitrile pharmacophore. *Journal of Medicinal Chemistry*, **53**(22), 7902–7917.

Gonzalez-Galvez, D., Lara, P., Rivada-Wheelaghan, O., et al. (2013). NHC-stabilized ruthenium nanoparticles as new catalysts for the hydrogenation of aromatics. *Catalysis Science and Technology*, **3**, 99–105.

Makosza, M. (2000). Phase-transfer catalysis. A general green methodology in organic synthesis. *Pure and Applied Chemistry*, **72**(7), 1399–1403.

Meshechkina, A. E., Mel'nik, L. V., Rybina, G. V., Srednev, S. S., and Shevchuk, A. S. (2012). Efficiency of phase-transfer catalysis in cyclopentene epoxidation with hydrogen peroxide. *Russian Journal of Applied Chemistry*, **85**(4), 661–665.

Reichart, B., Kappe, T., and Glasnov, T. N. (2013). Phase-transfer catalysis: mixing effects in continuous-flow liquid/liquid *O*- and *S*-alkylation processes. *Synlett*, **24**, 2393–2396.

Tsai, H. B. and Lee, Y. D. (1987a). Polyarylates (I): Investigation of the interfacial polycondensation reaction by UV. *Journal of Polymer Science*, **25**, 1505–1515.

Tsai, H. B. and Lee, Y. D. (1987b). Polyarylates. II. The molecular weight distribution measured by GPC. *Journal of Polymer Science*, **25**, 1709–1712.

Tsai, H. B. and Lee, Y. D. (1987c). Polyarylates III. Kinetics studies of interfacial polycondensation. *Journal of Polymer Science*, **25**, 2195–2206.

Tsai, H. B., Lee, Y. D., Jeng, J. T. (1988). Polyarylates V. The influence of phase transfer agents on the interfacial polycondensation. *Journal of Polymer Science*, **26**, 2039–2046.

Tsai, H. B., Jeng, J. T., and Tsai, R. S. (1990) Reaction kinetics of interfacial polycondensation of polyarylate. *Journal of Applied Polymer Science*, **39**, 471–476.

Wang, L., Hongxia Ma, H., Song, L., et al. (2014). Transfer hydrogenation of acetophenone in an organic-aqueous biphasic system containing double long-chain surfactants. *RSC Advances*, **4**, 1567–1569.

Wang, P., Lu, M., Zhu, J., Song, Y., and Xiong, X. (2011). Regioselective nitration of aromatics under phase-transfer catalysis conditions. *Catalysis Communications*, **14**, 42–47.

Wolinski, J. and Wronski, S. (2009). Interfacial polycondensation of polyarylate in Taylor-Couette-Reactor. *Chemical Engineering and Processing*, **48**, 1061–1071.

Yadav, G. D. and Badure, O. V. (2008). Selective engineering in *O*-alkylation of *m*-cresol with benzyl chloride using liquid–liquid–liquid phase transfer catalysis. *Journal of Molecular Catalysis A: Chemical*, **288**, 33–41.

Yadav, G. D. and Jadhav, Y. B. (2003). Kinetics and modeling of liquid–liquid phase transfer catalyzed synthesis of *p*-chlorophenyl acetonitrile: role of co-catalyst in intensification of rates and selectivity. *Journal of Molecular Catalysis A: Chemical,* **192**, 41–52.

Yadav, G. D. and Kadam, A. A. (2012). Atom-efficient benzoin condensation in liquid–liquid system using quaternary ammonium salts: pseudo-phase transfer catalysis. *Organic Process Research and Development*, **16**, 755–763.

Zhu, M. (2014). Integration of phase transfer catalysis into aqueous transfer hydrogenation. *Applied Catalysis A – General*, **479**, 45–48.

Zyuzin, I. (2013). Alkylation of 1,1-bis(methoxy-NNO-azoxy)alkanes under phase-transfer catalysis. *Russian Journal of Organic Chemistry*, **49**(4), 536–544.

Index